Flame Retardant Polymeric Materials

Series in Materials Science and Engineering

The series publishes cutting edge monographs and foundational textbooks for interdisciplinary materials science and engineering. It is aimed at undergraduate and graduate level students, as well as practicing scientists and engineers. Its purpose is to address the connections between properties, structure, synthesis, processing, characterization, and performance of materials.

Flame Retardant Polymeric Materials

A Handbook

Edited by
Yuan Hu
Xin Wang

CRC Press
Taylor & Francis Group
Boca Raton London New York

CRC Press is an imprint of the
Taylor & Francis Group, an **informa** business

CRC Press
Taylor & Francis Group
6000 Broken Sound Parkway NW, Suite 300
Boca Raton, FL 33487-2742

First issued in paperback 2020

© 2020 by Taylor & Francis Group, LLC
CRC Press is an imprint of Taylor & Francis Group, an Informa business

No claim to original U.S. Government works

ISBN-13: 978-1-138-29579-7 (hbk)
ISBN-13: 978-0-367-77926-9 (pbk)

Library of Congress Cataloging-in-Publication Data

Names: Hu, Yuan (Materials scientist), editor. | Wang, Xin (Materials scientist), editor.
Title: Flame retardant polymeric materials : a handbook / edited by Yuan Hu and Xin Wang.
Description: Boca Raton : CRC Press, [2020] | Series: Series in materials science and engineering | Includes bibliographical references and index.
Identifiers: LCCN 2019010040 | ISBN 9781138295797 (hardback)
Subjects: LCSH: Fire resistant polymers. | Fire resistant materials.
Classification: LCC TH1074.5 .F5776 2020 | DDC 628.9/223--dc23
LC record available at https://lccn.loc.gov/2019010040

**Visit the Taylor & Francis Web site at
http://www.taylorandfrancis.com**

**and the CRC Press Web site at
http://www.crcpress.com**

Contents

SECTION I   Fundamentals

SECTION II   Material Modification, Assembly, and Design

SECTION III   Flame Retardant Polymer Nanocomposites

Preface

Nowadays, the utilization of polymeric materials brings considerable conveniences to our daily life. However, most of the polymeric materials possess a fatal drawback, high flammability. There is a large number of property loss and casualties in polymer-related fire accidents every year on this planet. Therefore, more and more countries have set up laws and regulations to ensure that flame retardant treatment is compulsory for polymeric materials used in many fields such as buildings and construction, electrical and electronics, coatings and paints, wire and cables, etc. According to a report from US-based consultancy Grand View Research (Flame Retardant Market Analysis By Product, By Application, By End-Use, and Segment Forecasts, 2014–2025), global consumption of flame retardants reached a volume of 2.49 million tons in 2015 valued at $6.29 billion USD, which is predicted to exceed 4.0 million tons by 2025 valued at $11.96 billion USD with a compound annual growth rate of 4.9% from 2016 to 2025. The rapid development in flame retardant polymeric materials creates the need for an up-to-date monograph that introduces the latest advancements and applications in this field.

Over the past few decades, there are a number of classic monographs on flame retardant polymeric materials, including *Fire Retardant Materials* by Richard Horrocks and Dennis Price (2001), *Fire Retardancy of Polymers: New Applications of Mineral Fillers* by Michel Le Bras, Charles A. Wilkie, and Serge Bourbigot (2005), *Flame Retardant Polymer Nanocomposites* by Alexander B. Morgan and Charles A. Wilkie (2007), and *Fire Retardancy of Polymeric Materials* by Charles A. Wilkie and Alexander B. Morgan (2010). These books cover the broader aspects of flame retardants and their underlying scientific principles. However, some new arising flame retardant materials and techniques, such as graphene and derivatives, molybdenum disulfide, and layer-by-layer assembly have been developed. It is thereby imperative to introduce these latest scientific developments and technological advances to readers in this field.

This book focuses on key topics related to flame retardants of polymers, which are divided into four sections: Fundamentals; Material Modification, Assembly, and Design; Flame Retardant Polymer Nanocomposites; and Applications.

The first section consists of three chapters: introduction (Chapter 1), mechanisms and modes of action in flame retardancy of polymers (Chapter 2), and fire testing for the development of flame retardant polymeric materials (Chapter 3).

The second section then turns to areas of active research in terms of developing new methods and materials, including specific flame retardant systems (phosphorus-, silicon-, and boron-based flame retardants) (Chapters 4 through 6) and layer-by-layer assembly (Chapter 7).

In the third section, the use of several typical nano-fillers including carbon nanotubes and nanofibers (Chapter 8), graphene (Chapter 9), molybdenum disulfide and layered double hydroxides (Chapter 10), and polyhedral oligomeric silsesquioxane (Chapter 11) for reducing fire hazards of polymeric materials are reviewed and discussed.

The last section is designed to exhibit the latest advances in the current applications of flame retardant polymeric materials in the fields of construction insulating materials (Chapter 12), textiles (Chapter 13), wire and cable materials (Chapter 14) as well as fiber-reinforced composites (Chapter 15).

We hope that this book will be useful for professionals to develop new fire retardant materials. Also, we hope that it will be of value to graduate students who are interested in fundamentals and applications of flame retardant polymeric materials.

We should express great appreciation to all those who helped enable this book to be possible, especially the chapter authors who have spent their spare time to share their knowledge to the readers. We also would like to thank the reviewers who have taken time out of their busy lives to review each chapter. Finally, we would like to acknowledge those at Taylor & Francis for their professional organizing and editing work.

Yuan Hu

Xin Wang

Editors

Yuan Hu is head of the Department of Safety Science and Engineering of the University of Science and Technology of China (USTC) and Director of the Institute of Fire Safety Materials of USTC. He obtained his PhD from USTC in 1997. Prof. Hu has done research on fire safety materials for more than 20 years. His main research area includes: polymer/inorganic compound nanocomposites, new flame retardants and their flame retardant polymer, synthesis and properties of inorganic nanomaterials, combustion, and decomposition mechanism of polymer. Prof. Hu has published more than 400 papers (Science Citation Index) in research areas covering the range of fire safety of materials. Moreover, more than 40 invention patents have been authorized.

Xin Wang is an associate professor in the University of Science and Technology of China (USTC). He earned his PhD in Safety Science and Engineering from the USTC in 2013. Dr. Wang leads the high performance polymer nanocomposite group, setting technical direction and building partnerships with both industrial and academic communities, in order to promote the development of high performance flame retardant polymeric materials. Dr. Wang's research interests focus on synthesis of novel halogen-free flame retardants and preparation of layered nanomaterials and their use in flame retardant polymer nanocomposites. Dr. Wang has published more than 70 papers in the peer-reviewed international journals (citations 1980, H index = 36) in the field of flame retardant polymeric materials.

Contributors

Wei Cai
State Key Laboratory of Fire Science
University of Science and Technology of China
Hefei, P. R. China

Manfred Döring
Department of Polymer Synthesis
Fraunhofer Institute for Structural Durability
    and System Reliability LBF
Darmstadt, Germany

Sebastian Eibl
Bundeswehr Research Institute for Materials,
    Fuels, and Lubricants (WIWeB)
Erding, Germany

Bin Fei
Institute of Textiles and Clothing
The Hong Kong Polytechnic University
Hong Kong, P. R. China

Laurent Ferry
Center of Materials Research
Ecole des Mines d'Alès/Institut Mines Telecom
Alès, France

Lara Greiner
Department of Polymer Synthesis
Fraunhofer Institute for Structural Durability
    and System Reliability LBF
Darmstadt, Germany

Wenwen Guo
State Key Laboratory of Fire Science
University of Science and Technology of China
Hefei, P. R. China

Yuan Hu
State Key Laboratory of Fire Science
University of Science and Technology of China
Hefei, P. R. China

Katharina Kebelmann
Division Technical Properties of Polymeric
    Materials
Bundesanstalt für Materialforschung und –
    prüfung (BAM)
Berlin, Germany

Christian Lagreve
Center of Materials Research
Ecole des Mines d'Alès/Institut Mines Telecom
Alès, France

Hauke Lengsfeld
Epoxy Products
Schill + Seilacher "Struktol" GmbH
Hamburg, Germany

Xiaoyan Li
School of Materials Science and Technology
Tongji University
Shanghai, P. R. China

Lei Liu
School of Materials Science and Technology
Tongji University
Shanghai, P. R. China

Jose-Marie Lopez-Cuesta
Center of Materials Research
Ecole des Mines d'Alès/Institut Mines Telecom
Alès, France

Giulio Malucelli
Department of Applied Science and Technology
Politecnico di Torino
Alessandria, Italy

Alexander B. Morgan
Center for Flame Retardant Material Science
University of Dayton Research Institute
Dayton, Ohio

Ye-Tang Pan
High Performance Polymers and Fire Retardants
IMDEA Materials Institute
and
E.T.S. de Ingenieros de Caminos
Universidad Politécnica de Madrid
Madrid, Spain

Bernhard Schartel
Division Technical Properties of Polymeric
    Materials
Bundesanstalt für Materialforschung und –
    prüfung (BAM)
Berlin, Germany

Kelvin K. Shen
Rio Tinto Borax/U.S. Borax (Consultant)
Denver, Colorado

Lei Song
State Key Laboratory of Fire Science
University of Science and Technology of China
Hefei, P. R. China

De-Yi Wang
High Performance Polymers and Fire Retardants
IMDEA Materials Institute
Madrid, Spain

Junling Wang
State Key Laboratory of Fire Science
University of Science and Technology of China
Hefei, P. R. China

Xin Wang
State Key Laboratory of Fire Science
University of Science and Technology of China
Hefei, P. R. China

Zhengzhou Wang
School of Materials Science and Technology
Tongji University
Shanghai, P. R. China

Charles A. Wilkie
Department of Chemistry
Marquette University
Milwaukee, Wisconsin

Weiyi Xing
State Key Laboratory of Fire Science
University of Science and Technology of China
Hefei, P. R. China

Qiang Yao
Department of Polymers and Composites
CAS's Ningbo Institute of Materials Technology
    and Engineering
Ningbo, P. R. China

Bin Yu
Institute of Textiles and Clothing
The Hong Kong Polytechnic University
Hong Kong, P. R. China

Yan Zhang
State Key Laboratory of Fire Science
University of Science and Technology of China
Hefei, P. R. China

I

Fundamentals

1

Introduction

Yuan Hu and
Xin Wang

1.1 Fire Hazards of Polymeric Materials

Polymers can be divided into various categories based on sources (natural or synthetic polymers), based on physical/mechanical properties (plastics, rubbers, and fibers), and based on thermal properties (thermoplastic or thermosetting polymers). In this chapter, we classify polymers into hydrocarbon polymers, oxygen-, nitrogen-, silicon-, and halogen-containing polymers according to their chemical structure.

The hydrocarbon polymers without heteroatoms are further subdivided into aliphatic hydrocarbon polymers and aromatic hydrocarbon polymers. The aliphatic hydrocarbon polymers mainly include polyethylene, polypropylene, and ethylene propylene diene monomer, which are extensively used in the fields of packaging, and insulating structural components for motors, apparatuses, as well as electronic industry. The most important aromatic hydrocarbon polymers are polystyrene and styrenic copolymers like styrene-butadiene copolymer and methyl methacrylate butadiene styrene terpolymer. Polystyrene is widely applied as packaging and thermal insulating materials for buildings, while styrenic copolymers are rubbery materials as sheath of cables and wires, and tires.

The most important and extensively used oxygenated polymers include polyacrylics and polyesters. The major polyacrylic is polymethyl methacrylate, also known as plexiglass, which is widely used in the fields of motors and lighting industry and housing decoration. The most important polyesters are polyethylene terephthalate, polybutylene terephthalate, and polycarbonate, which are usually used as structural components in the field of airplanes, motors, and trains. Other oxygen-containing polymers include phenol-formaldehyde resin and polyformaldehyde.

The most commonly used nitrogen-containing polymers include polyamides, polyacrylonitrile, and polyurethanes. Polyamides, also known as nylon, are one of the most important materials in the fields of clothing, structural components of motors, and trains. Besides aliphatic polyamides, there are also several kinds of aromatic polyamides which possess outstanding thermal stability for specialist application as protective clothing. Polyurethanes are usually applied in the realm of packaging, soft furnishings, and thermal insulation materials. Polyacrylonitrile is a well-known synthetic fiber and also as precursor for synthesis of carbon fibers.

The main silicon-containing polymers are silicone rubbers including polydimethylsiloxane, polyoctylmethylsiloxane, etc. Silicone rubbers are widely used as sealing materials and high voltage insulators owing to their extremely high thermal stability, anti-aging property, outstanding electrical insulation, and excellent hydrophobicity.

The most important and extensively used halogen-containing polymers are polyvinyl chloride and polytetrafluoroethylene. Polyvinyl chloride can be further subdivided into rigid, semi-flexible, and flexible materials. Due to its highly tunable properties, polyvinyl chloride is extensively applied as wire and cable sheaths, protective clothing, and furniture fabrics. Polytetrafluoroethylene is famous for its exceptional thermal stability, high chemical resistance, and low friction coefficient, which is used as insulators and paintings for "non-stick" surfaces.

The chemical structure and the limiting oxygen index (LOI) of the most commonly used polymers are summarized in Table 1.1 (Tewarson 2005; Gilbert 2016). Hydrocarbon polymers and oxygenated polymers are composed of carbon, hydrogen, and oxygen, which have great tendency to burning in terms of relatively low LOI value. The aromatic groups in the hydrocarbon polymers can increase the LOI value. Generally, polymers with heteroatoms (nitrogen, silicon, chlorine, and fluorine) show higher LOI value than hydrocarbon polymers and oxygenated polymers. The higher the heteroatoms content in the polymer, the more difficult is the tendency to burning. Specifically, some polymers with heteroatoms like polytetrafluoroethylene cannot be ignited and will self-extinguish after removal of fire source. However, most of polymers are highly flammable, which potentially threatens people's life and property safety.

TABLE 1.1 Chemical Structures and LOI Values of Polymers

Type of Polymers	Unit	LOI (%)
Polyethylene	$-CH_2-CH_2-$	17
Polypropylene	$-CH_2-CH-$ $\quad\quad CH_3$	17
Polystyrene	$-CH_2-CH-$ (with phenyl ring)	18
Polymethyl methacrylate	$-CH_2-C-$ with CH_3 and ester group	17
Polyformaldehyde	$-CH_2-O-$	15
Polyethylene terephthalate	$-C(=O)-$(benzene ring)$-C(=O)-O-CH_2-CH_2-O-$	21
Polybutylene terephthalate	$-C(=O)-$(benzene ring)$-C(=O)-O-(CH_2)_4-O-$	18
Bis-phenol A polycarbonate	$-O-$(benzene ring)$-C(CH_3)_2-$(benzene ring)$-O-C(=O)-$	26

(Continued)

TABLE 1.1 (*Continued*) Chemical Structures and LOI Values of Polymers

Type of Polymers	Unit	LOI (%)
Phenol-formaldehyde resin		35
Polyamide 6		21–34
Polyamide 66		21–30
Polydimethylsiloxane		30
Polyvinyl chloride		23–43
Polytetrafluoroethylene		90

1.2 Polymer Combustion

The combustion reaction involves complicated chemical and physical processes, but it mainly depends on three factors: combustibles, combustive, and heat, which is also known as so-called fire triangle (Figure 1.1) (Laoutid et al. 2009). The combustive is generally the oxygen in the air, while combustibles are volatile products from decomposition of polymers. The diagrammatical illustration of polymer combustion is depicted in Figure 1.2 (Price et al. 2001). When polymers are exposed to an external heat

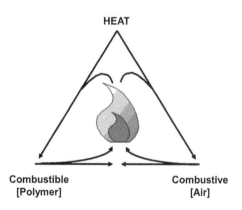

FIGURE 1.1 Principle of the combustion cycle. (Reprinted with permission from Laoutid, F. et al., *Mater. Sci. Eng. R-Rep.*, 63,100–125, 2009.)

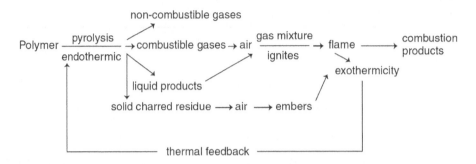

FIGURE 1.2   Diagrammatical illustration of polymer combustion. [Reprinted with permission from Price, D. et al., Introduction, in *"Fire Retardant Materials,"* A. R. Horrocks and D. Price (Eds.), CRC Press LLC, Cambridge, UK, p. 7, 2001.]

source to a certain level, the decomposition of polymers (also called depolymerization for polymers) occurs resulting in a large amount of combustible volatile products. These combustible volatile products are subsequently mixed with air and enter the gaseous phase to fuel the flame. The flame generates more heat and provides the feed-back loop that sustains the combustion process.

The decomposition of polymers is a crucial step for polymer combustion. Because combustion generally occurs in the gaseous phase, a solid polymer should decompose into small-molecular volatile products. Thermal decomposition of polymers is often triggered by chain scissions, which requires an input of heat. The heat provided to the polymers must exceed the bond dissociation energy between the covalently linked atoms. Table 1.2 collects some bond dissociation energy values for the most common bonds (Price et al. 2001). When an external heat source provides enough energy, the bond scissions induced thermal decomposition of polymers will start. Besides bond dissociation energy and heat, thermal decomposition behaviors of polymers are also highly dependent on the presence or absence of oxygen. Non-oxidizing thermal decomposition of polymers results in chain scissions by formation of free-radicals (equation 1.1) or by migration of hydrogen atoms to form more thermally stable molecules (equation 1.2) (Laoutid et al. 2009). In oxidizing thermal decomposition, the presence of oxygen reacted with polymers leads to various volatile products including carboxylic acids, alcohols, ketones, aldehydes, as well as highly reactive radicals like H· and OH· (Laoutid et al. 2009). Additionally, the

TABLE 1.2   Dissociation Energy of Some Common Bonds

Bond	Dissociation Energy (kJ/mol)	Bond	Dissociation Energy (kJ/mol)
C-H	340	H-I	297
C-C	607	C-F	553
C-O	1076	C-Cl	398
H-H	435	C-Br	280
H-F	569	C-I	209
H-Cl	431	C-P	515
H-Br	386	P-O	600

Source: Price, D. et al., Introduction, in *Fire Retardant Materials*, A. R. Horrocks and D. Price (Eds.), CRC Press LLC, Cambridge, UK, pp 12.

oxidizing thermal decomposition also generates crosslinking structures through recombination reactions of the macromolecular radicals. Thermal decomposition of polymers is also affected by physical factors, more specifically thermal conductivity.

$$R_1 - CH_2 - CH_2 - R_2 \rightarrow R_1 - CH_2 \bullet + \bullet CH_2 - R_2 \tag{1.1}$$

$$R_1 - CH_2 - CH_2 - CH_2 - R_2 \rightarrow R_1 - CH_2 = CH_2 + CH_3 - R_2 \tag{1.2}$$

1.3 Flame Retardation

As aforementioned, polymers combustion relates closely with three factors: combustibles, combustive, and heat. In order to achieve flame retardation of polymers, it is essential to interrupt one or more of the pathways between the three factors. This can be done in several approaches, for example, by endothermic reaction to cool down the polymers (heat absorption) or by releasing inflammable gases to dilute the flammable gases (lowering combustibles). Alternatively, it is also feasible to promote the char formation on the burning surface of polymers, thus blocking the mixing of the polymer decomposition volatiles with the air. Generally, flame retardation can act either in condensed phase or in gaseous phase, which can interfere with the various processes involved in polymer combustion (heating, decomposition, ignition, and propagation). The modes of action of flame retardant polymers are discussed in more detail in Chapter 2.

1.4 Research and Development of Flame Retardant Polymeric Materials

Because of the increasing concerns on fire safety of polymeric materials in use today (like wood, plastics, rubbers, textiles, etc.), research into the flame retardant polymeric materials has become one hot field of materials science study. The great research interest is clearly evidenced by the number of research publications over the last decade. A simple search with flame retardant or fire retardant as a keyword is conducted by using Institute for Scientific Information (ISI)-Web of Science, as shown in Figure 1.3.

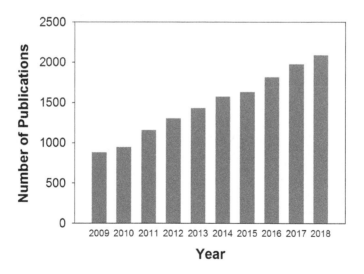

FIGURE 1.3 Publications number versus years searched using "flame retardant" or "fire retardant" as keywords in Institute for Scientific Information-Web of Science (up to December 2018).

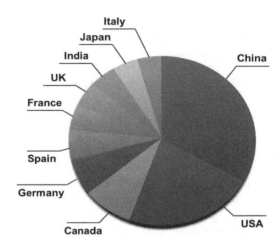

FIGURE 1.4 The top ten countries in publications number searched using "flame retardant" or "fire retardant" as keywords in Institute for Scientific Information-Web of Science (up to December 2018).

It can be seen that more than ten thousands of publications are published from 2009 to 2018 and rapid growth of publications year by year with more than two thousands publications in 2018. The statistics in the field of flame retardant polymeric materials is depicted in Figure 1.4. As can be seen, China contributes 33% publications in this field and hundreds of universities and institutes are devoted to scholarly research. The second and the third contributor are United States (US) and Canada, who contribute 22% and 9% publications, respectively. In the top ten nations, only China and India are developing countries. Several renowned flame retardant research organizations worldwide are listed in Table 1.3. The list is by no means comprehensive and specializes on the organizations that are often publishing papers on flame retardant polymeric materials. If you are interested in these organizations, you can learn more information through their websites.

TABLE 1.3 List of Several Renowned Flame Retardant Research Organizations

Name	Location	Website
National Fire Research Laboratory, National Institute of Standards and Technology (NIST)	USA	https://www.nist.gov/el/fire-research-division-73300
Underwriters Laboratories (UL)	USA	https://www.ul.com/
FM Global	USA	https://www.fmglobal.com/
BAM Federal Institute for Materials Research and Testing	Germany	https://www.bam.de/
Fraunhofer Institute for Structural Durability and System Reliability LBF	Germany	https://www.lbf.fraunhofer.de/en.html
Institute Centre for Materials Research and Innovation, The University of Bolton	UK	https://www.bolton.ac.uk/IMRI/
National Graduate School of Engineering Chemistry of Lille, Université Lille1	France	http://www.ensc-lille.fr/
Institute of Materials Physics and Engineering, the Polytechnic of Torino	Italy	http://www.disat.polito.it/
IMDEA Materials Institute	Spain	https://materials.imdea.org/
EMPA-Swiss Federal Laboratories for Materials Science and Technology	Switzerland	https://www.empa.ch/
College of Chemistry, The Sichuan University	China	http://chem.scu.edu.cn/EnglishSite
State Key Laboratory of Fire Science, The University of Science and Technology of China	China	https://en.sklfs.ustc.edu.cn/

In addition to academic research, industrial community is devoted largely to the development of new flame retardant materials for sale that meet the regulations. The major flame retardants manufacturers include Albemarle, BASF, Lanxess, Clariant, ICL, Chemtura, and Italmatch Chemicals. According to one report from US-based consultancy Grand View Research (GVR), global consumption of flame retardants reached 2.49 million tons in 2015, and it is expected to exceed 4.0 million tons in 2025 with a compound annual growth rate of 4.9%. In terms of value, the flame retardants market size is predicted to reach US$11.96 billion by 2025 from US$6.29 billion in 2015 (Global flame retardant market projected to reach US$11.96 billion by 2025 2017). Undoubtedly, the booming flame retardants market boosts the academic research in return.

1.5 Perspective of Flame Retardant Polymeric Materials

Taking all the changes underway for the political regulations, economic situations, and sociological concerns into account, it is quite difficult to precisely forecast what the perspective of flame retardant polymeric materials will be. However, some trends could be predicted about the future, which are listed as follows:

- Flame retardant polymeric materials with multi-functionality
- Combination or hybridization between organic and inorganic flame retardants
- More emphasis on smoke suppression rather than only caring heat suppression
- Correlation between bench-scale flame retardant test results and full-scale fire performance prediction
- Developing new flame retardant materials from bio-based resources.

In recent years, flame retardant polymeric materials are more and more applied in high technology fields such as electrical and electronics, automobile and airplane, new energy, etc. Owing to the specialized end-use purpose, multi-functionality is demanded. For instance, flame retardant polymers used in circuit boards require higher thermal conductivity to better release heat; as well, flame retardant polymers used in cable sheaths require antioxidant/UV resistance to serve a long time. However, balancing flame retardancy with other behaviors will continue to be a challenging task. Until now, polymer nanocomposite technology has been demonstrated to be effective to produce multi-functional polymeric materials with simultaneously improved flame retardancy and other behaviors. However, the industrial production of flame retardant polymer nanocomposites is severely limited by the high cost and processing technology, and it is still a long way to put flame retardant polymer nanocomposites into practice.

The current situation is that only a single flame retardant additive is used in most of the cases. The future is likely to combine or hybridize different additives to achieve synergistic flame retardant effect. For example, combination or hybridization between organic and inorganic flame retardants is beneficial to lowering more than one aspect of a fire risk scenario, with one species being used to reduce flame propagation or heat release, while another as smoke suppressant. Additionally, combination or hybridization is very likely to offer multi-functionality aforementioned, which will meet specialized end-use demands.

Traditional flame retardant technology focuses on reducing the heat release rate of polymeric materials during the combustion. However, poisonous smoke and gases rather than heat irradiation are the major reason for casualties during fire accidents. Carbon monoxide concentrations detected in real fire incidents achieve up to 7500 ppm (Troitzsch 2000), which probably cause a consciousness loss in 4 minutes (Irvine et al. 2000). With increasing concerns on higher fire safety materials, polymeric materials with high flame retardancy and low smoke emission are demanded. It is thereby urgent to develop novel fire safety technology. Several metal-based additives such as antimony oxides, basic iron (III) oxide, and zinc borate have been shown to be effective and frequently used smoke suppressant additives (Bourbigot et al. 1999; Carty & White 1998; Green 1996). Recently, metal or metal oxide nanomaterials or hybrids have also been shown to impart higher smoke suppressant characteristics to polymeric materials owing to their high aspect ratio (Dong et al. 2012; Fu et al. 2018; Zhou et al. 2014).

Today most of flame retardant polymers investigations are focused on small-scale or bench-scale flame-retardant testing, while their applications require full-scale flame retardant tests that provide insights into flammability of polymeric materials involved in a real fire. However, full-scale flame retardant tests generally require high expenses and large numbers of labor and materials, which are usually hard to generalize. Therefore, modeling efforts are ongoing to establish the correlation between bench-scale flame retardant tests results and full-scale fire performances. Some modeling prediction methods have been reported (Galgano et al. 2018; Lannon et al. 2018; Zhang et al. 2018; Zou et al. 2018), but these methods address very specific polymer type and very specific fire condition. There are no universal models for this type of prediction. Still, correlation between bench-scale and full-scale flame retardant performances will continue to be a section of future research, difficult though it will be.

Development of flame retardant polymeric materials from bio-based resources has attracted considerable interests due to increasing concerns over environmental issues and depletion of fossil resources. Until now, several kinds of bio-based resources including soybean oil (Basturk et al. 2013; Qiu et al. 2013; Sacristan et al. 2010), cardanol (Amarnath et al. 2018; Guo et al. 2019; Wang et al. 2017b), eugenol (Wan et al. 2016, 2015; Zhang et al. 2017), and vanillin (Wang et al. 2017a) have been frequently reported to prepare bio-based flame retardants, and polylactide and polybutylene succinate are the most commonly used bio-based polymers for flame retardant application (Bourbigot & Fontaine 2010; Suparanon et al. 2018; Wang et al. 2011). The ever-increasing demands for the better sustainability of polymeric materials will eventually induce the use of bio-based materials, and thereby the future emphasis will focus on exploiting cost-effective feedstocks for these bio-based flame retardants and derivative polymers.

Polymeric materials are very likely to be used more and more throughout society worldwide. With this increase, it seems that there will be more demand on flame retardant treatment of polymeric materials. This demand boosts huge research and development interest in both academic and industrial communities. Despite of ever-changing political regulations, economic situations, and sociological concerns, the unchanged theme in this field is pursuing better flame retardant materials without harms to people's health and environments. Flame retardant today for a better tomorrow!

References

Amarnath, N., Appavoo, D., and Lochab, B. 2018. Eco-friendly halogen-free flame retardant cardanol polyphosphazene polybenzoxazine networks. *ACS Sustainable Chemistry & Engineering* 6(1): 389–402.

Basturk, E., İnan, T., and Gungor, A. 2013. Flame retardant UV-curable acrylated epoxidized soybean oil based organic–inorganic hybrid coating. *Progress in Organic Coatings* 76(6): 985–992.

Bourbigot, S., and Fontaine, G. 2010. Flame retardancy of polylactide: An overview. *Polymer Chemistry* 1(9): 1413–1422.

Bourbigot, S., Le Bras, M., Leeuwendal, R. et al. 1999. Recent advances in the use of zinc borates in flame retardancy of EVA. *Polymer Degradation and Stability* 64(3): 419–425.

Carty, P., and White, S. 1998. A review of the role of basic iron(III) oxide acting as a char forming smoke suppressing flame retarding additive in halogenated polymers and halogenated polymer blends. *Polymers & Polymer Composites* 6(1): 33–38.

Dong, Y. Y., Gui, Z., Hu, Y. et al. 2012. The influence of titanate nanotube on the improved thermal properties and the smoke suppression in poly(methyl methacrylate). *Journal of Hazardous Materials* 209: 34–39.

Fu, X. L., Wang, X., Xing, W. Y. et al. 2018. Two-dimensional cardanol-derived zirconium phosphate hybrid as flame retardant and smoke suppressant for epoxy resin. *Polymer Degradation and Stability* 151: 172–180.

Galgano, A., Di Blasi, C., and Branca, C. 2018. Numerical evaluation of the flame to solid heat flux during poly(methyl methacrylate) combustion. *Fire and Materials* 42(4): 403–412.

Gilbert, M. 2016. Relation of structure to chemical properties. In *Brydson's Plastics Materials* (8th ed.), M. Gilbert, pp. 75–102. Oxford, UK: Elsevier.

Global flame retardant market projected to reach US$11.96 billion by 2025. 2017. Additives for Polymers 2017: 10–11.

Green, J. 1996. Mechanisms for flame retardancy and smoke suppression: A review. *Journal of Fire Sciences* 14(6): 426–442.

Guo, W. W., Wang, X., Gangireddy, C.S.R. et al. 2019. Cardanol derived benzoxazine in combination with boron-doped graphene toward simultaneously improved toughening and flame retardant epoxy composites. *Composites Part A: Applied Science and Manufacturing* 116: 13–23.

Irvine, D. J., McCluskey, J. A., and Robinson, I. M. 2000. Fire hazards and some common polymers. *Polymer Degradation and Stability* 67: 383–396.

Lannon, C. M., Stoliarov, S. I., Lord, J. M. et al. 2018. A methodology for predicting and comparing the full-scale fire performance of similar materials based on small-scale testing. *Fire and Materials* 42(7): 710–724.

Laoutid, F., Bonnaud, L., Alexandre, M. et al. 2009. New prospects in flame retardant polymer materials: From fundamentals to nanocomposites. *Materials Science & Engineering R-Reports* 63(3): 100–125.

Price, D., Anthony, G., and Carty, P. 2001. Introduction. In *Fire Retardant Materials*, ed. A. R. Horrocks and D. Price, pp. 1–28. Cambridge, UK: Woodhead Publishing.

Qiu, J. F., Zhang, M. Q., Rong, M. Z. et al. 2013. Rigid bio-foam plastics with intrinsic flame retardancy derived from soybean oil. *Journal of Materials Chemistry A* 1(7): 2533–2542.

Sacristan, M., Hull, T. R., Stec, A. A. et al. 2010. Cone calorimetry studies of fire retardant soybean-oil-based copolymers containing silicon or boron: Comparison of additive and reactive approaches. *Polymer Degradation and Stability* 95(7): 1269–1274.

Suparanon, T., Phusunti, N., and Phetwarotai, W. 2018. Properties and characteristics of polylactide blends: Synergistic combination of poly(butylene succinate) and flame retardant. *Polymers for Advanced Technologies* 29(2): 785–794.

Tewarson, A. 2005. Flammability of polymers. In *Plastics and the Environment*, ed. A. L. Andrady, pp. 403–489. Hoboken, NJ: John Wiley & Sons.

The Fire Bureau of the Ministry of Public Security. 2013. *Fire Statistical Yearbook of China-2012*. Beijing, China: The Personnel Press of China.

The Fire Bureau of the Ministry of Public Security. 2014. *Fire Statistical Yearbook of China-2013*. Beijing, China: The Personnel Press of China.

The Fire Bureau of the Ministry of Public Security. 2015. *Fire Statistical Yearbook of China-2014*. Kunming, China: Yun'nan people's Publishing House.

The Fire Bureau of the Ministry of Public Security. 2016. *Fire Statistical Yearbook of China-2015*. Kunming, China: Yun'nan people's Publishing House.

The Fire Bureau of the Ministry of Public Security. 2017. *Fire Statistical Yearbook of China-2016*. Kunming, China: Yun'nan people's Publishing House.

Troitzsch, J. H. 2000. Fire gas toxicity and pollutants in fires: The role of flame retardants. In *Flame Retardants 2000*, pp. 177–184. London, UK: Interscience Communications.

Wan, J. T., Gan, B., Li, C. et al. 2015. A novel biobased epoxy resin with high mechanical stiffness and low flammability: Synthesis, characterization and properties. *Journal of Materials Chemistry A* 3(43): 21907–21921.

Wan, J., Gan, B., Li, C. et al. 2016. A sustainable, eugenol-derived epoxy resin with high biobased content, modulus, hardness and low flammability: Synthesis, curing kinetics and structure–property relationship. *Chemical Engineering Journal* 284: 1080–1093.

Wang, S., Ma, S., Xu, C. et al. 2017a. Vanillin-derived high-performance flame retardant epoxy resins: Facile synthesis and properties. *Macromolecules* 50(5): 1892–1901.

Wang, X., Pan, Y. T., Wan, J. T. et al. 2014. An eco-friendly way to fire retardant flexible polyurethane foam: Layer-by-layer assembly of fully bio-based substances. *RSC Advances* 4(86): 46164–46169.

Wang, X., Song, L., Yang, H. Y. et al. 2011. Synergistic effect of graphene on antidripping and fire resistance of intumescent flame retardant poly(butylene succinate) composites. *Industrial & Engineering Chemistry Research* 50(9): 5376–5383.

Wang, X., Zhou, S., Guo, W. W. et al. 2017b. Renewable cardanol-based phosphate as a flame retardant toughening agent for epoxy resins. *ACS Sustainable Chemistry & Engineering* 5(4): 3409–3416.

Zhang, Q., Wang, Y. C., Bailey, C. G. et al. 2018. Quantifying effects of graphene nanoplatelets on slowing down combustion of epoxy composites. *Composites Part B: Engineering* 146: 76–87.

Zhang, X., Akram, R., Zhang, S. et al. 2017. Hexa(eugenol)cyclotriphosphazene modified bismaleimide resins with unique thermal stability and flame retardancy. *Reactive and Functional Polymers* 113: 77–84.

Zhou, K. Q., Zhang, Q. J., Liu, J. J. et al. 2014. Synergetic effect of ferrocene and MoS2 in polystyrene composites with enhanced thermal stability, flame retardant and smoke suppression properties. *RSC Advances* 4(26): 13205–13214.

Zou, G. W., Huo, Y., Chow, W. K. et al. 2018. Modelling of heat release rate in upholstered furniture fire. *Fire and Materials* 42(4): 374–385.

2

Mechanisms and Modes of Action in Flame Retardancy of Polymers

Yuan Hu
and Yan Zhang

2.1 Introduction

Polymer combustion is driven by thermally induced decomposition (pyrolysis) of solid polymer into smaller fragments, which then volatilize, mix with oxygen, and combust. The combustion process releases much heat, which reradiates onto the unburned polymer, thus continuing to drive pyrolysis and combustion until a lack of heat/fuel/oxygen causes the fire to extinguish. This is admittedly a simplistic explanation, but it holds basically true for just about all polymeric materials. Thermoplastic polymers have a tendency to drip and flow under fire conditions, which can lead to additional mechanisms of flame spread or propagation, whereas thermo-set polymers tend to not drip and flow, but instead produce pyrolysis gases from the surface of the sample directly into the condensed phase. A general schematic of this behavior is shown in Figure 2.1 (Morgan and Gilman 2013).

Flame retardant systems are intended to inhibit or stop the polymer combustion process. Herein, we present an abbreviated discussion and refer the reader to more detailed reviews published elsewhere. The main modes of action of flame retardant systems are reported and discussed below.

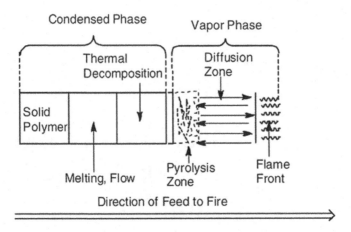

FIGURE 2.1   General schematic of polymer decomposition and combustion behavior. (Reprinted from Morgan, A.B. and Gilman, J.W., *Fire Mater.*, 37, 259–279, 2013. With permission.)

2.2 Physical Mode

The endothermic decomposition of some flame retardant additives induces a temperature decrease by heat consumption. This involves some cooling of the reaction medium to reduce the polymer combustion temperature. In this category, we can mention hydrated alumina or magnesium hydroxide, which start liberating water vapor at approximately 200°C and 300°C, respectively. Such a marked endothermic reaction is known to act as a "heat sink" (Bundersek et al. 2016; Cardenas et al. 2008; Nachtigall et al. 2006).

When the flame retardants decompose, with the formation of inert gases (H_2O, CO_2, NH_3, etc.), the combustible gas mixture is diluted, which limits the concentration of reagents thus reducing possibility of ignition (Wang et al. 2010; Wei et al. 2016).

In addition, some flame retardant additives lead to the formation of a protective solid or gaseous layer between the gaseous phase where combustion occurs and the solid phase where thermal degradation takes place. Such a protective layer limits the transfer of matter such as combustible volatile gases and oxygen. As a result, the amount of decomposition gases produced is significantly decreased. Moreover, the fuel gases can be physically separated from the oxygen, which prevents the combustion process being sustained (Bourbigot and Fontaine 2010).

2.2.1 Heat Sinks

Additives such as clays, talc, silica, or magnesia may simply represent non-combustible material which has to be heated up in order to bring the combustible part of the mass to ignition temperature. This mode of action generally requires very high loadings and is most satisfactory in elastomers which can tolerate such loadings without losing useful mechanical properties. It is mainly of value in combination with other flame retardants.

A more potent "heat sink" mode is that of additives which decompose with uptake of substantial heat (endothermic decomposition) and which do not release flammable vapors on doing so. Leading flame retardant additives with this property are aluminum hydroxide (alumina trihydrate, ATH, which begins to release water around 200°C), magnesium dihydroxide (MH, which releases water at about 300°C), basic magnesium calcium hydrate carbonate, (hydromagnesite), huntite-hydromagnesite, hydrated zinc borates, and to some extent, uncalcined clays (Hewitt et al. 2016; Laoutid et al. 2009, 2013, 2017; Nakamura et al. 2017; Rigolo and Woodhams 1992).

Calcium carbonate is somewhat too stable to be effective in this mode except in some formulations where it can provide a high temperature endothermic action (Wu et al. 2016). Mostly metal hydroxide and carbonate additives function in this way (Equations 2.1 and 2.2) (Morgan and Gilman 2013; Morgan and Wilkie 2014; Sawada et al. 1979).

Their thermal decomposition is endothermic and produces large amounts of non-combustible gasses, such as H_2O and CO_2. Not only is the water release endothermic, but also the water thus released takes up heat radiated from the flame on its way into the flame zone. In addition, it dilutes whatever fuel is also released from the pyrolysis of the polymer. The former (dilution effect) refers to the release of non-combustible vapors during combustion, diluting the oxygen supply to the flame or diluting the fuel concentration to below the flammability limit. Moreover, the residual solids after the water release has occurred still provide heat- and mass-transfer barriers. Besides heat sinks which are water-releasing and fuel-diluting, some other cases are of importance. Melamine is used as a flame retardant component, sometimes by itself in polyurethane elastomers, but more often in combinations with other flame retardants such as with chloroalkyl phosphates in flexible polyurethane foams. Part of its mode of action must be endothermic evaporation, endothermic dissociation (see below), and dilution of the fuel with a low-combustibility vapor.

$$2Al(OH)_3 \rightarrow Al_2O_3 + 3H_2O_{(g)} \tag{2.1}$$

$$4MgCO_3 \cdot Mg(OH)_2 \cdot 4H_2O \xrightarrow{\quad 4H_2O_{(g)}\quad} 4MgCO_3 \cdot Mg(OH)_2 \xrightarrow{\quad H_2O_{(g)}\quad} 4MgCO_3 \cdot MgO \xrightarrow{\quad 4CO_{2(g)}\quad} 5\,MgO \tag{2.2}$$

Yen et al. (2012) studied the flame retardancy of ethylene-vinyl acetate copolymer (EVA) in combination with metal hydroxide and nanoclay. It is suggested that the metal oxide layer on the burning surface is reinforced by the formation of silicate layer, which is both structured and compacted and acts as the insulation, and the newly formed layer responds to the synergistic effect of flame retardancy as well as smoke suppression observed in the EVA blends. The results obtained from cone calorimeter tests showed that EVA/MH composite containing 2 wt% of the nanoclay would give the best flame retardancy of the specimen. The mode of flame retardancy of the nanoclay may be ascribed to the formation of a stable barrier layer. Although EVA/ATH composite containing 1 wt% of the nanoclay presents the best results in cone calorimeter test, it is better to choose EVA/ATH with 2 wt% of the nanoclay, which results in lower limiting oxygen index (LOI) values. This is attributed to the fact that higher nanoclay content in EVA/ATH composite leads in a higher reinforced barrier layer, which improves performance in LOI and smoke density chamber tests.

Mineral fillers are very old technology, with definite references from the 1920s and some references suggesting that they may have been in use as far back as the seventeenth century. Regardless, they are a proven technology and are perceived to be very environmentally friendly. Further, under fire conditions, they tend to greatly lower smoke and reduce overall toxic gas emissions because the mineral filler is replacing flammable polymer fuel with non-flammable inorganic mass. These fillers also tend to be fairly inexpensive and can be readily coated with surfactants to make their use in polymers easier. These materials, however, suffer from two main drawbacks. The first is that they have a limited fire performance window. Specifically, once enough heat has consumed all the mineral filler and all the water/ CO_2 has been released, the metal oxide left behind provides no additional protection to the polymer. So, the mineral filler can delay ignition and slow initial flame growth, but it cannot stop it completely if enough constant external heat is applied. Another drawback is that, for mineral fillers to be effective, high loadings of mineral fillers are needed to flame retard the plastic, often at the expense of polymer mechanical properties.

2.2.2 Fuel Diluents

This is one of the least understood areas of action mode and indeed often overlooked despite the possibility of its great importance. Studies done with flame extinguishants (not in polymer systems) show a remarkable quantitative relation between flames inhibiting action and computed endothermic dissociation enthalpy and heat capacity (Ewing et al. 1994). In a study more related to polymers, the heat capacity effect of hydrogen bromide and hydrogen chloride was shown to be a main (although probably not exclusive) contributor to their mode of action (Lewin and Weil 2001). This may be a more important part of the action of hydrogen chloride and a lesser part of the action of hydrogen bromide. In both cases, fuel dilution by these non-combustible gases is important, as is the effect of an out flowing stream of non-combustible gas causing whatever fuel is also in the stream to be propelled past the flame zone before the fuel is fully oxidized.

A further mode of physical action of melamine (see prior section) is the reported dissociation of the intact melamine molecule in the vapor phase to form cyanamide, which may further dissociate endothermically to nitrogen-rich poor fuels (Braun et al. 2007; Gu et al. 2007).

As reported, the melamine polyphosphate shows some fuel dilution and a significant barrier effect. Using a combination of aluminum phosphinate and melamine polyphosphate results in some charring and a dominant barrier effect.

Chlorine-rich polymers such as polyvinyl chloride (PVC) or polyvinylidene chloride (PVDC), as well as chloroparaffins, also dissociate endothermically with release of HCl, a non-fuel (Georlette et al. 2000).

Some evidence was adduced that tris(dichloroisopropyl) phosphate, a widely used flame retardant in flexible polyurethane foam, may be working largely in a physical mode (at least in the upward burning configuration), by evaporating as the intact molecule prior to evolution of fuel (diisocyanate) from breakdown of the foam (Marklund et al. 2003). This phosphate is a poor fuel or even a non-fuel, and its outward flow of vapor from the heated foam prevents ignition of the foam as well as cooling the surface of the foam. This may not be the only mode of action in downward-burning tests, there is evidence of a surface barrier effect.

2.2.3 Barrier Effects

A gaseous "blanket" of non-combustible gases over the surface of the fire-exposed polymer, excluding oxygen from the surface, is often hypothesized as a flame retardant mechanism, particularly for halogen or halogen-antimony systems. The importance of excluding oxygen from the surface is surprisingly an unsettled question. In some fire configurations, the surface is extremely oxygen-depleted, and fire modeling is often done using a simplifying assumption that the pyrolysis of the polymer is anaerobic and that all the oxidation takes place in the flame zone. Some compelling evidence that this is not true was published by Celanese chemicals researchers who showed major surface and subsurface oxidation in the case of polypropylene. This could be an important mechanism where there is access of air to the burning surface, as in the important situation of upward burning where air is probably convected into the zone between the flame and the heated surface. Surface oxidation is often neglected in theoretical modeling, but needs more experimental investigation.

Another physical mode of action, often resulting from a chemical mode of action such as char formation, is formation on the surface of a fire-exposed polymer to prohibit heat and mass transfer. Charrable polymers will generally form such a barrier, especially if induced to do so by a char-inducing catalyst. This is a main mode of action of many phosphorus-containing flame retardants when they are used in polymers that have char-forming characteristics, mainly those polymers which have oxygen or nitrogen in their structure (Chao et al. 2015). The char-forming action is also often endothermic (at least in its earlier stages) and water-releasing, providing further contributions to flame extinguishment.

Importantly, a char barrier must be reasonably coherent and continuous. If it has too many holes or cracks, it can be ineffectual. It is believed that so-called "turbostratic" char which is composed of

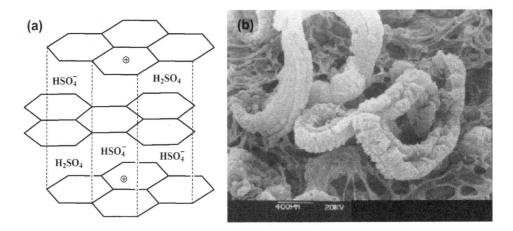

FIGURE 2.2   (a) Structure of expandable graphite intercalated by H_2SO_4 and (b) char layer of expandable graphite filled foam. (Reprinted from Modesti, M. et al., *Polym. Degrad. Stab.*, 77, 195–202, 2002. With permission.)

amorphous phase of carbon black and various precursors of the char with randomly distributed islands of graphitized carbon is the most efficient as a protective layer, because the amorphous part usually molten at the high temperature helps keeping flexibility and fluidity of the char, whereas graphite regions provide reinforcement and protect against oxidation (burn out, glowing). Little or no flame retardancy results from dispersing powdered charcoal or carbon black in a polymer. If the char is made to form in a poorly charring polymer by adding a powdered char-forming additive, it is important that the char be coherent, not particulate. It seems important that if a char-forming powder is added to a poorly charring polymer in order to form a char barrier, the char-forming additive should melt and flux together prior to charring.

An exception to the futility of adding powdered carbon is the great benefit of adding expandable graphite (acid-treated graphite) as shown in Figure 2.2 (Modesti et al. 2002). Although it is ineffectual as a fine powder, granular expandable graphite is very effective. It expands about 100 times more than its original volume and forms a thick mat of tangled twisted graphite fibers, and although this is not a continuous film, the expanded graphite layer is thick enough and formed rapidly enough to provide a heat-transfer barrier which is very effective in flame retardancy. Expandable graphite can be used where coarse black electrically conductive filler is acceptable.

Another postulate that is often made for the mode of action of non-volatile phosphorus flame retardants is that they coat the surface of the burning polymer with phosphoric or polyphosphoric acid (Cheng et al. 2016; Xu et al. 2017). This is qualitatively observable, for instance, by infrared or even by washing off and titrating the acid. It is surely an important part of the afterglow inhibition action of phosphorus flame retardants. The coating effect is suspected of being quite important, and may be involved in some of the strong effects of phosphorus synergism by some nitrogen compounds or metal salts, but quantitative studies of this mode of action are lacking. Coating with phosphoric acid is probably quite important in intumescent systems, where typical additives such as ammonium polyphosphate can generate polyphosphoric acid, which must coat the char. There, it may serve as a barrier to the diffusion of oxygen to the polymer or char surface or to some extent as a barrier of diffusion of the fuel to the flame. It is also known that phosphorus compounds can deactivate especially oxidizable sites on carbon.

Another significant family of barrier materials is silicaceous materials. The addition of silica by itself is ineffectual because it is too high melting. Small amounts of silica can aid or oppose other flame retardant actions by exerting melt-flow and wick effects. Silicic acid (hydrated silica) is effective to a commercial degree in viscose rayon (Visil®). Work at National Institute of Standards and Technology (NIST) also showed that silica gel provides flame retardant action through modification of polymer melt viscosity and building siliceous barrier. Very low melting alkali metal silicates are generally water-soluble, but are suitable for non-durable (disposable) flame retardant applications such as paper. Some moderately low melting borate-silicate glasses, available as powders (frits) seem to be useful contributors to the overall flame retardant effect of a multi-component formulation.

Organic silicon-containing products, such as some silicones, are useful, generally not as freestanding flame retardants, but as adjuvants. For example, phenyl methylsiloxanes appear to be useful flame retardant components in epoxy-based printed wiring boards, where another main contributor to the flame retardancy is based on a particular type of char-forming epoxy. Phenyl methylsiloxanes are also useful in polycarbonates where they seem to have two modes of action:

1. In the molten polymer they migrate quickly to the surface
2. They form siloxane cross-links which prevent dripping.

Some oligomeric methylsiloxanes are useful as flame retardant adjuvants, for example, Dow Corning DC-4071. It is more likely that they do not form a continuous siliceous barrier, but may work more by forming a radiant-heat-reflective layer on burning plastics.

Chao et al. (2015) synthesized a novel reactive-type phosphorus-nitrogen-silicon-containing flame retardant named after PSAP which have excellent flame retardancy and good compatibility with cycloaliphatic epoxy (Commercial name: ERL4221). The possible thermal degradation mechanism on charring can be proposed based on the thermogravimetric analysis/infrared spectrometry (TG-IR), SEM, and FT-IR tests: in the initial stage, when the modified flame retardant epoxy resin (FR-EP) is heated, the C-O-C groups from ERL4221 and formed between the hydroxyl of PSAP and ERL4221 were broken and released non-combustible gases such as CO_2 and H_2O during the combustion process. In addition, on increasing the temperature, the C-N-C structures in PSAP were decomposed accompanied by the formation of some amine groups and formyl groups simultaneously. Then, along with the release of NH_3, H_2O, and CO_2 the aminopropyl decomposes completely to produce a great amount of siloxanes containing active hydroxyl groups which promoted the formation of tight, smooth, compact, and stable char layer through a condensation reaction. Finally, the excellent flame retardant performance of FR-EP was achieved.

Low melting glasses, and especially devitrifying glasses which form a ceramic layer when strongly heated, also have been often proposed as barrier components, although commercial examples are few. Zinc alkali phosphate glasses from Dow Corning are examples of such glasses.

Borates also can be barrier formers. Particularly in combination with ATH, $Mg(OH)_2$ or other solid mineral additives, zinc borates can form a sintered mass which acts as a fire barrier. Boric acid in cotton batting fluxes when exposed to flame, releases water, provides an endotherm, and forms a protective glass. Sodium borate (borax) has a long history of flame retarding cellulose (wood, textiles, paper) and a substantial part of its action is attributed to formation of a glassy coating, which may be foamed due to the release of water vapor (Braun et al. 2007; Garba 1999).

Among them, zinc borates such as $2ZnO \cdot 3B_2O_3 \cdot 3.5H_2O$ are the most frequently used. Their endothermic decomposition (503 kJ/kg) between 290°C and 450°C liberates water, boric acid, and boron oxide (B_2O_3). The B_2O_3 formed softens at 350°C and flows above 500°C leading to the formation of a protective vitreous layer. In the case of oxygen-containing polymers, the presence of boric acid causes dehydration, leading to the formation of a carbonized layer. This layer protects the polymer from heat and oxygen. The release of combustible gases is thus reduced.

Layered clays, such as montmorillonite, can be made to separate into individual leaflets of about 1 nanometer thickness and orders-of-magnitude greater length and breadth. These are usually called "nanoclays." The separation into leaflets (exfoliation or delamination) is accomplished by introducing compounds or polymers which spread and cause the release of the leaflets. In a burning polymer, these nanoclays form surface barriers as the polymer burns away and contribute to flame retardancy or at least to reduce rate of heat release. The nanoclays have not been found to be "stand alone" flame retardants, but usually contribute to flame retardancy in combination with other retardants. In char-forming systems (see below), they often form good char-clay barriers on the burning polymer.

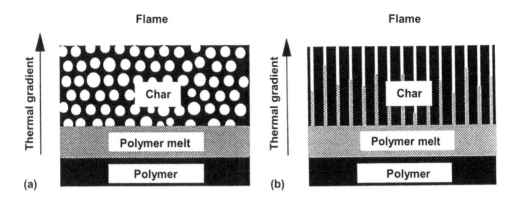

FIGURE 2.3 Diagrammatical illustrations of (a) ideal char structure, (b) poor char structure. (Reprinted from Price, D. et al., Introduction, in: *Fire Retardant Materials*, ed. Horrocks, A.R. and Price, D., p. 19, Cambridge, UK: Woodhead Publishing Ltd, 2001. With permission.)

In another study, the effects of intumescent flame retardant (IFR) incorporating organically modified montmorillonite (O-MMT) on the flame retardancy and melt stability of polylactic acid (PLA) were investigated (Li et al. 2009). The char layer protecting the polymer matrix plays a very important role in the performance of flame retardant. Figure 2.3 shows two char structures during combustion (Price et al. 2001). An ideal charring structure should have many integrated closed honeycomb pores; this structure is beneficial to form an adequate temperature gradient in the charring layer to protect the molten mass and matrix underneath (Figure 2.3a). In contrast, the structure shown in Figure 2.3b has a lot of channels and apertures allowing gas and molten mass of polymers to overflow and enter the flame region, thus it cannot isolate heat transfer from the matrix.

Over the past few decades, layered double hydroxide (LDH) has been incorporated into polymers as a flame retardant nanoadditive because of its high water content, non-toxicity, and layered structure. Wang et al. (2015) has synthesized a bio-based modifier (cardanol-BS) using cardanol from the renewable resource cashew nut shell liquid (CNSL), via ring-opening of 1,4-butane sultone (BS). Cardanol-BS modified-LDH was incorporated into epoxy resins (EPs) with different loadings. From the results, pristine LDH did not show so high an efficiency as modified-LDH in terms of the reduced peak heat release rate (PHRR), total heat release (THR), and total smoke production (TSP), and also the EP/LDH-6% composite exhibited no rating in the UL-94 vertical burning test. These findings supported that the flame retardant behavior increased with improved dispersion of nano filler in the polymer matrix. As shown in Figure 2.4, the well-dispersed modified-LDH nano fillers were beneficial to improving the quality of char residue, which effectively inhibited flammable volatiles escaping from the interiors and served as an effective thermal insulation layer to shield the underlying matrix from the exterior heat irradiation.

Except for the above layer materials, nano layer materials also played an important role these days. Graphite has a two-dimensional (2D) layered structure and is the most thermodynamically stable form among all of the carbon allotropes. Kim et al. (2014) used a newly developed ball-milling method for the synthesis of phosphorous modified graphene oxide (GO). And then the obtained graphene phosphonic acid (GPA) was coated on the Hanji paper. The physical causes of retardation are related to the cooling of the burning surface by endothermic reactions and vaporization. Chemical retardation is related to the thermal condensation of phosphonic acid into char formation on the surface of solid fuel (Hanji).

Besides, recent works have shown that covalent organic frameworks (COFs) nanosheets can also play a barrier effect during the combustion of polymer. Mu et al. (2017) synthesized melamine/o-phthalaldehyde covalent organic frameworks sheets and used them in the thermoplastic polyurethanes (TPU) matrix as flame retardant. The addition of COFs cannot only block heat radiation and release of gaseous pyrolysis products, but also increase graphitic degree and weight amount of char residues. The higher-quality char residues contribute to flame retardant of TPU through physical barrier effect.

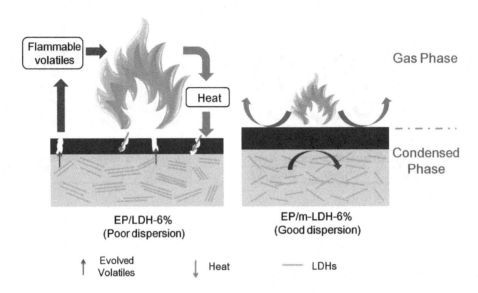

FIGURE 2.4 Schematic illustration of the flame retardant mechanism of cardanol-BS modified LDH in EP composites. (Reprinted from Wang, X. et al., *ACS Sustain. Chem. Eng.*, 3, 3281–3290, 2015. With permission.)

It is worth noting that barrier formation mechanisms, such as surface char formation, have an advantage over flame inhibition mechanisms because they do not cause an increase in smoke, carbon monoxide, and corrosive gases.

2.3 Chemical Mode

Flame retardancy through chemical modification of the fire process can occur in either the gaseous or the condensed phase. The free-radical mechanism of the combustion process can be stopped by the incorporation of flame retardant additives that preferentially release specific radicals (e.g., Cl• and Br•) in the gas phase. Flames are complex series of oxidation reactions, but the rate controlling step is a branching step, where atmospheric oxygen (O_2) reacts with hydrogen atom H• (derived from the fuel) to form O atom (a very reactive species) and hydroxyl radical (OH•, also a very reactive species). The O atom can attack a hydrogen molecule or a C–H structure to make OH radical and an H atom or carbon radical. Thus, one high-energy species yields two high energy species, a branching step which is critical to flame propagation. The main heat-producing step in a typical flame is OH• attacking CO to form CO_2 and H atom.

In the condensed phase, two types of chemical reactions triggered by flame retardants are possible: first, the flame retardants can accelerate the rupture of the polymer chains. In this case, the polymer drips and thus moves away from the flame action zone. Alternatively, the flame retardant can cause the formation of a carbonized (perhaps also expanded) or vitreous layer at the surface of the polymer by chemical transformation of the degrading polymer chains. This char or vitrified layer acts as a physical insulating layer between the gas phase and the condensed phase.

2.3.1 Flame Inhibition Effects

Halogen-containing additives (RX, where the R refers to C_nH_{2n+1} ($n \geq 0$) and X refers to the halogen atom) act by interfering with the combustion cycle in the gas phase, where the key combustion radicals (OH• and H•) are removed by decomposed halogenated species (Equations 2.3 through 2.6) thus effectively interfering in their oxidation. This modification of the combustion reaction pathway leads to a marked decrease in the exothermicity of the reaction, leading to a decrease of temperature, and therefore a reduction in the fuel produced. As a result, the effective flame retardant species is HX, which is replaced with the less reactive X• and regenerated by chemical equation (2.4). The difference of effectiveness of halogen-containing flame retardants has been attributed to the rate of Equations (2.5) and (2.6). Accordingly, the order of halogen effectiveness is F < Cl < Br ≈ I.

$$RX \rightarrow R^{\bullet} + X^{\bullet} \qquad (2.3)$$

$$X^{\bullet} + R'H \rightarrow R''^{\bullet} + HX \qquad (2.4)$$

$$HX + H^{\bullet} \rightarrow H_2 + X^{\bullet} \qquad (2.5)$$

$$HX + {}^{\bullet}OH \rightarrow H_2O + X^{\bullet}. \qquad (2.6)$$

Industrially speaking, producing organobromine compounds as flame retardants, turns out to be a very efficient chemistry, making it possible to produce large amounts of organobromine compounds cost effectively. This is because of the unique aspects of bromine and carbon chemistry that make these compounds easy to perbrominate, and so one small organic molecule can deliver a high payload of effective bromine to the fire. As indicated in the previous paragraph, the range of structures is widely varied for brominated flame retardants, and just some of the more common structures are shown in Figure 2.5 (Zhang et al. 2016). It should be noted that not all organobromine compounds make cost effective flame retardants. The flame retardant must be tailored to be compatible with the polymer, must have the right cost, and it must release its bromine under the right fire conditions—not too soon before the onset of polymer decomposition, but not too far after the polymer has begun to completely decompose. So, the structures in Figure 2.5 exist for a reason: they were found to work and have optimal properties for the polymers they are used in today. Halogenated flame retardants are sometimes used with synergists such as antimony oxide, zinc borate, or other phosphorus chemistry, as these other elements help to make the halogens more efficient in the vapor phase.

Of course, preventing access of oxygen (such as by a barrier or outgoing stream of gas) will prevent that step, but in most ventilated flames, oxygen is sufficiently available. The scarcer (rate-limiting) species, H•, can be scavenged in either of two ways. One important way is that H• can be made to react with some component (such as HBr) to give the less reactive H_2 and the relatively less reactive Br•. The hydrogen atom H• can also react with volatile antimony tribromide or trichloride to form HBr or HCl and the

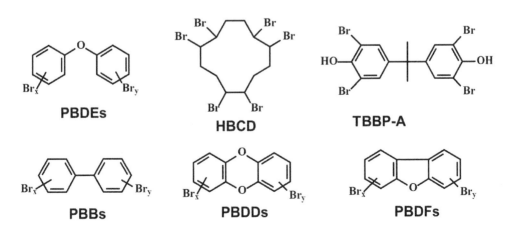

FIGURE 2.5 Bromine FR chemistry structures.

relatively less reactive $SbBr_2\bullet$ or $SbCl_2\bullet$ which can further undergo reactions with $H\bullet$ to give SbBr and SbCl, and ultimately Sb.

Antimony oxide is commonly used as a synergist with halogen-based flame-retardants. Antimony oxide works in the gas phase by facilitating the transfer of halogen and antimony into the gas phase for flame inhibition (Li et al. 2012). It has been proposed that antimony oxide gets converted to volatile antimony species, acting as an effective radical species that interrupt the combustion cycle. The sequence of reactions is proposed as the following steps (Zanetti et al. 2002).

$$Sb_2O_3 + 2HX \rightarrow 2SbOX + H_2O \tag{2.7}$$

$$5SbOX \rightarrow Sb_4O_5X_2 + SbX_{3(g)} \tag{2.8}$$

$$SbX_3^{\bullet\bullet} + H^\bullet \rightarrow SbX_2 + HX \tag{2.9}$$

$$SbX_2^{\bullet\bullet} + H^\bullet \rightarrow SbX_2 + HX \tag{2.10}$$

$$SbX^\bullet + H^\bullet \rightarrow Sb + HX \tag{2.11}$$

$$Sb + {}^\bullet OH \rightarrow SbOH \tag{2.12}$$

$$Sb + O^{\bullet\bullet} \rightarrow SbO \tag{2.13}$$

$$SbO + H^\bullet \rightarrow SbOH \tag{2.14}$$

$$SbOH + H^\bullet \rightarrow SbO + H_2 \tag{2.15}$$

$$SbOH + HO^\bullet \rightarrow SbO + H_2O. \tag{2.16}$$

Besides the halogens and antimony, there are other powerful vapor phase flame inhibitors. A broad comparison was done at NIST using measurement of the inhibition of premixed fuel-air flames. Interestingly, various phosphorus species were at the most active end of the scale, the halogens being roughly in the middle. Some relatively volatile phosphorus compounds are believed to exert their activity to a large degree in vapor phase flame inhibition; examples being dimethyl methylphosphonate (which is used in thermo-set resins), and triphenyl phosphate, probably lower-alkyl-substituted phenyl phosphates, and to some degree, tetraphenyl resorcinol diphosphate in styrenic plastics. The mode of action may be analogous to that of HBr, namely, phosphorus species, perhaps small fragments such as PO_2 and HPO_2, may scavenge hydrogen atoms and act as flame inhibitors.

Phosphorus compounds are unique in that they can be vapor phase or condensed phase flame retardants, depending upon their chemical structure and their interaction with the polymer under fire. Especially the effect of phosphorus oxides on recombination of $H\bullet$ and $OH\bullet$ species was investigated in presence of PH_3. Subsequent kinetic studies proposed that the radical phosphorus species such as $PO\bullet$, $PO_2\bullet$, and $HPO\bullet$ are key flame inhibitors (Green 1992). 9,10-dihydro-9-oxa-10-phosphaphenanthrene-10-oxide (DOPO) as a commercial one is notable for its growing use in epoxy circuit boards. In general, the sequence of reactions includes the following steps.

$$PO^\bullet + H^\bullet \rightarrow HPO \tag{2.17}$$

$$PO^\bullet + {}^\bullet OH \rightarrow HPO_2 \tag{2.18}$$

$$HPO + H^{\bullet} \rightarrow H_2 + PO^{\bullet} \tag{2.19}$$

$$^{\bullet}OH + H_2 + PO^{\bullet} \rightarrow H_2O + HPO \tag{2.20}$$

$$HPO_2^{\bullet} + H^{\bullet} \rightarrow H_2O + PO \tag{2.21}$$

$$HPO_2^{\bullet} + H^{\bullet} \rightarrow H_2 + PO_2 \tag{2.22}$$

$$HPO_2^{\bullet} + HO^{\bullet} \rightarrow PO_2 + H_2O. \tag{2.23}$$

Several other studies have demonstrated that sulfur species can also provide a degree of inhibition of H• and •OH in the flame. The catalytic cycle for flame inhibition by SO_2 was investigated (Rasmussen et al. 2007). However, it was found that the HSO_2/HSO_3 system is responsible for removal of H• and •OH species in flame.

Another way to scavenge the hydrogen atoms (H•) to quench the flame is to cause them to recombine to form relatively unreactive molecular hydrogen (H_2). It is a physical law that two fast-moving bodies, such as atoms, cannot recombine unless the momentum can be carried away, generally by a third body. This requires third body collisions which are infrequent, unless there is a surface present. Finely divided solids, such as smoke particles or a vapor-phase finely divided solid antimony species, or the white smoke from sublimed melamine, may serve as hydrogen atom recombination sites.

Many solid powders can also quench flames efficiently; this mode of flame extinguishment being used in solid fire extinguishers. Recombination sites may play a part in this mode of action, but endothermic volatilization of the solid is probably more important. It is possible that part of the action of melamine (which forms a white smoke) is of this type.

It has been reported that certain metallic compounds can also be effective as radical quenchers for flame inhibition. Studies based on a series of metallic inhibitors suggested a catalytic effect on radical recombination rates and effectiveness in reducing the burning velocity of the flame (Linteris et al. 2004, 2008). The pronounced effect of transition metal elements, such as Cr, Mn, Sn, U, and Ba on the recombination of H atoms in fuel has been reported (Bulewicz and Padley 1971). A mode of action was proposed based on the following equations, where X and M are any third body and metal center, respectively.

$$MO + H^{\bullet} \rightarrow MOH^{\bullet} \tag{2.24}$$

$$MOH^{\bullet} + X \rightarrow MOH + X^{\bullet} \tag{2.25}$$

$$MOH^{\bullet} + H^{\bullet} \rightarrow MO + H_2. \tag{2.26}$$

In the following work, Bulewicz and coworkers (1971) studied the catalytic effects of 21 metal species on recombination of chain-carrying flame radicals present at super-equilibrium levels in the post-flame zone of premixed, fuel rich, flames of $H_2/O_2/N_2$ at 1860 K. Figure 2.6 and Table 2.1 show the ratio of the catalyzed to un-catalyzed recombination rate for H atom caused by each of the metals, added at 1.3 mL/L. The cut-off value for this ratio was arbitrarily set to 1.1, and those above that value were described as having a strong catalytic effect (Cr, U, Ba, Sn, Sr, Mn, Mg, Ca, Fe, and Mo, in that order). Those species having lesser, but still measurable effect were Co, Pb, Zn, Th, Na, Cu, and La, while V, Ni, Ga, and Cl (included for comparison purposes) had no effect at these volume fractions. The possibility of heterogeneous recombination on particles was admitted, but at these low volume fractions, the authors argued for a homogeneous gas-phase mechanism involving H or OH reaction with the metal oxide or hydroxide (Linteris et al. 2008).

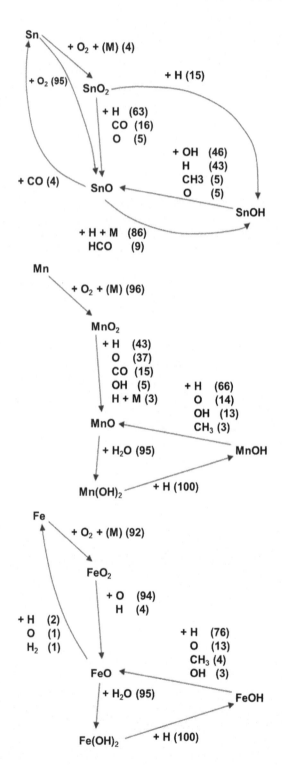

FIGURE 2.6 Reaction pathways for Sn, Mn, and Fe in a premixed methane-air flame ($\Phi = 1.0$, $X_{O2,\ ox} = 0.21$, $T_{in} = 353$ K). TMT, MMT, and $Fe(CO)_5$ present at (1963, 128, or 105) μL/L, respectively. The numbers in parentheses are the fractional consumption (percent) of the reactant molecule. (Reprinted from Linteris, G.T. et al., *Prog. Energy Combust. Sci.*, 34, 288–329, 2008. With permission.)

TABLE 2.1 Catalytic Efficiency of Different Metals in Promoting Radical Recombination

Strong Effect		Some Effect		No Effect	
Cr	2.8	Co	1.1	V	1
U	1.82	Pb	1.1	Ni	1
Ba	1.75	Zn	1.07	Ca	1
Sn	1.6	Th	1.06	Cl	1
Sr	1.35	Na	1.04		
Mn	1.3	Cu	1.04		
Mg	1.25	La	1.04		
Ca	1.25				
Fe	1.2				
Mo	1.16				

Source: Linteris, G.T. et al., *Prog. Energy Combust. Sci.*, 34, 288–329, 2008.
Values of k_{obs}/k_{uncat} ($T = 1860$ K; $X[M] = 1.3$ µL/L; $H_2/O_2/N_2 = 3/1/6$).

For example, a gas-phase mechanism for the iron-catalyzed radical recombination in flames was developed by Jensen and Jones (1974) and Rumminger et al. (1999), and later, expanded by Rumminger and Linteris (2000a). In both, the main catalytic cycle leading to radical recombination was:

$$FeOH + H \longleftrightarrow FeO + H_2$$
$$FeO + H_2O \longleftrightarrow Fe(OH)_2$$
$$Fe(OH)_2 + H \longleftrightarrow FeOH + H_2O$$
$$\text{(Net: } H + H \longleftrightarrow H_2)$$

This mechanism is schematically shown in Figures 2.7 and 2.8, in which the arrows connect reactants and products, and reaction partners are next to the arrow. Although the mechanism described in Rumminger and Kellogg (Kellogg and Irikura 1999; Rumminger and Linteris 2000b) also included other catalytic cycles besides the two shown in Figure 2.7, they were not found to be particularly important in methane–air flames, either pre-mixed or diffusion. In work with premixed $CO/N_2/O_2/H_2$ flames, however, the new O-atom cycle was found:

$$Fe + O_2 + M \longleftrightarrow FeO_2 + M$$
$$FeO_2 + O \longleftrightarrow FeO + O_2$$
$$FeO + O \longleftrightarrow Fe + O_2$$
$$\text{(Net: } O + O \longleftrightarrow O_2)$$

2.3.2 Char Formation

Polymers which form a high yield of carbonaceous char when strongly heated under air tend to be inherently flame retardant, examples being polyimides, polyaramides, liquid crystal polyesters, polyphenylene sulfide, polyarylenes, and many thermo-sets. In fact, it has long been known that the oxygen index correlates quite well to char yield for the higher charring polymers. One reason is that char represents carbonaceous material that didn't burn, thus the actual heat of combustion is lowered by the theoretical heat of combustion of the char. The second reason is that char represents a barrier to protect

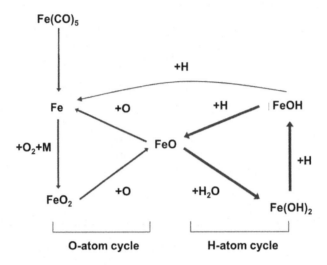

FIGURE 2.7   Schematic diagram of radical recombination reaction pathways is found to be important for CO/H$_2$/O$_2$/N$_2$ flames. Thicker arrows correspond to higher reaction flux. Reaction partners are listed next to each arrow. (Reprinted from Rumminger, M.D. et al., *Combust. Flame*, 120, 451–464, 2000. With permission.)

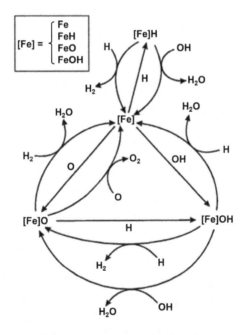

FIGURE 2.8   Schematic representation of different classes of reactions which may contribute to iron's super-efficient flame suppression ability through the catalytic recombination of radical species. (Reprinted from Kellogg, C.B. et al., *J. Phys. Chem. A*, 103, 1150–1159, 1999. With permission.)

the underlying material. Moreover, the formation of char in the case of oxygen-containing polymers is often accompanied by the release of water, or in the case of polycarbonates, the release of carbon dioxide.

Curiously, although the yield of char in relation to flame retardancy has been well studied, the rate of charring in relation to flame retardancy has not been much studied, although some studies indicate its likely importance.

Intumescent flame retardants get their name from their mode of flame retardancy during fire conditions. Specifically, they create a protective carbon foam under fire conditions; they rise up in response to heat (intumesce). This class of flame retardants is strictly condensed phase in its activity, and either provides its own carbon char or uses the polymer as a carbon char source. Intumescents are typically composed of three components that make the carbon char form. The first is an acid catalyst, which causes the carbon source (the second component) to cross-link and form a thermally stable form of carbon. The last component is the spumific or gas former, which causes the carbon source to become carbon foam. These three materials work together to make the intumescent work; by themselves, they provide some flame retardancy, but it is the combination, which provides the real protection when these materials are exposed to heat and flame. Examples of these chemical reactions are shown in Figure 2.9 (how the carbon char forms) and in Figure 2.10 (specific condensed phase reactions). Intumescent flame retardant are typically used to provide fire protection for fire barriers, steel, firewall holes, and applications requiring a high level of fire safety. Very commonly, the intumescents are incorporated into paint or barrier form, which is applied to another substrate, and the intumescent then protects that underlying material from thermal damage for some period of time. This versatility and their mode of flame retardant action mean that these materials are capable of providing very robust fire safety for highly demanding applications, and these features lead to their increasing use today. However, they do have some drawbacks, including water absorption issues (important if the underlying structure needs to be protected against corrosion or electrical short circuit) and low thermal stability. Intumescent systems work by activating well before the polymer obtains a chance to thermally decompose, and so most intumescent materials activate around 180°C–200°C, with some that now go up to 240°C before activating. This unfortunately eliminates their use in higher melting thermoplastics as they would activate during melt compounding into the thermoplastic. Therefore, intumescent materials, whereas they provide excellent fire protection, tend to be limited to lower temperature materials and fire protection barriers.

A substantial class of important polymers, such as polyesters and polyamides, containing oxygen and/or nitrogen, char poorly in the pure form, but can be induced to char by appropriate additives, most commonly, acid-generating additives. Additives that generate volatile acids such as hydrochloric are not especially effective because the acid is immediately volatilized. Less volatile is sulfuric acid, but it too is partially lost by volatilization. Sulfuric acid and its thermally decomposed derivatives such as ammonium sulfate or ammonium sulfamate are effective in polymers that char rapidly, such as cellulose, but are lost too soon in slow-charring polymers. The ideal acid for catalyzing char in burning oxygen- or nitrogen-containing polymers are phosphoric acids or polyphosphoric acids, which are substantially

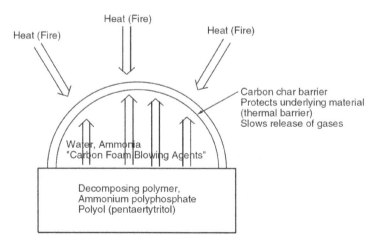

FIGURE 2.9 General schematic of intumescent formation. (Reprinted from Morgan, A.B. et al., *Fire Mater.*, 37, 259–279, 2013. With permission.)

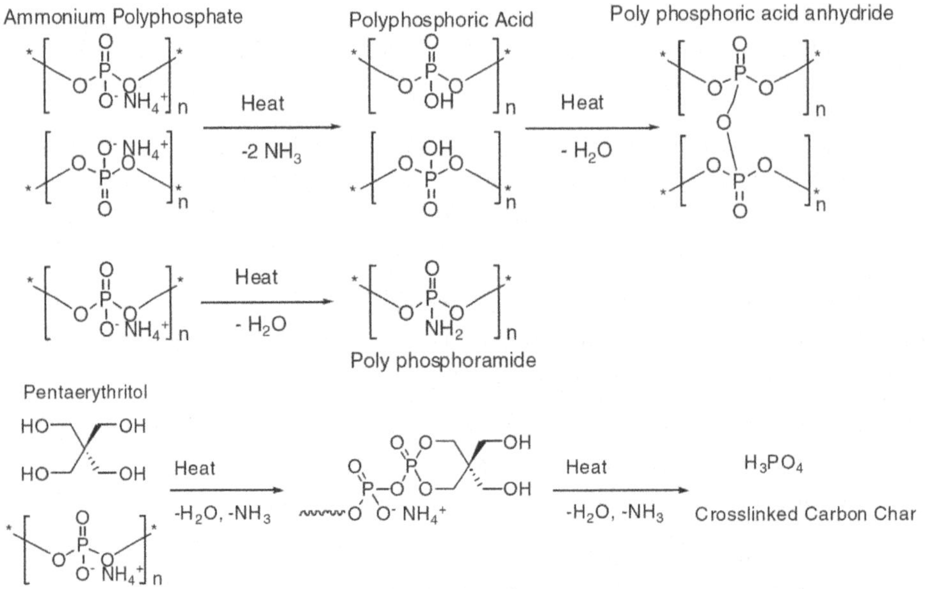

FIGURE 2.10 Intumescent flame retardant chemical reactions. (Reprinted from Morgan, A.B. et al., *Fire Mater.*, 37, 259–279, 2013. With permission.)

non-volatile. Since the acids are usually not suitable as polymer additives, thermally decomposable salts and esters of phosphorus acids are used. These represent one of the leading classes of flame retardants, used in a wide variety of charrable polymers. Further on, we will discuss how these same phosphorus compounds can be used in non-charrable or poorly charrable polymers. The leading char-inducing phosphorus compounds are ammonium phosphates (water soluble), ammonium polyphosphates (both water soluble and insoluble types), melamine phosphate, melamine pyrophosphate, melamine polyphosphate, ethylenediamine phosphate, piperazine pyrophosphate (insoluble polymer), and a wide variety of phosphate and phosphonate esters, which are discussed in other chapters.

For example, Zhao et al. (2016) used polyphosphoric acid (PPA) and melamine as raw materials to synthesize a macromolecular intumescent flame retardant tris(2-hydroxyethyl) isocyanurate polyphosphate melamine (TPM), which integrated char, acid, and gas sources. All these materials are non-toxic, and there is no corrosive and toxic substance in decomposed products, which conforms to the environmental friendly property of PLA. The possible pyrolysis route was carried out. First, tris(2-hydrooxyethyl) isocyanurate (THEIC) polyphosphate ester and melamine in TPM begin to degrade when it is heated to a certain temperature, then the ring-opening degradation of melamine commences, coupled with releasing of NH_3 at 300°C. From 350°C to 450°C, the scission of the phosphate ester bonds occurs via pyrolysis of THEIC polyphosphate ester to olefins and polyphosphoric acid. Further condensation reactions between polyphosphoric acids proceed to form a mixture of phosphorus degradation products containing P–O–P structure with different condensation degree and release some H_2O simultaneously (Twarowski 1993). These high viscous phosphorus-rich degradation products in a glass-like state can attach onto the surface of the matrix, which can serve as a protective barrier to hinder transfer of heat and combustible gas, and further catalyze the matrix into char. In addition, the olefins mentioned previously can develop various charring precursors and further cross-link aromatic structures with higher molecular mass, which is macroscopically presented as black carbonaceous skeleton. This carbonaceous layer can provide a good shield to effectively resist mass and heat transfer and retard degradation of underlying matrix.

Intumescent flame retardant is also adopted as a flame retardant coating to modified fabric. For example, Chen et al. (2015) prepared flame-retardant and self-healing superhydrophobic coatings on cotton fabric by a convenient solution-dipping method, which involves the sequential deposition of a trilayer of branched poly(ethylenimine) (bPEI), ammonium polyphosphate (APP), and fluorinated-decyl poly-hedral oligomeric silsesquioxane (F-POSS). One bilayer of bPEI/APP coating is sufficient to provide an efficient char layer on cotton fabric to extinguish flame because of the intumescent effect.

Polymers which char on heating often cross-link prior to char formation, and moreover, cross-linking between polymer chains is generally an important step during char formation. However, the opposite is not always true; polymers which are cross-linked (thermo-sets) or cross-link during decomposition do not always give extensive char. Charring usually depends on the type of cross-links, e.g., cross-links containing aromatic groups are always beneficial for charring, whereas aliphatic cross-links may decompose as easily as non-crosslinked polymer. Cross-linking, prior to and during char formation, generally reduces the rate and amount of volatile fuel and thus starves the flame. The flame retardants, such as phosphorus compounds, which increase char formation generally, lower the decomposition temperature of the polymer somewhat.

In the case of polyvinyl chloride, which loses its chlorine content as hydrogen chloride when exposed to fire, it is commercially useful to retard smoke formation by introducing solid catalysts, typically Zn-, Sn-, and/or Mo-based, which favor char formation (*via* dehydrochlorination) and lessen cracking to aromatic fragments which tend to form soot. Zinc borates, zinc stannates, and ammonium or melamine molybdates are often used for this purpose.

2.3.3 Drip Promotion or Drip Suppression

The flow properties of molten thermoplastics during ignition or flaming can play a large part in flame retardant modes of action. At the ignition step, melting or shrinking away from the flame can prevent ignition, this being a mode by which thermoplastic textiles can pass a flame test such as National Fire Protection Association (NFPA) 701 or the children's nightwear standard. Solids (such as infusible pigments or the inclusion of infusible fibers) that prevent the flow can aggravate flammability. The same effect may be seen in thermoplastics other than textiles.

A relatively weak flame retardancy standard such as UL-94 V-2, which allows even flaming drips, can usually be passed by such systems as foamed polystyrene or non-reinforced nylons. In the case of the foamed polystyrene, inclusion of a small amount of peroxide or other free-radical-generator which favors fragmentation, can favor the melt- flow-drip mode of extinguishment. In the case of nylons, melamine cyanurate is commonly added to favor non-flaming drip; the presence of a reinforcing solid can defeat this mode of extinguishment. On the other hand, melt flow leading to flaming drip is not allowed in some important test standards such as UL-94 V-0 or V-1. The most usual means to retard it is to include small (<0.5%) amounts of a finely divided shear-fibrillated polytetra fluoroethylene. Finely divided silica or clay can also be used for this purpose.

For example, flame-retardant-free and thermo-cross-linkable copolyesters have been synthesized by Zhao et al. (2014), and their flame retardation and anti-dripping behavior as a consequence of cross-linking during combustion were investigated. The large melt viscosity caused by cross-linked networks at high temperature played the most important role in anti-dripping of copolyesters. It is generally believed that the acromatization process occurs through a radical process. Reactions proposed include the formation of two phenylethynyl groups to one triphenylnaphthalene cross-links, which could play a positive effect on viscosifing and further on anti-dripping.

2.3.4 Additivity, Synergism, and Antagonism of Flame Retardant Combinations

Often, two or more flame retardants will act cooperatively to reach a flame retardancy goal. It is not meaningful to say that they are additive, since the relationship of flame retardant response vs. con-centration is not often linear, so "additivity" is not clearly defined in a quantitative sense.

Sometimes the effect is much larger than would be expected from consideration of what each additive would provide if by itself, and this is usually called "synergism." It is particularly clear where one of the additives, such as antimony oxide, has practically no effect by itself, but gives a large boost to a halogen additive. Many other cases of so-called synergism are not so clear. The topic has been reviewed in detail by one of the present authors.

Some particularly useful positive interactions ("synergism"), besides the antimony oxide-halogen case, are the following: phosphorus-containing textile finishes on cellulosic materials, which are greatly boosted by some, but not all, nitrogen compounds; tris(chloroalkyl) phosphates in flexible polyurethane foams, greatly aided by melamine; tetrabromophthalate esters in PVC, aided by tetrachlorophthalate esters; ATH plus $Mg(OH)_2$, often found better in combination in polyolefins; and combinations of certain chlorocarbons such as Dechlorane Plus® with iron oxide in polyamides. Dechlorane Plus can be also synergistic with decabromodiphenyl ether in polypropylene, which is indicative of Cl-Br synergism. These are discussed in the present volume in chapters on the individual polymer classes. A few demonstrable examples of phosphorus-halogen synergism have been noted, but it is not general.

In many cases, the reason for the beneficial combination is rather obvious, such as each component providing a different, non-competing mode of action, such as condensed phase charring together with vapor phase flame inhibition. More detail on useful combinations will be found in the chapters on each individual polymer classes.

Patentability of combinations of known ingredients requires that the combination be new, useful (beneficial), and non-obvious to "one of ordinary skill in the art." The compounder, having found a new and well-performing combination, may be able to get patent protection if it cannot be shown by the patent examiner that the "prior art" suggests such a combination, or better yet, where the prior art suggests that such as combination would be poor. It is sometimes found helpful in supporting patent prosecution arguments against obviousness, to provide comparative examples of the unsatisfactory performance of closely related compositions lying outside the range of the claimed compositions.

For example, a novel ternary nanostructure polyphosphazene nanotube (PZS)@ mesoporous silica (M-SiO_2)@bimetallic phosphide (CoCuP) was facilely fabricated using PZS as the template, where large amount of cetyltrimethylammonium bromide molecules were anchored to PZS via a similar layer-by-layer assembly strategy, and then uniform M-SiO_2 shells can be formed successfully by Hyeon's coating method (Qiu et al. 2017). Incorporation of the PZS@M-SiO_2@ CoCuP into TPU matrix increased the initial degradation temperature and char yield, implying the enhancement of thermal stability. In addition, introduction of 3 wt% PZS@M-SiO_2@CoCuP into TPU led to PHRR and THR increased by 58.2% and 19.4%, respectively. On one hand, random distributed PZS network structure acts as physical barrier to inhibit the escape of pyrolysis volatile, and the formation of cross-linked phosphorus oxynitride and carbonized aromatic networks during combustion also act an important role in condensed phase to enhance char formation, in accordance with formation of more compact and continuous char layer for TPU/PZS@M-SiO_2@CoCuP. Furthermore, CoCuP acts as transition metal catalyst to accelerate formation of P-rich carbonaceous char owing to its catalytic carbonization effect confirmed by X-ray photoelectron spectroscopy (XPS) analysis. The previous literatures have reported that the efficient solid acid silica has a large number of acid sites, which can easily catalyze the carbonization of degradation products in the presence of metal oxides. As a result, M-SiO_2 can promote the catalytic efficiency of CoCuP.

A few cases of antagonism in flame retardancy are known. A rather important case is the combining of antimony oxide-halogen systems with phosphorus flame retardants. Some PVC formulations, even commercial ones, have such combinations. There is evidence that the antimony oxide and phosphorus may tend to at least partially counteract each other, probably due to the formation of antimony phosphate which seems to be inert as a flame retardant.

Other cases occur with thermoplastics in cases where melt flow is an important part of the flame retardant mode of action. Two notable cases are polyethylene terephthalate flame retarded by phosphorus compounds (either built in or applied by the "thermosol" treatment). As little as 0.15% phosphorus content permits the fabric to pass the vertical flame tests. These additives or reactives work in large measure by favoring the melt-drip mode of extinguishment, and are defeated by solids such as silica from silicone processing oils or pigment printing with infusible pigments, or by laminating the polyester fabric with a cellulosic fabric that acts as a wick or as a scaffold to prevent melt flow.

Glass fiber reinforcement is antagonistic to the flame retardant action of melamine cyanurate in nylon which depends on non-flaming drips. The dripping is prevented by the glass.

Carbon nanotube as a pretty new flame retardant, Kashiwagi et al. (2005a, 2005b) reported that the in-situ formation of a continuous, network-structured protective layer from the tubes was critical for the optimization of flame retardancy of materials, because the layer thus acted as a thermal shield from energy feedback from the flame. However, Du and Fang (2011) found that the introduction of carbon nanotubes (CNTs) only enhanced thermal stability of materials in a certain temperature range, but caused a severe deterioration of flame retardancy due to the interaction of the network structure and the intumescent carbonaceous char.

In fact, there are complex interactions between CNTs and IFR especially on heating or in a fire. Firstly, the formation of CNTs network can, on one hand, play a role of barrier itself to protect the matrix; and on the other hand, act like a "skeleton frame" for the residues and improve the strength of carbonaceous char. Nevertheless, due to the severe tangle between nanotubes and the degradation of CNTs at high temperature, a breakdown of intumescent char residues occurs in combustion and which will deteriorate the fire performance. In a word, above analysis gave convincing proof to our hypothesis that the latter effect was predominant in the competition.

2.4 Summary

The multiple modes of flame retardant action, i.e., physical and chemical, condensed phase and vapor phase, are exhibited by many of the effective flame retardants in actual use. Detailed quantitative knowledge of the modes of action of each is often sparse and even contradictory. The compounder should use mode of action concepts cautiously, but they do serve to suggest effective choices and combinations. Moreover, experimental design and regression analysis using quantitative flammability measurements (as well as measurements of other properties) can be a useful tool in approaching optimum combinations.

References

Bourbigot, S., and Fontaine, G. 2010. Flame retardancy of polylactide: An overview. *Polymer Chemistry* 1: 1413–1422.

Braun, U., Schartel, B., Fichera, M. A. et al. 2007. Flame retardancy mechanisms of aluminium phosphinate in combination with melamine polyphosphate and zinc borate in glass-fibre reinforced polyamide 6, 6. *Polymer Degradation and Stability* 92: 1528–1545.

Bulewicz, E. M., and Padley, P. J. 1971. Catalytic effect of metal additives on free radical recombination rates in $H_2+O_2+N_2$ flames. *Symposium (International) on Combustion* 13: 73–80.

Bulewicz, E. M., Padley, P. J., Cotton D. H. et al. 1971. Metal-additive-catalysed radical-recombination rates in flames. *Chemical Physics Letters* 9: 467–468.

Bundersek, A., Japelj, B., and Music, B. et al. 2016. Influence of Al (OH)₃ nanoparticles on the mechanical and fire resistance properties of poly (methyl methacrylate) nanocomposites. *Polymer Composites* 37: 1659–1666.

Cardenas, M. A., García-López, D., Gobernado-Mitre, I. et al. 2008. Mechanical and fire retardant properties of EVA/clay/ATH nanocomposites-Effect of particle size and surface treatment of ATH filler. *Polymer Degradation and Stability* 93: 2032–2037.

Chao, P., Li, Y., Gu, X. et al. 2015. Novel phosphorus-nitrogen-silicon flame retardants and their application in cycloaliphatic epoxy systems. *Polymer Chemistry* 6: 2977–2985.

Chen, S., Li, X., Li, Y. et al. 2015. Intumescent flame-retardant and self-healing superhydrophobic coatings on cotton fabric. *ACS Nano* 9: 4070–4076.

Cheng, X. W., Guan, J. P., Tang, R. C. et al. 2016. Phytic acid as a bio-based phosphorus flame retardant for poly (lactic acid) nonwoven fabric. *Journal of Cleaner Production* 124: 114–119.

Du, B., and Fang, Z. 2011. Effects of carbon nanotubes on the thermal stability and flame retardancy of intumescent flame-retarded polypropylene. *Polymer Degradation and Stability* 96: 1725–1731.

Ewing, C. T., Beyler, C. L., and Carhart, H. W. 1994. Extinguishment of class B flames by thermal mechanisms; Principles underlying a comprehensive theory; Prediction of flame extinguishing effectiveness. *Journal of Fire Protection Engineering* 6: 23–54.

Garba, B. 1999. Effect of zinc borate as flame retardant formulation on some tropical woods. *Polymer Degradation and Stability* 64: 517–522.

Georlette, P., Simons, J., and Costa, L. 2000. Halogen-containing fire-retardant compounds. In *Fire Retardancy of Polymeric Materials*, 1st ed., ed. A. F. Grand and C. A. Wilkie, pp. 245–284. New York: Marcel Dekker.

Green, J. 1992. A review of phosphorus-containing flame retardants. *Journal of Fire & Flammability* 10: 470–487.

Gu, J. W., Zhang, G. C., Dong, S. L. et al. 2007. Study on preparation and fire-retardant mechanism analysis of intumescent flame-retardant coatings. *Surface and Coatings Technology* 201: 7835–7841.

Hewitt, F., Rhebat, D. E., Witkowski, A. et al. 2016. An experimental and numerical model for the release of acetone from decomposing EVA containing aluminium, magnesium or calcium hydroxide fire retardants. *Polymer Degradation and Stability* 127: 65–78.

Jensen, D. E., and Jones, G. A. 1974. Catalysis of radical recombination in flames by iron. *The Journal of Chemical Physics* 60: 3421.

Kashiwagi, T., Du, F., Douglas, J. F. et al. 2005a. Nanoparticle networks reduce the flammability of polymer nanocomposites. *Nature Materials* 4: 928–933.

Kashiwagi, T., Du, F., Winey, K. I. et al. 2005b. Flammability properties of polymer nanocomposites with single-walled carbon nanotubes: Effects of nanotube dispersion and concentration. *Polymer* 46: 471–481.

Kellogg, C. B., and Irikura, K. K. 1999. Gas-Phase hermochemistry of iron oxides and hydroxides: Portrait of a super-efficient flame suppressant. *The Journal of Physical Chemistry A* 103: 1150–1159.

Kim, M. J., Jeon, I. Y., Seo, J. M. et al. 2014. Graphene phosphonic acid as an efficient flame retardant. *ACS Nano* 8: 2820–2825.

Laoutid, F., Bonnaud, L., Alexandre, M. et al. 2009. New prospects in flame retardant polymer materials: From fundamentals to nanocomposites. *Materials Science and Engineering: R: Reports* 63: 100–125.

Laoutid, F., Lorgouilloux, M., Bonnaud, L. et al. 2017. Fire retardant behaviour of halogen-free calcium-based hydrated minerals. *Polymer Degradation and Stability* 136: 89–97.

Laoutid, F., Lorgouilloux, M., Lesueur, D. et al. 2013. Calcium-based hydrated minerals: Promising halogen-free flame retardant and fire resistant additives for polyethylene and ethylene vinyl acetate copolymers. *Polymer Degradation and Stability* 98: 1617–1625.

Lewin, M., and Weil, E. D. 2001. Mechanisms and modes of action in flame retardancy of polymers. In *Fire Retardant Materials*, ed. A. R. Horrocks and D. Price, pp. 31–68. Cambridge, UK: Woodhead Publishing Ltd.

Li, N., Xia, Y., Mao, Z. et al. 2012. Influence of antimony oxide on flammability of polypropylene/intumescent flame retardant system. *Polymer Degradation and Stability* 97: 1737–1744.

Li, S., Yuan, H., Yu, T. et al. 2009. Flame-retardancy and anti-dripping effects of intumescent flame retardant incorporating montmorillonite on poly (lactic acid). *Polymers for Advanced Technologies* 20: 1114–1120.

Linteris, G. T., Katta, V. R., and Takahashi, F. 2004. Experimental and numerical evaluation of metallic compounds for suppressing cup-burner flames. *Combustion and Flame* 138: 78–96.

Linteris, G. T., Rumminger, M. D., and Babushok, V. I. 2008. Catalytic inhibition of laminar flames by transition metal compounds. *Progress in Energy and Combustion Science* 34: 288–329.

Marklund, A., Andersson, B., and Haglund, P. 2003. Screening of organophosphorus compounds and their distribution in various indoor environments. *Chemosphere* 53: 1137–1146.

Modesti, M., Lorenzetti, A., Simioni, F. et al. 2002. Expandable graphite as an intumescent flame retardant in polyisocyanurate-polyurethane foams. *Polymer Degradation and Stability* 77: 195–202.

Morgan, A. B., and Gilman, J. W. 2013. An overview of flame retardancy of polymeric materials: Application, technology, and future directions. *Fire and Materials* 37: 259–279.

Morgan, A. B., and Wilkie, C. A. 2014. *The Non-halogenated Flame Retardant Handbook*, ed. A. B. Morgan and C. A. Wilkie, pp. 75–138. Hoboken, NJ: John Wiley & Sons.

Mu, X., Zhan, J., Feng, X. et al. 2017. Novel melamine/o-phthalaldehyde covalent organic frameworks nanosheets: Enhancement flame retardant and mechanical performances of thermoplastic polyurethanes. *ACS Applied Materials & Interfaces* 9: 23017–23026.

Nachtigall, S. M. B., Miotto, M., Schneider, E. E. et al. 2006. Macromolecular coupling agents for flame retardant materials. *European Polymer Journal* 42: 990–999.

Nakamura, K., Tsunakawa, M., Shimada, Y. et al. 2017. Formation mechanism of Mg-Al layered double hydroxide-containing magnesium hydroxide films prepared on Ca-added flame-resistant magnesium alloy by steam coating. *Surface and Coatings Technology* 328: 436–443.

Price, D., Anthony, G., and Carty, P. 2001. Introduction. In *Fire Retardant Materials*, ed. A. R. Horrocks and D. Price, pp. 1–28. Cambridge, UK: Woodhead Publishing Ltd.

Qiu, S., Wang, X., Yu, B. et al. 2017. Flame-retardant-wrapped polyphosphazene nanotubes: A novel strategy for enhancing the flame retardancy and smoke toxicity suppression of epoxy resins. *Journal of Hazardous Materials* 325: 327–339.

Rasmussen, C. L., Glarborg, P., and Marshall, P. 2007. Mechanisms of radical removal by SO_2. *Proceedings of the Combustion Institute* 31: 339–347.

Rigolo, M., and Woodhams, R. T. 1992. Basic magnesium carbonate flame retardants for polypropylene. *Polymer Engineering & Science* 32: 327–334.

Rumminger, M. D., and Linteris, G. T. 2000a. Inhibition of premixed carbonmonoxide-hydrogen-oxygen-nitrogen flames by iron pentacarbonyl. *Combust Flame* 120: 451–464.

Rumminger, M. D., and Linteris, G. T. 2000b. Numerical modeling of counter flow diffusion flames inhibited by iron pentacarbonyl. *Fire Safety Science*: 289–300.

Rumminger, M. D., Reinelt, D., Babushok, V. et al. 1999. Numerical study of the inhibition of premixed and diffusion flames by iron pentacarbonyl. *Combust Flame* 116: 207–219.

Sawada, Y., Yamaguchi, J., Sakurai, O. et al. 1979. Thermogravimetric study on the decomposition of hydromagnesite 4 $MgCO_3$ Mg $(OH)_2$ 4 H_2O. *Thermochimica Acta* 33: 127–140.

Twarowski, A. 1993. The influence of phosphorus oxides and acids on the rate of H+ OH recombination. *Combustion and Flame* 94: 91–107.

Wang, X., Kalali, E. N., and Wang, D. Y. 2015. Renewable cardanol-based surfactant modified layered double hydroxide as a flame retardant for epoxy resin. *ACS Sustainable Chemistry & Engineering* 3: 3281–3290.

Wang, Z. Y., Liu, Y., and Wang, Q. 2010. Flame retardant polyoxymethylene with aluminium hydroxide/melamine/novolac resin synergistic system. *Polymer Degradation and Stability* 95: 945–954.

Wei, P., Han, Z., Xu, X. et al. 2016. Synergistic flame retardant effect of SiO_2 in LLDPE/EVA/ATH blends. *Journal of Fire Sciences* 24: 487–498.

Wu, K., Zhu, K., Kang, C. et al. 2016. An experimental investigation of flame retardant mechanism of hydrated lime in asphalt mastics. *Materials & Design* 103: 223–229.

Xu, B., Li, W., Tu, D. et al. 2017. Effects of nitrogen-phosphorus flame retardants in different forms on the performance of slim-type medium-density fiberboard. *Bioresources* 12: 8014–8029.

Yen, Y. Y., Wang, H. T., and Guo, W. J. 2012. Synergistic flame retardant effect of metal hydroxide and nanoclay in EVA composites. *Polymer Degradation and Stability* 97: 863–869.

Zanetti, M., Camino, G., Canavese, D. et al. 2002. Fire retardant halogen-antimony-clay synergism in polypropylene layered silicate nanocomposites. *Chemistry of Materials* 14: 189–193.

Zhang, M., Buekens, A., and Li, X. 2016. Brominated flame retardants and the formation of dioxins and furans in fires and combustion. *Journal of Hazardous Materials* 304: 26–39.

Zhao, H. B., Liu, B. W., Wang, X. L. et al. 2014. A flame-retardant-free and thermo-cross-linkable copolyester: Flame-retardant and anti-dripping mode of action. *Polymer* 55: 2394–2403.

Zhao, X., Gao, S., and Liu, G. 2016. A THEIC-based polyphosphate melamine intumescent flame retardant and its flame retardancy properties for polylactide. *Journal of Analytical and Applied Pyrolysis* 122: 24–34.

3

Fire Testing for the Development of Flame Retardant Polymeric Materials

Bernhard Schartel
and Katharina
Kebelmann

3.1 Introduction

Modern civilization is no longer conceivable without polymeric materials derived from fossil fuels or sustainable resources, including applications in fields where flame retardancy is mandatory, such as transportation, construction, electrical engineering, and electronics (Troitzsch 2004). One of the main drawbacks of these all-purpose hydrocarbon-based materials is their intrinsic combustibility, which poses a significant and ever growing fire risk. The development of flame retardants and flame-retarded polymeric materials that meet fire safety regulations and pass decisive fire tests is a current challenge for materials research and development (Weil and Levchik 2016).

Based on highly flammable commodity polymers such as poly(propylene) (PP), polyethylene (PE), and poly(styrene) (PS), flammable technical polymers, such as polyamides (PA), bisphenol A polycarbonate (PC) and its blends, poly(butylene terephthalate) (PBT), and poly(urethane) (PU), and high-performance polymers such as epoxy resins are currently under development, particularly halogen-free, self-extinguishing thermoplastics, thermo-sets, fiber reinforced composites, fibers, and foams (Weil and Levchik 2016). Especially technical polymers applied in electrical engineering and thermo-sets in electronics, like computer housings and circuit boards, which must achieve the UL 94 V-0 classification, are of high interest due to the growing demand for these products. The same is true for materials in transportation such as for railway vehicles in Europe. The development of new flame retardants and flame-retarded materials is often based on optimizing well-known approaches, for instance, proposing new or optimized combinations with synergists, adjuvants and additives, and the modification of flame retardants or other additives such as microencapsulation to increase stability against hydrolysis or to avoid agglomeration. Further, the synthesis of new derivatives, by adjusting properties of flame retardants like molecular weight, thermal stability or compatibility with the polymer, as well as new developments like nanocomposites, flame retardants containing multiple flame retardant groups, flame retardants with complex, oligomeric, and polymeric structure, flame retardants based on renewable educts, and multi-functional flame retardants, show promising routes to fulfill the high demands on sustainability and efficiency (Wang 2017; Velencoso et al. 2018).

Thus, the R&D of flame-retarded materials considers essential variables such as components, their concentrations, distributions, and morphologies. Modifying the ingredients or their processing and manifold interactions expands our empirical understanding of the main influences on fire behavior to a multi-dimensional matrix. Flame retardant mechanisms and modes of action performed on a molecular or microscopic scale controls the reaction to fire of macroscopic specimens, so that they can pass a decisive fire test with respect to a distinct application. The fire testing of manifold polymeric materials with very different properties, application purposes, and fire protection goals becomes essential for R&D.

Fire tests exist on different scales, with each test simulating a distinct fire scenario to determine a distinct protection goal, characteristic for the desired application (Janssens 2008 2010). Fire tests measure a certain fire property as the reaction of a defined specimen under a specified fire scenario. The role played by the different material properties differs from one fire test to the next. Generally, fire scenarios and thus fire tests differ drastically in heat and mass transport, defined by characteristics such as applied heat flux, temperature, length scales, and ventilation (Wittbecker 2004; Schartel 2010). Fire scenarios can be distinguished into three main stages:

- Ignition: The start of flaming combustion or smoldering, typically occurring on a small length scale (cm) at ambient temperatures, well ventilated, and with low heat fluxes, apart from the local impact of the ignition source (flame, glow wire, cigarette, crib, etc.). Thus, the ignition source and how it is applied to the specimen become the main factors determining the fire scenario.
- Developing fire: Continuing flaming combustion occurs in this stage of fire growth, typically defined by an external heat flux $\dot{q}''_{ex} = 20\text{--}60 \text{ kW/m}^2$, larger length scales (dm to m), ambient temperatures above ignition temperature (700–900 K), and continuing ventilation. The heat fluxes and the heat release rate become the main factors determining the fire scenario and the fire property, respectively.
- Fully developed fire: Fully developed fire scenarios are characterized by a high external heat flux $\dot{q}''_{ex} > 40 \text{ kW/m}^2$, typical length scales of several meters, temperatures above auto ignition temperature (>900 K), and low ventilation controlled combustion.

Distinct fire properties belong to each fire scenario:

- Ignition:
  - Ignitability [time to ignition (t_{ig}), critical mass loss rate or heat flux for ignition, ignition temperature (T_{ig}), etc.]
  - Flammability (extinction behavior under ambient conditions after removing an ignition source, reaction to a small flame)
- Developing fire:
  - Flame spread (opposed flow and wind-aided flame spread, fire growth rate, etc.)
  - Heat release rate (HRR)
  - Total heat release (THR) and fire load (total heat evolved, THE = THR at the end of the test or at flameout)

- Fully developed fire:
  - Fire resistance (heat penetration, fire penetration, fire stability)
  - Fire load (THE)

Reaction to a small flame (flammability), flame spread, and fire penetration assess the performance of a certain specimen or components in a defined fire scenario established by a defined fire test, and are not material properties. Due to many existing test methods, many different terms describe a specimen's reaction to test conditions. Terms like "fireproof," "flash, flame, or fire resistant" and "self-extinguishing," but also "flammable" and "(in)combustible" often have different meanings and may lead to confusion (Schartel et al. 2016). In this chapter, "flammability" is used in the sense of "showing sustained flaming after removal of a small ignition source," and not in the general sense of "fire behavior." The manifold application of flame-retarded polymers leads to a broad range of protection goals for fire science and fire testing, including preventing sustained ignition (beginning of a fire), limiting the contribution to fire propagation (developing fire), or acting as a fire barrier (fully developed fire). As the protection goals and typical fire scenarios are different for various applications, a great variety of different fire tests are in use to monitor different fire properties of a defined specimen or its components. As the performance of a material in one fire test is often hard to transfer to another fire test, flame retardant materials are usually developed to pass a specific fire test.

This chapter abstains from describing each fire test apparatus or the testing procedures themselves in detail, since the corresponding international standards should be used to ensure accurate quality assurance. Instead, the main aim of this chapter is to arouse the reader's interest in sensible, targeted, and advanced uses of fire tests in the development of new materials. Typical and characteristic fire tests are selected and discussed, being aware that a lot of fire test are not mentioned. Furthermore, insight into how fire tests can be applied for the tailored development of flame-retarded materials is presented, although only in an overview, as a comprehensive review about what has been achieved so far is beyond the scope of this chapter.

3.2 Ignition

Ignition is the first stage of a fire, generally induced by either convection or irradiation, which leads to an increase in the surface temperature up to a critical temperature, accompanied by a critical fuel release rate, or by the direct effect of an open flame (Babrauskas 2003; Wittbecker 2004). It is of paramount interest to develop polymeric materials with properties such as low flammability to deter the start of a fire in critical situations and prevent sustained ignition. Furthermore, delaying sustained ignition aims to prevent the subsequent stages such as developing and fully developed fire. As typical ignition scenarios are spatially limited, bench-scale fire tests are commonly applied to simulate the fire scenario of ignition. Bench-scale fire tests are also preferred for laboratory applications such as comprehensive material testing and screening approaches (Schartel 2010). For successful and correct employment of those test methods, some important facts must be noted.

Each bench-scale fire test represents a specific fire scenario, and its results cannot be interpreted as intrinsic fire properties of the material. As for all fire tests, the fire behavior monitored is the response of a defined specimen to a specific fire scenario. It is important to apply the specimen dimensions speci-fied in the standards (ISO, IEC, ASTM, etc.), as fire test results are directly dependent on these, and on material thickness in particular. Establishing correct test conditions is crucial, including regular calibration of the instruments, proper specimen preconditioning, and mounting. Finally, the source of ignition, and how it is applied to the test specimen play a crucial role in small flame tests, and following the instructions regarding the size of the ignition source (Janssens 2008; Schartel et al. 2016), such as flame height, arrangement of flame and specimen, as well as how long the ignition source is applied, is essential (Blaszkiewicz et al. 2007). The application of the ignition source and the specimen setup for some common bench-scale tests are sketched in Figure 3.1. Testing vertical specimens is much more

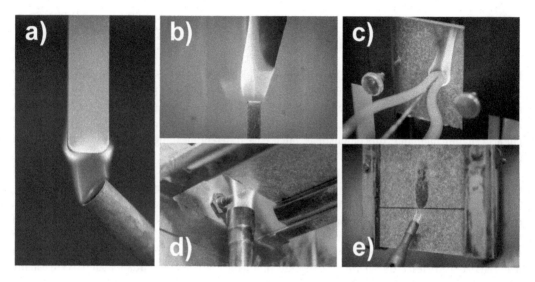

FIGURE 3.1   (a) UL 94 vertical setup, vertical specimen, small flame applied at the bottom; (b) oxygen index setup, vertical specimen, small flame applied from the top; (c) glow wire test setup, vertical specimen, glowing wire applied in the middle of the plate; (d) FMVSS 302 setup, horizontal specimen, flame applied at the edge; and (e) small burner test for building products, vertical specimen, flame applied at the surface.

demanding than horizontal ones and applying flames to the edges and the bottom much more so than at the surface or the top of the specimen. Most of the important fire tests, and thus the assessment of polymeric material within the ignition fire scenario, do not measure ignitability itself, but the reaction to an ignition source applied for a short time. Thus, the flammability of the specimen is usually assessed by monitoring the ability of the specimen to self-extinguish after removing the ignition source.

The most commonly applied small fire tests applied during the R&D of polymeric materials are the Underwriters Laboratories UL 94 test according to IEC 60695-11-10 and the oxygen index (OI) according to ISO 4589 (also known as the limiting oxygen index [LOI]) (Horrocks et al. 1989; Schartel 2010). Both methods evaluate the flammability of polymeric materials through their reaction to a small flame. The vertical UL 94 is the most important mandatory test in electronic and electrical engineering applications and for the general assessment of flame-retarded plastics. The UL 94 classifications V-2, V-1, and V-0 achieved for a certain specimen thickness (e.g., 3.2, 1.6, 0.8, or 0.4 mm) are used as the key information characterizing fire behavior on data sheets for polymeric materials. A defined number of specimens with dimensions of 127 × 12.7 mm are lighted twice from below by a defined gas burner flame, simulating an ignition source such as a candle or a match. The time until extinguishment is monitored as well as any burning dripping of the material. To achieve a V-0 classification, each of the five individual test bars must self-extinguish within 10 seconds (s) after the first and the second applications of the flame, the individual afterflame plus afterglow must not exceed 30 s, the total burning time of all five test bars must be less than 50 s, and no flaming drips that ignite a cotton bud are mandatory requirements. Since the V classifications demand immediate flame extinction in a few seconds, the vertical UL 94 test is quite demanding, as the results are based on falling below a critical flame spread in the direction of the flame. V-0 materials often show good performance in other and larger tests as well, e.g., for developing fires (Morgan and Bundy 2007). Requirements for a V-1 classification are self-extinguishment of each test bar within 30 s, an individual afterflame plus afterglow for less than 60 s, and an total burning time of less than 250 s and no flaming dripping by all five test bars. Materials with a V-2 classification meet the flaming time criteria of V-1, but exhibit flaming dripping that ignites a cotton bud. A complete characterization based on the UL 94 burning chamber test also includes a horizontal test for all materials that fall short of any V rating, where less flame retardant materials may extinguish after ignition and are classified as HB materials. This requires either self-extinguishment before burning reaches a reference mark on the specimen at 25 mm, or burning at a rate of less than 40 mm/min for specimens with 3–13 mm thickness, or at less than 75 mm/min for specimens less than 3 mm thick. As nearly all polymeric materials pass the criteria for an HB rating, this HB rating is hardly a sufficient material assessment. Nevertheless, immediate extinction or reduced burning rates can be used during development to assess trends in a quantitative manner.

There are also variants of the UL 94 test worth mentioning. For some electronic products, the classes 5VA and 5VB are desired according to IEC 60695-11-20, which entail an ignition source approximately five times more intense than the common vertical UL 94. Low-density foams (<250 kg/m^3) are tested according to ISO 9772 by using larger specimens of 150 × 50 mm in a horizontal setup, and classified by HF-1, HF-2, and HBF ratings according to burning time criteria; thin films not capable of supporting themselves in a horizontal position are tested according to ISO 9773 by rolling specimens around a mandrel in a vertical setup, and classified as VTM-0, VTM-1, or VTM-2. The criteria for VTM ratings are the same as for V ratings, but the VTM rating of a rolled film must not be considered as equivalent to a V rating.

The OI refers to the highest oxygen concentration at which a specimen of the material in a candle-like fire scenario self-extinguishes in less than 3 minutes, and at least less than 5 centimeters of the specimen is consumed by the fire. Furthermore, burning characteristics of the material are monitored, such as dripping, charring, glowing combustion, and afterglow. The standard contains six different specifications of specimen dimensions, particularly varying in thickness depending on the polymeric material, e.g., molding or cellular materials (specimen II: 80–150 × 10 × 10 mm), flexible films (specimen V: 140 × 52 × <10.5 mm), or sheets (specimen III: 80–150 × 10 × <10.5 mm). In most cases specimen IV, 80–150 × 6.5 × 3 mm for self-supporting molding or sheet materials for electronic purposes, is used, or specimen I, 80–150 × 10 × 4 mm for molding materials. The OI is quite often misunderstood as a material property, but it is not (Wharton 1981): for instance, varying a thickness from just 2–4 mm has been reported to shift the OI of carbon fiber reinforced epoxy resins from 27 vol.-% to 37 vol.-% (Camino et al. 1988; Schartel 2010). The test specimen is placed in a vertical chimney, which is continuously purged with a gas mixture of nitrogen and oxygen and ignited at the top with a gas burner flame. Due to the candle-like burning, the flame spread away from the flame is observed, and the scenario is generally less demanding than the UL 94 (Weil et al. 1992). This is why materials need OI values clearly greater than 21 vol.-%, often >26–29 vol.-% (Isaacs 1970) or in some cases >40 vol.-%, in order to pass a vertical classification in UL 94 (Schartel 2010).

Obviously, quite different fire scenarios arise during tests, with vertical versus horizontal setups, and various locations of the ignition source, with a flame located at the top of a vertical specimen during the OI versus on the bottom during the vertical and horizontal UL 94. Flame retardant phenomena that remove heat from the pyrolysis zone and thus reduce heat release and flame spread, and enhance self-extinguishment like dripping and melt flow, are influenced directly by the location of the ignition flame and the orientation of the sample. The same holds for protective residual layers. Thus, no general relation exists between the OI and the UL 94 tests when different materials or different flame retardant modes of action are compared (Braun et al. 2007; Schartel 2010).

There are specific protection goals to address each of the many applications of flame retardant polymers, requiring individual fire scenario circumstances. Thus, various small burner flame tests have been instituted by different countries and organizations in order to assess various fire ignition scenarios. A somewhat questionable fire test has been established for polymer materials in car interiors: the United States (US) Federal Motor Vehicle Safety Standard FMVSS 302 according to ISO 3795. This common test procedure assesses the fire risk of a 365 × 100 mm specimen with a thickness corresponding to components in the passenger compartment by monitoring flame spread after ignition. The specimen is fixed horizontally and subjected to a Bunsen burner flame for 15 s. The rate of flame spread must not exceed 101.6 mm/min over a distance of 254 mm. The measurement of the horizontal flame spread as the sole fire risk assessment of a material or component is generally not demanding, and, what is more, can be passed even by several materials known for their rather high flammability in vertical setups or fire load. Frankly, the value of the protection implied by passage of the FMVSS 302 fire test is unclear.

The flammability of construction materials, such as tests according to ISO 11925-2 *reaction to fire test for building products – ignitability when subjected to direct impingement of flame* or the German DIN 4102, is usually assessed by means of small flame tests either by surface flame impingement or by edge flaming using different, but similar specimens, e.g., 250 × 180 mm, 250 × 90 mm, 230 × 90 mm, or 190 × 90 mm. The classification distinguishes between flammable and easily flammable materials, whereas the class usually aspired to, "not easily flammable," demands additional fire tests that simulate a developing fire. An analogous idea is applied in the aviation sector, where a Bunsen burner type test contributes to assessing the self-extinguishment of most materials and parts used in the interior of airplanes approved by the US Federal Aviation Association, utilizing either a vertical or horizontal specimen (Federal Aviation Regulations [FAR], Part 25). A piloted flame is applied as the ignition source, and burn length, afterburn time, time of flaming dripping, and the burn rate are measured. Single cables are tested against a Bunsen burner in vertical set-up according to IEC 60332-1 and IEC 60332-2.

Besides applying a flame, tests with other ignition sources, i.e., radiation, glowing wire, and smoldering cigarettes, are performed to evaluate the ignitability of a polymeric material. In case of the glow wire test (GWT) according to IEC 60695-2-10 to IEC 60695-2-13, in electrical engineering, a heated resistance wire is pressed against the specimen for 30 s. Roughly summarized, if the specimen ignites for a duration of less than 5 seconds, the temperature of the wire is recorded as the glow wire ignition temperature (GWIT) for this material; if the specimen burns for less than 30 s the temperature is recorded as the glow wire flammability index (GWFI). This test simulates a specific fire scenario due to its very different ignition source compared to a small flame, and the results obtained do not correlate well with other fire tests. Nevertheless, it is commonly utilized along with the UL 94 during the development of materials for electrical engineering and electronic applications to identify the high fire risks typical for those applications. Also, the smoldering cigarette for the fire testing of upholsteries is a rather unique ignition source used to simulate the typical high fire risk scenario for the corresponding applications in furnishings.

3.3 Developing Fire

Developing fire is the next stage in the typical development of a fire after ignition, characterized by flaming combustion and flame spread involving a larger surface area. Compared to small scale tests used for ignition, the specimens used in these fire tests are usually larger in order to create a scenario closer to a typical developing fire. The main protection goal is the deceleration of flaming combustion, flame spread, fire growth, THR, and HRR. It is worth mentioning that the flame retardancy of polymeric materials does takes place mainly in these first two stages of fire, ignition and developing fire. The main protection goals may come down to reducing the reaction to small flame and the HRR.

To simulate the conditions of a developing fire, radiant heating systems are often applied. The fire testing of floorings as building products or in transportation is an excellent example for such fire tests. During the reaction-to-fire tests of floorings and the Standard Test Method for Critical Radiant Flux of Floor-Covering Systems, specimens 1050 × 230 mm and 1070 × 254 mm in size are mounted horizontally in a test chamber (ISO 9239-1, ASTM E648). To assess their burning behavior, specimens are irradiated with a gas-fired radiant panel, and the hot surface subsequently impinged by a pilot flame to initiate a flame spread. The Flooring Radiant Panel Test Apparatus assesses the flame spread based on classification of the critical heat flux below which no flame spread occurs, and the Critical Radiant Flux of Floor-Covering Systems test is based on the critical radiant flux at flameout.

To investigate the reaction to fire of building products and their contribution to the development of a fire, in the EU the Single Burning Item test (SBI) described in European Standard (EN) 13823 assesses the fire properties of building products surrounding a burning item in a room corner scenario on the intermediate scale. Two specimens, one with dimensions of 500 × 1500 mm and the other with dimensions of 1000 × 1500 mm are assembled in a corner setup and exposed to a sandbox burner as the well-defined flame ignition source. The results of the test lead to the EU building product classification A2, B, C, or D, based mainly on the fire growth rate (FIGRA), which is equal to the maximum value of HRR(t)/t (Sundström 2007). Furthermore, the flame spread, smoke production, and occurrence of dripping are monitored and classified. Thus, apart from flooring, the SBI is the most important fire test for building products in the EU.

Analogously, the Steiner (25-foot) tunnel test (ASTM E84) is the most important standard method in the US to assess the surface burning behavior of building materials made of polymers (e.g., ceilings, walls). Therefore, a specimen 510×7320 mm in area with a typical thickness of use is mounted horizontally, like a ceiling, with the surface facing the ignition source below (two gas burners, located 305 mm away from the specimen). During the test, the flame spread is observed, and a flame spread index and smoke developed index are calculated to describe the relative burning behavior of the materials.

An important specific protection goal for communication and energy cables is to limit vertical flame propagation. During a nearly full-scale test, a bundle of cables is mounted vertically and continuously exposed to a specified burner flame (EN 50399 and IEC 60332-3). Above the test chamber, the combustion gases are collected in a hood, allowing the measurement of heat release rate and smoke production, and the occurrence of flaming dripping is recorded.

In the aircraft industry, different test methods exist to appraise material properties and the fire behavior of manifold materials and products used in or for aircrafts. Specimens 150×150 mm in size of the most important parts appearing inside the passenger cabin are tested by a heat release test named after the institution where it was developed, the Ohio State University (OSUtest, FAR 25.853, ASTM E906). Here, the specimens are positioned vertically in an air stream under radiant heat flux and ignited by a pilot flame. The HRR is measured using the temperature (enthalpy) increase in the air stream leaving the device and is believed to give reasonable insight into the developing fire behavior of a material from a small specimen size at low cost and in a short time.

Particularly in R&D and in the investigation of flame retardant polymers, the cone calorimeter according to ISO 5660 has become the most popular bench-scale testing apparatus, providing comprehensive and meaningful insight into the fire properties of materials under a well-defined fire scenario of forced flaming conditions to simulate a developing fire (Babrauskas 1984; Schartel and Hull 2007). More recently, it has become a key fire test in EN 45545-2 *Fire test to railway components* for materials applied in railway vehicles in Europe. Like all of the more modern fire tests, the cone calorimeter is based on an oxygen consumption method (Hugget 1980; Janssens 1991) to calculate the HRR as the most important fire risk (Babrauskas and Peacock 1992). The instrumental setup corresponds to a bench-scale fire test, utilizing square specimens of only 100×100 mm with thicknesses up to 50 mm. It yields information on the fire performance of the material, allows the burning behavior of different polymeric materials to be compared, and elucidates the fire retardant modes of action. The developing fire scenario is simulated by exposing the specimen to an external radiant heat flux at 35 or 50 kW/m^2 and igniting it with a spark. The oxygen consumption, CO_2 and CO production, specimen mass, and smoke density are monitored. HRR, THR, and mass loss rate are calculated as well as peak heat release rate (PHRR), effective heat of combustion (EHC), CO yields, and the rate of smoke generation. Indices designed to depict the potential fire propagation or flame spread, such as the FIGRA = maximum of HRR(t)/t, PHRR/t_{PHRR}, PHRR/t_{ig}, fire performance index (FPI) = t_{ig}/PHRR, and the maximum average rate of heat emission (MARHE) are used for assessment (Schartel et al. 2017). The strength of the cone calorimeter, namely, its well-defined scenario that allows the kind of material behavior to be examined, is also its limitation, as the spark ignition, well ventilated fire, and horizontal position of the specimen eliminates the impacts of melt flow and dripping. Further, in the cone calorimeter, there is no real flame spread over the surface as is characteristic in developing fires, so the HRR is observed as function of the time when the pyrolysis front passes through the thickness of the specimen. Nevertheless, the HRR results obtained in the cone calorimeter are deemed acceptable for describing the fire hazard of polymeric materials. The HRR curve over time contains comprehensive information about the burning behavior of the specimen, as illustrated in Figure 3.2. Typical shapes of HRR curves are obtained and controlled by the burning behavior

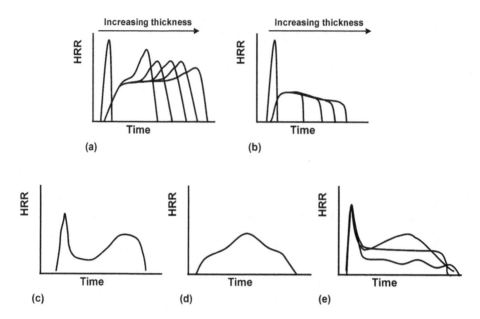

FIGURE 3.2 HRR curves in the cone calorimeter for typical burning behaviors of materials burning with no or only a small amount of residue, residue forming/charring materials, with protective layers or intumescent residues, and the impact of specimen thickness. (a) non-charring; (b) charring, residue forming; (c) charring; (d) charring, deforming; and (e) protection layer, intumescence.

of the material and the thickness of the specimen as the crucial parameters (Babrauskas 2002; Schartel et al. 2005; Schartel and Hull 2007).

Thermally thin specimens of non-charring polymers (Figure 3.2a) show early ignition and a spontaneous increase towards a high PHRR. The whole specimen heats up and pyrolyzes at once, with the HRR curve exhibiting one sharp peak; the PHRR becomes controlled by the THE. Intermediate sample thickness leads to an almost steady heat release rate increase for non-charring materials; the initial increase occurs only as a shoulder before reaching the PHRR. There is no steady state HRR in the form of a plateau-like HRR. The pyrolysis zone or the pyrolysis front velocity increases until a lack of fuel results in a decaying HRR. With increasing thickness, thermally thick specimens show a steady state HRR, visible as a plateau sometimes remaining nearly until the end of the test, often followed by a PHRR. The PHRR at the end is caused by the consumption of the specimen, yielding a thermally thin residual thickness.

Thermally thin samples of charring or residue forming polymers exhibit similar burning behavior to thermally thin non-charring ones: heating up and pyrolysis of the entire specimen at once (Figure 3.2b). Thermally thick samples show an initial increase resulting in a maximum value equal to the PHRR and followed by a plateau-like HRR or moderate decrease in HRR. The formation of the protective residual layer controls the PHRR as well as the moderate decrease afterwards. This behavior occurs not only for charring, but also for inorganic residues (ash) or inorganic-organic residues. Along with the plateau-like or decreasing HRR after the first maximum, the PHRR becomes a function of the protective properties of the fire residue.

Charring or residue-forming polymeric materials often form an unstable protective layer that bursts and cracks, leading to further pyrolysis and an increase in HRR towards the end of burning (Figure 3.2c). Apart from the destruction of residue, a second pyrolysis front may run through the specimen, or a thermally thin phenomenon at the end of burning may play a role. Such double peak behavior in HRR is typical for wood materials, for instance.

Many charring polymers tend to deform, leading to a slow increase and then decrease in HRR, and a low PHRR, often in the middle of the HRR curve (Figure 3.2d), due mainly to a decrease in the distance from the cone heater.

Materials that show the formation of a protective layer after the fuel is burned out of the top layer, such as a lot of intumescent polymers, but also glass fiber reinforced thermoplastics, exhibit a sharp peak of HRR at the beginning of burning followed by a low HRR (Figure 3.2e).

3.4 Fully Developed Fire

In the fully developed fire scenario, the crucial fire properties and main protection goals boil down to radically reducing the fire load in order to achieve incombustibility, or to increasing fire resistance. The latter is often discussed as a property of larger components or structures, as a way of assessing the fire stability or the stability against heat and fire penetration. It is obvious that we are talking about niche applications for polymers.

Non-combustibility tests, such as ISO 1182 in Europe, are performed on construction products of the highest "A" classification. A cylindrical specimen is placed in an electric furnace preheated to a high temperature, and the temperature rise within the furnace caused by the specimen and the mass loss are recorded. Apart from the incombustibility test, the bomb calorimeter is also applied to determine the calorimetric value, e.g., ISO 1716 in Europe. The bomb calorimeter is based on complete combustion without any charring in a high-pressure oxygen atmosphere. In sum, these tests can be passed only by materials that contain hardly any organic polymers, such as metal hydrate flame-retarded sheet molding compounds and bulk molding compounds, in which a few weight percent of polymer act as a glue between the inorganics.

Fire resistance testing is applied widely for building components on the large scale, e.g., column furnaces, wall furnaces, and ceiling furnaces, using gas burners to test the resistance over time, e.g., typically 30, 60, or 90 minutes, against the standard time temperature curve (also called standard fire temperature curve) such as EN 1363-1, ASTM E119-00a, or ISO 834-1 simulating a flashover to a fully developed fire in a room fire. Other time-temperature curves are used for more extreme scenarios such as tunnel fires or hydrocarbon curves for industrial fires. However, increasing fire resistance is also important for load bearing components in transportation, such as the burn through tests stipulated by the FAR25 regulations in the aviation sector, which apply the next generation burner (NexGen Burner, heat flux ~180 kW/m^2, T = 1093°C) on specimens 610 × 610 mm in size. Fire resistance is usually defined as property of a component, not of a material. Nevertheless, increasing the fire resistance of polymeric materials is a topic when the materials are used in load bearing structures.

For R&D purposes, investigations of fire resistance on the bench-scale are proposed, based on small furnaces using gas burners or electric heating (Tabeling 2004; Jimenez et al. 2016). Compared to using a cone calorimeter at high heat fluxes (Batholmai and Schartel 2007), approaches based on furnaces are much closer to the time-temperature scenarios of the reference fire resistance testing of components. Although it is an intrinsic problem for electric furnaces to simulate the rapid temperature increase at the beginning of the test, a modified electric muffle furnace precisely following the standard time-temperature profile up to 1000°C was developed by Morys et al. (2017a) and is proposed as advanced R&D tool to investigate the impact of the polymer binder on intumescent coatings, when combined with the online monitoring of thermal insulation and intumescence during measurement, and sophisticated investigation of the representative fire residue afterwards (Morys et al. 2017b).

Carbon and glass fiber reinforced polymer composites are increasingly used as lightweight materials in load bearing components in transportation, aviation, and shipping. Therefore, their fire resistance under fully developed fire scenarios is attracting ever greater interest. For R&D, bench-scale burn through tests have been proposed (Tranchard et al. 2015), as well as fire stability tests applying compression load and severe flame impingement (180 kW/m^2) on the bench-scale (Gibson et al. 2010, 2012) and intermediate-scale (specimens 500–1000 × 500 mm in size, compression loads up to 1 MN, NexGen Burner) (Hörold et al. 2013, 2017).

3.5 Screening

Modern flame retardant polymeric materials are multi-component systems consisting of manifold additives and fillers that influence the burning behavior. The development of novel flame-retarded materials must account for large numbers of variables—not only components such as the polymer, flame retardants, synergists, adjuvants, additives, and filler, but also their concentrations and dispersion. Further, each of the components is often varied in their characteristics, e.g., molecular weight, branching, size and size distribution, coating, or microencapsulation. This matrix of variations requires time, materials, and cost-efficient screening. The analysis of flame retardant polymers via intermediate-scale or large-scale equipment is expensive in terms of both time and materials, and most of the methods, including the analysis of large numbers of specimens, are not feasible for screening applications in R&D. Consequently, different assessments based on thermal analysis, bench-scale methods, and high through-put methods have been proposed, applying specimens down to a few milligrams in size. However, analytical methods based on such small specimens do not constitute fire scenarios. The replacement of temperature gradients as the key response of a specimen in fire with constant heating rates, and the elimination of feedback from the flame—and thus the coupling of the two chemical processes pyrolysis and combustion—limits the significance of the results (Schartel et al. 2016).

In a fire scenario, where a flaming combustion follows ignition, a polymer primarily decomposes under anaerobic pyrolysis conditions, due to both the limited oxygen diffusion into the specimen and the reduced oxygen concentration on the surface of the pyrolyzing material below the flame (Schartel et al. 2016). Just before ignition, thermo-oxidative decomposition occurs on the surface layer as well as in the residue during afterglow after flameout. What is more, thermo-oxidation plays an important role in unstable burning and smoldering, and thermo-oxidation is believed to occur in materials with very high specific surfaces like fibers and foams. On the other hand, some of the most common flammability tests, such as UL 94 and OI, start with a pilot flame and thus pyrolysis under anaerobic conditions, and monitor flame front velocity and the extinguishment of the specimen. A well-established thermo-analytical method to analyze the decomposition of a polymer under anaerobic and/or aerobic conditions is thermogravimetric analysis (TGA), or TGA coupled with evolved gas analysis (TGA-FTIR, TGA-MS, and TGA-GCMS). To examine pyrolytic decomposition, milligrams of the material are heated in a controlled nitrogen atmosphere at constant heating rates of 10 or 20 K/min, and the weight loss recorded over time and temperature (Schartel 2010; Schartel et al. 2016). For systems that do not contain halogen groups, and are thus not controlled by flame inhibition, van Krevelen established a powerful relation between the OI of polymers and their char yield (μ) in wt.-% gained as residue of the specimen after analysis with TGA in nitrogen, summarized in the following equation (van Krevelen 1975; Schartel et al. 2017):

$$OI = 17.5 + 0.4\mu. \tag{2.1}$$

Although formulated in the last century, the equation is still a powerful tool for assessing systems in which charring is the most important flame retardancy mechanism. Furthermore, it makes clear that the most important characteristic to look at in TGA results is the char yield, not decomposition temperatures or activation energy.

OI and UL 94 are used regularly to screen the fire behavior of flame retardant polymers. However, strictly speaking, both tests deliver information about the flammability in an ignition scenario. The relevance of OI for other fire tests is particularly questionable (Weil et al. 1992). Accelerated methods capable of analyzing many specimens in a short time are interesting for screening applications in the field of R&D. The Microscale Combustion Calorimeter (MCC), also known as the pyrolysis combustion flow calorimeter (PCFC), is probably the most successful screening tool today, as it measures the "flammability characteristics of plastic and other solid materials [by using] controlled thermal decomposition of milligram specimens and combustion of the pyrolysis gases at high temperature in excess oxygen to simulate the process of flaming combustion" (Lyon and Walters 2015). The measurements deliver the heat release rate obtained by the oxygen consumption method, as a function of temperature at which the volatiles are produced by anaerobic pyrolysis (Lyon and Walters 2004; Lyon et al. 2007; Schartel et al. 2007). However, the PCFC, consisting of the pyrolysis of a 5 mg specimen and a combustion furnace at 900°C ensuring the complete combustion of the volatiles, does not simulate important chemical and physical features typical for flame-retarded polymeric materials, such as flame-to-polymer heat transfer, incomplete combustion due to flame inhibition, the influence of changed melt flow and dripping behavior, or the formation of a residual protective layer during pyrolysis. Nevertheless, it is strong for not only systematic screening, but also for the quantitative assessment of carbonaceous charring and fuel dilution. On the other hand, the effect of inert fillers is often overestimated, since their dilution in the condensed phase is monitored only in wt.-%, but not vol.-%. Kraemer and Stoliarov proposed the Milligram-scale Flame Calorimeter (MFC) (Raffan-Montoya et al. 2015), which also uncoupled pyrolysis from combustion, determining the heat release rate via oxygen combustion, but realizing this combustion in a laminar, nearly axisymmetric diffusion flame, so that flame inhibition can be monitored.

One proposal for enhancing screening options for laboratory applications was the rapid cone calorimeter (Gilman et al. 2003; Davis et al. 2010), consisting of a cone calorimeter equipped with a conveyor belt feeding a constant supply of reduced-size specimens into the burning zone. This approach was adopted by Rabe et al. (2017) in the development of the rapid mass calorimeter, leading to a further reduction in the time and effort required for maintenance and calibration (Rabe and Schartel 2017a, 2017b), and comprising a mass loss calorimeter in which the balance has been replaced by a linear motion unit equipped with two specimen holders and a thermo-pile chimney.

All kinds of bench-scale versions of larger tests have been proposed to reduce the time and specimen sizes needed for testing, such as a simple and economical test module to put wires and cables into a standard cone calorimeter, simulating the EN 50399 for screening applications (Gallo et al. 2017). A "Mini SBI" test has been proposed by several groups (De Corso and Castrovinci 2015; Bachelet et al. 2017), and a small-scale Steiner tunnel by Bourbigot et al. (2013).

3.6 Other Fire Hazards

A major danger that prevents people from escaping fires is generated smoke, in particular smoke production due to incomplete combustion of the burning materials, and toxic products, such as CO and, for materials containing nitrogen, HCN (Stec and Hull 2010). It is of crucial interest for R&D to minimize the production of smoke and toxic gases to extend escape times, especially in fields like transportation, in which increasing the escape time is a major or the most important protection goal. CO and smoke production depend not only on the properties of the combustible polymeric material, but also strongly on the fire scenario, such as the source of ignition, ventilation, temperature, irradiation, flame properties, and presence of quenching effects (Stec et al. 2008). Not only is the correlation of the smoke and CO production between different fire scenarios and fire tests difficult, but even their rankings can change when a set of materials is compared in different fire tests (Hull et al. 2005). However, measuring the light-obscuring properties of evolved smoke is an established method to assess the fire hazard quantitatively with respect to preventing people from escaping, as is coupling fire tests with calibrated Fourier transform infrared spectroscopy of the evolved gases to assess toxicity.

The NBS smoke chamber, or smoke density chamber according to ISO 5659-2 or ASTM E662, is the most established test method to determine the specific optical density of smoke produced by pyrolyzing, smoldering, or combusting a specimen. Small variations of the test parameters are applied for materials used in the passenger compartments of aircrafts, railways (EN 45545) and ships (International Marine Organization Fire Test Procedure Code (IMO FTPC) Part 2). Generated smoke is collected in a closed chamber and measured by monitoring the attenuation of a light beam, expressed by the specific optical smoke density (D_s). Quantification of toxic combustion gases, CO_2, CO, HF, HCl, HBr, HCN, SO_2, and NO_2, is mandatory in the fire protection regulations for the interiors of railway vehicles in Europe (EN 45545-2), and CO, HF, HCl, HBr, HCN, SO_2, and NO_x in the International Marine Organization requirements for ships, although different standards accept different concentrations. Several standard fire tests assessing the HRR are equipped with optical density measurements within their exhaust systems, e.g., to monitor the transmitted intensity of a laser beam. Typical examples are the cone calorimeter and the SBI. The latter offers the classification of smoke production, whereas the data of the cone calorimeter can be used for rough estimation. However, in practice they are not used as alternative to the smoke density chamber. All kinds of fire tests have been proposed in combination with FTIR gas analysis to investigate the production of toxic gases.

More recently, soot particles contained in the smoke generated by the burning of polymeric materials have been analyzed in depth, for instance, by Stec and co-workers (2013). The total amount of soot collected in a filter connected to a smoke chamber was analyzed by gravimetrical determination, and a multi-stage cascade impactor provided a mass-based particle size distribution. Results showed that the burning conditions directly influence the size of particulate matter within the smoke.

Generally, preventing ignition, self-extinguishing during early stages of fire, and reduced flame spread induced by flame retardants reduce smoke production and the production of toxic gases in real fires. Nevertheless, reducing combustion efficiency through flame inhibition drastically increases the yields of CO, smoke, and toxic gases. Thus, flame-retarded systems that reduce both fire growth and the production of smoke and toxic gases are in demand for transportation applications such as aviation, shipping, and railway vehicles. Common approaches include combined applications of flame retardants and smoke suppressants and changes to the flame retardant or the polymer material itself.

Furthermore, corrosion due to highly corrosive fire effluents such as HBr are critical fire hazards, particularly for electrical and electronic devices and installations, industrial machinery, and large facilities and buildings, causing enormous economic damage. Thus, flame retardants containing halogen, for instance, are strictly avoided in applications like submarines. The potential for corrosion is assessed in tests such as the corrosion test apparatus according to IEC 60754, or the cone corrosometer according to ISO 11907.

Another important fire hazard is smoldering (Ohlemiller 1985; Babrauskas 2003; Rein 2016), characterized by heterogeneous, exothermic, flameless combustion on the surface of a solid porous fuel. It propagates by conduction, convection, and radiation of the heat released during the reaction. The typical propagation rates of smoldering are lower than flame spread rates. Several materials can sustain smoldering, such as cellulose, wood, cotton, but also synthetic foams. Weaker ignition sources, such as cigarettes, hot materials, or glowing wires, typically lead to smoldering. Furthermore, smoldering can be the first stage of a fire, acting as kind of ignition source for flaming combustion. Thus, it provides a pathway to flaming for heat sources too weak to produce a flame directly. Smoldering fires are believed to be one of the leading causes of fires in residential areas. Smoldering typically produces much higher yields of toxic gases and smoke. Since smoldering is based on exothermic thermo-oxidative decomposition rather than endothermic pyrolysis, different materials and different flame retardant approaches are needed as well as specific smoldering fire tests.

3.7 Uncertainty

The determination and reduction of uncertainty is an intrinsic task in fire testing. The repeatability of fire test results within the same laboratory, as well as reproducibility when comparing different laboratories, is affected by various factors including the uncertainty of the analytical equipment quantifying the reaction of the specimen to the test method, the intrinsic burning characteristics of the material, variations between the specimens for each material, and the impact of the uncertainties of test conditions (fire scenario) including operator's actions (Janssens 2007; Schartel et al. 2016). Combinations of several factors occur in most experimental setups, leading to somewhat amplified uncertainties which are complicated to determine, much less to eliminate. For example, materials with difficult intrinsic characteristics like strong deformation (collapse, bending, intumescence, melt flow, etc.) are sensitive to test conditions, including specimen mounting and flame application, and reduce the repeatability of test results. Additionally, as only a low number of replications of tests are conducted in fire testing, they are generally insufficient to perform appropriate statistical analysis. Thus, it becomes an essential premise to apply representative and homogenous specimens (i.e., morphology, ingredients, homogeneity, and thickness), to use defined specimen sizes, and to carefully follow the described test setups (i.e., specimen mounting) and instructions (i.e., specimen preconditioning, specimen size). In terms of flame-retarded polymeric materials, proper processing becomes an important factor influencing the results; for instance, extrusion on a large extruder followed by injection molding is accepted as standard for flame-retarded thermoplastics, whereas mixing a batch using a Brabender and a hydraulic press to prepare the specimen usually yields different fire performance with greater uncertainty. In terms of fire testing, even less obvious steps like proper preconditioning can make a huge difference for some materials. Safronava et al. (2015) recently demonstrated the extremely strong influence on fire performance exerted by the water content in the specimen.

In flammability and reaction to fire assessments, such as the OI, UL 94, and SBI tests, it is imperative to provide an appropriate ignition or heat source, always applying the same intensity, duration, and location of application, and even the same retraction technique. Furthermore, the specimen size and the subjective visual observation of ignition, flame height, and flameout by the operator affect the results of the tests. To maintain reproducible test conditions, proper maintenance, calibration, and following the instructions precisely are essential and often subject to neglect.

The problem may be larger than admitted during daily fire testing or research. Figure 3.3 shows two remarkable examples illustrating four cone calorimeter measurements. Cone calorimeter results are commonly believed to show a reproducibility of <10%, although all comprehensive studies show that the uncertainty is very different for different materials and heat fluxes applied (Scudamore et al. 1991). In particular, the uncertainty increases due to a very large uncertainty in time to ignition, when irradiations are applied close to the critical irradiation needed for ignition. Also flame-retarded polymeric materials with a low HRR show larger relative uncertainty due to the material independent absolute uncertainty of the HRR base line. When nearly identical specimens of polymeric materials are used, showing burning behavior with excellent reproducibility, repeatability within a fire laboratory is usually clearly lower. Uncertainties close to or less than 2% can be achieved when the material's contribution to uncertainty approaches 0, and the uncertainty of the fire scenario is minimized by performing all calibration and maintenance tasks properly. However, Figure 3.3a features a sort of worst case, namely, a comparison of the two sample holder options commonly applied in the cone calorimeter: measuring with the retainer frame and measuring without the frame in an aluminum tray. Due to this difference in sample mounting, the HRR curves of the epoxy resin measured obviously differ strongly in some key parameters, even in their basic shape. Only the initial HRR increase and the basic shape of the HRR during the main combustion stage are similar, whereas the PHRR obtained with the aluminum tray is clearly higher than the one obtained with the retainer frame. What is more, only the HRR curve of the sample placed in the retainer frame reveals a subsequent burning stage based on the pyrolysis of material previously protected by the frame, characterized by non-stable flames. Thermo-oxidation occurs, so that the residue at the end of burning is also much less than for the fire test in the open tray.

For many studies, the material itself is a dominant source of uncertainty, as illustrated in Figure 3.3b. Three HRR curves of the same PC/acrylonitrile butadiene styrene copolymer (ABS) blend differ crucially from each other. The materials show strong deformation before ignition and during burning as well. Due to this behavior, t_{ig}, time to PHRR (t_{PHRR}) and PHRR show significant uncertainties; however, reasonable repeatability was observed when the same material was re-measured 2 years later. The third strongly differing HRR was measured for the same PC/ABS, but using a different production batch of the same commercial PC. The PC had exactly the same ingredients, composition, and processing, and the polymer analytics revealed no differences at all. Nevertheless, the specimen deformed differently, leading to a different HRR.

Analytical methods that use mg-sized specimens, such as TGA measurements, target certain material properties. They focus on well-defined setups, e.g., a controlled heating rate and homogeneous temperature within the specimen. Nevertheless, a broad range of parameters must be considered to minimize

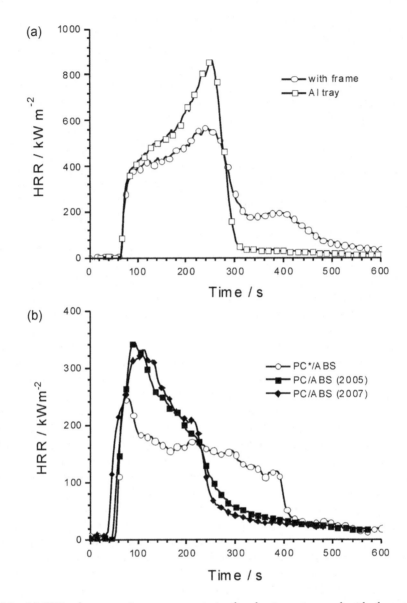

FIGURE 3.3 (a) HRR of epoxy resin measurements in the aluminum tray and with the retainer frame. A decreased PHRR and additional subsequent unstable burning (thermo-oxidative decomposition) of mainly the material previously protected by the retainer tray occurred and (b) merely using a different production batch PC* of the commercial PC component of the PC/ABS blend resulted in different deformation and thus a very different HRR curve compared to usual varyations demonstrated by measurements of different years (2005 and 2007).

uncertainty and deliver unambiguous results, such as the heating rate, suitable texture and size of the specimen, comparison of only the same specimen mass, the crucible applied, flow rates, impurities in the atmosphere, as well as the decomposition behavior of the materials investigated. When analyzing flame retardant polymers, attention must be paid to macroscopic phenomena that can influence the results, such as thermally insulating residues due to intumescence; deformation, which may lead to insufficient thermal contact between material and crucible; and diffusion barriers. These phenomena may influence the TGA results as kinds of artifacts, such that the results become very dependent on the mass and form of the specimen, be it powder, resin, or a thin film.

3.8 Assessing Fire Behavior and Flame Retardant Modes of Action

For the R&D of flame retardants and flame-retarded polymers, the quantitative assessment of fire risks plays the most important role. Furthermore, some of the most common fire tests deliver only pass or failure, without offering any insight into the modes of action. However, when it comes to scientifically based development and research, identifying the main modes of action may play an essential role, and further contribute to their quantitative assessment. Measurements with the cone calorimeter allow a quantitative assessment of not only the fire risks such as HRR, PHRR, and THE arising from the burning material in developing fires, but also of the modes of action. Measuring heat and mass, and thus effective heat of combustion and char/residue yield, the cone calorimeter is a strong tool to quantify the effect of inorganic residue and carbonaceous char enhancement in the condensed phase, and flame inhibition and fuel dilution in the gas phase. Furthermore, the general impact of intumescence or the formation of a residual protection layer may be assessed with respect to reducing the heat release rate.

The cone calorimeter measures the EHC as a product of the effective heat of combustion of the volatiles h_c^0 and the combustion efficiency of the flame χ. The cone calorimeter simulates a well ventilated developing fire, thus $\chi = 1$ for most of the polymers. Fuel dilution reduces h_c^0; flame inhibition the χ. Since the THE is the total mass loss multiplied by the EHC, it is proportional to χh_c^0. Charring and fuel dilution in the condensed phase are taken into account by the factor $(1 - \mu)$, where μ is the char yield between 0 and 1. Thus the total mass loss is the specimen multiplied by $(1 - \mu)$ and the THE is:

$$THE \sim \chi h_c^0 (1 - \mu) m_0$$

This THE can be used to quantify the flame retardant effects of charring, fuel dilution, and flame inhibition when comparing the results of a flame-retarded polymer to the results of the polymer (Brehme et al. 2011; Deng et al. 2017; Rabe et al. 2017). It is also possible to compare the THE results of the cone calorimeter and Microscale Combustion Calorimeter tests to calculate and thus assess the flame inhibition action of a flame retardant in terms of χ (Lyon et al. 2006; Schartel et al. 2007).

Assuming that the HRR is also:

$$HRR(t) \sim \theta(t) \chi h_c^0 (1 - \mu),$$

the results for the THE, or for the EHC and μ, respectively, can be used to determine the protective layer effect of the fire residue in terms of the empirical factor $\theta(t)$.

Flame retardancy modes of action and flame retardancy mechanisms can also be identified due to other parameters in the cone calorimeter, by other fire tests, and by means of analytical methods. For instance, the increase in smoke and CO yield during the burning stage proves incomplete combustion as well as the reduced EHC. However, CO and smoke production are determined not only by the combustion, but also by the pyrolysis of the polymer. Pyrolysis, often thermo-gravimetric experiments coupled with evolved gas analysis or with residue analysis, is quite often applied to identify flame retardancy modes of action. Nevertheless, this is hardly correct, as these analytical methods are not fire tests, but merely give insight into the pyrolysis, such as phosphorus release or carbonaceous charring, and thus the mechanisms behind the flame retardant modes of action. Great care is demanded, and multi-methodical approaches and reliable quantification are required, e.g., the fate and behavior of the complete amount of phosphorus should be checked and quantified in the gas and the condensed phases. Significant changes in the amounts and ratios of the volatile products often occur, but because the conservation of atoms equals the conservation of effective heat of combustion, this indicates a reduction in heat release only when it is accompanied by a corresponding change in carbonaceous char.

Depending much more on its properties than on the amount, every fire residue also works as a protective layer, often called a "barrier to heat and mass transport." While this is certainly true, it should be noted that in most cases hardly any relevant mass is captured in the form of volatiles inflating the fire residue like a balloon. The main effect is always the change in heat transfer. For layered silicate polymer nanocomposites, a reduced HRR and PHRR has been observed in the cone calorimeter due to the formation of an inorganic-carbonaceous residual protective layer. It has been shown for some systems in detail that heat shielding, the increase in reradiation by the hot surface, is the only important factor (Schartel and Weiß 2010; Wu et al. 2012).

To examine flame retardant mechanisms appearing in the condensed phase and initiating the formation of carbonaceous char or inorganic residue, further analysis of fire residues is carried out (Levchik and Wilkie 2000; Camino and Delobel 2000). Images obtained by scanning electron microscopy (SEM) have become standard procedure, revealing details of the morphology; SEM with energy dispersive X-ray spectroscopy (SEM/EDX) adds information on the chemical composition of residues (Sturm et al. 2012). Elemental analysis, attenuated total reflection (ATR)-FTIR, Raman spectroscopy, and solid state NMR are used successfully to address the formation of carbonaceous char, metal phosphates, silicates, borates, or other glasses to understand the chemistry behind the fire residue and protective layer. Analysis of the amount, morphology, and chemical structure of residue helps us to understand how and how much fuel is kept away from combustion in the flame and to explain the potential of the residual protective layer (Braun et al. 2007; Bourbigot et al. 2000; Brehme et al. 2011).

The burning of a polymer is a physico-chemical process strongly influenced by the coupling of the flame in the gas phase with the pyrolysis in the condensed phase. Several physical phenomena, such as the heat absorption of the materials, have a major impact on ignition properties and fire behavior. Back in the 1970s, Hallman et al. proposed that the role of the reflectance-absorptance characteristic of the polymer surface and dependent on the radiation sources must be considered for ignition (Hallman et al. 1974). To measure the absorption of radiative heat flux in the cone calorimeter, Linteris used four different sample platelets of various thicknesses (0.4, 0.6, 0.9, 1.4 mm) placed under the cone heater at $50 \ kW/m^2$ irradiation (Linteris et al. 2011). A heat flux meter was positioned below the platelets to measure the irradiation transmitted through the sample. Carbon nanoparticles, such as carbon nanotubes, were reported to show strong changes in heat absorption at greater depths, reducing the t_{ig} (Kashiwagi et al. 2004).

Higher thermal conductivity leads to a decelerated increase in the surface temperature, delayed decomposition, and release of volatile fuel, and thus to later ignition. Thus, in carbon nanoparticle polymer nanocomposites, it has been proposed that the increased heat absorption within the top layer, reducing the t_{ig}, competes with the increased thermal conductivity increasing the t_{ig} (Dittrich et al. 2013a, 2013b).

The tendency of polymeric materials to easy ignition, resulting in extensive flame spread, has led to testing with the application of coatings in order to shift the time of ignition to a higher order of magnitude as a powerful flame retardant approach. Not only sacrificial and intumescent coatings have been proposed (Bourbigot et al. 2004), but also heat reflective coatings (Schartel et al. 2012). Three-layer systems, consisting of an adhesive, an IR mirror, and a protective covering, reduce heat absorption by two orders of magnitude and thus increase the time of ignition sufficiently. Shifting t_{ig} by a factor of five to ten also efficiently decreases the MARHE.

Melt flow and dripping are crucial phenomena influencing fire properties, particular in reaction to fire tests such as the vertical UL 94. In general, melt flow and dripping can be both detrimental and beneficial. Flaming dripping provides an additional ignition source, increases flame spread, and may result in pool fires. On the other hand, melt flow and dripping are used as efficient flame retardant modes of action, as they remove mass and heat from the actual pyrolysis zone. It turns out that well-adjusted melt flow and dripping behavior is crucial for passing many fire tests. For instance, both immediate non-dripping self-extinguishing behavior and immediate non-flaming dripping self-extinguishing behavior are exploited in successful commercial materials classified as V-0 in UL 94. The role of dripping in the UL 94 has been addressed by several groups recently, delivering a much better understanding of the interaction and competition between gasification, charring, and dripping in UL 94 fire behavior

(Wang et al. 2010; Kandola et al. 2014; Matzen et al. 2015). Flame retardants in thermoplastics are often combined with special polytetrafluoroethylene fibers, introducing flow limits at low shear rates to achieve V-0 non-dripping (Levchik 2007; Kempel et al. 2015). Radical generators enhancing decomposition to liquid products in the condensed phase, alone or combined with flame retardants causing flame inhibition, are used to flame-retard non-charring foams such as PS foams, foils, and thermoplastics to achieve the UL 94 classifications V-2 or non-flaming dripping V-0 (Eichhorn 1964; Aubert et al. 2007; Beach et al. 2008; Pawelec et al. 2012).

3.9 Conclusion

Fire behavior is not a material property, but the reaction of a defined specimen in a specific fire scenario. Thus, while fire tests are not the only tool for performing R&D of flame retardants and flame retardant polymeric materials, passing a certain fire test is often the actual goal. The fire scenarios and fire properties addressed by the various fire tests are quite different. The reaction to a small ignition source in ignition tests, the flame spread in developing fires, and fire resistance against a fully developed fire are often related to each other only to a very limited extent. Different material properties control fire behavior: not only the usual suspects determining the fuel amount and quality of fuel released, such as char yield and effective heat of combustion of the volatiles, but also a zoo of additional parameters dependent on phenomena like heat absorption and dripping.

Since fire performance depends on complex interaction between material properties, as well as between test specimens and the fire scenario, proper fire testing demands efforts with respect to quality management, specimen preparation, and performing the fire test according to established standards, including proper know-how about mounting, preconditioning, and accounting for uncertainty.

Tailored fire testing and advanced data evaluation enable the screening, identification, and quantification of the flame retardant modes of action and the investigation of special phenomena.

Fire testing delivers information on the relevant fire performance in the R&D of flame retardant polymeric materials. What is more, with some additional effort, fire testing can yield supplementary data that can improve our understanding and assess the flame retardant modes of action as a starting point to deduce the routes and guidelines for future optimizations and development strategies.

References

Aubert, M., Roth, M., Pfaendner, R. et al. 2007. Azoalkanes: A novel class of additives for cross-linking and controlled degradation of polyolefins. *Macromolecular Materials and Engineering* 292: 707–714.

Babrauskas, V. 1984. Development of the cone calorimeter: A bench-scale heat release rate apparatus based on oxygen consumption. *Fire and Materials* 8: 81–95.

Babrauskas, V. 2002. Heat release rates. In: DiNenno, P.J., Drysdale, D., Beyler, C.L. et al. *The SFPE Handbook of Fire Protection Engineering*. 3rd ed., Quincy, MA: National Fire Protection Association. Sect. 3, chap. 1: 3-1-3-37.

Babrauskas, V. 2003. *Ignition Handbook*. Issaquah, WA: Fire Science Publishers.

Babrauskas, V., and Peacock, R. 1992. Heat release rate: The single most important variable in fire hazard. *Fire Safety Journal* 18: 255–272.

Bachelet, P., Sarazin, J., Duquesne, S. et al. 2017. Scale reduction of SBI: Correlation methodology between the standard and the small scale test. *FRPM 2017 Fire Retardant Polymeric Materials*. Manchester, UK, July 3–6.

Batholmai, M., and Schartel, B. 2007. Assessing the performance of intumescent coatings using bench-scaled cone calorimeter and finite difference simulations. *Fire and Materials* 31: 187–205.

Beach, M. W., Rondan, N. G., Froese, R. D. et al. 2008. Studies of degradation enhancement of polystyrene by flame retardant additives. *Polymer Degradation and Stability* 93: 1664–1673.

Blaszkiewicz, M., Bowman, P., and Masciantonio, M. 2007. Understanding the repeatability and repro-ducibility of UL 94 testing. In: *18th Annual Conference on Recent Advances in Flame Retardancy of Polymeric Materials*. Stamford, CT: BCC.

Bourbigot, S., Bachelet, P., Samyn, F. et al. 2013. Intumescence as method for providing fire resistance to structural composites: Application to poly(ethylene terephtalate) foam sandwich–structured composite. *Composite Interfaces* 20: 269–277.

Bourbigot, S., Le Bras, M., Dabrowski, F. et al. 2000. PA-6 clay nanocomposite hybrid as char forming agent in intumescent formulations. *Fire and Materials* 24: 201–208.

Bourbigot, S., Le Bras, M., Duquesne, S. et al. 2004. Recent advances for intumescent polymers. *Macromolecular Materials and Engineering* 289: 499–511.

Braun, U., Schartel, B., Fichera, M.A. et al. 2007. Flame retardancy mechanisms of aluminium phos-phinate in combination with melamine polyphosphate and zinc borate in glass-fibre reinforced polyamide 6,6. *Polymer Degradation and Stability* 92: 1528–1545.

Brehme, S., Schartel, B., Goebbels, J. et al. 2011. Phosphorus polyester versus aluminium phosphinate in poly(butylene terephthalate) (PBT): Flame retardancy performance and mechanisms. *Polymer Degradation and Stability* 96: 875–884.

Camino, G., and Delobel., R. 2000. Intumescence. In: Grand, A. F., Wilkie C. A. *Fire Retardancy of Polymeric Materials*. New York: Marcel Dekker Publications. Chap. 7: pp. 217–243.

Camino, G., Costa, L., Casorati, E. et al. 1988. The oxygen index method in fire retardance studies on polymeric materials. *Journal of Applied Polymer Science* 35: 1863–1876.

Davis, R. D., Lyon, R. E., Takemori, M. T. et al. 2010. High throughput techniques for fire resistant mate-rials development. In: Wilkie, C.A., Morgan A. B. *Fire Retardancy of Polymeric Materials*. Boca Raton, FL: CRC Press. Chap. 16: pp. 421–452.

De Corso, A., and Castrovinci, A. 2015. Mini single burning item [miniSBI]: A unique lab-scale device to pre-screen reduced specimen for the EN 13823. In: *Fire Resistance in Plastics*. Cologne, Germany, December 8–10.

Deng, C., Yin, H., Li, R. M. et al. 2017. Modes of action of a mono-component intumescent flame retar-dant MAPP in polyethylene-octene elastomer. *Polymer Degradation and Stability* 138: 142–150.

Dittrich, B., Wartig, K. A., Hofmann, D. et al. 2013a. Carbon black, multiwall carbon nanotubes, expanded graphite and functionalized graphene flame retarded polypropylene nanocomposites. *Polymers for Advanced Technologies* 24: 916–926.

Dittrich, B., Wartig, K. A., Hofmann, D. et al. 2013b. Flame retardancy through carbon nanomaterials: Carbon black, multiwall carbon nanotubes, expanded graphite, multi-layer graphene and gra-phene in polypropylene. *Polymer Degradation and Stability* 98: 1495–1505.

Eichhorn, J. 1964. Synergism of free radical initiators with self-extinguishing additives in vinyl aromatic polymers. *Journal of Applied Polymer Science* 8: 2497–2524.

Gallo, E., Stöcklein, W., Klack, P. et al. 2017. Assessing the reaction to fire of cables by a new bench-scale method. *Fire and Materials* 41: 768–778.

Gibson, A. G., Browne, T. N. A., Feih, S. et al. 2012. Modeling composite high temperature behavior and fire response under load. *Journal of Composite Materials* 46: 2005–2022.

Gibson, A. G., Otheguy Torres, M. E., Browne, T. N. A. et al. 2010. High temperature and fire behav-iour of continuous glass fibre/polypropylene laminates. *Composites Part A: Applied Science and Manufacturing* 41: 1219–1231.

Gilman, J. W., Bourbigot, S., Shields, J. R. et al. 2003. High throughput methods for polymer nanocom-posites research: Extrusion, NMR characterization and flammability property screening. *Journal of Materials Science* 38: 4451–4460.

Hallman, J. R., Welker, J. R., and Sliepcevich, C. 1974. Polymer surface reflectance-absorptance charac-teristics. *Polymer Engineering & Science* 14: 717–723.

Hörold, A., Schartel, B., Trappe, V. et al. 2013. Structural integrity of sandwich structures in fire: An intermediate-scale approach. *Composite Interfaces* 20: 741–759.

Hörold, A., Schartel, B., Trappe, V. et al. 2017. Fire stability of glass-fibre sandwich panels: The influence of core materials and flame retardants. *Composite Structures* 160: 1310–1318.

Horrocks, A. R., Tune, M., and Price, D. 1989. The burning behaviour of textiles and its assessment by oxygen-index methods. *Textile Progress* 18: 1–186.

Hugget, C. 1980. Estimation of rate of heat release by means of oxygen consumption measurement. *Fire and Materials* 4: 61–65.

Hull, T. R., Wills, C. L., Artingstall, T. et al. 2005. Mechanisms of smoke and CO suppression from EVAcomposites. In: Le Bras, M., Wilkie, C.A., Bourbigot, S. et al. *Fire Retardancy of Polymers New Application of Mineral Fillers*. Cambridge, UK: Royal Society of Chemistry. Chap 28: pp. 372–385.

Isaacs, J. 1970. The oxygen index flammability test. *Journal of Fire and Flammability* 1: 36–47.

Janssens, M. 2007. Uncertainty of fire test results. In: *11th International Fire Science and Engineering Conference*, Interflam 2007, Egham, UK, September 3–5.

Janssens, M. 2008. Challenge in fire testing: A tester's viewpoint. In: Horrocks, A. R., Price, D. *Advances in Fire Retardant Materials*. Cambridge, UK: Woodhead Publishing. Chap. 10: pp. 233–254.

Janssens, M. 2010. Challenge in fire testing: A tester's viewpoint. In: Wilkie, C.A., Morgan A. B. *Fire Retardancy of Polymeric Materials*. Boca Raton, FL: CRC Press. Chap. 14: pp. 349–386.

Janssens, M. L. 1991. Measuring rate of heat release by oxygen consumption. *Fire Technology* 27: 234–249.

Jimenez, M., Bellayer, S., Naik, A. et al. 2016. Topcoats versus durability of an intumescent coating. *Industrial & Engineering Chemistry Research* 55: 9625–9632.

Kandola, B. K., Ndiaye, M., and Price, D. 2014. Quantification of polymer degradation during melt dripping of thermoplastic polymers. *Polymer Degradation and Stability* 106: 16–25.

Kashiwagi, T., Grulke, E., Hilding, J. et al. 2004. Thermal and flammability properties of polypropylene/carbon nanotube nanocomposites. *Polymer* 45: 4227–4239.

Kempel, F., Schartel, B., Marti, J. M. et al. 2015. Modelling the vertical UL 94 test: Competition and collaboration between melt dripping, gasification and combustion. *Fire and Materials* 39: 570–584.

Levchik, S. V. 2007. Introduction to flame retardancy and polymer flammability. In: Morgan, A. B., Wilkie, C. A. *Flame Retardant Polymer Nanocomposites*. Hoboken, NJ: John Wiley & Sons. Chap. 1: pp. 1–29.

Levchik, S., and Wilkie, C. A. 2000. Char formation. In: Grand, A. F., Wilkie, C. A. *Fire Retardancy of Polymeric Materials*. New York: Marcel Dekker Publications. Chap. 6: pp. 171–215.

Linteris, G., Zammarano, M., Wilthan, B. et al. 2011. Absorption and reflection of infrared radiation by polymers in fire-like environment. *Fire and Materials* 36: 537–553.

Lyon, R. E., and Walters, R. N. 2004. Pyrolysis combustion flow calorimetry. *Journal of Analytical and Applied Pyrolysis* 71: 27–46.

Lyon, R. E., and Walters, R. N. 2015. Practical aspects of microscale combustion calorimetry. In: *Proceedings of the Fire and Materials 2015 Conference*. Greenwich, UK: Interscience Communication, pp. 104–115.

Lyon, R. E., Walters, R. N., and Gandhi, S. 2006. Combustibility of cyanate ester resins. *Fire and Materials* 30: 89–106.

Lyon, R. E., Walters, R. N., and Stoliarov, S. I. 2007. Screening flame retardants for plastics using microscale combustion calorimetry. *Polymer Engineering & Science* 47: 1501–1510.

Matzen, M., Kandola, B., Huth, C. et al. 2015. Influence of flame retardants on the melt dripping behaviour of thermoplastic polymers. *Materials* 8: 5621–5646.

Morgan, A. B., and Bundy, M. 2007. Cone calorimeter analysis of UL 94 V-rated plastics. *Fire and Materials* 31: 257–283.

Morys, M., Illerhaus, B., Sturm, H. et al. 2017a. Revealing the inner secrets of intumescence: Advanced standard time temperature oven (STT Mufu+)—μ-computed tomography approach. *Fire and Materials* 41: 927–939.

Morys, M., Illerhaus, B., Sturm, H. et al. 2017b. Size is not all that matters: Comparing the influence of different binders on the performance and morphology of intumescent coatings. *Journal of Fire Sciences* 35: 284–302.

Ohlemiller, T. J. 1985. Modeling of smoldering combustion propagation. *Progress in Energy and Combustion Science* 11: 277–310.

Pawelec, W., Aubert, M., Pfaendner, R. et al. 2012. Triazene compounds as a novel and effective class of flame retardants for polypropylene. *Polymer Degradation and Stability* 97: 948–954.

Rabe, S., and Schartel, B. 2017a. The rapid mass calorimeter: A route to high throughput fire testing. *Fire and Materials* 41: 834–847.

Rabe, S., and Schartel, B. 2017b. The rapid mass calorimeter: Understanding reduced-scale fire test results. *Polymer Testing* 57: 165–174.

Rabe, S., Chuenban, Y., and Schartel, B. 2017. Exploring the modes of action of phosphorus-based flame retardants in polymeric systems. *Materials* 10: 455.

Raffan-Montoya, F., Ding, X., Stoliarov, S. I. et al. 2015. Measurement of heat release in laminar diffusion flames fueled by controlled pyrolysis of milligram-sized solid samples: Impact of bromine- and phosphorus-based flame retardants. *Combustion and Flame* 162: 4660–4670.

Rein, G. 2016. Smoldering combustion. In: Hurley, M.J. *SFPE Handbook of Fire Protection Engineering.* 5th ed. New York: Springer, Society of Fire Protection Engineers. Chap. 19: pp. 581–603.

Safronava, N., Lyon, R. E., Crowley, S. et al. 2015. Effect of moisture on ignition time of polymers. *Fire Technology* 51: 1093–1112.

Schartel, B. 2010. Uses of fire tests in materials flammability development. In: Wilkie, C. A., Morgan, A. B. *Fire Retardancy of Polymeric Materials.* Boca Raton, FL: CRC Press, 2010. Chap. 15: pp. 387–420.

Schartel, B., and Hull, T. R. 2007. Development of fire-retarded materials—Interpretation of cone calorimeter data. *Fire and Materials* 31: 327–354.

Schartel, B., and Weiß, A. 2010. Temperature inside burning polymer specimens: Pyrolysis zone and shielding. *Fire and Materials* 34: 217–235.

Schartel, B., Bartholmai, M., and Knoll U. 2005. Some comments on the use of cone calorimeter data. *Polymer Degradation and Stability* 88: 540–547.

Schartel, B., Beck, U., Bahr, H. et al. 2012. Short communication: Sub-micrometer coatings as an IR mirror: A new route to flame retardancy. *Fire and Materials* 36: 671–677.

Schartel, B., Pawlowski, K. H., and Lyon, R. E. 2007. Pyrolysis combustion flow calorimeter: A tool to assess flame retarded PC/ABS materials? *Thermochimica Acta* 462: 1–14.

Schartel, B., Wilkie, C. A., and Camino, G. 2016. Recommendations on the scientific approach to polymer flame retardancy: Part 1—Scientific terms and methods. *Journal of Fire Sciences* 34: 447–467.

Schartel, B., Wilkie, C.A., and Camino, G. 2017. Recommendations on the scientific approach to polymer flame retardancy: Part 2—Concepts. *Journal of Fire Sciences* 35, 3–20.

Scudamore, M., Briggs, P., and Prager, F. 1991. Cone calorimetry—A review of tests carried out on plastics for the Association of Plastic Manufacturers in Europe. *Fire and Materials* 15: 65–84.

Stec, A. A., and Hull, T. R. 2010. *Fire Toxicity.* Cambridge, UK: Woodhead Publishing.

Stec, A. A., Hull, T. R., Lebek, K. et al. 2008. The effect of temperature and ventilation condition on the toxic product yields from burning polymers. *Fire and Materials* 32: 49–60.

Stec, A. A., Readman, J., Blomqvist, P. et al. 2013. Analysis of toxic effluents released from PVC carpet under different fire conditions. *Chemosphere* 90: 65–71.

Sturm, H., Schartel, B., Weiß, A. et al. 2012. SEM/EDX: Advanced investigation of structured fire residues and residue formation. *Polymer Testing* 31: 606–619.

Sundström, B. 2007. *The Development of a European Fire Classification System for Building Products— Test Methods and Mathematical Modelling.* Lund, Sweden: Fire Safety Engineering and Systems Safety.

Tabeling, F. 2004. Zum Hochtemperaturverhalten dämmschichtbildender Brandschutzsysteme auf Stahlbauteilen. Dissertation, Gottfried Wilhelm Leibniz Universität Hannover.

Tranchard, P., Samyn, F., Duquesne, S. et al. 2015. Fire behaviour of carbon fibre epoxy composite for aircraft: Novel test bench and experimental study. *Journal of Fire Sciences* 33: 247–266.

Troitzsch, J. 2004. *Plastics Flammability Handbook.* 3rd ed., Munich, Germany: Carl Hanser Verlag.

van Krevelen, D. 1975. Some basic aspects of flame resistance of polymeric materials. *Polymer* 16: 615–620.

Velencoso, M. M., Battig, A., Markwart, J. C. et al. 2018. Molecular firefighting—how modern phosphorus chemistry can help solve the challenge of flame retardancy. *Angewandte Chemie International Edition* 57: 10450–10467.

Wang, D. Y. 2017. *Novel Fire Retardant Polymers and Composite Materials.* Cambridge, UK: Woodhead Publishing.

Wang, Y., Zhang, F., Chen, X. et al. 2010. Burning and dripping behaviors of polymers under UL 94 vertical burning test conditions. *Fire and Materials* 34: 203–215.

Weil, E. D, Hirschler, M. M., Patel, N. et al. 1992. Oxygen index: Correlations to other fire tests. *Fire and Materials* 16: 159–167.

Weil, E. D., and Levchik, S. V. 2016. *Flame Retardants for Plastics and Textiles.* 2nd ed., Munich, Germany: Carl HanserVerlag.

Wharton, R. 1981. Correlation between the critical oxygen index test and other fire tests. *Fire and Materials* 5: 93–102.

Wittbecker, F. W. 2004. Methodology of fire testing. In: Troitzsch, J. *Plastics Flammability Handbook.* 3rd ed., Munich, Germany: Carl HanserVerlag. Chap. 8: pp. 209–221.

Wu, G. M., Schartel, B., Bahr, H. et al. 2012. Experimental and quantitative assessment of flame retardancy by the shielding effect in layered silicate epoxy nanocomposites. *Combustion and Flame* 159: 3616–3623.

II

Material Modification, Assembly, and Design

4

Phosphorus-Containing Flame Retardants: Synthesis, Properties, and Mechanisms

Qiang Yao and
Charles A. Wilkie

4.1 Introduction

Phosphorus-containing flame retardants (PFR) are an important class of flame retardants, with a long history to mitigate the risk of fire. As early as the beginning of the nineteenth century, simple inorganic ammonium phosphate was used as a coating to prevent the burning of theater curtains made from natural materials (Granzow 1978). With the advent of modern polymeric materials, the impetus to flame-retard some of the highly flammable materials, particularly polyurethane, resulted in the development of chloro-alkyl phosphates that have good handling properties, compatibility, and high efficiency (Papa 1972). Since the early 1990s, polymer alloys such as polycarbonate/acrylonitrile-butadiene-styrene copolymers (PC/ABS) and high impact polystyrene/poly(phenylene oxide) (HIPS/PPO) have gained importance in office equipment and business machine housings, and organophosphorus flame retardants, such as bisphosphates of low volatility, have been introduced to flame-retard these engineering resins (Gosens et al. 1993; Bright et al. 1997). Recently, the movement toward non-halogenated flame retardants has brought a significant push to develop phosphorus-containing flame retardants. To retain the high performance of polymers such as polyamide and polyesters, inorganic-organic hybrid phosphorus-containing flame retardants have been developed, and gradually take the market for brominated flame retardants (Jenewein et al. 2002). With the miniaturization of the electronic components, properties such as critical tracking index (CTI) and glow wire index temperature (GWIT) have become increasingly important (Markarian 2005). Since phosphorus-containing flame retardants are naturally more amenable to these requirements than brominated flame retardants, the former will continue to experience fast growth. However, the concern of the flame retardants' impact on the health and environment has pushed some traditional PFRs out of market (Quednow and Püttmann 2009; Van der Veen and de Boer 2012). The future of phosphorus flame retardants will depend on the balance of the need of fire safety and their impact on the health and environment.

In this chapter, the characteristics of phosphorus will be given first and serve as the basis for the synthesis and mechanism of phosphorus-containing flame retardants. The modes of action of phosphorus-containing flame retardants will then be presented with focus on the factors affecting the efficiency. This is followed by the synthesis and applications that include some common synthetic methodologies to prepare phosphorus-containing flame retardants and the selected commercially important applications. Finally, the trends of phosphorus-containing flame retardants and their future will be envisaged. The focus of the chapter is on electric and electronic applications.

4.2 Characteristics of Phosphorus

Phosphorus has an electron configuration of s^2p^3 [Ne]. It can form hyper-valent compounds through a three-center four-electron bonding mechanism (Musher 1969), and oxidation states range from −3 to +5, depending on the substituents. The oxidation state of phosphorus in the ordinary types of phosphorus compounds can be calculated simply by assigning +1 to phosphorus when it is connected to oxygen through a single bond, +2 through a double bond, and −1 when a carbon or hydrogen is directly attached to phosphorus.

According to the Sanderson scale (Sanderson 1983), phosphorus has an electronegativity of 2.515 which is lower than oxygen's 3.654, but fairly close to carbon's 2.746. This difference explains the higher reactivity of P–O bonds than P–C bonds toward polar reactions such as nucleophilic substitutions on P. However, under thermo-lysis conditions and in the absence of eliminations and rearrangement, P–C bonds tend to undergo homolysis easier than P–O bonds because the former have lower bond dissociation energies (BDE) than the latter (Table 4.1). Nevertheless, the substituents can exert a large effect on the BDE of P–C/P–O bonds and thus change the preference to which bond is cleaved. For simple substituents, ab initio computations found that BDE (P–OPh) < BDE (P–CH$_3$) < BDE (P–Ph) < BDE (P–OCH$_3$) (Hemelsoet et al. 2010).

An important characteristic of phosphorus is its high affinity toward oxygen. Elemental white phosphorus spontaneously reacts with oxygen to produce transient species such as PO or PO$_2$ radicals before transforming to stable derivatives (Andrews and Withnall 1988). The complete combustion of phosphorus produces a large amount of heat with an enthalpy of formation of about 2970 kJ/mol (Holmes 1962; Hartley et al. 1963), and yields the highly thermally stable, P$_4$O$_{10}$. P$_4$O$_{10}$ is a powerful desiccant and produces non-volatile phosphoric acid or polyphosphates upon hydrolysis. These intermediates and ultimate stable products generated by the oxidation of phosphorus are fundamental to the flame retardancy of PFRs.

4.3 Types of PFRs

Phosphorus-containing flame retardants take a variety of chemical structures. However, because of the propensity of oxidation, the most prevalent phosphorus-containing flame retardants possess the P=O double bond. Figure 4.1 lists some common types of phosphorus-containing flame retardants.

TABLE 4.1 Mean Bond Dissociation Energy of P–X Bond

Bond	P–H	P–Cl	P–Me	P–Ph	P–O	P=O
BDE (kJ/mol)	322	331	272	297	385	586

Source: Maiti, S. et al., *Prog. Polym. Sci.*, 18, 227–261, 1993; Hartley, S. et al., *Quart. Rev. Chem. Soc.*, 17, 204–223, 1963.

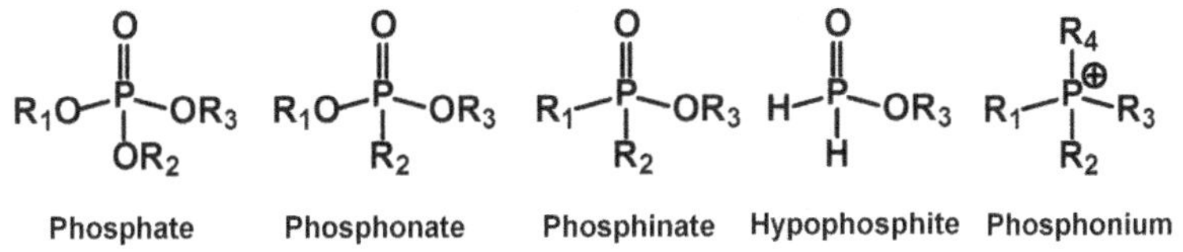

FIGURE 4.1   Common types of phosphorus-containing flame retardants.

4.4 PFR Flame Retarding Mechanisms

The combustion of polymeric materials is extremely complicated and involves the degradation of polymers with the generation of fuels, oxidation of fuels in the combustion zone to release heat, and heat feedback to polymers to propagate the generation of fuels. The phosphorus-containing flame retardants can interrupt any of these processes to break down this auto-catalytic cycle and protect polymers from burning. They can quench the flame via chemical and/or physical pathways. They may act in the vapor phase or the condensed phase depending on the type and location of phosphorus-containing flame retardants or their degradation products.

4.4.1 General Mechanisms

From molecules to radicals, phosphorus compounds act as flame retardants in a variety of forms. As molecules, phosphorus-containing flame retardants with labile P–O bonds often interact with condensation polymers such as polyamides, polyesters, and polycarbonates to change the degradation of polymers. For example, due to the strong electron withdrawing effect of sulfone group, the P–O bond of polysulfonyldiphenylene phenylphosphonate (PSPP) is greatly polarized and is easily subject to the polar reaction. When PSPP is mixed with polyesters, the transesterification reaction effectively reduces the molecular weight of the latter and accelerates the dripping of polyesters (Wang et al. 1996). Dripping not only removes the heat from the degradation zone, it also reduces the supply of fuels to the gas phase. As a result, good flame retardancy of polyesters by phosphonate is achieved.

Besides direct involvement in the degradation of polymers, phosphorus-containing flame retardants can also alter the generation and types of fuels from polymers through their degradation products, phosphoric acids, or polyphosphoric acids (Kishore and Mohandas 1982; LeVan 1984). These acids are strong enough to catalyze the dehydration, cyclization, and aromatization of polymers or char-forming agents to carboneous chars. Through the formation of char, smaller amounts of fuels from polymers are generated. Char also shields the underlining polymers from the heat and lessens further degradations. In addition, polyphosphoric acids are thermally stable enough to form a coating on the surface of polymer matrix. This coherent layer works in a similar way as chars to physically shield the polymers from the combustion zone.

Some phosphorus-containing flame retardants are chemically and thermally stable enough to vaporize into the combustion zone. After they enter the vapor phase, they react in the same ways as does elemental phosphorus. As a matter of fact, phosphorus-containing flame retardants can be viewed as the stable elemental phosphorus with the substituents imparting good handling property, correct volatility, and chemical inertness to phosphorus. Most of phosphorus-containing flame retardants or their primary decomposition products eventually transform to PO or PO_2 radicals in the combustion zone regardless of their original structures (Siow and Laurendeau 2004; Shmakov et al. 2005). The fundamental role played by PO or PO_2 radicals is to scavenge H or OH radicals through a catalyzed radical coupling chemistry (Figure 4.2) (Hastie and Bonnell 1980; Twarowski 1995).

In order for Rxn 1 to occur, the bond energy of H to O=P–O• must be high so that Rxn 1 can effectively compete with the radical branching reaction (Rxn 5). However, it should not be too high so that

$$H + PO_2 + M \longrightarrow HOPO + M \quad \text{(Reaction 1)}$$
$$H + HOPO \longrightarrow H_2 + PO_2 \quad \text{(Reaction 2)}$$
$$OH + PO_2 \longrightarrow HOPO_2 \quad \text{(Reaction 3)}$$
$$H + HOPO_2 \longrightarrow H_2O + PO_2 \quad \text{(Reaction 4)}$$

FIGURE 4.2 The catalyzed radical recombination by PO_2 radical. M is a third body.

Rxn 2 would be endothermic and the reverse reaction would be favored. By using rate constants given in Reference Twarowski (1996), it is estimated that the ratio of the rate constant of Rxn 1/Rxn 5 ≈ 914 at a flame temperature of 1500 K; i.e., Rxn 1 is favored over Rxn 5. On the other hand, Rxn 2 has very low activation energy close to 0 kJ/mol, and the rate constant of the forward reaction is high. The recombinations of H with H and OH radicals through Rxn 2 to Rxn 4 lead, respectively, to H_2 and H_2O, stop the oxidation process, and regenerate the PO_2 active species. Thus, even a small amount of PO_2 can cause the combustion to become unsustainable.

$$H + O_2 \longrightarrow OH + O \quad \text{(Reaction 5)}$$

Besides radical scavenging in the vapor phase, physical effects, such as heat capacity and dilution of flame retardants might be important, especially at high concentration of flame retardants. The physical effect has been shown in many brominated flame retardants (BFR)/fuel systems (Noto et al. 1998), and it should be no surprise that PFRs would behave similarly in the gas phase.

4.4.2 Complexities of the Mechanisms

Since the phosphorus in the form of either molecules or radicals can interrupt any step of the burning cycle, several mechanisms often operate either simultaneously or sequentially. The dominant mechanism may change from one to another even for the same PFR depending on the structures of polymeric materials or concentrations of flame retardants.

For example, it has been demonstrated that dimethyl methylphosphonate (DMMP) shows a highly efficient gas phase action at low concentration; however, the fuel effect of the hydrocarbon moiety of the DMMP molecule actually promotes combustion in lean flames and at high loadings (Babushok et al. 2016). Levchik et al. (1996b) found that bis(m-aminophenyl)methylphosphine oxide changes from the condensed phase to a gas phase fire retardant action with an increased loading level in the same epoxy resin. Also, Seefeldt and his colleagues demonstrated that aluminum diethylphosphinate (ADEP) works quite differently in semi-aromatic polyamides (SAPA) and aliphatic polyamides (APA). ADEP shows both the condensed phase action and flame inhibition in SAPA (Seefeldt et al. 2013), but mostly flame inhibition in APA (Braun et al. 2007).

Because flame retardancy is the collective results of many factors, singling out one factor becomes difficult. This adds extra difficulty to the study of the mechanism. What is more, the tools used to probe the mechanism could also influence the categorization of mechanisms. For example, the widely used cone calorimeter is an excellent tool to discriminate the gas phase action from the condensed phase action in the absence of dripping. However, since it completely eliminates the dripping and may force the phosphorus species to go into the gas phase, the gas phase action could be overestimated. On the other hand, the Microscale Combustion Calorimetry (MCC) tends to misjudge flame inhibition because the gas-phase reactions are essentially forced to completion (Raffan-Montoya et al. 2015).

Besides the tools, the conditions used for particular equipment might have an influence also. This is especially true in thermal analysis. In the thermogravimetric analysis (TGA) study, triphenyl phosphate (TPP) has often been considered to act solely in the gas phase when blended with polypropylene; however, Korobeinichev et al. (2013) showed that TPP actually chemically interacts with the degradation of polypropylene at a very high heating rate and high temperatures, close to those at ignition and combustion of the polymer. Further study indicates that TPP increases the flammability of polymers by its action in the condensed phase (Korobeinichev et al. 2016). This discrepancy is likely because the rate of the evaporation of TPP is slower than that of its decomposition at the high heating rate, and consequently some of TPP decomposes in the condensed phase before it escapes.

4.4.3 Factors Affecting Efficiencies of Flame Retardancy

The efficiency of phosphorus-containing flame retardants is affected by their own chemical structures as well as many variables such as polymer's structures and the presence or absence of the second component in the polymers. A fine tuning of the structure can bring a tremendous effect on the efficiency. This effect is pronounced even for the homologs of PFRs.

For example, despite a low phosphorus content, 2,5-di-tertbutylphenyl diphenyl phosphate (DDP) has been shown to be as effective as triphenyl phosphate for polycarbonate, but 2-tertbuylphenyl diphenyl phosphate is less effect than either, even though it has a higher phosphorus content than DDP (Yoon et al. 2013). The high efficiency of DDP is attributed to its high diffusivity at elevated temperatures. In polyamide 6/6, aluminum diisobutylphosphinate gives a UL 94 V-0 rating at 1.6 mm with a loading level as low as 10%, while aluminum diethylphosphinate fails completely at the same concentration (Yao et al. 2010). The excellent flame retardancy of the former is attributed to its high volatility at the correct temperature, and also possibly to its lower enthalpy of decomposition through facile ß-hydrogen elimination to P(=O)–H, a tautomeric structure to P–OH which is similar to HOPO in Rxn 1.

The oxidation number of phosphorus has been noted to affect the efficiency of PFRs. Mariappan et al. (2013) found that triphenyl phosphate was a better flame retardant in polyurea than triphenyl phosphite and triphenyl phosphine oxide, but triphenyl phosphite performed best in epoxy resins.

In the gas phase, since the phosphorus-containing flame retardants need to decompose to active PO or PO_2 radicals, the easier are these radicals produced, the higher is the efficiency of the PFR. Phosphorus compounds with the activation energy of decomposition higher than 377 kJ/mol have been shown to be ineffective in the gas phase (Babushok and Tsang 1999).

In addition, the efficiency of gas phase action of phosphorus-containing flame retardants decreases with the increased flame temperatures. MacDonald et al. (2001) showed that the effectiveness of DMMP in a stoichiometric adiabatic flame decreases by 90% for a mere 300 K increase in flame temperature. This is a significant result. Since the flame temperatures are not the same for all combusting polymers, this would lead to a different efficiency in polymers even for the same gas phase PFRs. Further, the efficiency of PFRs is much more sensitive to the flame temperature than those of BFRs. It can be estimated that the inhibition effectiveness of CF_3Br is lowered less than 50% by a 300 K temperature increase from an adiabatic temperature of 2000 K to 2300 K (Saso et al. 1999). Therefore, the high sensitivity of PFRs on the flame temperature might be one reason that the gas phase PFRs do not work as a universal inhibitor as BFRs do in a variety of polymers. From the study of temperature dependence, there seems to exist an optimum working temperature range for gas phase PFRs. Below a certain temperature, PFR is slow to generate active scavenging radicals, such as PO_2 radical. But at a very high temperature, the radical scavenging ability is reduced.

The efficiency of PFRs in the gas phase is also concentration dependent. That is, the low concentration of PFRs has a higher efficiency than the high concentration of PFRs. Increasing the amount will not increase their chemical effectiveness. This means that only a small amount of gas phase PFRs is needed.

Thus, the saturation effect and the temperature dependence of PFRs hint that a high efficiency of flame retardancy could be achieved by the combination of gas phase action and condensed phase action or flame inhibition plus physical process such as gas phase dilution. These strategies have been used to improve the efficiency of both BFRs (Babushok and Tsang 1999) and PFRs. For example, Weil et al. (1996) found that a combination of gas phase phosphate flame retardants with char-forming agents works very well in ABS.

The efficiency of PFRs can also be affected by the other additives present in the polymer, such as glass fiber which can bring the wick effect and reduces the efficiency of PFRs. On the other hand, the use of synergists can boost the performance of PFRs. For example, aluminum diethylphosphinate is well known to be ineffective for polyamides, but the combination with melamine polyphosphate or cyanurate is a superior flame retardant (Braun et al. 2007, 2008). Phosphorus-phosphorus synergism has also been documented (Despinasse and Schartel 2013). However, the combination may not work as the synergist in all polymers and sometimes they act even as antagonists (Levchik et al. 1996a).

In summary, for PFRs to gain the maximum efficiency, they need to work at a right time with a right amount and a right partner to limit the generation and supply of fuels.

4.5 Synthesis and Applications

The diversity of the PFR structures dictates the need for a multiplicity of synthetic methods. Many methods have been reported; however, except for a few processes, harsh reaction conditions, the use of expensive reagents, multiple-steps, or low yields prevent their commercialization. As a matter of fact, the number of industrially viable synthetic methods is very limited for economic productions of PFRs. Most PFRs have been made by utilizing P–Cl or P–H functional groups, respectively, for introducing P–O or P–C bonds in one or two steps. Some unusual PFRs can be prepared through the chemical modification of the auxiliary groups on phosphorus.

4.5.1 Synthesis

4.5.1.1 P–O

$POCl_3$ is the basic raw materials for many phosphates. The P–Cl bond is highly polarized and subjected to attack by nucleophiles such as –OH, epoxide, and amine groups. The driving force here is the formation of a more stable P–O bond and in some cases a gain in entropy by producing volatile HCl as well.

An early commercialized phosphate is tris (1-chloro-2-propyl) phosphate (TCPP). It is made from the insertion reaction of $POCl_3$ on propylene oxide. This is an excellent flame retardant for polyurethane foams (Papa 1972). The synthesis of bisphosphates such as bisphenol A bis(diphenyl bisphosphate) and resorcinol bis(diphenyl bisphosphate) also begin with $POCl_3$. Depending on the desirable molecular weights and the availability of chlorophosphates, two different processes have been developed (Figure 4.3) (Knauer 1894). In method (a), a bischlorophosphate intermediate is formed first; it is then capped with mono-alcohol to furnish the bisphosphate. In this method, Rxn 6 needs to be carried out with excess

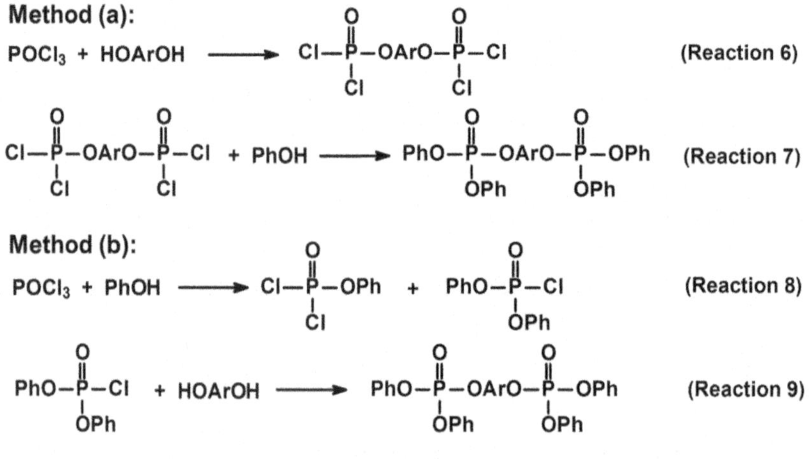

FIGURE 4.3 Synthesis of bisphosphates.

$POCl_3$ in order to suppress the formation of oligomers. In method (b), a monochlorophosphate intermediate is isolated in the first step (Rxn 8); it subsequently reacts with diol to produce bisphosphate. Rxn 8 is usually stopped at the stage of dichloro- and monochlorophosphates to minimize the formation of triphosphate; however, if a sterically hindered phenol is used, the reaction can be controlled to produce only monochlorophosphate (Campbell et al. 2002). To accelerate the reaction, Lewis acids such as aluminum chloride or magnesium chloride are often employed as catalysts.

Besides aromatic alcohols, simple aliphatic alcohols can react with $POCl_3$ to make alkyl phosphates. However, since the alkyl phosphate is an alkylating agent, the alkyl group is subjected to attack by the by-product HCl to generate phosphoric acid that further reacts with chlorophosphates to give pyrophosphate. Thus, tetraethyl pyrophosphate, which is highly toxic, is a common by-product in the synthesis of triethyl phosphate.

To avoid the alkylation, sterically hindered aliphatic alcohols can be used instead. Two common sterically hindered alcohols are neopentyl glycol and pentaerythritol. Due to the presence of four hydroxyl groups, pentaerythritol has a rich chemistry and the pentaerythritol phosphates are excellent char-forming agents (Telschow 1999). However, these phosphates usually have low thermal stability or are easily hydrolyzed (Wroblewski and Verkade 1996).

Besides the chemistry directly based on the esterification of $POCl_3$, the transesterification can be exercised indirectly to make P–O bonds. For example, (poly)methylphosphonates have been produced from the transesterification of diphenyl methylphosphonate and diols (Figure 4.4) (Honig and Weil 1977; Freitag 2006; Piotrowski et al. 2009). Similarly, triphenyl phosphate is reacted with diols to produce phosphate oligomers (Deanin and Ali 1994).

Inorganic phosphates flame retardants can be prepared from the condensation of amine salts of phosphoric acids. Melamine polyphosphate has been made by this route (Cichy et al. 2003).

4.5.1.2 P–N

In theory, phosphoramides or phosphonamides can be made from amines and P–Cl; however, because the reaction generates HCl that neutralizes the amine reactant, acid-scavengers, typically an organic amine such as triethylamine, are required. At large scale, it has been found that the by-product triethylammonium chloride is difficult to completely remove from the product and its presence, even at a trivial level, can significantly deteriorate the physical properties of the flame retardant composites (Worku et al. 2003). To overcome the economic disadvantages and the impurities, phosphoramides might be prepared through amminolysis of phosphates or phosphonates. For example, poly(methylphosphoramide) have been obtained from the transesterification of methylphosphonate with diamines (Figure 4.5) (Honig and Weil 1977).

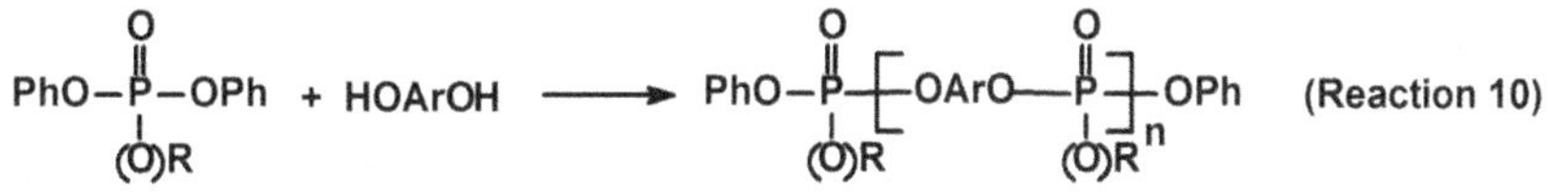

(Reaction 10)

FIGURE 4.4 Transesterification of phosphonates or phosphates with diols.

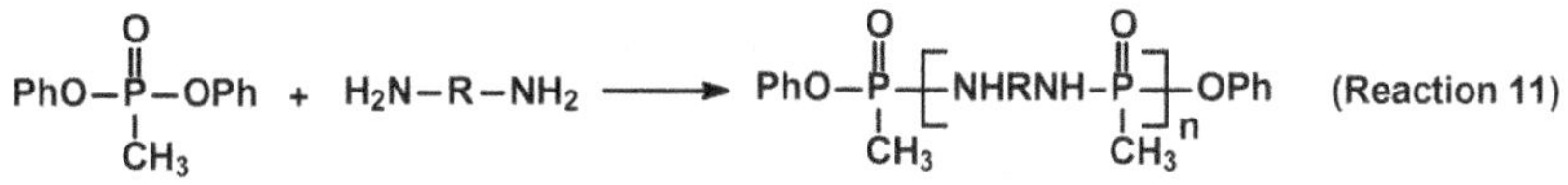

(Reaction 11)

FIGURE 4.5 Aminolysis of diphenyl methylphosphonate with diamines.

4.5.1.3 P–C

Due to good thermal and chemical stabilities, phosphonates or phosphinates that possess one or two P–C bonds have become increasingly important for engineering resins and textiles. Consequently, several general synthetic methods have been developed to construct P–C bonds. Currently, the most common ones are based on the Friedel-Crafts reaction of aromatics with PCl_3, P–H addition to unsaturated functional groups or coupling with alkyl halides, and the Arbuzov reaction of phosphites.

The Friedel-Crafts reaction of aromatics with PCl_3 has been known since 1873 when Michaelis discovered that benzene reacts with PCl_3 in the vapor phase to produce dichlorophenylphosphine (Michaelis 1873). This reaction can be run in the liquid phase under milder conditions. Under the influence of a stoichiometric amount of aluminum chloride, phosphorus trichloride readily reacts with benzene to produce dichlorophenylphosphine (Michaelis 1879; Buchner and Lockhart Jr 1951). When ionic liquids are used, only a catalytic amount of $AlCl_3$ needed (Wang and Wang 2003).

Dichlorophenylphosphine is a useful intermediate, reacting with acrylic acid to make 2-carboxyethyl(phenyl)phosphinic acid (CEPPA) (Birum and Jansen 1978). CEPPA can replace part of terephthalic acid and be polymerized with ethylene glycol to obtain flame retardant polyesters (Kleiner et al. 1976).

PCl_3 is also a starting material to make phosphites which can be subsequently converted to the corresponding phosphonates via the Arbuzov reaction (Bhattacharya and Thyagarajan 1981). The Arbuzov reaction demands high temperature, but the yield is often high. For example, diphenyl methylphosphonate (DPMP) has been prepared with a yield of 95% from triphenyl phosphite and dimethyl methylphosphonate (DMMP) at a temperature above 210°C (Levchik et al. 2008).

In an unusual way, PCl_3 reacts with o-phenylphenol to produce the intermediate (I) in the presence of only a catalytic amount of $ZnCl_2$ (Figure 4.6). In this reaction, a P–C bond is formed by an intramolecular Friedel-Crafts reaction. The intermediate (I) is easily converted to DOPO after hydrolysis. 9,10-dihydro-9-oxa-10-phosphaphenanthrene-10-oxide (DOPO) is a platform chemical from which many derivatives have been synthesized by utilizing its P–H bond (Saito 1972).

The P–H bond is relatively weak with bond energy of 322 kJ/mol. It can generate the phosphorus-centered radicals through hydrogen abstraction by other radicals. The phosphorus-centered radicals such as phosphinyl or phosphonyl radicals are electrophilic and can add to electron-rich alkenes (Hamilton and Williams 1960). As the result of forming the more stable C–H bond, the overall free radical addition reaction of P–H to alkenes is strongly exothermic and the reaction usually goes to the completion for simple alkenes (Fossey et al. 1995). For example, diethylphosphinate is readily made from the free radical reaction of sodium hypophosphite to ethylene (Figure 4.7). For sterically hindered alkenes, chain transfer agents such as thiols have been demonstrated to help bring the reaction to completion (Yao et al. 2017).

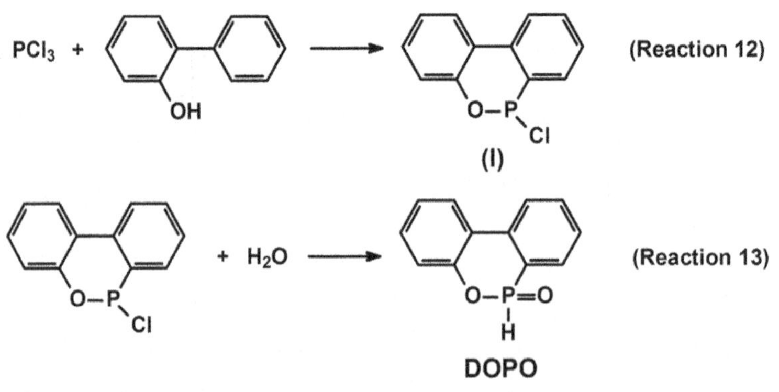

FIGURE 4.6 Synthesis of DOPO.

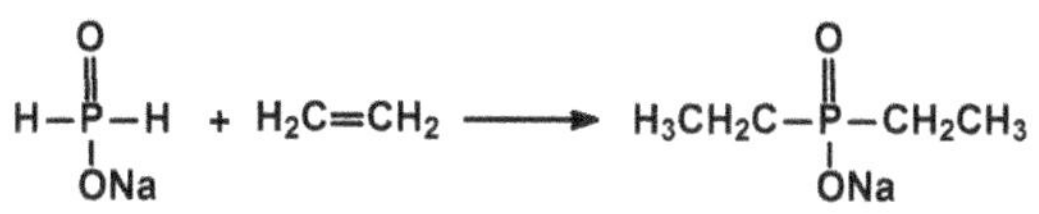

FIGURE 4.7 Synthesis of dialkylphosphinate.

Since the simple alkenes have low boiling points, the free radical addition reactions of P–H to alkenes need to be carried out under pressure. To overcome this inconvenient reaction condition, alcohols or esters which are liquids at room temperature have been proposed to replace the alkenes. Through dehydration or decarboxylation, alkenes are generated *in-situ* and the reactions occur as in the case of the alkenes. Diisobutylphosphinate has been prepared by this way (Yao and Levchik 2012).

For the electron deficient alkenes, the Michael reaction is employed to generate the P–C bond from the P–H bond (Figure 4.8). This reaction is usually catalyzed by strong bases or Lewis acids. In the case of H-phosphonates, the role of the catalyst is to shift the tautomeric equilibrium to phosphites, which are nucleophilic and add to the Michael acceptors (Rulev 2014).

Besides alkenes, carbonyl groups can also be used to react with P–H bond. Tetrakis hydroxymethyl phosphonium chloride or sulfate is thus made from PH_3 and formaldehyde in the presence of HCl (Hoffman 1921; Reeves et al. 1955) or sulfuric acid, respectively. In addition, P–H adds to epoxy to open the ring and form a P–C bond (Figure 4.9) (Wang and Lin 1999).

Another reaction to generate a P–C bond from P–H bond is through the Michaelis-Becker reaction (Cohen et al. 2003). 6H-dibenz[c, e][1,2]oxaphoshorin,6,6'- (1,4-ethanediyl)bis-,6,6'-dioxide (II) has been prepared from DOPO and dichloroethane by this reaction (Rxn 17 in Figure 4.10) (Angell et al. 2016). It is a convenient method, but because the reaction requires the use of strong bases, many side reactions can occur, and thus the yields are often low to mediocre. To overcome these shortcomings, a new synthetic method for compound (II) has been developed (Rxn 18 in Figure 4.10) (Yao et al. 2015).

4.5.1.4 Derivatives Not Directly Involving P (Nitration/Alkylation of Benzene Ring)

Some phosphorus-containing flame retardants can be prepared by modifying the functional groups on the phosphorus. For example, triphenyl phosphate has been reported to convert to polyphosphate (Katti and Krishnamurthy 1984) and alkylated phenyl phosphate through Friedel-Crafts reactions. In a similar way, methyldiphenylphosphine oxide undergoes nitration followed by the reduction to amines (Hergenrother et al. 2005).

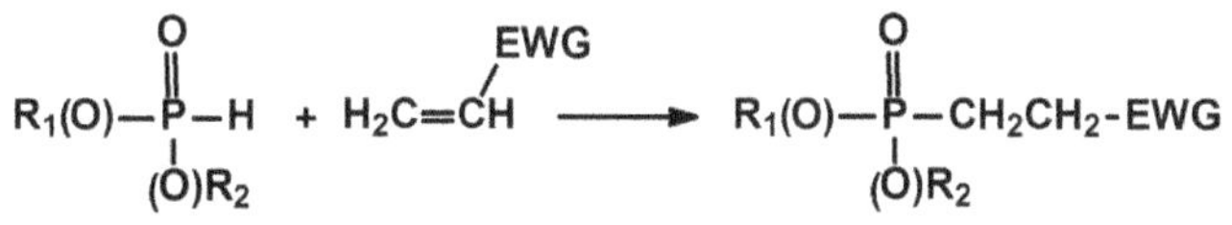

FIGURE 4.8 The Michael addition of P–H to electron deficient alkenes.

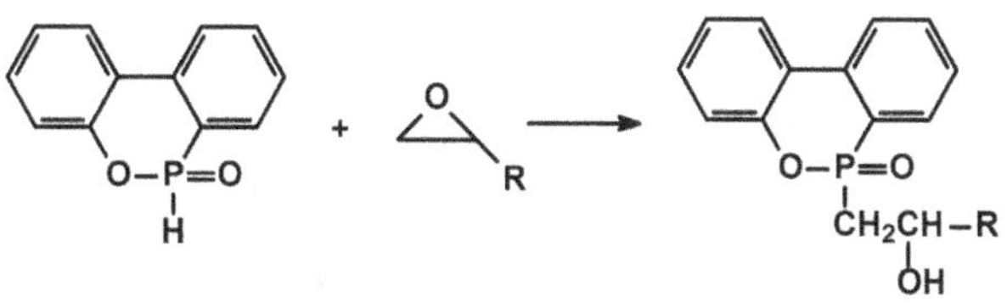

FIGURE 4.9 The addition of P–H to epoxides.

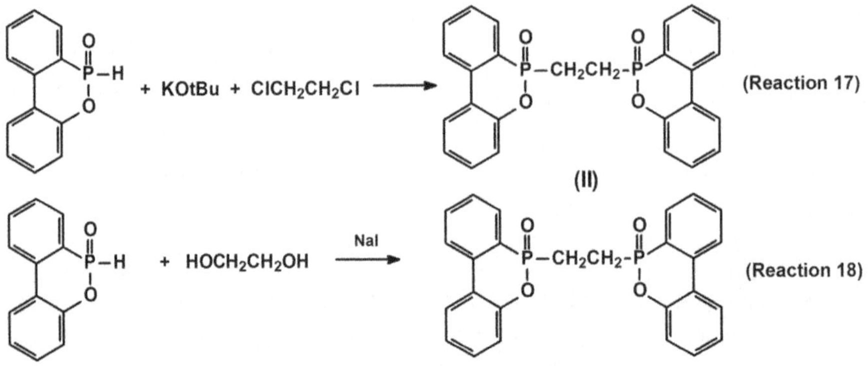

FIGURE 4.10 Preparation of compound (II).

4.5.2 Applications

Phosphorus-containing flame retardants have been used in many types of polymeric materials, and a number of excellent reviews can be found in the literature (Weil and Levchik 2004; Levchik et al. 2005; Levchik and Weil 2006; Levchik 2014). For simplicity, only selected flame retardant composites with focus on the electronic and electric field are presented below.

4.5.2.1 PC/ABS

PC/ABS alloy is widely used in the electronic enclosures due to its excellent physical properties such as heat resistance, dimensional stability, moldability, and impact strength at low temperatures. Bisphosphate flame retardants have been the gold standard for flame retardant PC/ABS since the mid-1990s. To suit the requirements of the hydrolytic stability, physical form, thermal property, and/or economy, several major bisphosphates have been commercialized (Figure 4.11) (Hergenrother et al. 2005).

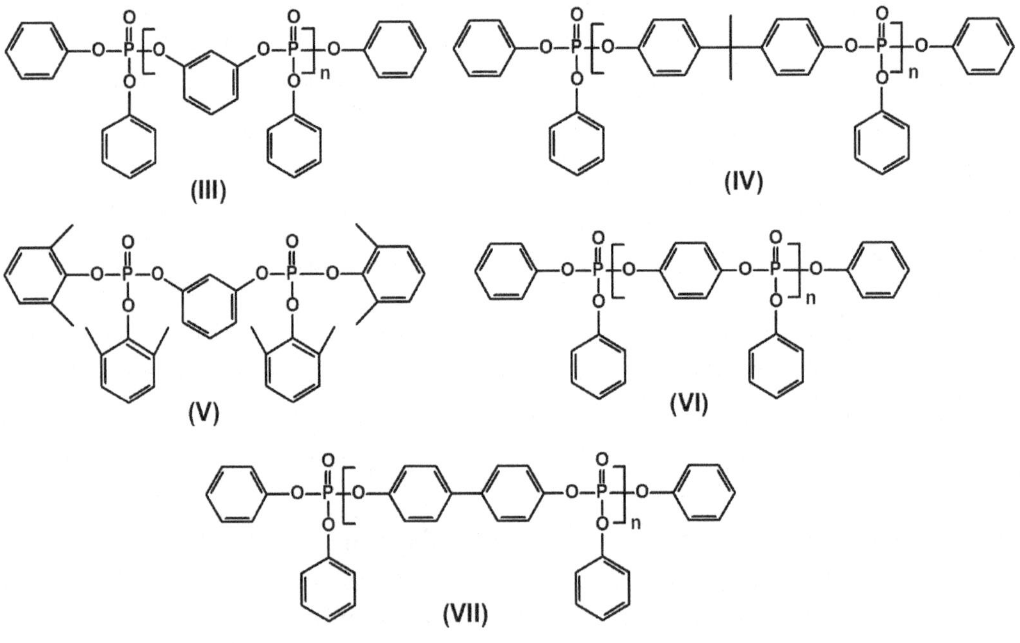

FIGURE 4.11 Chemical structures of several commercialized bisphosphates.

Compound (III) is the original bisphosphate to flame retardant PC/ABS and HIPS/PPO; however, due to the presence of the electron withdrawing group of phosphate in the meta position, the P–O bond connecting to the resorcinyl group is further polarized. Subsequently, compound (III) is highly susceptible to hydrolysis (Levchik et al. 2000). With time, compound (III) gradually has lost its market share to compound (IV) which has better hydrolytic stability owing to the electron donating alkyl group in the para-position of the bridging unit. Compound (V) possesses sterically hindered groups in the ortho-position of phenoxy groups and thus shows the distinctly better hydrolytic stability than compounds (III) and (IV) (Levchik 2014). Compound (VI) is a solid, which is preferred by some compounders for easy handling, and its relative higher phosphorus content gives it an edge over solid compound (V) in terms of loading. To increase the heat resistance of flame retardant PC/ABS, compound (VII) can be used (Eckel et al. 2003).

4.5.2.2 PA/PBT

There are not many phosphorus-containing flame retardants for polyamides and polybutylene terephthalate, especially glass fiber reinforced grade. The glass fiber reinforced polyamide (PA) and polybutylene terephthalate (PBT) require a high processing temperature at which many phosphorus-containing flame retardants either decompose or vaporize. Also since PA and PBT are condensation polymers, they can transesterify phosphorus compounds containing labile P–O bonds during the extrusion or injection. In addition, flame retardant PA and PBT are primarily used in the electronic and electric industry where physical properties are highly valued in addition to flame retardancy. Phosphorus-containing flame retardants that have a plasticizing action on PA and PBT inevitably lead to a lower heat distortion temperature (HDT) of flame retardant composites. A high HDT value is usually preferred especially in thin-walled electronic products. Consequently, there are only a few phosphorus-containing flame retardants that have been successfully applied in PA and PBT.

Red phosphorus has been used as a flame retardant in polyamide and polybutylene terephthalate (Levchik and Weil 2000). Excellent electric property can be obtained for red phosphorus flame retardant polyamide 6,6 (Jou et al. 2001). However, in terms of the efficiency based on the phosphorus content, red phosphorus is only mediocre since usually more than 6% phosphorus is needed in the flame retardant formulation. Additionally, for safety and to overcome the red color, red phosphorus is often encapsulated with lightly colored inorganics or thermo-setting resins (Albanesi and Rinaldi 1984; Sakon et al. 1991; Peerlings et al. 2004).

Metal salts of hypophosphorous acid, such as calcium hypophosphite, work together with melamine cyanurate as flame retardant in PBT (Engelmann and Wartig 2007). However, these salts are not particularly thermally stable, and can degrade to release phosphine during processing (De Campo et al. 2013), thus it is not suitable for polyamide 6,6.

In polyamide 6,6, aluminum dialkylphosphinates, particularly ADEP, have been the choice of non-halogenated flame retardants (Kleiner et al. 1998; Dietz et al. 2004). Although ADEP alone does not work well, its combination with melamine polyphosphate (MPP) (Schlosser et al. 2001), melamine cyanurate, or aluminum salt of a phosphorus acid (Krause et al. 2013) shows remarkable efficiency in polyamide. At a loading level of 15%–17%, the combination of ADEP and MPP enable PA6 or PA66 to achieve a UL 94 V-0 rating at 1.6 mm.

In the search of a single component flame retardant, Yao et al. (2010) discovered that aluminum diisobutylphosphinate (ADBP) is unique among all the dialkyl phosphinate. Although ADBP has a lower phosphorus percent than ADEP, the former is highly efficient and polyamide 6 or polyamide 66 containing 12.5% ADBP can easily gain a UL-94 V-0 rating. It is believed that the excellent temperature match of the volatilization of ADBP with the decomposition of polyamides and the tendency to generate the PO_2 radical account for the distinctive efficiency of ADBP.

For the high temperature polyamides, ADEP works very well on its own. At 15% loading, ADEP enables glass fiber reinforced poly(hexamethylene terephthalamide-co-2-methylpentamethylene terephthalamide) (PA 6T/DT) to achieve a UL-94 V-0 rating (Seefeldt et al. 2013).

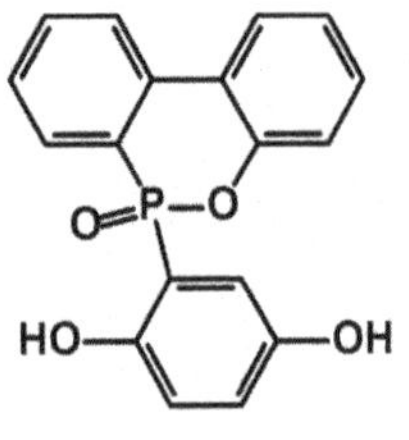

FIGURE 4.12   Chemical structures of compounds (VIII) and (VIIII).

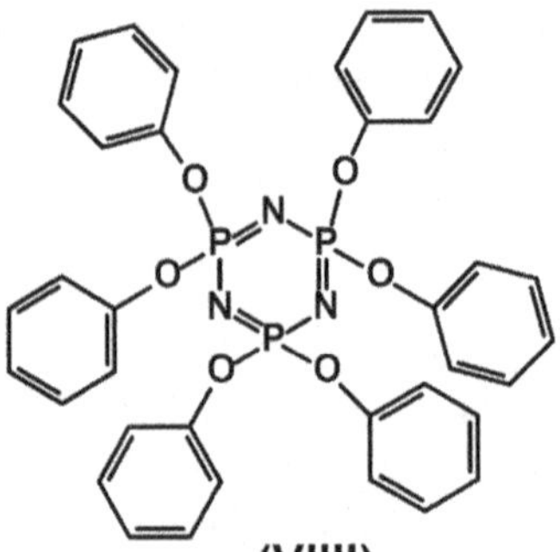

4.5.2.3 Epoxy

DOPO and its derivatives have been the main non-halogenated flame retardants in epoxy resins. Since DOPO has a reactive P–H group, DOPO itself can be used as a flame retardant by reacting with the epoxy resins (Wang and Lin 1999; Ito and Miyake 2001; Kazuo et al. 2005). To overcome the limitations imposed by the mono-functional group of DOPO which reduces the crosslinking density and thus lowers glass transition temperature of flame retardant epoxy, DOPO derivative with di-functional groups such as compound (VIII) (Figure 4.12) have been developed (Wang and Shieh 2001; Katsuyuki and Kenji 2002). These derivatives work as the chain extenders through their functional groups as well as flame retardants via phosphorus. Although the reactive phosphorus-containing flame retardants are constructed to retain the good thermal property of the epoxy resins, the prepared epoxy containing these reactive PFRs suffers from the high viscosity and dilution by epoxy resins of low viscosity is needed.

Besides the reactive DOPO and its derivatives, the additive type of inert DOPO derivatives is also reported to give excellent flame retardancy as well as good physical property. Notably, high glass transition temperature of flame retardant laminates can be obtained with compound (II) in conjunction with melamine polyphosphate (White et al. 2013).

Reactive or additive cyclic phosphazenes have been proposed for the epoxy resins. From the commercial point of view, additive cyclic phosphazenes such as compound (VIIII) (Figure 4.12) is preferred as epoxy molding compound for sealing semi-conductor (Shinsuke et al. 2011). Compared to the inert DOPO derivatives, compound (VIIII) has better solubility in common organic solvents, which is a big advantage due to its easy mixing into the base epoxy resins.

4.6 Concluded Remarks and Trends of PFRs

With the increase of the awareness of health and environmental problems associated with halogenated flame retardants, phosphorus-containing flame retardants have been promoted as an alternative and experienced fast growth; however, their future will largely depend on the need for fire safety and the health issues they bring to the world.

Specifically, the driving forces for them to grow or diminish are: (1) changes in regulations: while the law about the fire safety is becoming stringent in general, some applications such as upholstered furniture has seen the relaxation of the regulation on the use of flame retardants. (2) Performance needs: with the increasing demands on the physical property and the recyclability of flame retardant composites, PFRs face the strong competition from other types of flame retardants. PFRs need possess excellent thermal and chemical stability, and their effects on the physical properties of the flame retardant composites should be minimal. Only a very few of PFRs current in the market meet the increasing requirements of industry. (3) Cost: phosphorus chemistry is extremely versatile; however, most of the synthetic methods are not economically advantageous. Also the waste in the production of PFRs is usually associated with a high phosphorus residue and by-products such as HCl which require costly treatment. The development of industrially viable economic process, especially for the new PFRs, is critical. (4) Health effects: the health profiles of phosphorus compounds are highly unpredictable. Some of them are essential to the function of the human body, but others are extremely toxic. As a result, certain chloroalkyl phosphate flame retardants have been subjected to the law scrutiny and restrictions.

Apart from the driving forces of phosphorus-containing flame retardants and on the more fundamental side, factors governing the efficiency of phosphorus-containing flame retardants need to be systematically studied and understood. The current design of new phosphorus-containing flame retardants is still largely based on the empirical evidence, and is only applicable in some cases. Even in the well-studied gas phase PFRs, how to connect the gas phase activity obtained from the study of the small molecule PFRs with the actual PFRs when physical and chemical changes take place first in the condensed phase is a challenge. Also many research tools studying the mechanism are limited by the low heating rate, which has been considered to be far away from the actual fire conditions and could lead to misinterpretation. Good research tools and methods need to be developed.

On the market side, the electronic and electric sector such as electric cars, mobile electronics, and LED lightings will witness the fastest growth of phosphorus-containing flame retardants in the near future. Notably PFRs will find increasing use in the low K dielectric applications such as high data rate ethernet switching and printed circuit boards due to its versatile structures.

The past several years have seen a great increase in regulations for flame retardants. It is unlikely that this will change and it seems reasonable to expect that there may be new regulations that come into play in the future. It will require an active cadre of talented synthetic chemists to continue to develop new and better phosphorus flame retardants.

References

Albanesi, G., and Rinaldi, G. 1984. Process for stabilizing by encapsulation red phosphorus to be used as flame retardant of polymeric materials and product so obtained. US Patent 4,440,880.

Andrews, L., and Withnall, R. 1988. Matrix reactions of oxygen atoms with P4. Infrared spectra of P4O, P2O, PO and PO2. *Journal of the American Chemical Society* 110(17): 5605–5611.

Angell, Y. L., White, K. M., Angell, S. E. et al. 2016. DOPO derivative flame retardants. US Patent 9,522,927.

Babushok, V., Linteris, G., Katta, V. et al. 2016. Influence of hydrocarbon moiety of DMMP on flame propagation in lean mixtures. *Combustion and Flame* 171: 168–172.

Babushok, V., and Tsang, W. 1999. Influence of phosphorus-containing fire suppressants on flame propagation. *Fire Research and Engineering, Third International Conference Proceedings*, Chicago, IL, Society of Fire Protection Engineers, Bethesda, MD, pp. 257–267.

Bhattacharya, A. K., and Thyagarajan, G. 1981. Michaelis-Arbuzov rearrangement. *Chemical Reviews* 81(4): 415–430.

Birum, G. H., and Jansen, R. F. 1978. Production of 2-carboxyethyl (phenyl) phosphinic acid. US Patent 4,081,463.

Braun, U., Bahr, H., Sturm, H. et al. 2008. Flame retardancy mechanisms of metal phosphinates and metal phosphinates in combination with melamine cyanurate in glass-fiber reinforced poly (1, 4-butylene terephthalate): The influence of metal cation. *Polymers for Advanced Technologies* 19(6): 680–692.

Braun, U., Schartel, B., Fichera, M. A. et al. 2007. Flame retardancy mechanisms of aluminium phosphinate in combination with melamine polyphosphate and zinc borate in glass-fibre reinforced polyamide 6, 6. *Polymer Degradation and Stability* 92(8): 1528–1545.

Bright, D. A., Dashevsky, S., Moy, P. Y. et al. 1997. Resorcinol bis (diphenyl phosphate), a non-halogen flame-retardant additive. *Journal of Vinyl and Additive Technology* 3(2): 170–174.

Buchner, B., and Lockhart Jr., L. B. 1951. An improved method of synthesis of aromatic dichlorophosphinesl. *Journal of the American Chemical Society* 73(2): 755–756.

Campbell, J. R., Howson, P. E., and Reitz, M. L. 2002. High purity sterically hindered diaryl chlorophosphates and method for their preparation. US Patent Application 2002/0013488.

Cichy, B., Luczkowska, D., Nowak, M. et al. 2003. Polyphosphate flame retardants with increased heat resistance. *Industrial & Engineering Chemistry Research* 42(13): 2897–2905.

Cohen, R. J., Fox, D. L., Eubank, J. F. et al. 2003. Mild and efficient Cs_2CO_3-promoted synthesis of phosphonates. *Tetrahedron Letters* 44(47): 8617–8621.

De Campo, F., Murillo, A., Li, J. et al. 2013. Flame retardant polymer compositions comprising stabilized hypophosphite salts. European Patent 2,678,386.

Deanin, R., and Ali, M. 1994. Aromatic organic phosphate oligomers as flame-retardants in plastics. *Abstracts of Papers of the American Chemical Society* 208: 129.

Despinasse, M. C., and Schartel, B. 2013. Aryl phosphate–aryl phosphate synergy in flame-retarded bisphenol A polycarbonate/acrylonitrile-butadiene-styrene. *Thermochimica Acta* 563: 51–61.

Dietz, M., Horold, S., Nass, B. et al. 2004. New environmentally friendly phosphorus based flame retardants for printed circuit boards as well as polyamides and polyesters in E&E applications. *Proceedings of International Congress and Exhibition on Electronics Goes Green*, Berlin, Germany, pp. 771–776.

Eckel, T., Wittmann, D., Zobel, M. et al. 2003. Flame-resistant thermostable polycarbonate ABS moulding materials. US Patent 6,566,428.

Engelmann, J., and Wartig, D. 2007. Halogen-free flameproof polyester. US Patent 7,169,838.

Fossey, J., Lefort, D., and Sorba, J. 1995. Free Radicals in Organic Chemistry, John Wiley & Sons, Chichester, p.92.

Freitag, D. 2006. Preparation of polyphosonates via transesterification without a catalyst. US Patent Application 2006/0020104.

Gosens, J. C., Pratt, C. F., Savenije, H. B. et al. 1993. Polymer mixture having aromatic polycarbonate, styrene I containing copolymer and/or graft polymer and a flame-retardant, articles formed therefrom. US Patent 5,204,394.

Granzow, A. 1978. Flame retardation by phosphorus compounds. *Accounts of Chemical Research* 11(5): 177–183.

Hamilton, L. A., and Williams, R. H. 1960. Synthesis of compounds having a carbon-phosphorus linkage. US Patent 2,957,931.

Hartley, S., Holmes, W., Jacques, J. et al. 1963. Thermochemical properties of phosphorus compounds. *Quarterly Reviews, Chemical Society* 17(2): 204–223.

Hastie, J. W., and Bonnell, D. 1980. *Molecular Chemistry of Inhibited Combustion Systems*, National Bureau of Standards, Washington, DC.

Hemelsoet, K., Van Durme, F., Van Speybroeck, V. et al. 2010. Bond dissociation energies of organophosphorus compounds: an assessment of contemporary ab initio procedures. *The Journal of Physical Chemistry A* 114(8): 2864–2873.

Hergenrother, P. M., Thompson, C. M., Smith Jr., J. G. et al. 2005. Flame retardant aircraft epoxy resins containing phosphorus. *Polymer* 46(14): 5012–5024.

Hoffman, A. 1921. The action of hydrogen phosphide on formaldehyde. *Journal of the American Chemical Society* 43(7): 1684–1688.

Holmes, W. 1962. Heat of combustion of phosphorus and the enthalpies of formation of P_4O_{10} and H_3PO_4. *Transactions of the Faraday Society* 58: 1916–1925.

Honig, M. L., and Weil, E. D. 1977. A convenient synthesis of diaryl methylphosphonates and transesterification products therefrom. *The Journal of Organic Chemistry* 42(2): 379–381.

Ito, M., and Miyake, S. 2001. Flame-retardant resin composition and semiconductor sealant using the same. US Patent 6,180,695.

Jenewein, E., Kleiner, H. J., Wanzke, W. et al. 2002. Synergistic flame protection agent combination for thermoplastic polymers. US Patent 6,365,071.

Jou, W., Chen, K., Chao, D. et al. 2001. Flame retardant and dielectric properties of glass fibre reinforced nylon-66 filled with red phosphorous. *Polymer Degradation and Stability* 74(2): 239–245.

Katsuyuki, A., and Kenji, T. 2002. Flame-Retardant Thermosetting Resin Composition. JP Patent 3,268,498.

Katti, K., and Krishnamurthy, S. 1984. Studies of phosphazenes. XIX. New polyorganophosphazenes derived from the friedel–crafts reactions of hexa (aryloxy) cyclotriphosphazenes with polyhaloalkanes. *Journal of Polymer Science: Polymer Chemistry Edition* 22(11): 3115–3128.

Kazuo, I., Chiaki, A., Toshihiko, K., and Naritsuyo, T. 2005. Epoxy Resin Composition Including Phosphorus. JP Patent 3,613,724.

Kishore, K., and Mohandas, K. 1982. Action of phosphorus compounds on fire-retardancy of cellulosic materials: A review. *Fire and Materials* 6(2): 54–58.

Kleiner, H. J., Budzinsky, W., and Kirsch, G. 1998. Low-flammability polyamide molding materials. US Patent 5,773,556.

Kleiner, H. J., Finke, M., Bollert, U. et al. 1976. Flame retarding linear polyesters and shaped articles thereof. US Patent 3,941,752.

Knauer, W. 1894. Ueber die o-Chlorphosphine der zweiatomigen Phenole. *Berichte der deutschen chemischen Gesellschaft* 27(2): 2565–2572.

Korobeinichev, O., Gonchikzhapov, M., Paletsky, A. et al. 2016. Counterflow flames of ultrahigh-molecular-weight polyethylene with and without triphenylphosphate. *Combustion and Flame* 169: 261–271.

Korobeinichev, O., Paletsky, A., Kuibida, L. et al. 2013. Reduction of flammability of ultrahigh-molecular-weight polyethylene by using triphenyl phosphate additives. *Proceedings of the Combustion Institute* 34(2): 2699–2706.

Krause, W., Bauer, H., Sicken, M. et al. 2013. Flame retardant-stabilizer combination for thermoplastic polymers. US Patent Application 2013/0190432.

LeVan, S. L. 1984. Chemistry of fire retardancy. In *The Chemistry of Solid Wood*, ed. R. Rowell, pp. 531–574. Washington, DC: American Chemical Society.

Levchik, S. 2014. Chapter 2: Phosphorus-based FRs. In *Non-Halogenated Flame Retardant Handbook*, eds. A. B. Morgan and C. A. Wilkie, pp. 17–74. Beverly, MA: Scrivener Publishing LLC.

Levchik, G., Levchik, S., and Lesnikovich, A. 1996a. Mechanisms of action in flame retardant reinforced nylon 6. *Polymer Degradation and Stability* 54(2–3): 361–363.

Levchik, S., Piotrowski, A., Weil, E. et al. 2005. New developments in flame retardancy of epoxy resins. *Polymer Degradation and Stability* 88(1): 57–62.

Levchik, S. V., Bright, D. A., Moy, P. et al. 2000. New developments in fire retardant non-halogen aromatic phosphates. *Journal of Vinyl and Additive Technology* 6(3): 123–128.

Levchik, S. V., Camino, G., Luda, M. P. et al. 1996b. Epoxy resins cured with aminophenylmethylphosphine oxide 1: Combustion performance. *Polymers for Advanced Technologies* 7(11): 823–830.

Levchik, S. V., Dashevsky, S., Weil, E. et al. 2008. Oligomeric, hydroxy-terminated phosphonates. US Patent 7,449,526.

Levchik, S. V., and Weil, E. D. 2000. Combustion and fire retardancy of aliphatic nylons. *Polymer International* 49(10): 1033–1073.

Levchik, S. V., and Weil, E. D. 2006. Flame retardants in commercial use or in advanced development in polycarbonates and polycarbonate blends. *Journal of Fire Sciences* 24(2): 137–151.

MacDonald, M., Gouldin, F., and Fisher, E. 2001. Temperature dependence of phosphorus-based flame inhibition. *Combustion and Flame* 124(4): 668–683.

Maiti, S., Banerjee, S., and Palit, S. K. 1993. Phosphorus-containing polymers. *Progress in Polymer Science* 18(2): 227–261.

Mariappan, T., Zhou, Y., Hao, J. et al. 2013. Influence of oxidation state of phosphorus on the thermal and flammability of polyurea and epoxy resin. *European Polymer Journal* 49(10): 3171–3180.

Markarian, J. 2005. Flame retardants for polyamides-new developments and processing concerns. *Plastics, Additives and Compounding* 7(2): 22–25.

Michaelis, A. 1873. Ueber aromatische Phosphorverbindungen. Zweite Mittheilung. *Berichte der deutschen chemischen Gesellschaft* 6(1): 816–819.

Michaelis, A. 1879. Ueber ein homologes des phosphenylchlorids. *Berichte der deutschen chemischen Gesellschaft* 12(1): 1009–1009.

Musher, J. 1969. The chemistry of hypervalent molecules. *Angewandte Chemie International Edition in English* 8(1): 54–68.

Noto, T., Babushok, V., Hamins, A. et al. 1998. Inhibition effectiveness of halogenated compounds. *Combustion and Flame* 112(1–2): 147–160.

Papa, A. J. 1972. Flame retardation of polyurethane foams in practice. *Industrial & Engineering Chemistry Product Research and Development* 11(4): 379–389.

Peerlings, H., Wagner, M., and Podszun, W. 2004. Microencapsulation of red phosphorus. US Patent 6,765,047.

Piotrowski, A. M., Weil, E. D., Yao, Q. et al. 2009. Process for preparing diaryl alkylphosphonates and oligomeric/polymeric derivatives thereof. US Patent 7,541,415.

Quednow, K., and Püttmann, W. 2009. Temporal concentration changes of DEET, TCEP, terbutryn, and nonylphenols in freshwater streams of Hesse, Germany: possible influence of mandatory regulations and voluntary environmental agreements. *Environmental Science and Pollution Research* 16(6): 630–640.

Raffan-Montoya, F., Ding, X., Stoliarov, S. I. et al. 2015. Measurement of heat release in laminar diffusion flames fueled by controlled pyrolysis of milligram-sized solid samples: Impact of bromine-and phosphorus-based flame retardants. *Combustion and Flame* 162(12): 4660–4670.

Reeves, W. A., Flynn, F. F., and Guthrie, J. D. 1955. Laboratory preparation of tetrakis-(hydroxymethyl)-phosphonium chloride. *Journal of the American Chemical Society* 77(14): 3923–3924.

Rulev, A. Y. 2014. Recent advances in Michael addition of H-phosphonates. *RSC Advances* 4(49): 26002–26012.

Saito, T. 1972. Cyclic organophosphorus compounds and process for making same. US Patent 3,702,878.

Sakon, I., Sekiguchi, M., and Kanayama, A. 1991. Method for producing red phosphorus flame retardant and nonflammable resinous composition. US Patent 5,041,490.

Sanderson, R. T. 1983. Electronegativity and bond energy. *Journal of the American Chemical Society* 105(8): 2259–2261.

Saso, Y., Ogawa, Y., Saito, N. et al. 1999. Binary CF_3Br-and CHF_3–inert flame suppressants: Effect of temperature on the flame inhibition effectiveness of CF_3Br and CHF_3. *Combustion and Flame* 118(3): 489–499.

Schlosser, E., Nass, B., and Wanzke, W. 2001. Flame-retardant combination. US Patent 6,255,371.

Seefeldt, H., Duemichen, E., and Braun, U. 2013. Flame retardancy of glass fiber reinforced high temperature polyamide by use of aluminum diethylphosphinate: Thermal and thermo-oxidative effects. *Polymer International* 62(11): 1608–1616.

Shinsuke, H., Hiroyuki, S., and Haruaki, T. 2011. Epoxy Resin Molding Material for Sealing Electronic Parts and Electronic Parts. JP patent 4,635,882.

Shmakov, A., Korobeinichev, O., Shvartsberg, V. et al. 2005. Inhibition of premixed and nonpremixed flames with phosphorus-containing compounds. *Proceedings of the Combustion Institute* 30(2): 2345–2352.

Siow, J. E., and Laurendeau, N. M. 2004. Flame inhibition activity of phosphorus-containing compounds using laser-induced fluorescence measurements of hydroxyl. *Combustion and Flame* 136(1–2): 16–24.

Telschow, J. E. 1999. Pentaerythritol phosphate derivatives as flame retardants for polyolefins. *Phosphorus, Sulfur, and Silicon and the Related Elements* 144(1): 33–36.

Twarowski, A. 1995. Reduction of a phosphorus oxide and acid reaction set. *Combustion and Flame* 102(1–2): 41–54.

Twarowski, A. 1996. The temperature dependence of H+ OH recombination in phosphorus oxide containing post-combustion gases. *Combustion and Flame* 105(3): 407–413.

Van der Veen, I., and de Boer, J. 2012. Phosphorus flame retardants: Properties, production, environmental occurrence, toxicity and analysis. *Chemosphere* 88(10): 1119–1153.

Wang, C. S., and Lin, C. H. 1999. Synthesis and properties of phosphorus-containing epoxy resins by novel method. *Journal of Polymer Science Part A: Polymer Chemistry* 37(21): 3903–3909.

Wang, C. S., and Shieh, J. Y. 2001. Phosphorus-containing dihydric phenol or naphthol-advanced epoxy resin or cured. US Patent 6,291,626.

Wang, Y. Z., Zheng, C. Y., and Wu, D. C. 1996. Flame-retardant action of polysulfonyldiphenylene phenylphosphonate on PET. *Acta Polymerica Sinica* 4: 439–445.

Wang, Z. W., and Wang, L. S. 2003. Preparation of dichlorophenylphosphine via Friedel–Crafts reaction in ionic liquids. *Green Chemistry* 5(6): 737–739.

Weil, E. D., and Levchik, S. 2004. Current practice and recent commercial developments in flame retardancy of polyamides. *Journal of Fire Sciences* 22(3): 251–264.

Weil, E. D., Zhu, W., Patel, N. et al. 1996. A systems approach to flame retardancy and comments on modes of action. *Polymer Degradation and Stability* 54(2–3): 125–136.

White, K. M., Angell, Y. L., Angell, S. E. et al. 2013. DOPO-derived flame retardant and epoxy resin composition. US Patent 8,536,256.

Worku, A. Z., De Schryver, D. A., Landry, S. D. et al. 2003. Flame retardant polymer compositions. US Patent 6,617,379.

Wroblewski, A. E., and Verkade, J. G. 1996. 1-Oxo-2-oxa-1-phosphabicyclo [2.2.2] octane: A new mechanistic probe for the basic hydrolysis of phosphate esters. *Journal of the American Chemical Society* 118(42): 10168–10174.

Yao, Q., and Levchik, S. V. 2012. Process for the alkylation of phosphorus-containing compounds. US Patent 8,097,747.

Yao, Q., Levchik, S. V., and Alessio, G. R. 2010. Phosphorus-containing flame retardant for thermoplastic polymers. US Patents 7,807,737.

Yao, Q., Wang, J., and Mack, A. G. 2015. Process for the preparation of DOPO-derived compounds and compositions thereof. US Patent 9,012,546.

Yao, Q., Zhao, Y. Y., and Tang, T. B. 2017. Preparation method and application of dialkyl phosphonic acid compound. Chinese Patent Application CN106279268(A).

Yoon, D., Jung, H. T., Kwon, G. et al. 2013. Dynamics and mechanism of flame retardants in polymer matrixes: Experiment and simulation. *The Journal of Physical Chemistry B* 117(28): 8571–8578.

Recent Advances in Silicon-Containing Flame Retardants

Weiyi Xing and
Junling Wang

5.1 Introduction

Nowadays, polymer materials have been widely used in human daily life because of their remarkable properties, low weight, and ease of processing. However, these materials are also known for their high flammability. During the combustion of polymer, a great deal of heat can be released, accompanied by the generation of considerable toxic gases and smoke. Thus, it is of great significance to improve the flame retardancy of polymer for extending their use to most applications.

In general, there are two kinds of flame-retardant additives: halogen and halogen-free. Halogen flame retardants are confirmed to have high flame-retardant efficiency, owing to its free radical scavenging mechanism. However, considering its proven or suspected adverse effects on the environment, halogen flame retardants are being phased out and forbidden for use. Then, the development of effective and environmentally friendly flame retardant is very necessary. As a result, flame-retardant elements of P, N, and Si have come into the notice of researchers. A lot of phosphorus-, nitrogen-, and silicon-containing flame retardants have been developed, and their influence on polymer flame retardancy has been investigated.

This chapter is chiefly focusing on the recent advances in silicon-containing flame retardant. This type of flame retardants is mainly divided into silicone, silica, organosilane, and silicate. Silicone has gained extensive attentions from researchers because of their excellent thermal stability and high heat resistance, with very limited release of toxic gases during thermal decomposition. In present study, various types of nanosilica, mesoporous silica, and silica gel were added to polymers to investigate their flame-retardant effectiveness and mechanism of flame retardancy. The silica particles may accumulate near the polymer surface, acting as a thermal insulation layer. Moreover, the accumulation of silica particles may reduce the combustible gas concentration near the surface. As is well-known, the silane coupling agents possess a good capability of bonding the fillers and polymer matrix as well as a great contribution to forming a compact and dense char structure during combustion. In addition, the flame-retarded Si–O–Si network structure can be formed through the sol–gel reaction of silane coupling agents. Layered silicates (also called phyllosilicates) also show potentials in flame-retardant area, owing to its layer structure. These compounds are rendered organophilic by replacement of the metal ions that normally balance the negative charges with organic cations carrying an aliphatic chain, typically alkylammonium or alkyl phosphonium salts. These four types of silicon-containing compounds show significant flame-retardant functions on polymer materials. This chapter aims to describe most of the recent work related to applications of these compounds as flame retardants and introducing their flame-retardant mechanism on polymer systems.

5.2 Silicon-Containing Flame Retardants

5.2.1 Silicone

Silicone can be used as flame-retardant agents through direct blending within the polymer matrix, incorporation into porous fillers, or by synthesizing block/graft copolymers including silicone segments. Pape et al. reported that the addition of silicone powder could improve the flame retardancy of various polymer systems including polycarbonate (PC), polystyrene (PS), polypropylene (PP), polyamide (PA), and epoxy (EP) (Pape and Romenesko 1997). After its low incorporation at 1–5 wt%, the obviously reduced peak heat release rate (PHRR), decreased levels of carbon monoxide generation and smoke evolution were observed. Silicone powder was composed of polydimethylsiloxane (PDMS) and fumed silica. This powder has been modified with organofunctional groups (amino, epoxy, or methacrylate) to enhance its compatibility with polymer matrix. Integration with other conventional flame retardant, such as halogen/antimony, ammonium polyphosphate, and magnesium hydroxide, also gave improvements in the flame retardancy of polymer.

5.2.1.1 Polycarbonate (PC)

Iji et al. employed several types of silicone polymers as flame retardant for PC (Iji and Serizawa 1998). Here, the used silicone varied in their chain structure (linear or branched type), the nature of functional groups in the chain (e.g., methyl, phenyl, mixture of them), and the type of end-groups (e.g., methyl, phenyl, hydroxyl, methoxy, vinyl). Branched silicone polymers containing a mixture of methyl and phenyl groups along the chain and end-capped by methyl groups enhanced the limit oxygen index (LOI) values and final residue most effectively (Figure 5.1). They found that the superior flame-retardant effect of the

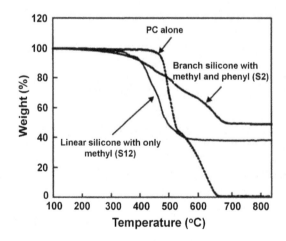

FIGURE 5.1 Thermogravimetric analyses of PC/silicone derivative blends. (Reprinted from Iji, M. and Serizawa, S., *Polym. Adv. Technol.*, 9, 593–600, 1998. With permission.)

silicone derivatives (branched silicone with methyl and phenyl) is attributed to their excellent dispersion in matrix and migration on the surface during combustion. Owing to its migration, a uniform and highly flame-resistant barrier was formed above the underlying polymer. The char residue was composed of polysiloxane and condensed aromatic compounds. Furthermore, the efficiency of the branched silicone was increased with the decreasing molecular weight. The decreases in molecular weight resulted in its low viscosity, which further accelerates the migration to sample surface. Nodera et al. investigated the different flame-retardancy effects of nano- and micro-sized silica in PC–PDMS block copolymer (Nodera and Kanai 2006a). The thermal stability and char residue for PC–PDMS block copolymer were slightly enhanced by adding the nano-sized silica. The maximum loss rate decreased with the increasing nano-sized silica content because of the good sealing efficiency of char residue. Moreover, the incorporation of micro-sized silica had little effect on the sealing efficiency of residue because micro-sized silica cannot cover on enough of the surface of the residue. Their further works showed that monodisperse nano-sized silica (50 nm, in powder form) had a positive effect on the flame-retardant mechanism of PC–PDMS (Nodera and Kanai 2006b). Moreover, the colloidal silica (nano-sized silica with a median particle size of 20 nm dispersed in liquid polyethyleneglycol) had higher flame retardancy than the silica of 50 nm. PC–PDMS block copolymer was mainly distributed on the surface of the nano-sized silica through physical interactions, which was responsible for the improvement in flame retardancy. In fact, the size of silica was almost the same as the size of PDMS domains effective in flame retardancy (domain size ~50 nm). Furthermore, micro-sized silica had a little effect on the sealing efficiency of char residue. Thus, it was supposed that the PDMS on the surface of micro-sized silica had a behavior similar to large PDMS domain ineffective in flame retardancy. Therefore, the tested samples with micro-sized silica or aggregated nano-sized silica have lower LOI values. Based on such these results, they proposed a conceptual model of silica distributions in PDMS which influence the flame-retardancy efficiency of PDMS (Figure 5.2). When the content of nano-sized silica is less than the PDMS content, silica particles were perfectly dispersed within the matrix and fully covered with PDMS (Figure 5.2b). The monodisperse silica was effective in the flame-retardant enhancement of polymer. In addition, PDMS located on the surface of nano-sized silica was thought to exhibit higher thermal stability than PDMS alone and generate high sealing char residues. With the increases of the nano-sized silica content, the aggregated silica can be observed and its surface was still covered with PDMS (Figure 5.2d). At large nano-sized silica content, the uncovered aggregated silica particles and single silica particle contents increased (Figure 5.2e). Hirai and Ishii used both PDMS and TiO_2 which surfaces were treated with siloxane as flame retardant for PC resin (Hirai and Ishii 2003). After its additions, PC resin achieved V-0 rate in UL-94 flammability test. Barren et al. (2000) employed TiO_2 treated with PDMS as flame retardant in PC and the remarkable enhancements in flame retardancy were obtained. Lupinski et al. added up to 9 wt% of PDMS-treated fillers such as TiO_2, clay or carbon black into PC resin for improving its flame retardancy. As a result, the V-0 rate in UL-94 test was observed (Lupinski and Hawkins 1992).

5.2.1.2 Polycarbonate/Polyacrylonitrile–Butadiene–Styrene (PC/ABS)

Silicone additives can be modified by introduction of hetero-atom-containing molecules that may have synergistic effect in flame-retardant action. Zhong et al. used flame-retardant named after DPA–SiN containing phosphorous, nitrogen, and silicon elements simultaneously in PC/ABS resin (Zhong et al. 2007b). Incorporation of 30 wt% of DPA–SiN into PC/ABS improved the LOI value from 21% to 27%. At this loading level, the HRR, total heat release rate (THR), and effective heat of combustion (EHC) was reduced about 50%. The addition of DPA–SiN changed the thermal behavior of PC/ABS via forming more stable char layer during thermal decomposition. The morphology of char residue indicated that a smooth and compact char layer was obtained at 800°C, whereas char layer could swell well to form a barrier to resist the transfer of heat and mass during fire. Another type of flame retardant named after DVN containing Si, P, and N for PC/ABS had also been developed by them (Zhong et al. 2007a). In PC/ABS, DVN had flame-retardant effect both in condensed and gaseous phase. Incorporation of 30 wt% DVN into PC/ABS has improved thermal stability and flame retardancy. The LOI values of PC/ABS

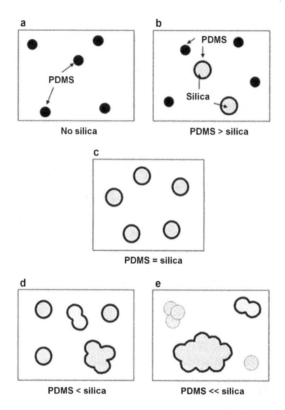

FIGURE 5.2 Schematic distributions of PDMS in the PC–PDMS block copolymers containing nano-sized silica.: (a) No silica; (b) PDMS > silica; (c) PDMS = silica; (d) PDMS < silica; (e) PDMS << silica. (Reprinted from Nodera, A. and Kanai, T., *J. Appl. Polym. Sci.*, 101, 3862–3868, 2006. With permission.)

were enhanced from 21.2% to 27.2%. The morphology of chars obtained after the LOI test were observed by SEM (Figure 5.3). The outer surface was smooth and plain, while the internal surface exhibited cell structure. The swollen internal structure, dense, and smooth outer surface provided a good barrier to the heat and mass transfer. The char residue could retard the overflow of the flammable volatiles at high temperature. The thermal decomposition performance of PC/ABS/DVN occurred through decomposition of phosphorus-containing groups to hydrate the char source-containing groups during heating. Then, a continuous and protective carbon–silica layer was formed. SiO_2 compounds reacted easily with phosphate to yield silicophosphate.

5.2.1.3 Ethylene Butyl Acrylate (EBA)

Davidson et al. reported the improvement in the flame retardancy of acrylate-based copolymers in wire and cable applications by adding chalk and silicone agents (Davidson and Wilkinson 1992). 5 wt% of tri-methylsilyl chain-ended PDMS gum containing nominally 0.2 mol% vinyl groups and 30 wt% of stearate-coated calcium carbonate were added into 65 wt% of EBA copolymer (noted as CaSiEBA), resulting in the increased LOI values from 18% to 34%. The flame retardancy of CaSiEBA has been investigated by other groups in detail and its applications in flame-retarded polyolefin were also performed. Hermansson et al. (2003) observed that CaEBA (sample only with calcium carbonate) shows a collapsed char residue during burning, while CaSiEBA exhibits an intumescent structure with a compact and smooth char, even though there is no obvious differences between the LOI values of CaEBA and SiEBA (sample only with PDMS) (Table 5.1). Ester pyrolysis of EBA facilitated the reaction between chalk and carboxylic acid which leads to the formation of gases and ionomers. The volatile species diluted the combustible gases concentration

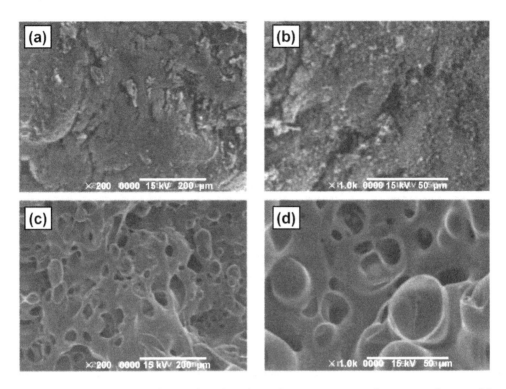

FIGURE 5.3 The outer (a, b) and internal (c, d) surfaces of PC/ABS/DVN-30 after LOI test. (Reprinted from Zhong, H. et al., *Fire Mater.*, 31, 411–423, 2007. With permission.)

TABLE 5.1 Effect of Chalk and Silicone Additions on Flame Retardancy of EBA Copolymers

Sample	LOI (%)	PHRR (kW/m^2)	TTI (s)	Mass Loss (%)
EBA	18	1304	77	98.1
SiEBA	19	1044	84	98.7
CaEBA	20	656	102	71.7
CaSiEBA	30.5	326	148	65.9

Source: Hermansson, A. et al., *Fire Mater.*, 27, 51–70, 2003.

and also caused the melt to effervesce, resulting in the generation of an intumescent structure. This intumescent structure acted as a heat insulating layer and protected the material underneath from burning. Hermansson et al. (2005) also suggested that the reinforcement of intumescent structure through an insulation effect is triggered by temperature. Moreover, the effect of chalk and silicone distributions to flame-retardant efficiency was also studied by them (Hermansson et al. 2006). A homogeneous distribution of chalk particles led to a larger contact area between the particles and polymer matrix, which is positive for the interactions and the chemical mechanisms occurring between the solid and melt phase. Homogeneous dispersion of silicone elastomer was also important, as it facilitates the formation a uniform thickness of SiO$_2$ at the surface that withstands the pressure of volatile gases and the initial crack propagation. In fact, more extensive mixing was beneficial for the dispersion of chalk particles and silicone elastomer, as well as flame-retardancy improvement. Huhtala et al. found that the partial substitution of calcium carbonates with organic montmorillonite (OMMT) nanofillers reduced dripping properties of burned materials (Huhtala and Motha 2008). Kramer et al. used ethylene–acrylate-based copolymer containing chalk and silicone as flame-retardant agent for PP. In all formulations, the acrylate content was 8 wt% in the EBA copolymer with the chalk of 30–40 wt% and the silicone of 5–7 wt%. These three samples showed similar degradation process in nitrogen, proceeding principally in a single step which led to significant mass losses above 400°C with a maximum rate at 460°C–470°C. In air, CaSiEBA-PP and CaSiEBA polymers displayed decomposition in three temperature regions (a minor mass loss at 250°C and two major steps at 355°C and 430°C, respectively). The subsequent mass loss above 600°C was attributed to the degradation of chalk filler. Compared with CaSiEBA, the higher initial mass loss of CaSiEBA-PP at 250°C indicated the earlier decomposition of PP. Ignition of CaSiEBA-PP occurred as the surface layer reached temperatures between 350°C and 400°C. Moreover, the infrared spectroscopy analysis suggested that PP shows a advanced decomposition as compared to EBA, and its pyrolysis provided a significant contribution to the ignitable gases. Therefore, the combustion of CaSiEBA-PP self-accelerated and generated a rapid pyrolysis on the polymer surface.

5.2.1.4 Polypropylene

Modified silicone-based flame-retardant of PSiN containing phosphorus, nitrogen, and silicon has been synthesized and added into PP matrix (Li et al. 2006). Li et al. (2005) also reported that PSiN could enhance the flame-retardant and thermal stability of PP. When 30 wt% PSiN was loaded, LOI value of PP/PSiN was enhanced from 17% to 26%, and the char yield at 800°C was improved from 0 to 27 wt%. The phosphorus component in PSiN facilitated the char formation of PP and the silicon component improved the thermal stability of char owing to the generation of thermal insulating silica layer. Phosphorus-containing compounds degraded at relatively low temperature to form a protective phosphorus–carbon layer with weak phosphonate bonds. The nitrogen-containing compounds decomposed to generate ammonia or other molecules. This incombustible ammonia gas diluted the concentration of the oxygen and combustion volatiles as well as swelled the carbon layer. Chen et al. prepared flame-retardant PP composites by melt blending method starting from ammonium polyphosphate (APP), pentaerythritol (PER), and hydroxyl silicone oil (HSO) (Chen and Jiao 2009). Cone calorimeter experiments showed that a synergistic effect occurred when HSO was combined with APP and PER in PP composites.

Wang et al. investigated the synergistic effects of β–cyclodextrin-containing silicone oligomer (CDS) on intumescent flame-retardant (IFR) PP system (Wang and Li 2010). IFR/PP system containing 6.25 wt% CDS showed the highest LOI value of 35.0% with a UL-94 V-0 rating. The experimental results demonstrated that the combination of CDS and charring-forming agent (CFA) presents synergistic effects in the flame retardancy and char formation.

A novel silicone-containing macromolecular charring agent (Si-MCA) was synthesized, and its synergistic effect with APP on flame-retardant PP composites was also investigated by Lai et al. (2015). Synthetic route of Si-MCA was shown in Figure 5.4. When the content of APP and Si-MCA were 18.7 and 6.3 wt%, respectively, the composite showed the LOI value of 33.5% with V-0 rating. When the temperature was higher than 300°C, the phosphoric acid released from APP could catalyze the esterification of the Si-OH of Si-MCA, and thus a precursor of intumescent char (-P-O-Si-) was formed. After that, lots of incombustible gases (NH_3, H_2O, and CO_2) were released and swelled the precursor char. As a result, a multi-cellular swollen char was generated and covered on the polymer. This residue could effectively slow down the heat and mass transfer, thus protected the substrate material from burning.

Lai et al. (2016) also combined the *N*-alkoxy hindered amine (NOR 116) with novel IFR (APP and Si-MCA) to flame-retard PP. When the content of NOR 116 was 0.3 wt%, the LOI value of sample was increased from 35.0% to 42.5%, with V-0 rating in UL-94 test. Meanwhile, the HRR, THR, average mass loss rate (av-MLR), and smoke production rate (SPR) of composite were also significantly reduced.

5.2.1.5 Epoxy Resin (EP)

Liu et al. compared nano-silica, diglycidylether terminated-polydimethylsiloxane (PDMS–DG), and tetraethoxysilane (TEOS) as silicon-source additive for phosphorus EP (Liu and Chou 2005). They observed that nano-silica is failed to migrate to surface, forming a protective silicon layer.

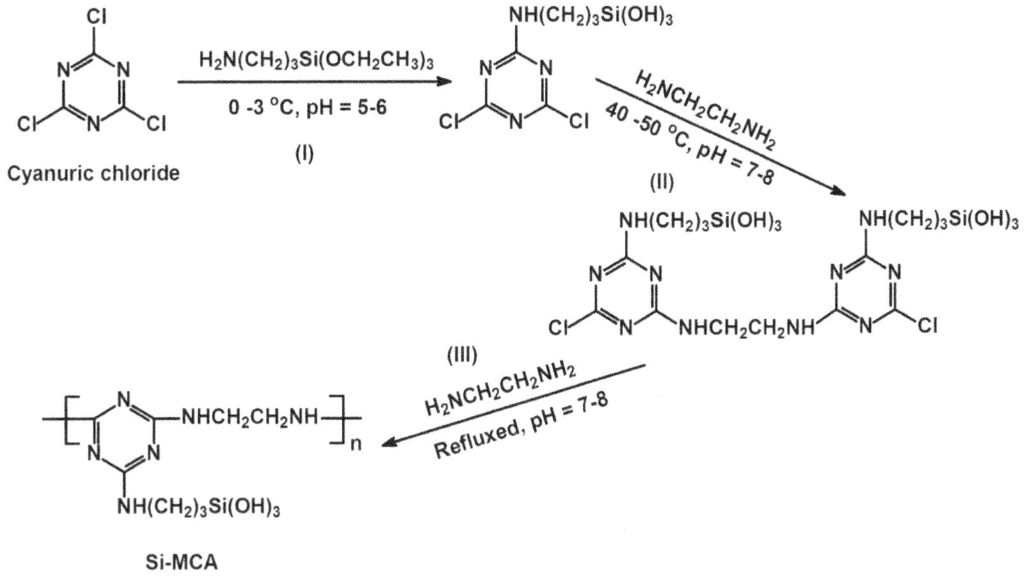

FIGURE 5.4 Synthetic route of Si-MCA.

In contrast, PDMS–DG and TEOS showed significant performance in silicon migration and protective layer formation. However, self-degradation of PDMS–DG during the heating period resulted in some silicon loss in the condensed phase. Thus, formation of epoxy–silica hybrid structure through sol–gel reactions by using TEOS was a good approach for achieving phosphorus–silicon synergism of flame retardation in EP. Hsiue et al. (2000) found that the synergistic effect of phosphorus–silicon on EP fire resistance can be further strengthened by using modified siloxane reagents instead of silanes. They mixed 44.3 wt% of bis(3-glycidyloxy)phenylphosphine oxide (BGPPO) with 34.6 wt% of diamine aminopropyl-terminated polydimethylsiloxane (PDMS–NH$_2$) in place of the triglycidyloxy phenyl (TGPS), resulted in a high LOI value of 45%. Furthermore, the LOI values increased with the increasing of silicon content. This highly synergistic flame-retardant efficiency of siloxane with phosphorus mainly originated from the formation of a continuous silica layer. The combination of EP emulsion and self-crosslinked silicone acrylate (SSA) emulsion as mixed binder for preparing water-borne intumescent fire resistive coating was proposed by Wang and Yang (2010). SSA could react with EP and increased cross-linking degree of mixed binder. Then, the binder with high cross-linking degree could enhance anti-oxidation performance and char residue of coatings at high temperature. The epoxy-based intumescent fire resistive coating with 11.2 wt% SSA showed higher decomposition temperature than its counterpart without SSA, indicating that increasing SSA content was favorable to the fire protection of the coatings. A novel silicone derivative of 9,10-dihydro-9-oxa-10-phosphaphenanthrene-10-oxide (DOPO)-SiD was synthesized via the reaction 1,3,5,7-tetramethyl-1,3,5,7-teravinylcyclotetrasiloxane (D4Vi) with DOPO and employed as flame retardant for cycloaliphatic EP (Li et al. 2014a). DOPO-SiD was found to effectively prevent the cycloaliphatic EP from dripping during combustion. Song et al. (2014) synthesized a novel dicyclic silicon-/phosphorus hybrid (SDPS) and investigated its performance on flame retardancy of EP. To construct an intumescent system, SDPS was cooperatively used with mono (4, 6-diamino-1,3,5-triazin-2-aminium) mono (2,4,8,10-tetraoxa-3,9-di-phosphaspiro [5.5] undecane-3,9-bis (olate) 3, 9-dioxide) (SPDM), which was a well-established P–N intumescent flame retardant. The combination of SPDM and SDPS resulted in a relatively high LOI value and a relatively low PHHR for EP. Another phosphorous-silicone-nitrogen ternary flame retardant of [(1,1,3,3-tetramethyl-1,3-disilazanediyl)di-2,1-ethane-diyl]bis(diphenylphosphine oxide) (PSiN) was synthesized and added into EP (Li et al. 2014b). The preparation scheme for PSiN was given in Figure 5.5. LOI values of EP composites

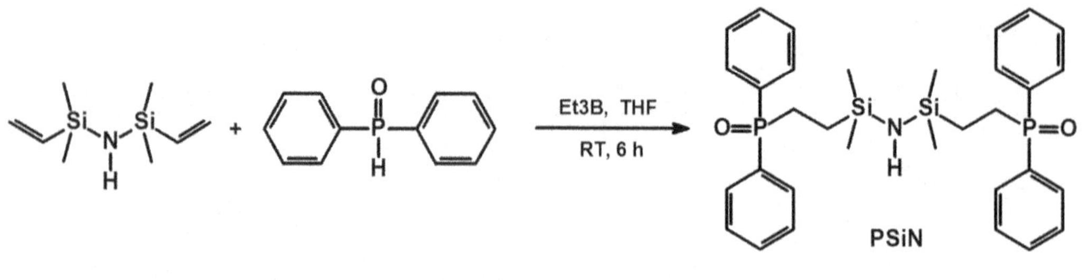

FIGURE 5.5   Diagrammatical illustration of preparation process of PSiN.

increased with the increasing of PSiN contents. An UL-94 V-0 rating was achieved when 20 wt% PSiN was incorporated. Characterization of char residue implied that silica tended to migrate to the surface of polymer, acting as protective layer. Nitrogen components turned into gas in the combustion process. Phosphorus components played a role of flame retardant in both condensed phase and gas phase.

5.2.1.6 Low-Density Polyethylene (LDPE)

Silicone elastomer-containing chalk was also used to improve flame retardancy of LDPE (Hermansson et al. 2003). LOI and cone calorimeter tests were performed on a formulation of 12.5 wt% of silicone elastomer, 30 wt% of chalk, and 57.3 wt% of LDPE. The LOI value increased from 18% to 24.5%, while cone calorimeter test showed a remarkable improvement in flame retardancy in terms of PHRR, time to ignition (TTI), and mass loss (Table 5.2). The integration of silicone and chalk facilitated the formation of stable intumescent char residue, which acted as heat and mass barrier. Silicone alone has also been studied to enhance thermal stability of LDPE. Increasing thermal stability of LDPE by PDMS was explained by cross-linking formation between radicals from degradation product of LDPE and vinyl groups in PDMS backbone (McNeill and Mohammed 1995). Owing to the immiscibility of LDPE–PDMS blends, ethylene–methyl acrylate (EMA) copolymer was introduced as a chemical compatibilizer (Santra et al. 1993). EMA reacted with PDMS rubber during melt-mixing at 180°C to form EMA-grafted PDMS rubber (EMA-g-PDMS) in situ. Jana et al. investigated the thermal stability of LDPE-PDMS blends (50:50) compatibilized with various proportions of EMA (Jana and Nando 2003). EMA-g-PDMS acted as virtual bridges between the two components holding the continuous (LDPE) and dispersed (PDMS) phases together. Thus, the formation of intra- as well as inter-molecular cross-linking between LDPE and PDMS matrix enhanced the mechanical strength and thermal stability of blend.

5.2.2 Silica

5.2.2.1 Silica Nanoparticles

The mild processing and characterization of silica/epoxy (SiO_2/EP) nanocomposite was conducted by Chen and co-workers (Chen and Morgan 2009). In this paper, a milder processing procedure was applied to prepare SiO_2/EP nanocomposites, and the dispersion of SiO_2 nanoparticles in matrix was investigated. The results showed that the well-dispersed state of SiO_2 nanoparticles could be achieved after milder processing. After the milder processing procedure, the glass transition temperature (T_g) of

TABLE 5.2   TGA, LOI, and Cone Calorimeter Results of LDPE and CaSiLDPE

Material	Peak HRR (kW/m²)	TTI (s)	Mass Loss (%)	LOI (%)
LDPE	1420	76	99.4	18
CaSiLDPE	320	95	69.4	24.5

Source: Hermansson, A. et al., *Fire Mater.*, 27, 51–70, 2003.

nanocomposite with 5 wt% silica loading did not change, and the drop in T_g was minimal when SiO_2 loading up to 15 wt%, but some effects of the self-polymerization of EP on T_g were noted with 30 wt% loading of SiO_2. Thermal analysis and flammability tests suggested that nano-SiO_2 had only an inert filler effect (dilution of fuel) on the flammability reduction and char yield increase. The effect of SiO_2 nanoparticles on the thermal stability and curing kinetics of SiO_2/EP nanocomposite was investigated by Arabli and co-workers (Arabli and Aghili 2015). After the inclusion of SiO_2 nanoparticles, the weight loss rates of the nanocomposite were reduced, suggesting the enhanced thermal stability. The thermal stability of SiO_2/EP nanocomposite increased with increasing SiO_2 content. When the temperature was elevated, the SiO_2 nanosilica tended to move toward the surface, forming a thermal insulating material and protecting the underlying polymers. Therefore, adding SiO_2 nanoparticles into the polymer improved the polymer flame retardancy and thermal resistance. The similar introduction of nano-SiO_2 into EP was also reported by Roenner et al. (2017). The flammability and mechanical properties of EP nanocomposite were evaluated after the additions of nano-SiO_2. In the single edge notch bending test, the addition of 36% nano-SiO_2 particles doubled the toughness and increased the flexure modulus by 50% (Figure 5.6). Flammability of samples was estimated via time to ignition at constant irradiation. Adding up to 36% nano-SiO_2, the time to ignition increased by 38% although a sharp decrease was observed around 24% SiO_2 addition. The increased time to ignition is mostly due to a higher thermal diffusivity, increased inert content, as well as a strengthening of char residue, which acts as a mass barrier for degraded products.

Wang et al. (2006) investigated the effect of acrylic polymer and nanocomposite with nano-SiO_2 on thermal degradation and fire resistance of ammonium polyphosphate (APP)-dipentaerythritol-melamine (APP-DPER-MEL) coating. The interaction of APP, DPER, MEL, and acrylic resin between 300°C and 450°C led to the formation of an intumescent char structure. As the carbonization agent and binder of flame-retardant coating, the acrylic resin played an important role during thermal degradation and char formation of coating. The thermal degradation of acrylic resin was connected with its

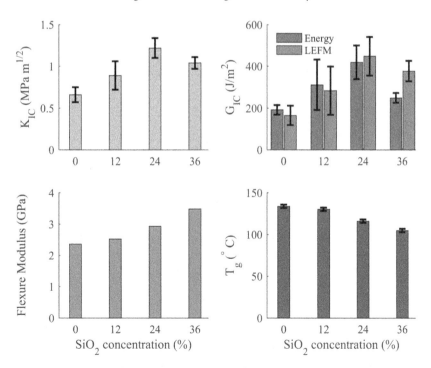

FIGURE 5.6 Overview of the change in toughness, flexure modulus, and glass transition temperature with nano-SiO_2 concentration. (Reprinted from Roenner, N. et al., *Fire Saf. J.*, 91, 200–207, 2017. With permission.)

molecular structure and molecular weight. The melting and thermal decomposition of APP and acrylic resin (3F-1 type) with benzene rings and high molecular weight were at the same temperature range, thus a good synergistic effect could be observed. However, the melting and decomposition temperature range of F-963 acrylic resin with low molecular weight and lack of benzene rings was lower than that of APP. Then, APP could not effectively catalyze the acrylic resin, which is adverse to char formation and chemical interaction. The flame-retardant coating with acrylic resin (3F-1 type) showed a compact and coherent char structure, while the flame-retardant coating with acrylic resin (F-963 type) only formed a loose char structure.

In order to obtain better polymer performance, the modified nano-SiO$_2$ was synthesized and employed to polymer matrix. Yamauchi et al. (2009) synthesized flame-retardant-immobilized SiO$_2$ nanoparticles and added it into EP. The brominated flame-retardant, poly(2,2',6,6'-tetrabromobisphenol-A) diglycidyl ether (PTBBA), was successfully immobilized onto SiO$_2$ nanoparticles through the reaction between terminal amino groups of hyperbranched polyamidoamine (PAMAM)-grafted SiO$_2$ (SiO$_2$-PAMAM) and epoxy groups of PTBBA. The flame-retardant property of EP was considerably improved by the addition of SiO$_2$-PAMAM-PTBBA. Theil-Van Nieuwenhuyse et al. (2013) synthesized phosphorylated SiO$_2$/PA 6 (PA 6) nanocomposites by in situ sol-gel method in molten conditions and investigated their flame retardancy. This synthesis was based on the hydrolysis-condensation reactions of diethylphosphatoethyltriethoxysilane (SiP) as a functional inorganic precursor in combination with or without TEOS dispersed in the molten PA 6 matrix. It was found that the phosphorylated SiO$_2$ causes a dramatic decrease in PHRR (Figure 5.7).

Dong et al. (2013) investigated the effect of DOPO immobilized SiO$_2$ (SiO$_2$-DOPO) nanoparticles in the intumescent flame-retarded PP composites. Keeping the IFR content at 25 wt%, different loadings of SiO$_2$ and SiO$_2$-DOPO nanoparticles were incorporated into the flame-retardant composites. The PP/IFR (25 wt%) composite revealed a LOI value of 28.5% and rated V-2 grade in UL-94 test. By incorporating 0.5 wt% and 1 wt% SiO$_2$-DOPO nanoparticles, the LOI values of composite increased to 31.5% and 32.1%, respectively, with UL-94 V-0 rating. The DOPO immobilized SiO$_2$ nanoparticles showed better synergistic effect on the flame retardancy than neat SiO$_2$ nanoparticles. Moreover, with the incorporation of 1.0 wt% of SiO$_2$-DOPO, a compact char residue was formed without pore cracking. The enhanced properties of polylactic acid (PLA) with SiO$_2$ nanoparticles were reported by Lv et al. (2016). Surface-grafting modification was used in this study by grafting 3-glycidoxypropyltrimethoxysilane (KH-560) onto the

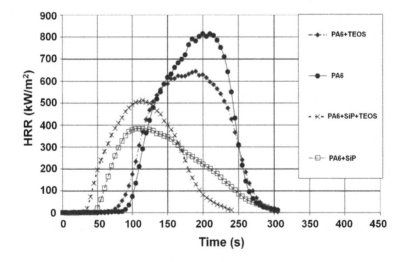

FIGURE 5.7   HRR curves for pure PA 6 and in situ nanocomposites versus time at 35 kW/m^2. (Reprinted from Theil-Van Nieuwenhuyse et al., *Polym. Degrad. Stab.*, 98, 2635–2644, 2013. With permission.)

surface of SiO_2 nanoparticles. When the contents of silica particles increased, the weight loss ratios of the nanocomposites decreased. The main reason for this phenomenon was that the KH-560-treated SiO_2 particles showed a high degree of homogeneity in the PLA matrix, resulting in better thermal insulation performance. Consequently, the flame-retardant properties of PLA were improved greatly by adding modified SiO_2 nanoparticles. Zhu et al. (2016) also studied the influence of modified nano-SiO_2 on the flame resistance of IFR-cellulosic fabrics. Pure cotton fabric (CF) without any flame retardants and additives was highly combustible with low LOI value of 18.4%. With the addition of 20 wt%, intumescent flame retardants consisted of APP-PER-melamine, the LOI of IFR-CF increased to 29.3%. The co-addition of 4 wt% nano-SiO_2 further increased the LOI value of the IFR-cellulosic fabrics to 31.5%.

5.2.2.2 Mesoporous Silica

Mesoporous silica SBA-15 synthesized from Pluronic P123 and tetraethoxysilane was used as a synergistic agent to improve flame retardancy of PP/IFR system (Li et al. 2011). LOI, UL-94 rating, and thermogravimetric analysis (TGA) analyses were used to evaluate the synergistic effect of SBA-15 on PP/IFR system. PP/IFR system could reach V-0 with loading of SBA-15 ranging from 0.5 to 3 wt%, while the sample without SBA-15 had no rating in UL-94 test. The LOI value increased from 25.5% for PP/IFR to 32.2% for PP/IFR with 1 wt% SBA-15. TGA results showed that the maximum decomposition temperature increased from 455°C to 488°C owing to the addition of 1 wt% SBA-15, indicating the improved thermal stability. Meanwhile, adding SBA-15 promoted the char formation of PP/IFR blends. SEM images of char residues suggested the PP/IFR blends with SBA-15 could form smaller size of carbonaceous microstructures in char layer than the PP/IFR, which effectively protected the underlying matrix from combustion.

Wang et al. (2013b) investigated the influence of mesoporous fillers (MCM-41 and SBA-15) on the flammability and tensile behavior of PP-IFR system. The PP composite containing 25 wt% IFR consisted of APP:PER (2:1 by weight) exhibited no rating in UL-94 tests. By contrast, the co-addition of 1 wt% MCM-41 or SBA-15 into PP/25 wt% IFR composite showed UL-94 V-0 rating, which demonstrated effective flame-retardant synergism between MCM-41 or SBA-15 with IFR in PP. They also reported the application of mesoporous MCM-41 as synergistic agent in flame-retardant natural rubber (NR) composite (Wang et al. 2013a). Microencapsulated ammonium polyphosphate by melamine-formaldehyde-APP resin was employed as flame-retardant additive to NR together with MCM-41. The LOI values of NR/MCMF-APP composites were enhanced as compared to that of the NR/APP composites at the equivalent loading. When MCM-41 was further added into the NR/MCMF-APP composite, the LOI values were further improved accompanying with V-0 rating in UL-94 test. The char residue-containing MCM-41 displayed more homogenous and compact structure than the samples without MCM-41, indicating that MCM-41 was effective in retarding the degradation of polymer and improving its flame retardancy. Wei et al. (2013) synthesized a novel hybrid organic-inorganic mesoporous silica named after DM and used it as flame retardant for PC/ABS. The PC/ABS composite containing 6 wt% triphenyl phosphate and 2 wt% DM exhibited a LOI value of 28% and V-0 rating in UL-94 test.

5.2.2.3 Silica Gel

Ni et al. (2011) prepared gel-silica/ammonium polyphosphate (MCAPP) core-shell microspheres and investigated its flame-retardant effect in polyurethane (PU) composites. The LOI values of PU/MCAPP composites were higher than those of PU/APP composites. For instance, the LOI value of PU composite with 15 wt% MCAPP was 33%, while the sample containing 15 wt% APP was only 29%. When the MCAPP was heated, the SiO_2 gel on the surface of APP could release water vapor, which would dilute the concentration of air. Meanwhile, the shell of SiO_2 gel restrained the volatile products to make the materials swell to form a stable intumescent char residue. Moreover, the composite with 12.5 wt% MCAPP achieved V-0 in UL-94 test. Xu et al. (2017) investigated the flame retardancy and thermal behavior of PC composites filled with α-zirconium phosphate@gel-silica (ZS). The as-prepared ZS particles exhibited an obvious coating layer on the surface of the original α-ZrP, while the α-zirconium phosphate (α-ZrP) particles showed lamellar structure with a smooth and clear surface. The LOI value of the PC composite with co-addition of 3 wt% ZS and 3 wt% silicone resin (SR) were increased significantly to 32.3% and a UL-94 V-0 rating was also achieved. By comparison, the PC composite with co-addition of 3 wt% α-ZrP and 3 wt% SR displayed UL-94 V-1 rating, because the absence of gel-silica coating cannot restrain the melt dripping in UL-94 tests. Zhao et al. (2015) prepared the granular silica aerogel/polyisocyanurate rigid foam (PIR) composites and investigated their flame-retardant and thermal insulation properties. The LOI value of PIR composite foam increased from 29.4% (0 wt% aerogel) to 34.6% (8 wt% aerogel) and thermal conductivity was decreased from 0.0346 W/(m·K) (0 wt% aerogel) to 0.0233 W/(m·K) (8 wt% aerogel). Meanwhile, the compressive strength and the specific strength were improved by 136% and 92.2%, respectively, compared to those of the control foam.

5.2.3 Silicate

Owing to its natural abundance and high aspect ratio, montmorillonite (MMT) has drawn extensive attentions from researchers and became the most promising candidate of layered silicates. To improve its dispersion in polymer, the modification with organic compounds has been performed. Wang et al. (2004) reported the preparation and characterization of flame-retardant acrylonitrile–butadiene–styrene/montmorillonite (ABS/MMT) nanocomposites. The used organophilic montmorillonite (OMT) were prepared from MMT by ion exchange reaction using hexadecyl trimethyl ammonium bromide in water. The PHRR of ABS/5%OMT, ABS/18% decabromodiphenyl oxide (DB) and ABS/15%DB/3% antimony oxide (AO) was 28%, 50%, and 68% lower than that of pure ABS, respectively. With co-addition of 5 wt% OMT with DB and/or AO, the PHRR of ABS/18%DB/5%OMT and ABS/15%DB/3%AO/5%OMT nanocomposites were reduced by 68% and 78%, respectively, compared with that of pure ABS, indicating that addition of OMT can further enhance the flame retardancy of polymer.

Shi et al. (2009) prepared the ethylene-vinyl acetate/montmorillonite (EVA/MMT) nanocomposites and investigated their flame-retardant as well as mechanical properties. Exfoliated EVA/MMT nanocomposites were prepared by a master-batch process using polymer-modified layered silicate instead of small molecule surfactant-modified clays. The control sample EVA-0, which contained 20% PVAc (the used cationic vinyl acetate copolymer) showed a heat release capacity (HRC) value of 603 J/(g·K), while the HRC of EVA/0.84 wt% exfoliated MMT was 581 J/(g·K). Increasing exfoliated MMT loading led to further reduction in HRC. Furthermore, EVA nanocomposite with exfoliated MMT displayed higher yield strength, modulus, and toughness than its counterpart with intercalated MMT. Huang et al. (2010) prepared a phosphorus–nitrogen-containing flame-retardant (PNFR) functionalized montmorillonites (PNFR-MMT) and applied it in PU. Both T_{inital} and T_{max} of PU/20%PNFR-MMT were higher than those of PU/1%PNFR and PU/4%MMT, which indicated the synergistic effect between PNFR and MMT. Cone calorimeter experiments showed that PNFR-MMT can significantly reduce the flammability in terms of PHRR. Compared to PU, the PHRR of PU/20%PNFR-MMT was reduced by 25%. Xin et al. (2017) reported a novel triazine-rich polymer wrapped MMT (PTAC–MMT) and its application in flame-retardant polybutylene terephthalate (PBT). The PBT composite containing 1.67 wt% PTAC–MMT and 8.33 wt% aluminum diethylphosphinate (AlPi) showed a relatively high LOI value of 36.4% and achieved UL-94 V-0 rating. By contrast, the PBT/10 wt% AlPi composite exhibited no rating in UL-94 tests, suggested that PTAC–MMT imparted excellent anti-dripping effect to PBT.

He et al. (2017) investigated the effect of DOPO/MMT nanocomposite on flame retardancy of EP. The LOI value of pure EP, EP/DOPO, and EP/MMT composites was measured to be 23.0%, 31.2%, and 26.0%, respectively. By comparison, the LOI value of EP/DOPO+MMT (physically mixture of DOPO and MMT) and EP/DOPO-MMT was determined to be 33.0% and 33.4%, respectively, higher than that of EP/DOPO and EP/MMT. These results imply that DOPO and MMT show some synergistic effect in the flame retardancy of EP. Moreover, EP/DOPO + MMT showed V-1 rating in UL-94 test, while EP/DOPO – MMT exhibited V-0 rating.

5.2.4 Silane

Cheng et al. investigated the UV-curing behavior and properties of tri/di(acryloyloxyethyloxy) phenyl silane used for flame-retardant coatings (Cheng and Shi 2010). The silicon-containing acrylates, tri(acryloyloxyethyloxy) phenyl silane (TAEPS) and di(acryloyloxyethyloxy) methyl phenyl silane (DAEMPS) were prepared by using the transetherification of phenyltrimethoxyl silane and dimethoxy-methylphenyl silane with 2-hydroxylacrylate, respectively. The LOI values increased along with the increase of TAEPS or DAEMPS content. It is concluded that the addition of TAEPS or DAEMPS can efficiently enhance the flame retardancy of a commercial epoxy acrylate oligomer (EB600, molar weight: 500 g/mol). These silicon-containing compounds worked mainly in condensed phase. As incorporating TAEPS or DAEMPS into EB600, the char residue increased which shielded the underlying polymers from attack by flame.

Zhang et al. (2009) used vinyl-tris-(methoxydiethoxy) silane as an effective and eco-friendly flame retardant for electrolytes in lithium ion batteries. Figure 5.8 showed the flame propagation and burning-up time profiles of the electrolyte with different amounts of VTMS. The bare electrolyte was highly flammable with short flame propagation time and short burning-up time, which are 4.8 seconds(s) and 47.5 s, respectively. With increasing the VTMS content, both the flame propagation time and the burning-up time increased (Figure 5.8a). With the addition of 20 vol% vinyl-Tris-(methoxydiethoxy) silane (VTMS), the flame propagation time increased to 24.3 s, almost five times that of the bare electrolyte. The burning-up time showed similar trend (Figure 5.8b), which indicated that VTMS was a promising flame retardant for the electrolytes of lithium ion batteries.

Gao et al. (2011) studied effect of polysiloxane and silane-modified silica (M-SiO_2) on flame retardancy of IFR/PP system. Compared to 20 wt% IFR/PP composite (no rating in UL-94 tests), the co-addition of 5.5 wt% polysiloxane or M-SiO_2 into IFR-PP could pass UL-94 V-0 classification. By comparison to the 5.5 wt% M-SiO_2/20 wt% IFR/PP composite, the 5.5 wt% polysiloxane/20 wt% IFR/PP composite showed better flame retardancy in terms of much lower PHRR and THR observed in cone calorimeter tests. Furthermore, the incorporation of polysiloxane imparted higher water resistance to IFR-PP than M-SiO_2. A UL-94 V-0 rating material with a LOI of over 34% was still retained after the water treatment.

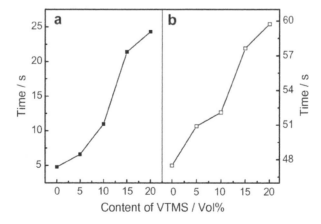

FIGURE 5.8 The flame propagation (a) and burning-up time (b) profiles of 1.0 mol/L $LiPF_6$-EC/EMC/DMC (1:1:1, v/v/v) electrolyte containing different amounts of VTMS. (Reprinted from Zhang, H. et al., *Electrochem. Commun.*, 11, 526–529, 2009. With permission.)

5.3 Combination with Other Flame Retardant Additives

5.3.1 Carbon Nanotubes

Zhang et al. (2012) investigated the flame retardancy of blob-like multi-walled carbon nanotubes/silica nanospheres (SiO_2-g-MWNTs) hybrids in polymethyl methacrylate (PMMA). The preparation process was diagrammatically shown in Figure 5.9. Compared with PMMA, all the PMMA nanocomposites showed an obvious reduction in HRR. The PHRR was reduced from 699 kW/m^2 for PMMA to 496 kW/m^2 for PMMA/pristine-MWNTs, to 471 kW/m^2 for PMMA/polyacrylic acid (PAA)-g-MWNTs, and to 452 kW/m^2 for PMMA/SiO_2-g-MWNTs, respectively. The PMMA/SiO_2-g-MWNTs showed the lowest PHRR and average heat release rate (AHRR), which was attributed to a more compacted layer formed by the presence of SiO_2-g-MWNTs. Hapuarachchi et al. employed the MWNTs and sepiolite nanoclays as flame retardants for PLA (Hapuarachchi and Peijs 2010). The maximum drop (–59.8%) in HRC was observed in the formulation of PLA with 10 wt% sepiolite and 10 wt% MWNT, which was better than PLA with 10 wt% sepiolite (–29.5%) and PLA with 10 wt% MWNT (–50.5%). The superior flame retardancy was attributed to the formation of a tighter char residue via the bridging of MWNTs and sepiolite clay during decomposition. Kuan et al. (2010) firstly prepared the functionalized CNTs (vinyltriethoxysilane [VTES]—CNTs) and further investigated its effect on the flame retardancy and thermal stability of EP. The LOI values increased with increasing the VTES—CNTs amount. With 1 wt% VTES—CNTs, the resultant EP composite achieved a LOI of 23% and UL-94 V-1 rating; when the content of VTES—CNTs was beyond 3 wt%, the LOI value was higher than 25% and UL-94 V-0 rating materials were obtained. The LOI and UL-94 results demonstrated a good flame-retardant effect of VTES—CNTs on EP.

5.3.2 Graphene Oxide and Graphene

Qian et al. (2013) prepared silicon nanoparticle decorated graphene composites and investigated reinforcement on the fire safety and mechanical properties of polyurea nanocomposites. Evidently, the PHRR of polyurea nanocomposites was significantly reduced with the incorporation of silicon nanoparticle decorated graphene. Wang et al. (2015) synthesized a novel nanosilica/graphene oxide hybrid (m-SGO)

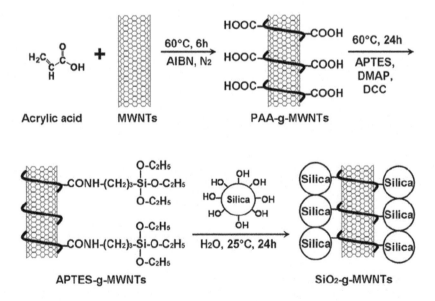

FIGURE 5.9 Schematic illustration of the synthesis procedure of SiO_2-g-MWNTs. (Reprinted from Zhang, T. et al., *Polym. Degrad. Stab.*, 97, 1716–1723, 2012. With permission.)

and applied it in flame-retarding EP composites. Interestingly, this sandwich-like structure can transform into silica nanosheets during the combustion process, and thus not only may inherit high flame-retardant efficiency of inorganic layered clays, but also may endow the resultant nanocomposites with special functionalities and outstanding mechanical properties. Li et al. (2015) also synthesized a silica nanospheres/graphene oxide (SGO) hybrid and employed it as flame retardant for phenolic foams (PF). The incorporation of SGO into PF foam led to a reduction in PHRR and THR, as well as an enhanced thermal stability of the PF foams. Moreover, the addition of 1.5 phr SGO into the PF foam increased the flexural and compressive strengths by 31.6% and by 36.2%, respectively.

5.3.3 Layered Double Hydroxides

Jiang et al. (2014) synthesized the mesoporous silica@Co–Al layered double hydroxide (m-SiO$_2$@Co–Al LDH) spheres via layer-by-layer method and investigated their effects on the flame retardancy of EP. The addition of 2 wt% m-SiO$_2$@Co–Al LDH resulted in a 39.3% decrease in PHRR, a 36.2% decrease in THR, a 23.8% decrease in total smoke release (TSR), and a 16.7% decrease in maximum average heat rate emission (MAHRE). These results suggested that the EP/m-SiO$_2$@Co–Al LDH nanocomposite presents better flame retardancy than EP/m-SiO$_2$ and EP/Co–Al LDH. The incorporation of m-SiO$_2$@Co–Al LDH resulted in a significant improvement of char yield (Figure 5.10). Raman spectra indicated a higher graphitization degree of char was formed due to the catalysis effect of m-SiO$_2$@Co–Al LDH. Moreover, it was found that a much more continuous and cohesive char surface was formed for EP/m-SiO$_2$@Co–Al LDH nanocomposite.

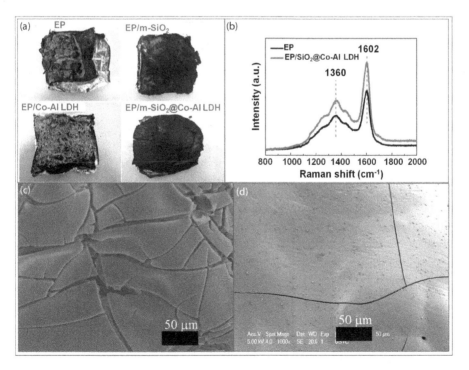

FIGURE 5.10 Digital photos (a) and Raman spectra (b) of the residues from EP and its nanocomposites, SEM images of residues of EP (c), and EP/m-SiO$_2$@Co–Al LDH (d). (Reprinted from Jiang, S.D. et al., *ACS Appl. Mater. Interfaces*, 6, 14076–14086, 2014. With permission.)

5.3.4 Molybdenum Disulfide

Zhou et al. (2018) reported the in-situ-growth of zero-dimensional (0D) silica nanospheres on 2D molybdenum disulfide (MoS_2) nanosheets toward reducing fire hazards of EP. Pure EP burned quickly with a PHRR value of 758 kW/m². There was no obvious variation in PHRR value by incorporating 2 wt% MoS_2 into EP matrix. In contrast, the incorporation of 2 wt% MoS_2-SiO_2 into EP could significantly slow down the combustion process. Specifically, the PHRR value was reduced to 512 kW/m² corresponding to a 32.5% reduction. The THR value of EP/MoS_2-SiO_2 also decreased obviously, which demonstrated that more EP chains participated in the carbonization process owning to the catalysis of MoS_2-SiO_2 hybrids.

5.4 Conclusions

The fire safety of polymeric materials can be significantly enhanced through addition of silicone-containing flame-retardant additives. These incorporated fillers always made great contribution to char formation, which protected the underlying matrix from combustion. Moreover, inorganic silicon dioxide layer with high thermal insulating property was also generated above polymer, preventing the heat transfer. Silicone and silane additives showed great compatibility with polymer, but silica and silicate compounds were immiscible with matrix, which was adverse to the flame retardancy improvement. Then, surface modification for these two fillers is necessary. In order to acquire higher flame-retardant efficiency, their combination with other flame-retardant fillers, such as carbon nanotubes, graphene oxide or graphene, layered double hydroxide, and molybdenum disulfide, were also exploited. Overall, the silicon-containing flame retardants have multiple advantages such as environmental friendliness, excellent flame-retardant efficiency, and high thermal stability which will be a promising alternative to phosphorus-containing compounds.

References

Arabli, V., and Aghili, A. 2015. The effect of silica nanoparticles, thermal stability, and modeling of the curing kinetics of epoxy/silica nanocomposite. *Advanced Composite Materials* 24(6): 561–577.

Barren, J. P., Chen, F. F. S., and Osborn, A. J. 2000. Polycarbonate resin blends containing titanium dioxide. US Patents 6133360A.

Chen, C., and Morgan, A. B. 2009. Mild processing and characterization of silica epoxy hybrid nanocomposite. *Polymer* 50(26): 6265–6273.

Chen, X., and Jiao, C. 2009. Synergistic effects of hydroxy silicone oil on intumescent flame retardant polypropylene system. *Journal of Polymer Research* 16(5): 537–543.

Cheng, X. E., and Shi, W. F. 2010. UV-curing behavior and properties of tri/di (acryloyloxyethyloxy) phenyl silane used for flame-retardant coatings. *Progress in Organic Coatings* 69(3): 252–259.

Davidson, N. S., and Wilkinson, K. 1992. Flame retardant polymer composition. US Patents 5091453.

Dong, Q., Liu, M., Ding, Y. et al. 2013. Synergistic effect of DOPO immobilized silica nanoparticles in the intumescent flame retarded polypropylene composites. *Polymers for Advanced Technologies* 24(8): 732–739.

Gao, S., Li, B., Bai, P. et al. 2011. Effect of polysiloxane and silane-modified SiO_2 on a novel intumescent flame retardant polypropylene system. *Polymers for Advanced Technologies* 22(12): 2609–2616.

Hapuarachchi, T. D., and Peijs, T. 2010. Multiwalled carbon nanotubes and sepiolite nanoclays as flame retardants for polylactide and its natural fibre reinforced composites. *Composites Part A: Applied Science and Manufacturing* 41(8): 954–963.

He, X., Zhang, W., and Yang, R. 2017. The characterization of DOPO/MMT nanocompound and its effect on flame retardancy of epoxy resin. *Composites Part A: Applied Science and Manufacturing* 98: 124–135.

Hermansson, A., Hjertberg, T., and Sultan, B. A. 2003. The flame retardant mechanism of polyolefins modified with chalk and silicone elastomer. *Fire and Materials* 27(2): 51–70.

Hermansson, A., Hjertberg, T., and Sultan, B. A. 2005. Linking the flame-retardant mechanisms of an ethylene-acrylate copolymer, chalk and silicone elastomer system with its intumescent behaviour. *Fire and Materials* 29(6): 407–423.

Hermansson, A., Hjertberg, T., and Sultan, B. A. 2006. Distribution of calcium carbonate and silicone elastomer in a flame retardant system based on ethylene–acrylate copolymer, chalk and silicone elastomer and its effect on flame retardant properties. *Journal of Applied Polymer Science* 100(3): 2085–2095.

Hirai, Y., and Ishii, K. 2003. Polycarbonate resin composition and its molded articles, US Patents 6664313B2.

Hsiue, G. H., Liu, Y. L., and Tsiao, J. 2000. Phosphorus-containing epoxy resins for flame retardancy V: Synergistic effect of phosphorus–silicon on flame retardancy. *Journal of Applied Polymer Science* 78(1): 1–7.

Huang, G., Gao, J., Li, Y. et al. 2010. Functionalizing nano-montmorillonites by modified with intumescent flame retardant: Preparation and application in polyurethane. *Polymer Degradation and Stability* 95(2): 245–253.

Huhtala, J., and Motha, K. 2005. Flame retardant polymer composition comprising nanofillers. European Patent 1512718.

Iji, M., and Serizawa, S. 1998. Silicone derivatives as new flame retardants for aromatic thermoplastics used in electronic devices. *Polymers for Advanced Technologies* 9(10–11): 593–600.

Jana, R., and Nando, G. 2003. Thermogravimetric analysis of blends of low-density polyethylene and poly (dimethyl siloxane) rubber: The effects of compatibilizers. *Journal of Applied Polymer Science* 90(3): 635–642.

Jiang, S. D., Bai, Z. M., Tang, G. et al. 2014. Synthesis of mesoporous silica@ Co–Al layered double hydroxide spheres: Layer-by-layer method and their effects on the flame retardancy of epoxy resins. *ACS Applied Materials & Interfaces* 6(16): 14076–14086.

Kuan, C. F., Chen, W. J., Li, Y. L. et al. 2010. Flame retardance and thermal stability of carbon nanotube epoxy composite prepared from sol–gel method. *Journal of Physics and Chemistry of Solids* 71(4): 539–543.

Lai, X., Qiu, J., Li, H. et al. 2016. Thermal degradation and combustion behavior of novel intumescent flame retardant polypropylene with N-alkoxy hindered amine. *Journal of Analytical and Applied Pyrolysis* 120: 361–370.

Lai, X., Yin, C., Li, H. et al. 2015. Synergistic effect between silicone-containing macromolecular charring agent and ammonium polyphosphate in flame retardant polypropylene. *Journal of Applied Polymer Science* 132(10): 41580.

Li, J., Wei, P., Li, L. et al. 2011. Synergistic effect of mesoporous silica SBA-15 on intumescent flame-retardant polypropylene. *Fire and Materials* 35(2): 83–91.

Li, Q., Jiang, P., Su, Z. et al. 2005. Synergistic effect of phosphorus, nitrogen, and silicon on flame-retardant properties and char yield in polypropylene. *Journal of Applied Polymer Science* 96(3): 854–860.

Li, Q., Jiang, P., and Wei, P. 2006. Synthesis, characteristic, and application of new flame retardant containing phosphorus, nitrogen, and silicon. *Polymer Engineering & Science* 46(3): 344–350.

Li, X., Wang, Z., and Wu, L. 2015. Preparation of a silica nanospheres/graphene oxide hybrid and its application in phenolic foams with improved mechanical strengths, friability and flame retardancy. *RSC Advances* 5(121): 99907–99913.

Li, Y. J., Gu, X. Y., Zhao, J. et al. 2014a. Flame retardancy effects of phosphorus-containing compounds and cationic photoinitiators on photopolymerized cycloaliphatic epoxy resins. *Journal of Applied Polymer Science* 131(7): 40011.

Li, Z. S., Liu, J. G., Song, T. et al. 2014b. Synthesis and characterization of novel phosphorous-silicone-nitrogen flame retardant and evaluation of its flame retardancy for epoxy thermosets. *Journal of Applied Polymer Science* 131(24): 40412.

Liu, Y. L., and Chou, C. I. 2005. The effect of silicon sources on the mechanism of phosphorus–silicon synergism of flame retardation of epoxy resins. *Polymer Degradation and Stability* 90(3): 515–522.

Lupinski, J. H., and Hawkins, C. M. 1992. Flame retardant polycarbonate compositions. US Patent 5153251.

Lv, H., Song, S., Sun, S. et al. 2016. Enhanced properties of poly (lactic acid) with silica nanoparticles. *Polymers for Advanced Technologies* 27(9): 1156–1163.

McNeill, I. C., and Mohammed, M. H. 1995. Thermal analysis of blends of low density polyethylene, poly (ethyl acrylate) and ethylene ethyl acrylate copolymer with polydimethylsiloxane. *Polymer Degradation and Stability* 50(3): 285–295.

Ni, J., Chen, L., Zhao, K. et al. 2011. Preparation of gel-silica/ammonium polyphosphate core-shell flame retardant and properties of polyurethane composites. *Polymers for Advanced Technologies* 22(12): 1824–1831.

Nodera, A., and Kanai, T. 2006a. Flame retardancy of a polycarbonate–polydimethylsiloxane block copolymer: The effect of the dimethylsiloxane block size. *Journal of Applied Polymer Science* 100(1): 565–575.

Nodera, A., and Kanai, T. 2006b. Flame retardancy of polycarbonate–polydimethylsiloxane block copolymer/silica nanocomposites. *Journal of Applied Polymer Science* 101(6): 3862–3868.

Pape, P. G., and Romenesko, D. J. 1997. The role of silicone powders in reducing the heat release rate and evolution of smoke in flame retardant thermoplastics. *Journal of Vinyl and Additive Technology* 3(3): 225–232.

Qian, X., Yu, B., Bao, C. et al. 2013. Silicon nanoparticle decorated graphene composites: Preparation and their reinforcement on the fire safety and mechanical properties of polyurea. *Journal of Materials Chemistry A* 1(34): 9827–9836.

Roenner, N., Hutheesing, K., Fergusson, A. et al. 2017. Simultaneous improvements in flammability and mechanical toughening of epoxy resins through nano-silica addition. *Fire Safety Journal* 91: 200–207.

Santra, R., Samantaray, B., Bhowmick, A. et al. 1993. In situ compatibilization of low-density polyethylene and polydimethylsiloxane rubber blends using ethylene–methyl acrylate copolymer as a chemical compatibilizer. *Journal of Applied Polymer Science* 49(7): 1145–1158.

Shi, Y., Kashiwagi, T., Walters, R. N. et al. 2009. Ethylene vinyl acetate/layered silicate nanocomposites prepared by a surfactant-free method: Enhanced flame retardant and mechanical properties. *Polymer* 50(15): 3478–3487.

Song, S., Ma, J., Cao, K. et al. 2014. Synthesis of a novel dicyclic silicon-/phosphorus hybrid and its performance on flame retardancy of epoxy resin. *Polymer Degradation and Stability* 99: 43–52.

Theil-Van Nieuwenhuyse, P., Bounor-Legare, V., Bardollet, P. et al. 2013. Phosphorylated silica/polyamide 6 nanocomposites synthesis by in situ sol–gel method in molten conditions: Impact on the fire-retardancy. *Polymer Degradation and Stability* 98(12): 2635–2644.

Wang, G., and Yang, J. 2010. Influences of binder on fire protection and anticorrosion properties of intumescent fire resistive coating for steel structure. *Surface and Coatings Technology* 204(8): 1186–1192.

Wang, H., and Li, B. 2010. Synergistic effects of β-cyclodextrin containing silicone oligomer on intumescent flame retardant polypropylene system. *Polymers for Advanced Technologies* 21(10): 691–697.

Wang, N., Mi, L., Wu, Y. et al. 2013a. Enhanced flame retardancy of natural rubber composite with addition of microencapsulated ammonium polyphosphate and MCM-41 fillers. *Fire Safety Journal* 62: 281–288.

Wang, N., Zhang, J., Fang, Q. et al. 2013b. Influence of mesoporous fillers with PP-g-MA on flammability and tensile behavior of polypropylene composites. *Composites Part B: Engineering* 44(1): 467–471.

Wang, R., Zhuo, D., Weng, Z. et al. 2015. A novel nanosilica/graphene oxide hybrid and its flame retarding epoxy resin with simultaneously improved mechanical, thermal conductivity, and dielectric properties. *Journal of Materials Chemistry A* 3(18): 9826–9836.

Wang, S., Hu, Y., Song, L. et al. 2004. Preparation and characterization of flame retardant ABS/ montmorillonite nanocomposite. *Applied Clay Science* 25(1–2): 49–55.

Wang, Z., Han, E., and Ke, W. 2006. Effect of acrylic polymer and nanocomposite with nano-SiO$_2$ on thermal degradation and fire resistance of APP–DPER–MEL coating. *Polymer Degradation and Stability* 91(9): 1937–1947.

Wei, P., Tian, G., Yu, H. et al. 2013. Synthesis of a novel organic-inorganic hybrid mesoporous silica and its flame retardancy application in PC/ABS. *Polymer Degradation and Stability* 98(5): 1022–1029.

Xin, F., Guo, C., Chen, Y. et al. 2017. A novel triazine-rich polymer wrapped MMT: Synthesis, characterization and its application in flame-retardant poly (butylene terephthalate). *RSC Advances* 7(75): 47324–47331.

Xu, L., Lei, C., Xu, R. et al. 2017. Synergistic effect on flame retardancy and thermal behavior of polycarbonate filled with α-zirconium phosphate@gel-silica. *Journal of Applied Polymer Science* 134(19): 44829.

Yamauchi, T., Yuuki, A., Wei, G., Shirai, K., Fujiki, K., Tsubokawa, N. 2009. Immobilization of flame retardant onto silica nanoparticle surface and properties of epoxy resin filled with the flame retardant-immobilized silica. *Journal of Polymer Science* 47(22): 6145–6152.

Zhang, H., Xia, Q., Wang, B. et al. 2009. Vinyl-tris-(methoxydiethoxy) silane as an effective and ecofriendly flame retardant for electrolytes in lithium ion batteries. *Electrochemistry Communications* 11(3): 526–529.

Zhang, T., Du, Z., Zou, W. et al. 2012. The flame retardancy of blob-like multi-walled carbon nanotubes/ silica nanospheres hybrids in poly (methyl methacrylate). *Polymer Degradation and Stability* 97(9): 1716–1723.

Zhao, C., Yan, Y., Hu, Z. et al. 2015. Preparation and characterization of granular silica aerogel/ polyisocyanurate rigid foam composites. *Construction and Building Materials* 93: 309–316.

Zhong, H., Wei, P., Jiang, P. et al. 2007a. Thermal degradation behaviors and flame retardancy of PC/ ABS with novel silicon-containing flame retardant. *Fire and Materials* 31(6): 411–423.

Zhong, H., Wu, D., Wei, P. et al. 2007b. Synthesis, characteristic of a novel additive-type flame retardant containing silicon and its application in PC/ABS alloy. *Journal of Materials Science* 42(24): 10106–10112.

Zhou, K., Tang, G., Gao, R. et al. 2018. In situ growth of 0D silica nanospheres on 2D molybdenum disulfide nanosheets: Towards reducing fire hazards of epoxy resin. *Journal of Hazardous Materials* 344: 1078–1089.

Zhu, F. L., Xin, Q., Feng, Q. Q. et al. 2016. Influence of nano-silica on flame resistance behavior of intumescent flame retardant cellulosic textiles: Remarkable synergistic effect? *Surface and Coatings Technology* 294: 90–94.

6

Recent Advances in Boron-Based Flame Retardants

Kelvin K. Shen

6.1 Introduction

Borates function as flame retardants, smoke suppressants, afterglow suppressants, and anti-tracking agents in both halogen-containing and halogen-free polymers. The use of borates as flame retardants was reviewed extensively (Shen et al. 2010; Shen 2014a, 2014b). Recently, there have been a large number of literature/patents claiming the use of borates as flame retardants in a variety of polymer systems. This chapter will cover recent developments on the use of borates as flame retardants, as well as some reports/patents that were not presented in previous reviews. Due to a large number of new publications, this chapter will only present papers/patents that have commercial potential and technical novelty. Due to limitation of the size of the chapter, it is not intended to be inclusive for all publications.

In halogen-containing systems, boron-based flame retardants can provide the following benefits:

- Synergists of halogen sources
- Synergists of antimony oxide
- Can either partially or completely replace antimony oxide
- Smoke suppressant
- Afterglow suppressant
- HCl/HBr suppressant
- Improve aged elongation properties of polyolefins
- Improve thermal stability of aromatic bromine/Sb system
- Provide anti-tracking (i.e., Comparative Tracking Index—CTI) and anti-arcing properties of base polymer.

In halogen-free systems:

- Reduce rate of heat release
- Promote char/residue formation

- Stabilize the char and prevent dripping
- Inhibit oxidation of the char (afterglow suppression)
- Most borates release significant amount of water to provide flame retardancy
- Promotes sintering between filler particles—effective flux and glass former for ceramification
- Provide anti-tracking and anti-arcing properties of base polymer
- Display synergy with nitrogen, phosphorus, silicon compounds
- Buffer the fire retardant system and stabilize the package during processing.

6.2 Properties of Boron-Based Products and Applications

6.2.1 Boric Acid (H_3BO_3) and Boric Oxide (B_2O_3)

Boric acid or orthoboric acid (commercially available as *Optibor®*) is a triclinic crystal that is soluble in water (5.46 wt% 25°C), alcohols, and glycerin. It is a weak acid and has a pH of 4 (saturated solution at room temperature). Upon heating from 75°C to around 125°C, it loses part of its water of hydration to form metaboric acid (HBO_2). The metaboric acid can be further dehydrated to boric oxide at around 260°C–270°C.

$$2H_3BO_3 \rightarrow 2HBO_2 + 2H_2O \rightarrow B_2O_3 + H_2O$$

Boric oxide, also known as anhydrous boric acid, is a hard glassy material, which softens at about 325°C and melts at about 450°C–465°C. It is produced commercially by fusion of boric acid. Through this procedure, it generally contains up to 0.5% water. It can, however, absorb water and revert back to boric acid; however, this normally does not affect its fire retardancy performance. Boric oxide and boric acid are not recommended for use in non-polar hydrocarbon polymers, because boric acid (or boric acid formed via boric oxide hydration by moisture) may lead to migration to the polymer surface. Some examples of boric acid/boric oxide usage as flame retardants are illustrated below.

Boric acid, working mostly in the condensed phase, is used extensively as a flame retardant in cellulosic products. For example, it is used in cotton batting for inhibiting afterglow combustion, in cellulosic insulation for inhibiting flame spread/afterglow combustion, and various wood/paper product products for flammability control (Shen 2014a). In many applications, boric acid is used in combination with sodium borate.

6.2.1.1 Coatings

Epoxy: Boric acid (BA) in conjunction with ammonium polyphosphate (APP) is used extensively in epoxy intumescent coating for hydrocarbon fires (Hanafin and Bertrand 2000). It helps to form a strong char, results in smallest decrease of viscosity as the epoxy decomposes at a high mechanical strength, and a high degree of expansion (Jimenez et al. 2006). The effects of APP and BA on thermal degradation of epoxy intumescent coating containing expandable graphite and melamine was also reported recently (Ullah et al. 2017). The effect of BA and kaolin in the same coating system was also reported (Ullah et al. 2011).

Inorganic Coating: It was reported that a combination of BA, potassium carbonate, and surfactant in water, can provide not only flame retardancy (ASTM E84 Class I), but also biocidal properties of gypsum board with paper binder (Curzon and Smoot 2010). When a fabric was treated with an aqueous solution consisting of boric acid, sodium borate, sodium silicate, guanylurea phosphate, and magnesium chloride, it can pass the Federal Aviation Regulation (FAR Part 25) Vertical Flammability Test (Part I) and OSU Rate of Heat Release Test (Part IV, OSU-Ohio State University) (Khadbai et al. 2014).

Wood Products Coating: For wood products used in construction applications, they normally have to pass two fire tests. First is the ASTM E119 fire endurance test that measures the ability of the product of maintaining its load bearing capacity during a fire. The second test is the ASTM E2768 that requires a 30-minute (min) burn in the E84 surface burning test set-up. Recently, Koppers Chemicals claims the use of fire retardant packages impregnate and applied a coating to the surface of a wood product to achieve maximum flame distance of 1.2 meter (code requirement <3.2 meter in E84 for 30 min). The impregnation package (via vacuum or pressure) consists of monoammonium phosphate, boric acid, and ethylenediamine in water, and the coating consists of typical water-based intumescent coating with resin/APP/pentaerythritol/melamine (Zhang et al. 2017).

6.2.1.2 Phenolics

Boric acid is commonly used in phenolics to impart thermal stability (stabilizing both the char and the phenolics) and fire retardancy. For example, the use of boric acid and aluminum trihydroxide (ATH) in phenolics for sandwich panel construction was reported (Nobuyoki et al. 2002). A flame retardant fiberglass reinforced plastics (FRP) cover for conveyors was also reported. The cover is made of phenolic resin (100 parts) and boric acid (50–100 parts). It does not deform for 20 min in the cone calorimeter test (ISO 5660-1/50 kW/m^2) and has a peak heat release rate (PHRR) of 13 kW/m^2 and total heat release of 6 MJ/m^2 (Saratani 2011).

6.2.1.3 Polyetherimide and Polyimide

A very low level of BA (around 0.25% by weight) in polyetherimide (PEI), and glass-filled PEI can reduce PHRR by almost 50% in the FAR Part 25 OSU Heat Release test for aircraft application (Bassett 1992). Recently, it was reported that addition of B$_2$O$_3$ nanoparticles to polyimide (PI) can significantly improve the thermal and mechanical properties of the B$_2$O$_3$ (generated from BA in situ)/polyimide composite films (Jian et al. 2017).

6.2.1.4 Polyphenylene Ether

Boric acid (or boric oxide) in conjunction with polyvinylidene difluoride (PVDF) is reported to be an effective fire retardant in polyphenylene ether (PPE) to reduce smoke production drastically (Claesen et al. 1993). For example, a combination of PPE (80%), pentaerythritol tetrabenzoate (PETB) (20%), boric oxide (<1%), and PVDF (2%) can result in V-0 (1.6 mm) and a smoke density (Ds in ASTM E-662) of 87 (4 min). The benefits of having boric oxide and PVDF were also illustrated in systems such as PPE/high impact polystyrene (HIPS), polycarbonate, polyether sulfone, and polyphenylene sulfide. In a subsequent patent, the former GE group also reported that boric acid (1%) alone without PVDF can provide a drastic smoke reduction in PPE/polystyrene (PS) and PPE/PETB (Hoeks et al. 1997).

6.2.1.5 Polypropylene

The use of a low level of boric acid in polypropylene (PP) containing a novel intumescent flame retardant (IFR) based on bis (2,6,7-trioxa-1-phosphabicyclo [2,2,2] octane-1-oxo-4-hydroxymethyl) phenylphosphonate was reported by Peng et al. (2008). They demonstrated that BA has a synergistic effect with the IFR. For example, addition of 1.0 wt% of BA to 29 wt% of IFR in PP gives an limiting oxygen index (LOI) of 34.2% and UL-94 V-0 rating, whereas 30% loading of IFR alone gives an LOI of only 30.3%. In addition, the incorporation of BA can improve the thermal stability and promote the formation of a carbonaceous charred layer.

6.2.1.6 Polystyrene Foam

There has been considerable commercial interest in developing halogen-free flame retardant PS foam for thermal insulation applications (Wagner et al. 2016). It was reported that PS bead was first treated with boric acid and pre-formed in a heated rotary mixer. The pre-formed beads were further treated with boric acid/sodium silicate/phenolics and subjected to molding around 100°C–150°C. The product has low flammability and generates no black smoke during combustion (Fujimori 2002). Subsequently, it was reported the use of boric acid/ATH/phenolics/red phosphorus can achieve LOI in the range of 30%–38%

and PHRR in the range of 15–52 kW/m^2 in the Cone Calorimeter test at 50 kW/m^2 (Fujimori 2006). More recently, XFLAM in Australia reported the use of a combination of phenolic resole, expandable thermoplastic microspheres, and boric acid (silane treated) to treat pre-expanded PS beads (Bracegirdle et al. 2016, 2017). The final PS foam produced has an average specific mass loss rate of around 1.27 g/m^2·s, whereas the untreated PS foam has the mass loss rate of 9.81 g/m^2·s (ISO 17554 at 50 kW/m^2).

6.2.2 Metallo-Borates

6.2.2.1 Alkali Metal borates

Borax Decahydrate (Na$_2$O·2B$_2$O$_3$·10H$_2$O), also called borax, is slightly soluble in cold water (4.71% by wt at 20°C) and very soluble in hot water (30% at 60°C). It has a pH of 9.24 (1% solution at ambient temperature) and exhibits excellent buffering property.

Borax Pentahydrate (commercially available as *Neobor*®) is the most common form of sodium borate used in a variety of industries. Its advantages vs. borax lie in the lower transportation, handling, and storage cost of a more concentrated product. It starts to dehydrate at about 65°C, loses all water of hydration when heated above 320°C, and fuses when heated above 740°C. In water, it hydrolyzes to give a mildly alkaline solution with excellent buffering properties.

Due to their low dehydration temperature and water solubility, sodium borates (except the anhydrous sodium borate) are normally only used as flame retardants in cellulose insulation, wood timber, textiles, urethane foam, and coatings. Thus, in cellulosic material and wood products, it is commonly used in combination with boric acid, which is an effective smoldering inhibitor. In addition, a combination of boric acid and sodium borate can also result in significantly higher water solubility.

Anhydrous Borax (Na$_2$O·2B$_2$O$_3$) commercially known as *Dehybor*® and often called AB, can absorb surface moisture. With a melting point of 742°C, it is an excellent flux and glass former. Thus, it is an effective additive for ceramification during polymer combustion in applications such as wire and cable, sealants, etc. Its potential as a flame retardant in polymers has not been fully explored. Interestingly enough, a combination of ammonium polyphosphate, ammonium phosphate, borax, anhydrous borax (≤45 microns), and melamine was claimed for use in urethane panels to meet the German Institute for Standardization (German standard DIN) 4102, Part 1, B-1 (Schmittmann et al. 1984). They specifically reported that the combination of borax and anhydrous borax yields surprisingly effective results.

Recently, it was reported that reduced graphene oxide (rGO) and sodium metaborate (Na$_2$O·B$_2$O$_3$·xH$_2$O) synergistically display flame retardancy in sawdust pellets (Nine et al. 2017).

Disodium Octaborate Tetraborate (Na$_2$O·4B$_2$O$_3$·4H$_2$O)—This unique form of sodium borate (DOT), known as *Polybor*®, is an amorphous material and thus can be dissolved into water rapidly (solubility 9.7 wt% at room temperature and 21.9% at 30°C). It is particularly effective in reducing the flammability of wood/cellulose/paper products. For example, a mixture comprising this sodium borate (4.55%) with a sodium silicate (0.75%), as well as other aluminum silicate additive is used for protection against fire and wood-destroying organisms for outdoor application (Herve 2013). Weyerhaeuser developed a coating consisting of acrylic latex (16%–30% dry wt), sodium borate (26–45), expandable graphite (13–25), and isocyanate (13–25) for oriented strand board (OSB) used in the construction industry (Parker et al. 2014). This formulation meets the 30 min. ASTM E2768 with a flame spread less than 3.2 meter. Highly flame retardant rigid polyurethane foam was prepared by vacuum or pressure impregnation with the sodium borate solution. The heat release with 150 and 400 wt% of the sodium borate in the Cone Calorimeter test were 8.8 and 4.1 MJ/m^2, respectively (Tsuyumoto et al. 2011). In a single processing step, polysaccharide-based fire-retardant coatings were applied to flexible polyurethane foam via squeezing and releasing several times in the FR solutions that contain boron-based FRs and/or a clay char former. The efficacy of this technology was validated in full-scale fire tests. For example, full-scale fire tests of furniture containing 1.5% starch and 11.5% sodium polyborate has achieved a 75% lower PHRR than when a standard foam was used (120 kW/m^2 vs. 350–580 kW/m^2) (Davis et al. 2015).

6.2.2.2 Alkaline Earth Metal Borates

6.2.2.2.1 *Calcium Borates (xCaO·yB₂O₃·zH₂O)*

6.2.2.2.1 Calcium Borates ($xCaO \cdot yB_2O_3 \cdot zH_2O$)

A variety of calcium borates can be prepared by reacting calcium hydroxide and boric acid. Synthetic gowerite ($CaO \cdot 3B_2O_3 \cdot 5H_2O$) and calcium metaborate ($CaO \cdot B_2O_3 \cdot 4H_2O$ or $CaO \cdot B_2O_3 \cdot 6H_2O$) are commercially available mostly for the glass industry. Among all of the known calcium borates, the natural mineral colemanite ($2CaO \cdot 3B_2O_3 \cdot 5H_2O$) is the most well known in the field of flame retardants.

In halogen-containing systems, calcium borates are generally not as effective as zinc borates, because zinc ion is a stronger Lewis acid than calcium ion. Most calcium borates have low dehydration temperatures (around 115°C–200°C) except colemanite, which has a dehydration temperature of 290°C–300°C. It is mostly used in rubber modified roofing membrane at loadings in the range of 12%–14% (Grube et al. 1992; Khan et al. 2000). The use of colemanite in ethylene-vinyl acetate (EVA) or ethylene-methacrylate (EMA) in conjunction with ATH or magnesium hydroxide (MDH) was recently reported (Cavodeau et al. 2017). It was shown that colemanite in EVA/ATH does not act as a synergistic agent, but seems to improve fire properties on its own.

Synthetic calcium borates with lower dehydration temperatures have been used in fire-retardant grade sealants and caulks. Hitachi reported the use of synthetic calcium borate ($2CaO \cdot B_2O_3 \cdot H_2O$) in epoxy filled with silica for semi-conductor encapsulation. This calcium borate's dehydration temperature is in the range of 200°C–400°C (Ishii et al. 2006). Calcium metaborate is claimed to be advantageously incorporated into frits as fire retardant in fire doors particularly for those based on sodium silicate. It is believed that calcium borosilicate will be formed upon heating in this composition (Crompton 1992).

A calcium borate whisker ($2CaO \cdot B_2O_3 \cdot H_2O$) was prepared by a hydrothermal method. It releases water at about a remarkably high temperature (370°C). Its flame retardancy performance is not yet being evaluated (Song et al. 2012). Recently, it was demonstrated that the $2CaO \cdot B_2O_3 \cdot H_2O$ nanoflake displays good flame retardancy with wood powder as shown by the LOI test (Geng et al. 2017).

6.2.2.2.2 *Magnesium Borates (xMgO·yB₂O₃·zH₂O)*

6.2.2.2.2 Magnesium Borates ($xMgO \cdot yB_2O_3 \cdot zH_2O$)

Due to its high charge to size ratio, the Mg^{2+} cation has a strong tendency to include water in its coordination sphere. Thus, most synthetic magnesium borates contain non-hydroxyl water which can cause them to have low dehydration temperature. For use in plastics, these magnesium borates have to be fully or partially pre-dehydrated. Kyocera reported the use of an unspecified magnesium borate in silica-filled epoxy/phenolic for electronic packaging. The addition of 25% magnesium borate resulted in a V-0 (1.6 mm) formulation with good moldability and high temperature reliability (Haruomi 2003).

Preparation of magnesium borate ($2MgO \cdot B_2O_3 \cdot 1.5H_2O$) nano-rod and nano-wire were reported by Zhang et al. (2017). LOI of compressed pure wood dust, wood dust containing non-nano magnesium borate (20 wt%), nano-rod (20%), and nano-wire (20%) are 23.8, 25.3, 29.0, and 29.3, respectively.

6.2.2.3 Transition Metal Borates and Miscellaneous Metal Borates

6.2.2.3.1 *Zinc Borates (xZnO·yB₂O₃·zH₂O)*

6.2.2.3.1 Zinc Borates ($xZnO \cdot yB_2O_3 \cdot zH_2O$)

Among all of the boron-containing fire retardants used in polymers, zinc borate has the most commercial importance. Major commercial zinc borates and their applications were reviewed previously (Shen 2000, 2014a; Shen et al. 2008) (Table 6.1).

Products—Rio Tinto Minerals/U.S. Borax first patented and commercialized a unique form of zinc borate with a molecular formula, known in the trade as *Firebrake®* ZB (Nies and Hulbert 1972). A recent single crystal X-ray crystallography study showed that *Firebrake* ZB has a structure of $Zn[B_3O_4(OH)_3]$ (i.e., with a mineral formula of $2ZnO \cdot 3B_2O_3 \cdot 3H_2O$) (Schubert et al. 2003). In contrast to previously known zinc borates, this zinc borate is stable to 290°C–300°C. It should be noted that this borate does not have interstitial water molecules (Figure 6.1). Due to their low dehydration temperature, the usage of ZB-223 and ZB-237 are mostly limited to halogen-containing polymers such as flexible polyvinyl chloride (PVC). Due to the demand of high production throughput and thin-walled electrical parts,

TABLE 6.1 Major Commercial Zinc Borates

Formula	Approx. Starting Dehydration Temp. (°C)	Trade Name
$2ZnO \cdot 3B_2O_3 \cdot 7H_2O$	170	ZB-237
$2ZnO \cdot 2B_2O_3 \cdot 3H_2O$	200	ZB-223
$2ZnO \cdot 3B_2O_3 \cdot 3.5H_2O$	290	*Firebrake*®ZB, ZB2335, ZB-467
$2ZnO \cdot 3B_2O_3$	NA	*Firebrake*®500
$4ZnO \cdot B_2O_3 \cdot H_2O$	>415	*Firebrake*®415

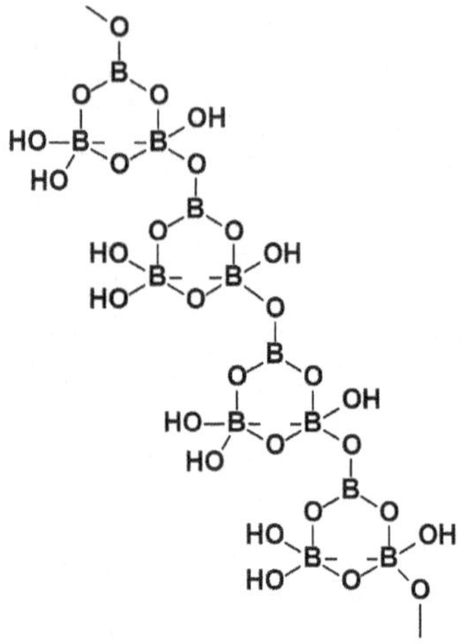

FIGURE 6.1 Molecular structure of Firebrake ZB (zinc atoms complexing with oxygen atoms are not displayed).

engineering plastics are processed at increasingly higher temperatures. To meet this market demand, U.S. Borax also developed an anhydrous zinc borate, *Firebrake*®500 ($2ZnO \cdot 3B_2O_3$), that is stable to at least 500°C and *Firebrake*®415 ($4ZnO \cdot B_2O_3 \cdot H_2O$) that is stable to >415°C.

In recent years, there are many nano-sized zinc borates with different morphologies such as whisker, flake, and disc reported. Some may have commercial potential.

6.2.2.3.1.1 Epoxy   In halogen-containing epoxy, *Firebrake* ZB is an effective flame retardant and smoke suppressant, particularly with an aliphatic halogen source (Shen 2002).

In halogen-free epoxy, for example, it has been used as a sole flame retardant in semi-conductor encapsolant application (Kamikooryma et al. 1994). In epoxy intumescent coatings for hydrocarbon fires, zinc borate is commonly used in conjunction with APP, silica, ATH, etc. for strong char formation (also see Section 6.2.3.1 of Ammonium Pentaborate). Interestingly enough, a combination of zinc borate and antimony trioxide was demonstrated to have good flame retardancy and remarkable smoke reduction in epoxy-based IFR coating (Zhang 2016a).

The use of a combination of ATH and zinc borate in epoxy for aerospace applications was reported. In Cone Calorimeter tests, it was demonstrated that zinc borate promotes the formation of a strong char *via* sintering with alumina and results in substantial decrease of PHRR and average Heat Release Rate (aveHRR) of epoxy resin (Formicola et al. 2009). Synergy was observed when the total loading reaches 20% or higher at 1:1 ratio of zinc borate: ATH.

Dow reported the use of a combination of ATH, zinc borate, and melamine in epoxy structural adhesives for bonding metals. The adhesive strongly resists ignition and does not interfere significantly with mid-frequency direct current (DC) welding performance (Eagle et al. 2016). TAD (Figure 6.2) is a reaction product of triallylisocyanurate and 9, 10-dihydro-9-oxa-10-phosphaphenanthrene-10-oxide (DOPO).

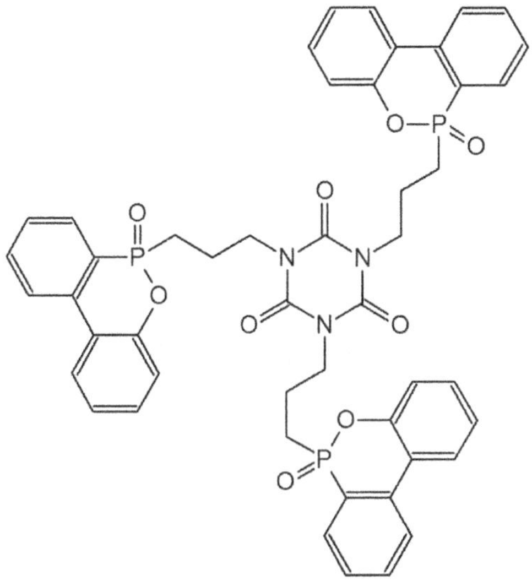

FIGURE 6.2 TAD-reaction product of triallylisocyanurate and DOPO.

Zinc borate or boric oxide (1.5%) in epoxy/4,4′-diamino diphenyl sulfone containing TAD (4.5%) can efficiently improve the UL-94 performance (V-0 at 3.2 mm). However, boron phosphate in this case displays a negative effect on UL-94 test (Tang et al. 2017).

Boron compounds are known to inhibit thermal oxidation of carbon species. Interestingly enough, in a carbon fiber reinforced epoxy containing ATH or MDH, it was demonstrated that addition of zinc borate (almost independent of the loading level) can minimize the respirable carbon fiber formation during polymer combustion (Eibl 2017).

6.2.2.3.1.2 Phenolics Phenolics—Zinc borate in combination with MDH and boehmite was reported for use in glass fiber-reinforced phenolics that have high heat resistance and V-0 rating. Most of all, it can achieve CTI—level 2 rating for packaging requirement of motor parts (Weng 2013). The zinc borate and ATH combination was also used in phenolics for molding electrical parts.

6.2.2.3.1.3 Polyamides Zinc borates (*Firebrake* ZB and *Firebrake* 500) have been used extensively in both halogen-containing and halogen-free glass-reinforced PA. Depending on the processing temperature, *Firebrake* 500 has been used in PA, particularly in high temperature PA (PPA) and PA 66.

> *Halogen-Containing PA*: A combination of brominated polystyrene, zinc borate, and antimony trioxide can provide good CTI and V-0 performance. For an antimony-free formulation, one can still maintain V-0 rating with the use of zinc borate, but at slightly higher bromine content. In high temperature PA, *Firebrake* 500 can replace antimony trioxide completely.
>
> *Halogen-free PA*: Zinc borate in PA 66 containing Red P cannot only displays synergy in fire retardancy, but also imparts corrosion resistance toward metals such as copper (Bonin and LeBlanc 1995). Zinc borate can trap trace amounts of phosphine (PH_3) derived from red phosphorus. Partial replacement of melamine pyrophosphate with zinc borate in glass-reinforced polyamide 66 cannot only maintain V-0 (1.6 mm), but also increase the CTI from 250 to 600 volts (Martens et al. 1997).

Aluminum diethylphosphinate (DEPAL) offered by Clariant is the dominant fire retardant on the market for halogen-free PAs and polyesters. DEPAL does not melt before it starts to decompose/vaporize at temperatures >350°C (Hoerold and Schacker 2007). In 25% glass-reinforced PA 66, OP1312 (a blend of DEPAL, MPP, and zinc borate) can achieve V-0 at 1.6 mm with 16% total loading and 5VA 18% loadings.

TABLE 6.2	PA 66 with Fiber Glass—UL-94 and Glow Wire Performances

	Examples (wt.%)		
Components	1	2	3
Polyamide 66	50	50	50
Glass Fibers	30	30	30
Exolit OP1312[a]	20	—	—
Exolit OP1314[a]	—	20	—
Exolit OP 1400[b]			20
Properties			
UL-94 (0.4 mm)	V-0 (33 s)	V-0 (35 s)	V-0 (30 s)
UL-94 (0.4 mm) (168 hr/70°C)	V-0 (26 s)	V-0 (40 s)	V-0 (44 s)
UL-94 (0.8 mm)	V-0 (24 s)	V-0 (24 s)	V-0 (31 s)
UL-94 (0.8 mm) (168 hr/70°C)	V-0 (44 s)	V-0 (39 s)	V-0 (24 s)
GWIT (0.8 mm)	725°C	700°C	700°C
GWIT (1.5 mm)	750°C	700°C	700°C
GWIT (3.0 mm)	800°C	750°C	750°C
GWFI (0.8–3.0 mm)	960°C	960°C	960°C

[a] Exolit OP1312 and 1314 contain DEPAL, MPP, and zinc borate.
[b] Exolit OP1400 contains two proprietary phosphorus compounds.

Recently, it was reported that a combination of DEPAL and aluminum ethylphosphinate in the presence of MPP/*Firebrake* ZB (or *Firebrake* 500) provides not only good flame retardancy, but also no discoloration, no extrudation, and low melt index (i.e., no degradation during processing) (Hoerold and Hill 2015). It was demonstrated that the OP1312 package in PA 66/glass fiber (GF) outperforms the brominated system in terms of peak heat release, time to ignition, and fire growth index (Isitma et al. 2010). The zinc borate in OP1312 is also a thermal stabilizer. OP1312 in PA is highlighted by a low loading level (compared to about 27% with brominated polystyrene/Sb_2O_3 system), low material density, and a high CTI (>600 volts), and a good glow wire ignition temperature GWIT (Narayan 2018) (Table 6.2). It was also reported that this zinc borate can also improve corrosion resistance of DEPAL in high temperature PPA (Yin 2009). Clariant recently reported that PA 66 containing DEPAL (11.4%), MPP (5.7%), talc (10%), *Firebrake* 500, and glass fiber (20%) can achieve UL-94 V-0 at 0.6 mm and GWIT of 775°C at 1 mm and 800°C at 2 mm (Nass et al. 2007). The addition of MPP and zinc borate to DEPAL in PA causes the change of mode of action from gas phase to a mixed gas/condensed phase. The formation of boron phosphate and aluminum phosphate in the condensed phase was demonstrated (Braun et al. 2007). Hexaphenoxycyclophosphazene (HPCP) (Figure 6.3) is a very thermally and hydrolytically stable phosphorus/nitrogen (P/N) flame retardant. In PA 66 containing OP1312 (DEPAL/MPP/zinc borate), addition of HPCP can significantly improve UL-94 classification in particular for 0.4 mm (7 days, 70°C) (Konig 2014).

6.2.2.3.1.4 Polycarbonates	In polycarbonates (PCs), a combination of core-shell polymer (5.75%), BDP [bis-phenol-A bis(diphenyl phosphate)] (12.5%), polytetrafluoroethylene/styrene-acrylonitrile (PTFE-SAN) (0.8%), and the zinc borate (1.48%) provides UL-94 V-0 (1.5 mm), 5VA (2.3 mm), and good physical properties (Rogunova 2009). Synergism of the zinc borate and polysiloxane in PC were reported for achieving V-0 at 1.6 mm (Yang et al. 2013). Zinc borate in PC containing BDP and with a core/shell modifier can lower the PHRR and provides a drastic smoke reduction per ASTM E662 (Li 2016).

Adding various metal borates (calcium borate, magnesium borate, and zinc borate) to PC/silicone rubber/BDP was evaluated. These borates can change the pyrolysis and flame retardancy action. Smoke suppression, LOI, as well as UL-94 classification are improved. Zinc borate performed better than magnesium borate and calcium borate (Wawrzyn et al. 2014).

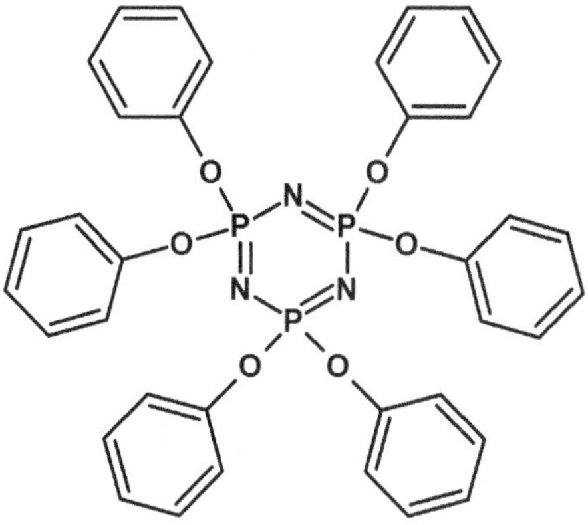

FIGURE 6.3 Chemical structure of hexaphenoxycyclophosphazene.

6.2.2.3.1.5 Polyimide/Polyetherimide/Silicone Polyetherimide In addition to boric acid/boric oxide, the use of nano-zinc borate as a flame retardant in polyimide was observed to enhance thermal stability of the nanocomposite (Koytepe et al. 2009). A flame retardant formulation of silicone polyetherimide containing zinc borate (*Firebrake* 415, 3%) and PTFE (1%) can achieve Flame Spread Index (ASTM E162) of 6 at 3.2 mm and with no dripping (Table 6.3). It is suitable for transportation industry applications (Bhandari and Gallucci 2013).

6.2.2.3.1.6 Polyolefins

  Halogen-Containing System: When used in conjunction with a halogen-based flame retardant, *Firebrake* zinc borate can partially replace antimony oxide (30%–40%) and still maintain the same fire test performance. In addition, it can improve aged elongation properties, increase char formation, and decrease smoke generation. The B_2O_3 moiety in zinc borate can also provide after-glow suppression (Shen et al. 2010).

  Halogen-Free System: Depending on application, high loadings of ATH or MDH are generally required (30–70 wt%) in polyolefins. Recent developmental efforts have focused on co-additives of ATH or MDH with the aim of reducing total loading and developing stronger char/residue formation (Shen 2014b).

TABLE 6.3 Silicone Polyimide Compositions for Transportation

	Examples (parts by wt.)		
Components	1	2	3
Silicone-Polyetherimide (Siltem 1600)	100	96.0	95.3
Zinc Borate (Firebrake 415)	—	3.0	3.0
Properties			
ASTM E-162 (3.2 mm)	Fail	Fail	Pass
Flame Spread Index (Is)	15	6	6
Dripping (Y/N)	Y	Y	N
Comments	Continuous Flaming Drip	6 drips	No Dripping
UL-94 (1.6 mm)			
Flame Out Time (sec.)	>15 s	<10 s	<8 s
Dripping	Y	V-0	V-0

- Zinc borate reduces heat release rate and smoke evolution in most metal hydroxide-containing polymers such as EVA or EVA/polyethylene (PE). The main benefit of the zinc borate is promoting the formation of a strong char/ceramic residue that prevents burning drips, delays oxidative pyrolysis, and protects underlying polymers
- The use of co-additives such as silicone/silica, melamine polyphosphate, phosphate ester, red phosphorus, ammonium polyphosphate, melamine cyanurate, and nanoclay can augment the fire test performance of zinc borate/metal hydroxide system
- Partial substitution of MDH with the fine grade of zinc borate (5–10 wt %) in EVA (28% VA) resulted in not only HRR reduction, but also better electrical properties/physical properties (except tensile strength) (Shen et al. 2006). In addition, there was a significant torque reduction during compounding, an indication of improvement in processability
- Partial replacement of ATH with zinc borate in EVA reduces carbon monoxide yield under fuel rich conditions in the Purser furnace (Hull et al. 2002).

Addition of MDH (240 phr), zinc borate (30), and silicone powder (30) to EVA, modified with propylene/ethylene/butene co-polymer (80:20 ratio), can meet vertical wire burn (VW-1) and good physical properties in wire and cable applications (Noda et al. 2014). Brucite (a natural magnesium hydroxide) surface was deposited with fibrous nano-zinc borate ($12ZnO \cdot 3B_2O_3 \cdot H_2O$). This hybrid material (brucite to the zinc borate ratio is 39.5:5.5) in EVA at 45% loading can meet V-0 (3.0 mm). In addition, compared to brucite alone in EVA, it can increase the tensile strength by 20% (Wang et al. 2014). *Polypropylene*: In halogen-containing PP, zinc borate can partially replace antimony trioxide and provide good afterglow control. In halogen-free system, the use of a combination of silicone, *Firebrake* ZB, ATH, and a zinc or sodium ionomer in PP achieves low HRR and smoke generation (Paul 2015). The preparation of a neutralized intumescent flame retardant by reacting pentaerythritol, phosphoryl trichloride, and melamine was reported. The addition of a small amount of *Firebrake* 500 (anhydrous zinc borate with 2 wt% at the expense of neutralized intumescent flame retardant) in PP leads to a dramatic improvement in the fire test performances (LOI and V-0 rating). The zinc borate promotes the formation of a cohesive intumescent barrier (Fontaine et al. 2008).

The use of a combination of MDH, ATH, zinc borate, expandable graphite, and optionally antimony trioxide in polyethylene or high density polyethylene (HDPE) was also reported for manufacturing radio frequency transparent pallets that give low rate of heat release in the ASTM E1354 test (Muirhead 2011).

6.2.2.3.1.7 Polyphenylene Ether In a cross-linked PPE/PS printed wire board application (Table 6.4), a combination of phosphate ester (PX200) and zinc borate (*Firebrake* ZB-Fine) can provide a V-0 at 1.6 mm, good dielectric constant, insulation resistance, and soldering heat resistance with a glass transition temperature (Tg) of 165°C (Satoshi and Kazuhiko 2009).

6.2.2.3.1.8 PVC Zinc borate is widely used commercially in many PVC applications such as wall covering, wire and cable, coated fabrics, carpet backing, etc.

Rigid PVC: Firebrake ZB can outperform antimony trioxide in rigid PVC as evidenced by LOI.
Flexible PVC: Depending on the plasticizer, filler, and halogen content, *Firebrake* ZB can either partially or completely replace antimony oxide. In addition, it can function as a smoke suppressant and afterglow suppressant (Shen et al. 2006).

6.2.2.3.1.9 Elastomers (SBR, SEBS, SEPS, Silicone) Zinc borates have commonly been used in halogen and halogen-free elastomers (such as ethylene-propylene-diene-monomer(EPDM), ethylene-propylene (EP), styrene-butadiene rubber (SBR), neoprene, and silicone). The applications include conveyor belting, flooring, wire and cable, electrical insulation, door seals, etc. (Shen et al. 2006). The use of a combination of ATH, silica, zinc borate, and zeolite in styrene-ethylene-butylene-styrene (SEBS) and styrene-ethylene-propylene-styrene (SEPS) can achieve good dielectric breakdown strength, track and erosion resistances, in particular physical properties such as 650% elongation (Wu et al. 2017).

TABLE 6.4 Printed Wire Board Based on PPO/PS

	Examples (wt.%)		
Components	1	2	3
PPO (MA-modified)	50	50	50
Triallyl Isocyanurate	46	46	46
Polystyrene (GP)	4	4	4
Bis(1,2-pentabromophenyl)ethane	20	—	—
Sb_2O_3	4	—	—
Phosphate Esters (PX-200)	—	15	15
Silica (7 microns)	15	15	—
Firebrake ZB-Fine	—	—	15
Peroxide	6	6	6
Properties			
UL-94 (1.6 mm)	V-0	V-0	V-0
Peel Strength (KN/M)	1.2	1.1	1.0
Insulation Resistance (Ω)	1×10^{14}	4×10^{14}	4×10^{14}
Glass Transition (°C)	170	160	165
Dielectric Constant (MHz)	3.6	3.6	3.5
Soldering Heat Resistance (260°C)	Good	Good	Good

In silicone, zinc borate is known to be an effective flame retardant, particularly in the presence of other co-additives such as ATH, calcium carbonate, etc. It is believed that boric oxide from zinc borate may react with silica (derived from silicone combustion) to form borosilicate. The use of a combination of mica (25%)/glass frit containing no borate (1.5%)/MDH (10%) in peroxide-cured silicone to produce a self-supporting ceramic material at 1050°C. Addition of zinc borate (1.5%) resulted in not only better mechanical properties such as less shrinkage at high temperatures, but also much better fire resistance properties as evidenced by the mass loss measurement (Alexander et al. 2010). The composition is suitable for partitions, wall linings, fire door inserts, door seals, or cables.

6.2.2.3.1.10 Unsaturated Polyesters/Vinyl Esters *Firebrake* zinc borate can be used in halogen-containing polyester. For example, in Hetron 92 AT (an unsaturated polyester with chloredic anhydride and 3%–5% Sb_2O_3), the zinc borate can replace antimony trioxide completely and result in good transparency and still meet the Class I in the E84 tunnel test. Unlike antimony trioxide, *Firebrake* ZB has a refractive index similar to that of organic polymer; as a result, they do not induce opacity to polymer products. In halogen-free systems, it was reported that a combination of DEPAL and *Firebrake* zinc borate can achieve V-0 at 1.5 mm thickness and LOI 40% (Knop et al. 2005).

6.2.2.3.1.11 Wood Plastic Composites Zinc borate is an effective wood preservative, known in the trade as *Borogard*® ZB. In PVC/wood composite, zinc borate can also provide some flame retardancy in PVC/wood composite or HDPE/wood composite containing intumescent flame retardants. Most important, it can provide effective smoke suppression. Interestingly enough, it was also reported that the zinc borate in HDPE/wood composite can provide substantial reduction in UV degradation of the resin (Manning et al. 2010).

6.2.2.3.2 Aluminum Borate ($9Al_2O_3 \cdot 2B_2O_3$)

A variety of transition and other metal borates are reported in the literature. These include aluminum borate, silver borate, iron borate, copper borate, nickel borate, strontium borate, lead borate, zirconium borate, etc. In a polybutylene terephthalate (PBT) containing brominated polystyrene and antimony trioxide, aluminum borate whiskers ($9Al_2O_3 \cdot 2B_2O_3$) were used to replace glass fiber resulting in much improved CTI (from 275 to 375 volts) and some flexural modulus improvement and still maintain the V-0 (0.8 mm) performance (Hironaka 2001). Aluminum borate can also replace glass fiber in a non-halogen melamine cyanurate-based PA 66 to achieve excellent mechanical strength, CTI, and better UL-94 test performance (Takeda and Kirikoshi 2002).

6.2.2.3.3 *Manganese Borates (MnO·2B$_2$O$_3$·3H$_2$O)*

Manganese borate is commercially available as a 60:40 blend with calcium sulfate from societa chimica larderello (SCL) for printing ink application. Although MnO has a catalytic effect on IFR flame retardancy, the use of manganese borate as a flame retardant has not been reported.

6.2.3 Boron-Nitrogen Containing Compounds

6.2.3.1 Ammonium Pentaborate [(NH$_4$)$_2$O·5B$_2$O$_3$·8H$_2$O)]

Upon thermal decomposition, ammonium pentaborate (APB) first gives off a large amount of water starting at about 120°C. At about 200°C, it starts to give off ammonia and, at about 450°C, it is all converted to boric oxide which is a glass-former. APB functions as both inorganic blowing agent and a glass-forming fire retardant.

$$\left(NH_4\right)_2O \cdot 5B_2O_3 \cdot 8H_2O \rightarrow 2NH_3 \uparrow + 8H_2O \uparrow + 5B_2O_3$$

APB solution can be sprayed on paper or the paper can be dipped into the solution to yield fire retardant products. In addition to the cellulose, it has also been used as a flame retardant in polyurethane foam and epoxy intumescent coatings. APB's usage in polymers, however, is limited by its high water solubility (10.9%) and low dehydration temperature.

APB is an effective intumescent flame retardant in thermoplastic polyurethane. Partially dehydrated APB at 5–10 phr loading in thermoplastic polyurethane (TPU) can provide 7- to 10-fold improvement in a burn-through test (Myers et al. 1985). The use of APB in conjunction with APP, zinc borate, silica, etc. in a flame-retardant epoxy-intumescent coating was claimed for protection of steel against hydrocarbon fires (Nugent et al. 1992). The use of a combination of APB and ZrO$_2$ at 20% loading (1:1 ratio) in a polyisocyanurate can reduce HRR by 81% in a Cone Calorimeter test at 70 kW/m^2 (Gilman 1998). Interestingly, it was reported that APB in urea formaldehyde adhesive for plywood cannot only increase gel time, but also decrease formaldehyde emission (Gao et al. 2015).

6.2.3.2 Boron Nitride (BN)

Hexagonal boron nitride (h-BN) can be prepared from reaction of boric acid and urea or melamine. For example, pyrolysis of melamine diborate can yield hexagonal boron nitride. It is commonly referred to as "white graphite" because of its platy hexagonal structure similar to graphite. Under high pressure and at 1600°C, h-BN is converted to cubic BN, which has a diamond-like structure. h-BN is stable in inert or reducing atmosphere to about 2700°C and in oxidizing atmosphere to 850°C. It is an excellent thermal conductor and has been frequently cited as functional filler for fire retardant encapsulants for electrical/electronic applications. Single-walled BN nanotubes were recently reported. They are far more resistant to oxidation than carbon nanotubes (Bengu and Marks 2001). BN nanotube (0.01 wt%) was reported to improve thermal resistance of polybenzimidazole (Hiroaki et al. 2008). It was also claimed that a boron nitride-coated calcium borate in combination with spherical alumina can be used in epoxy printed circuit boards having a high glass transition temperature, peel strength, moisture/heat resistance, and UL-94 V-0 performance at 0.4 mm (Ohtsuka et al. 2013).

The use of exfoliated h-BN nano-sheet as a high performance binder-free fire resistant coating for wood was claimed. It was demonstrated that the oxidation resistance property of h-BN wood coating is stable to 900°C in air (Liu et al. 2017).

6.2.3.3 Guanidinium Borate [x[C(NH$_2$)$_3$]O·yB$_2$O$_3$·zH$_2$O]

Several guanidinium borates have been reported in the literature. For example, guanidinium tetraborate [C(NH$_2$)$_3$]$_2$O.2B$_2$O$_3$.4H$_2$O] can be prepared by reacting guanidinium carbonate with boric acid in hot water or by reacting guanidinium chloride with borax in cold water (Weakley 1985). Guanidium borate or its precursors (BA plus guanidine) are mostly used for flame retardant cellulosic product. For example, Sanyo in Japan reported the combination of calcium imidosulfonic acid and guanidinium borate or guanidinium sulfamate provides excellent flame retardancy in cellulosic materials (Yoshiyuki and Yashushi 1997). A combination of guanidium borate and imidazolium borate was reported for treating cotton fabrics to raise LOI from 18.8% to 24.6% and 25.8%, respectively, at 5 wt% loading (Dogan 2014).

6.2.3.4 Melamine Diborate [(C$_3$H$_8$N$_6$)O·B$_2$O$_3$·2H$_2$O)]/(C$_3$H$_6$N$_6$·2H$_3$BO$_3$)

Melamine diborate (MB, Figure 6.4), known in the fire retardant trade as melamine borate, is a white powder, which can be prepared readily from melamine and boric acid. It is partly soluble in water and acts as an afterglow suppressant and char promoter in cellulosic materials. Budenheim Iberica claims that, in a 1:1 combination with ammonium polyphosphate, MB (10%–15%) can be used for phenolic bound non-woven cotton fibers. In general, MB can be used as a char promoter in intumescent systems for various polymers including polyolefins or elastomers. However, its low dehydration temperature (about 130°C) limits its application in thermoplastics that are processed at above 130°C. MB is also reported to suppress afterglow combustion in flame proofing textiles with ammonium polyphosphate or monoammonium phosphate to meet the German DIN 53,459 and Nordtest NT-Fire 002 (Thijssen and Laser 1992). In EVA (60% VA), it was found that EVA-ATH-MB is protected by both the gas and condensed phase mechanism, whereas MB contributes to the formation of a protective layer due to the formation of boric oxide, boron nitride, and BNO at the surface. The boron species can decrease thermal conductivity in the ignition temperature range leading to delay of time to ignition relative to melamine or melamine phosphate (Hoffendahl et al. 2015).

6.2.4 Boron-Phosphorus Containing Compounds

6.2.4.1 Boron Phosphate (BPO$_4$)

Boron phosphate, an inorganic polymer of empirical formula BPO$_4$, can be prepared by heating phosphoric and boric acids to calcining temperature (Petric et al. 2010). It is a white infusible solid that vaporizes at 1450°C–1462°C without decomposition. As opposed to the tri-valency in boric oxide, most of the boron atoms in boron phosphate are tetra-coordinated (i.e., as a BO$_4^-$ group, Figure 6.5). Boron phosphate is a low cost, low toxicity flame retardant. An amorphous form of boron phosphate is commercially available from Budenheim and Italmatch. Commercially, it is mostly used as an additive in polyphenylene ether alloy. In a PA-PPE alloy, a combination of boron phosphate (5%) and silicone oil (2%) resulted in a V-0 (1.6 mm) rating (Shaw 1998). The use of boron phosphate in conjunction with

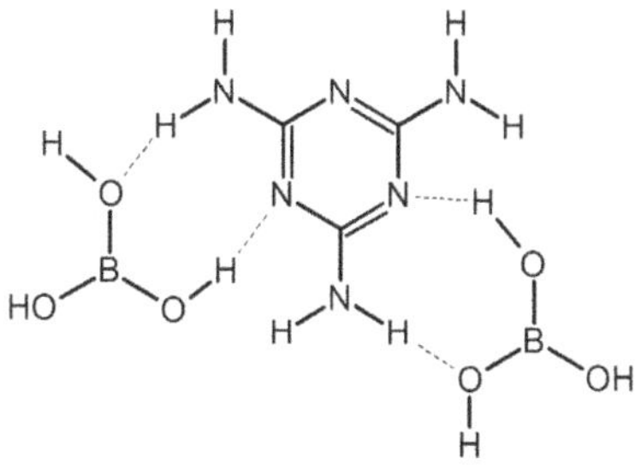

FIGURE 6.4 Chemical structure of melamine diborate.

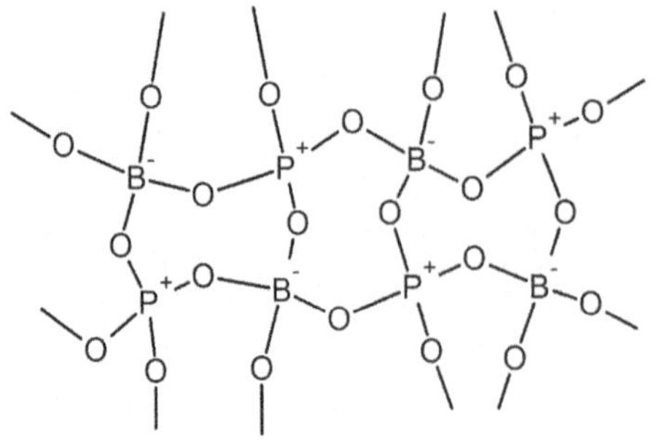

FIGURE 6.5   Chemical structure of boron phosphate.

zinc borate prevents the UL-94 5VA burn-through in PPE/HIPS/SEBS containing triphenyl phosphate (Yin et al. 2003). A patent by Nexans claimed the use of zinc borate or boron phosphate in electrical insulation which is resistant to high temperature such as 1100°C (Demay et al. 2003). The use of boron phosphate as a flame retardant in proton-conducting polyimide for fuel cell membrane application was reported (Cakmakci and Gungor 2013). Recently, it was reported that crystalline boron phosphate can function as a solid acid. In epoxy, it can catalyze the pyrolysis of epoxy at a lower temperature, reduce the release of flammable gas, and promote char formation (Zhou et al. 2014; Li et al. 2017).

6.2.4.2 Metal Borophosphates

BASF reported the preparation of different metal borophosphates (Ewald and Hibst 2014). These compounds were found to exhibit high thermal stability and are useful as flame retardants for plastics. Partial replacement of DEPAL with the supposedly less expensive zinc or magnesium borophosphate can still maintain the UL-94 V-0 performance. The advantage of this additive is that it contains both boron and phosphorus. For easy processing, it could replace both melamine polyphosphate and zinc borate.

6.2.4.3 Ammonium Borate Phosphate

It was claimed that boiling a mixture of boric acid and diammonium phosphate can result in a solid that has a flame retardancy effect in unsaturated polyester (Sperber 2017). The solid product was not analyzed and could be some sort of "ammonium borate phosphate" that has never been reported.

6.2.4.4 Melamine Borate Phosphate

Melamine borate phosphate (MBP) was prepared by solvothermal method from melamine/boric acid/phosphoric acid (Figure 6.6). It was used to improve flame retardancy of phenol formaldehyde (PF) foam toughened with polyethylene glycol (PEG) (Hu et al. 2017). MBP (4 phr) can raise the LOI from 37.5% to 45.5%, and the Cone Calorimeter test of PHRR, THR, TSR and mean CO and CO_2 yields of the treated sample are all lower than that of untreated PF foam. In addition, the use of MBP results in increase in flexural strength and compressive strength.

6.2.5 Boron-Silicon Containing Compounds

The use of borosilicate glass/ceramic frits and borosiloxane were previously reviewed (Shen 2014a, 2014b). Borosilicate glass is a range of glasses based on boric oxide, silica, and a metal oxide. It has excellent thermal shock resistance and chemical resistance. Recently, synergistic flame retardant effects of borosilicate hollow glass microsphere and magnesium hydroxide in EVA was reported (Liu et al. 2014). Trovotech in Germany reported that the use of melamine cyanurate (MC, 4%) and their special porous amorphous glass particle (4%) can achieve a non-dripping V-0 in polyamide 6, whereas typical

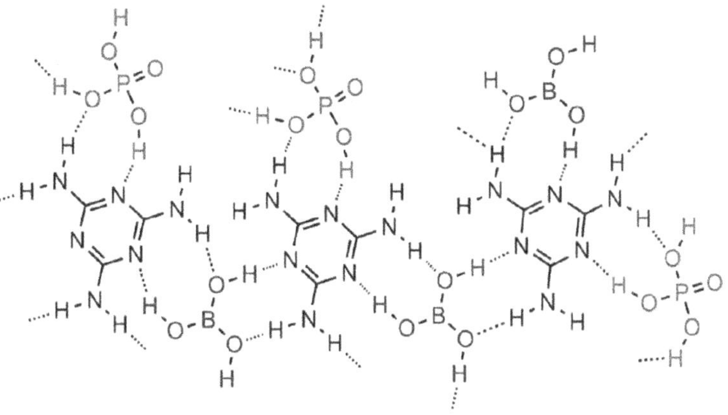

FIGURE 6.6 Proposed structure of melamine borate phosphate. (Reprinted from Hu, L. et al., *Polym. Plast. Technol. Eng.*, 56, 678–686, 2017. With permission.)

polyamide 6 formulation requires about 10–12 loading of MC to achieve V-0 with flaming drips (Hans-Juergen and Ferner 2015). These porous borosilicate glass particles can sinter at heat treatment temperature around 360°C–400°C. This sintering action is believed to be responsible for the flame retardancy.

6.2.6 Boron-Carbon Containing Compounds

6.2.6.1 Arylboronic Acid [ArB(OH)$_2$]

Unlike the B–O–C bond in boric acid ester, boronic acids with the B–C bond are hydrolytically stable. There have been many publications on the use of boronic acid derivative as flame retardants. They are known to release water on thermolysis, thereby leading to the formation of boroxines or boronic anhydrides. For example, chemical modification of PS by introducing boronic acid functionality (–B(OH)$_2$) to predominately the para-position improves fire retardancy. It was reported that the boronated PS is significantly less flammable than the unmodified PS. The virgin PS has a LOI of 18.3% vs. 25.3% for the boronated PS (degree of substitution 9.2%). The char yield also increased from <1% to 7% at 600°C. It was postulated that the boronic acid groups are active in the condensed phase by assisting intermolecular crosslinking, forming the 6-membered boroxine ring, and promoting the formation of a protective char layer (Armitage et al. 1996). It was reported that 10 mole% of a boronic acid derivative (two boronic acid/pinacol moieties attached to dimethyl terephthalate) can reduce total heat release reduction by 20% in a flexible polyurethane formulation (Benin et al. 2014).

Phenylboronic acid was also successfully bonded to polyethyleneimine. This polymeric product was applied on cotton fabrics at 33.8% loading to achieve self-extinguishing. Further improvement of the coating-washing durability was achieved by a novel formaldehyde-free crosslinking treatment (a phosphazene acrylate derivative) The new coating on cotton can achieve LOI values 29.6% and 23.2% before and after repeated launderings, respectively (30 wt% loading) (Chan et al. 2017). The use of a combination of polydopamine and a reaction product of azo-boronic acid and bis-phenol-A in polylactic acid (PLA) was also reported. LOI of 23.7% and V-0 were achieved at 10 wt% loading (Tawiah et al. 2017). A combination of the aromatic boronic acid derivative 2,4,6,-tris(4-boronic-2-thiophene)-1,3,5-triazine and MDH was thoroughly evaluated in an epoxy resin. The synergistic combination at 10 phr each can achieve V-0 at 3.2 mm, LOI 32.5%, and significantly better of PHRR and THR in Cone Calorimeter test at 35 kW/m^2 (Zhang et al. 2016a, 2016b).

6.2.6.2 Boric Acid Esters [B(OR)$_3$]

In general, boric acid esters (also called boresters) derived from alcohols and boric acid are hydrolytically unstable for general use in plastics/elastomers, wood, paper, or cotton. However, pressure treating of wood with trimethylborate [B(OCH$_3$)$_3$, bp 68°C–69°C] is reported to be effective in rendering it fire resistant. In this case, once trimethylborate penetrates the wood, it will revert back to boric acid in the presence of moisture.

6.2.6.3 Boron Carbide (B$_4$C)

Boron carbide is produced industrially by the carbo-thermal reduction of B$_2$O$_3$ in an electric arc furnace. It is a black powder and has a melting point of 2445°C. It was reported (Kobayashi et al. 1995) that the addition of boron carbide (10–15 wt%) in a variety of intumescent coatings containing ammonium polyphosphate, and blowing agent resulted in improving weight retention, compression strength, and peel strength during fire. More important, it can retard oxidative weight loss of the foamed layer at 1000°C or even higher. The effect of boron carbide nanoparticles on fire resistance of carbon fiber/epoxy composite was evaluated (Rallini et al. 2013). The flame resistance was evaluated through residual mechanical properties after the exposure of the specimens to a direct torch with a heat flux of 500 kW/m^2. It was demonstrated that boron carbide improves the thermal stability of the composite and inhibits the thermal oxidation of the carbon fiber.

6.3 Conclusions

Boron-based flame retardants are multi-functional fire retardants and can function as flame retardants, smoke suppressants, afterglow suppressants, anti-tracking/anti-arcing agents, etc. They function as flame retardants in the condensed phase. They can be used in both halogen-containing and halogen-free polymer systems.

Borates have been commercially successful in both plastics and rubbers. The fire test performance of borates can be enhanced with the use of co-additives such as metal hydroxides, and N-, P-, and Si-containing compounds. Borates alone can also function as flame retardant in char-forming polymers such as polyethersulfone, polyimide, etc.

Future FR development will be focused on phosphorus, N/P compounds, silicon-based, and boron-based compounds. Taking advantage of synergistic/beneficial effects of interaction among additives, future endeavors will also be focused on developing a flame retardant package instead of using just one or two flame retardant additives.

References

Alexander, G., Graeme, A., Cheng, Y. et al. 2010. Fire resistant silicone polymer compositions. U.S. Patent 7,652,090.

Armitage, P., Ebdon, J. R., Hunt, B. J. et al. 1996. Chemical modification of polymers to improve flame retardance-I. The influence of boron containing groups. *Polymer Degradation and Stability* 54: 387–393.

Bassett, W. H. 1992. Non-halogen, flame retardant engineering thermoplastics. *Third Annual BCC Conference on Flame Retardancy*, Stamford, CT.

Benin, V., Gardelle, B., and Morgan, A. B. 2014. Heat release of polyurethanes containing potential flame retardants based on boron and phosphorus chemistries. *Polymer Degradation and Stability* 106: 108–121.

Bengu, E., and Marks, L. D. 2001. Single-walled nanostructures. *Physical Review Letters* 86: 2385.

Bhandari, Y. J., and Gallucci, R. R. 2013. Silicone polyimide compositions with improved flame retardance. U.S. Patent 8,349,933.

Bonin, Y., and LeBlanc, J. 1995. Fire retardant, noncorrosive polyamide composition. U.S. Patent 5,466,741.

Bracegirdle, B., Clarke, S., and McMillan, S. 2017. Foam composite. WO 2017136878.

Bracegirdle, B., Monti, M., Smyth, G. et al. 2016. Foam composite. U.S. Patent Appl. 20160053065.

Braun, U., Schartel, B., Ficherra, M. A. et al. 2007. Flame retardancy mechanisms of aluminum phosphinate in combination with melamine polyphosphate and zinc borate in glass-fiber reinforced polyamide 66. *Polymer Degradation and Stability* 92: 1528–1545.

Cakmakci, E., and Gungor, A. 2013. Preparation and characterization of flame retardant and proton conducting boron phosphate/polyimide composites. *Polymer Degradation and Stability* 98: 927–933.

Cavodeau, F., Viretto, A., Otazaghine, B. et al. 2017. Influence of colemanite on the fire retardancy of ethylene-vinyl acetate and ethylene-methyl methacrylate copolymer. *Polymer Degradation and Stability* 144: 401–410.

Chan, S. Y., Si, L., Lee, K. I. et al. 2018. A novel boron-nitrogen intumescent flame retardant coating on cotton with improved washing durability. *Cellulose* 25: 843–857.

Crompton, G. 1992. Materials for and manufacture of fire and heat resistant components. U.S. Patent 5,082,494.

Claesen, C. A., Lohmeijier, J. H. G., Boogers, M. P. et al. 1993. Polymer containing fluorinated polymer and boron compound. U.S. Patent 5,182,325.

Curzon, J. L., and Smoot, T. W. 2010. Flame retardant and microbe inhibiting method and compositions. U.S. Patent 7,767,010.

Davis, R., Li, Y., Gervasio, M. et al. 2015. One-pot, bioinspired coatings to reduce flammability of flexible polyurethane foams. *ACS Applied Materials & Interfaces* 7: 6082–6092.

Demay, J., Lejeune, M., Gardelein, M. et al. 2003. Insulating composition for a security electric cable. U.S. Patent Appl. 20030199623.

Dogan, M. 2014. Thermal stability and flame retardancy of guanidium and imidazolium borate finished cotton fabrics. *Journal of Thermal Analysis and Calorimetry* 118: 93–98.

Eagle, G. G., Lutz, A., and Jialanella, G. L. 2016. Flame retardant structural epoxy resin adhesives and process for bonding metal members. U.S. Patent 9,346,983.

Eibl, S. 2017. Potential for the formation of respirable fibers in carbon fiber reinforced plastic materials after combustion. *Fire and Materials* 41: 808–816.

Ewald, B., and Hibst, H. 2014. Borophosphate, borate phosphate, and metal borophosphate as flame proofing additives for plastics. U.S. Patent 8,841,373.

Fontaine, G., Bourbigot, S., and Duquesne, S. 2008. Neutralized flame retardant phosphorus agent: Facile synthesis, reaction to fire in PP and synergy with zinc borate. *Polymer Degradation and Stability* 93: 68–76.

Formicola, A. D., De Fenzo, A., Zarrelli, M. et al. 2009. Synergistic effects of zinc borate and aluminum trihydroxide on flammability behavior of aerospace epoxy system. *Express Polymer Letters* 3: 376–384.

Fujimori, T. 2002. Foamed polystyrene products and method for their production. U.S. Patent 6,344,267.

Fujimori, T. 2006. Process for producing foam. WO Patent Appl. 2006043435.

Gao, W., Du, G., and Kamdem, D. P. 2015. Influence of ammonium pentaborate (APB) on the performance of urea formaldehyde (UF) adhesives for plywood. *Journal of Adhesion* 91: 186–196.

Geng, Y. J., and Liu, Z. H. 2017. Preparation and thermodynamic characterization of $2CaO \cdot B_2O_3 \cdot H_2O$ nanomaterials with enhanced flame retardant properties. *Colloids and Surfaces A: Physical and Chemical Aspects* 522: 563–568.

Gilman, J. 1998. New non-halogenated fire retardant for commodity and engineering polymers. WO 98/47980.

Grube, L. L., and Frankoski, S. P. 1992. Fire retardant bitumen. U.S. Patent 5,110,674.

Hanafin, J., and Bertrand, D. C. 2000. Low density light weight intumescent coating. U.S. Patent 6,096,812.

Hans-Juergen, V., and Ferner, U. 2015. Flame retardant composition for thermoplastic polymers consisting of porous, amorphous glass powder and melamine cyanurate. U.S. Patent Appl. 2015/0018453.

Haruomi, H. 2003. Epoxy resin composition and electrical/electronic part equipment. JP 2003096272.

Herve, P. 2013. Method for the treatment of wood by silicate borate pairing. U.S. Patent Appl. 2013/0319286.

Hiroaki, K., Yoshio, B., Chunyi, Z. et al. 2008. Heat resistant composite composition and method for making the same. Jpn. Kokai Tokkyo Koho JP 2008007699.

Hironaka, H. 2001. Reinforced and flame retarded thermoplastic resin composition and process for producing the same. U.S. Patent 6,221,947.

Hoeks, T. L., Claesen, C. A., Lohmeijer, J. H. G. et al. 1997. Thermoplastic composition with boron compound. U.S. Patent 5,648,415.

Hoerold, S., and Hill. M. 2015. Flameproof agent stabilizer combination for thermoplastic and duroplastic polymers. U.S. Patent 9,068,061.

Hoerold, S., and Schacker, O. 2007. Flame retardant and stabilizer combined for thermoplastics polymers. U.S. Patent, 7,255,814.

Hoffendahl, C., Fontaine, G., Duquesne, S. et al. 2015. The combination of aluminum trihdrate and melamine borate as fire retardant additives for elastomeric ethylene vinyl-acetate. *Polymer Degradation and Stability* 115: 77–88.

Hu, L., Wang, Z., and Zhao, Q. 2017. Flame retardant and mechanical properties of toughened phenolic foams containing a melamine phosphate borate. *Polymer-Plastics Technology and Engineering* 56: 678–686.

Hull, T. R., Quinn, R. E., Areri, I. G. et al. 2002. Combustion toxicity of fire retarded EVA. *Polymer Degradation and Stability* 77: 235–242.

Ishii, T., Kokaku, M., Nagsi, A. et al. 2006. Calcium borate flame retardation system for epoxy molding compound. *Polymer Engineering & Science* 46: 799–806.

Isitma, N. A., Gunduz, H. O., and Kaynak, C. 2010. Halogen-free flame retardants that outperform halogenated counterparts in glass fiber reinforced polyamides. *Journal of Fire Sciences* 28: 87–98.

Jian, S., Liu., S., Chen, L. et al. 2017. Nano-boria reinforced polyimide composition with greatly enhanced thermal and mechanical properties via in-situ thermal conversion of boric acid. *Composite Communications* 3: 14–17.

Jimenez, M., Duquesne, S., and Bourbigot, S. 2006. Characterization of the performance of an intumescent fire protective coating. *Surface & Coatings Technology* 201: 979–987.

Kamikooryma, Y., Hirose, M., and Koshibie, S. 1994. Production of flame resistant epoxy resin composition. JP 06,107,914.

Khadbai, A., Sparling, J. D., Anderson, J. B. et al. 2014. Method of preparation and applications for flame retarding compositions. U.S. Patent 8,814,998.

Khan, A. G., Bruns, J. R., Rains, R. D. et al. 2000. Roll roofing membrane. U.S. Patent 6,134,856.

Knop, S., Sicken, M., and Hoerold, S. 2005. Flame retardant thermoset compositions. U.S. Patent Appl. 2005/0101708.

Kobayashi, N., Yoshida, K., Sagawa, K. et al. 1995. Intumescent fire resistant coating, fire Resistant material, and process. U.S. Patent, 5,401,793.

Konig, A. 2014. Flame retardant thermoplastic molding composition. U.S. Patent 8,653,168.

Koytepe, S., Vural, S., and Seçkin, T. 2009. Molecular design of nano-metric zinc borate containing polyimide as a route to flame retardant materials. *Materials Research Bulletin* 44: 369–376.

Li, X. 2016. Flame retardant polycarbonate. U.S. Patent [Appl.] 2016/0032097.

Li, Y., Wang, Y., Yang, X. et al. 2017. Acidity regulation of boron phosphate flame retardant and its catalyzing carbonization mechanism in epoxy resin. *Journal of Thermal Analysis and Calorimetry* 129: 1481–1494.

Liu, J., Kutty, R. G., Zheng, Q. et al. 2017. Hexagonal boron nitride nanosheet as high-performance binder-free fire resistant wood coatings. *Small* 13: 1602456.

Liu, L., Hu, J., Zhuo, J. et al. 2014. Synergistic flame retardant effects between hollow glass microsphere and magnesium hydroxide in ethylene-vinyl-acetate. *Polymer Degradation and Stability* 104: 87–94.

Manning, M., Gnatowski, M. J., Mah, C. et al. 2010. Performance enhancement in the stabilization of organic materials. U.S. Patent 7,691,932.

Martens, M. M., Kasowski, R., Cosstick, R. et al. 1997. Polyamide or polyester compositions. WO 97/23565.

Muirhead, S. A. 2011. Flame retardant-containing shipping container. U.S. Patent Appl. 20110187022.

Myers, R. E., Dicksons, E. D., Licursi, E. et al. 1985. Ammonium pentaborate: An intumescent flame retardant for thermoplastic polyurethanes. *Journal of Fire Science* 3: 432–449.

Narayan, S. 2018. Organophosphinate flame retardants for versatile polymer application. *29th Annual Conference on Recent Advances in Flame Retardancy of Polymeric Materials*, Stamford, CT, May 20–23.

Nass, B., Hoerold, S., and Schaker, O. 2007. Polymeric molding compositions based on thermoplastic polyamide. U.S. Patent Appl. 20070072967.

Nies, N., and Hulbert, P. 1972. Zinc borate of low hydration and method for preparing same. U.S. Patent 3,649,172.

Nine, J. M., Tran, D. N. H., Kabiri, S. et al. 2017. Graphene-borate as an efficient fire retardant for cellulosic materials with multiple and synergistic modes of action. *ACS Applied Materials & Interfaces* 9: 10160–10168.

Nobuyoki, T., Yasuaki, N., Takashi, H. et al. 2002. Resol-type phenolic resin compound for manufacturing phenolic. Japanese Patent 2002241530.

Noda, K., Yamaguchi, M., Komya, K. et al. 2014. Flame-retardant resin composition, method, for producing same, molded body of same, and electric wire. EP Patent 2805994A1.

Nugent, R. M., Ward, T. A., Greigger, P. P. et al. 1992. Flexible intumescent coating composition. U.S. Patent 5,108,832.

Ohtsuka, H., Ueyama, D., and Sogame, M. 2013. Resin composition, and prepreg and laminate sheet containing the same. TW 201302885.

Parker, E. M., Winterowd, J. G., and Robak, G. 2014. Water resistant low flame spread intumescent fire retardant coating. U.S. Patent Appl. 2014/0295164.

Paul, S. 2015. Flame retardant additive for polymers, free of halogens, antimony oxide, and phosphorus containing substance. U.S. Patent 8,975320.

Peng, H. Q., Wang, D. Y., Zhou, Q. et al. 2008. A novel charring agent containing caged bicyclic phosphate and its application in intumescent flame retardant polypropylene systems. *Journal of Industrial and Engineering Chemistry* 14: 589–595.

Petric, M, Grozav, M, and Ilia, G. 2010. Boron phosphate flame-retardant for certain resins. *Revista de Chimie* 61: 1183–1185.

Rallini, M., Natali, M., Kenny, J. M. et al. 2013. Effect of boron carbide nanoparticles on the fire reaction and fire resistance of carbon fiber/epoxy composites. *Polymer* 54: 5154–5165.

Rogunova, M. 2009. Impact resistant, flame retardant thermoplastic molding composition. US Patent 20090143513.

Saratani, G. 2011. Cover for conveyor. Japan Patent 2011026454.

Satoshi, M., and Kazuhiko, O. 2009. Polyphenylene ether resin composition, prepreg containing the same, laminated plate, copper-clad laminated plate and printed circuit board. Japanese Patent 4255759.

Schmittmann, H. B., and Their, A. 1984. Fire retardant and compounds based thereon. U.S. Patent 4,438,028 (1984).

Schubert, D. M., Alam, F., Visi, M. Z. et al. 2003. Structural characterization and chemistry of the industrially important zinc borate, $Zn[B_3O_4(OH)_3]$. *Chemistry of Materials* 15: 866–871.

Shaw, J. P. 1998. Flame retardant polyamide-polyphenylene ether compositions. U.S. Patent 5,714,550.

Shen, K. K. 2000. Zinc borates- 30 years of successful development as multifunctional fire retardants. *Polymer Materials Science & Engineering* 83: 64–67.

Shen, K. K. 2002. Zinc borates as multifunctional fire retardants in epoxies. *Thirteenth Annual BCC Conference on Flame Retardancy*, Stamford, CT.

Shen, K. K. 2006. Overview of flame retardancy and smoke suppressant in flexible PVC. *Society of Plastics Engineering Vinyltech Conference*, Atlanta, GA.

Shen, K. K. 2014a. Boron-based flame retardants in non-halogen based polymers. In *Non-halogenated Flame Retardant Handbook*, ed. A. B. Morgan and C. A. Wilkie, Chapter 6, pp. 201–235. Hoboken, NJ: Scrivener Publishing/Wiley.

Shen, K. K. 2014b. Review of recent advances on the use of boron-based flame retardants. In *Polymer Green Flame Retardants*, ed. C. D. Papaspyrides and P. Kiliaris, Chapter 11, pp. 367–388. Amsterdam, the Netherlands: Elsevier.

Shen, K. K., Kochefahani, S. H., and Jouffret, F. 2010. Boron-based flame retardants and fire retardancy. In *Fire Retardancy of Polymeric Materials*, 2nd ed., ed. C. A. Wilkie and A. B. Morgan, Chapter 9, pp. 207–237. Boca Raton, FL: CRC Press (Taylor & Francis Group).

Shen, K. K., Kochesfahani, S. H., and Jouffret, F. 2008. Zinc borates as multifunctional polymer additives. *Polymers for Advanced Technology* 19: 469–474.

Shen, K.K., Olson, E., Amigouet, P. et al. 2006. Recent advances on the use of metal hydroxides and borates as fire retardants in halogen-free polyolefins. *The 17th Annual BCC Conference on Flame Retardancy*, Stamford, Connecticut.

Song, S., Nai, X., Li, W. et al. 2012. Hydrothermal synthesis of calcium borate whiskers. *Advanced Material Research* 399–401: 693–697.

Sperber, D. S. 2017. Flame-retardant formulations and methods relating thereto. U.S. Patent Appl. 2017/0066970.

Takeda, T., and Kirikoshi, K. H. 2002. U.S. Patent Appl. 2002/0002228.

Tang, S., Qian, L., and Dong, Y. 2017. Synergistic flame retardant and mechanism of boron compounds on phosphaphenathrene containing epoxy resins. *Proceedings of FRPM 2017*, Manchester, UK, July 6: paper C-MF-1.

Tawiah, B., Yu, B., and Fei, B. 2017. Synthesis and application of synergistic azo-boron/polydopamine flame retardants in poly(lactic acid). *Proceedings of FRPM 2017*, Manchester, UK, July 3: paper FRH-1.

Thijssen, S., and Laser, J. 1992. Flameproof product. U.S. Patent 5,082,727.

Tsuyumoto, I., Onoda, Y., Hashizume, F. et al. 2011. Flame retardant rigid polyurethane foams prepared with amorphous sodium polyphosphate. *Journal of Applied Polymer Science* 122: 1707–1711.

Ullah, S., Ahmad, F., and Megat-Yusoff, P. 2011. Effect of boric acid with kaolin clay on the thermal degradation of intumescent fire retardant coating. *Journal of Applied Sciences* 11(21): 3645–3649.

Ullah, S., Ahmad, F., and Schariff, A. M. 2017. Effects of ammonium polyphosphate and boric acid on the thermal degradation of an intumescent fire retardant coating. *Progress in Organic Coatings* 109: 70–82.

Wagner, J., Deglmann, P., Fuchs, S. et al. 2016. A flame retardant synergism of organic disulfide and phosphorus compounds. *Polymer Degradation and Stability* 129: 63–76.

Wang, X., Pang, H., Chen, W. et al. 2014. Controllable fabrication of zinc borate hierarchical nanostructure on brucite surface for enhanced mechanical properties and flame retardant behaviors. *ACS Applied Materials & Interfaces* 6: 7223–7235.

Wawrzyn, E., Schartel, B., Karrasch, A. et al. 2014. Flame-retarded bisphenol A Polycarbonate/silicon rubber/bisphenol A bis(diphenyl phosphate): adding inorganic additives. *Polymer Degradation and Stability* 106: 74–87.

Weakley, T. J. R. 1985. Guanidinium tetraborate. *Acta Crystallographica* C41: 377–379.

Weng, G. 2013. Phenolic resin molding compound. CN Patent 103087465.

Wu, F., Don, X., Hillbourg, H. et al. 2017. Electrical insulating material and method for preparing insulating material element. U.S. Patent Appl. 2017/0250001.

Yang, S., Lv, G., Wang, Q. 2013. Synergism of polysiloxane and zinc borate flame retardant polycarbonate. *Polymer Degradations and Stability* 98: 2795–2800.

Yin, M., Kitamura, T., Ishiwa, K. et al. 2003. Flame retardant composition and article. WO 03/004560.

Yin, Y. 2009. Flame resistant semi-aromatic polyamide resin composition and article. U.S. Patent 2009/0030124.

Yoshiyuki, U., and Yashushi, N. 1997. Flame retardant composition. Jpn. Kokai Tokkyo Koho JP1997227870.

Zhang, F., Chen, P., Wang, Y. et al. 2016a. Smoke suppression and synergistic flame retardancy properties of zinc borate and antimony trioxide in epoxy-based intumescent fire retardant coating. *Journal of Thermal Analysis and Calorimeter* 123: 1319–1327.

Zhang, J., Horton, J., and Gao, X. H. 2017. Methods of conferring fire retardancy to wood and fire-retardant wood products. U.S. Patent 9,669,564.

Zhang, L., and Liu, Z. H. 2017. Preparation of $2MgO \cdot B_2O_3 \cdot 1.5H_2O$ nanomaterial and evaluation of their flame retardant properties. *Journal of Thermal Analysis and Calorimeter* 129: 715–719.

Zhang, T., Liu, W., Wang, M. et al. 2016b. Synergistic effect of an aromatic boronic acid derivative and magnesium hydroxide on the flame retardancy of epoxy. *Polymer Degradation and Stability* 130: 257–263.

Zhou, Y., Feng, J., Peng, H. et al. 2014. Catalytic pyrolysis and flame retardancy of epoxy with solid acid boron phosphate. *Polymer Degradation and Stability* 110: 395–404.

7

Layer-by-Layer Assembly: A Novel Flame-Retardant Solution to Polymeric Materials

Giulio Malucelli

7.1 Introduction

During the last 2 decades, several surface engineered methods have been designed and applied to different types of polymeric materials, in order to provide these latter with functional features (comprising flame retardancy, hydrophobicity, oleophobicity, antibacterial/biocide properties, etc), without affecting the overall behavior of the bulk. In fact, the surface of any polymeric material (in form of sheets, fabrics, thin or thick films) is the first to interact with the surrounding environment: therefore, it is sufficient to modify the polymer surface in order to provide the whole underlying material with the required characteristics. In addition, the bulk addition of different specific additives often shows such drawbacks as the required high loading owing to low effectiveness due to bulk dilution, detrimental effects on polymer mechanical properties, etc. (Malucelli et al. 2014; Malucelli 2016a). Specifically referring to flame retardancy, different surface engineered methods have been designed and successfully applied to plastics and fabrics: among them, plasma treatments, sol-gel processes, and layer-by-layer (LbL) methods have demonstrated their suitability for providing the treated substrates with flame-retardant features.

It is noteworthy that all these methods are quite efficient flame-retardant treatments, as polymer flammability is a typical surface property: in particular, the polymer surface represents the critical zone in the polymer combustion scenario depicted in Figure 7.1.

In fact, the heat approaching the polymer surface is transferred to the polymer bulk; from this latter, volatile products originating from the thermal degradation diffuse toward the polymer surface and the gas phase, hence feeding the flame. Therefore, a key role in either polymer ignition or combustion is played by the polymer surface, since its physico-chemical features significantly influence the flux of

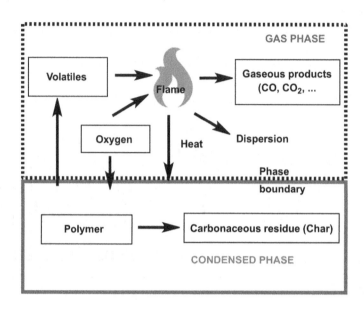

FIGURE 7.1 Polymer combustion cycle.

combustible volatiles toward the gas phase. In this context, any surface treatment that activates during the early stages of combustion because of the application of a flame or the exposure to a heat flux could be significantly beneficial in terms of fire protection: in fact, the formation, during the early stages of combustion, of an inorganic (ceramic), organic (intumescent), or hybrid organic/inorganic thermally stable surface layer, acting as a barrier to the exchange of mass and/or heat, is able to confer flame-retardant features to the underlying substrate.

In particular, by combining advancements in polymer surface engineering with nanotechnology, it could be possible, as it will be described in the next paragraphs, to design a pioneering low impact strategy to fire retardancy, the effectiveness of which is much higher than in the case of the protection created when polymer combustion has already started, as it typically occurs with bulk addition of "traditional" fire retardant additives.

This chapter is aimed at providing the reader with a general overview of the aforementioned surface engineered treatments; then, a detailed description of the layer-by-layer approached is carried out, highlighting the current achievements and limitations and presenting some possible further progresses for the next future.

7.2 Flame-Retardant Surface Engineering Methods

In the next paragraphs, three surface approaches, namely sol-gel treatments, nanoparticle absorption, and plasma treatments will be discussed from a general point of view, highlighting their main peculiarities within flame retardancy. Then, the layer-by-layer strategy will be thoroughly reviewed.

7.2.1 Sol-gel Treatments

Although the sol-gel chemistry approach has been used from the 1950s, especially for producing ceramic materials, only quite lately it has been exploited for conferring flame-retardant features to textiles. The basic principle of this surface engineered approach refers to the exploitation of the *in-situ* formation of sol-gel derived ceramic phases (i.e., ceramic particles or coatings) on fabrics treated with suitable precursors. These ceramic phases can act as a thermal shield during the exposure of the fabrics to a flame or to a heat flux. In particular, the resulting ceramic barrier is able to limit the transfer of oxygen and heat, hence, remarkably reducing the development of combustible volatile species that support further textile degradation phenomena; in addition, this ceramic barrier promotes the conversion of the polymer substrate into *char*, i.e., an aromatic stable carbonaceous residue (Alongi and Malucelli 2012; Alongi et al. 2015).

The sol-gel method can be considered as an example of soft-chemistry bottom-up strategy, which utilizes hydrolysis and condensation reactions occurring in a liquid medium where selected alkoxy precursors (namely, tetramethoxysilane, tetraethoxysilane, aluminum isopropoxide, titanium tetraisopropoxide) are located. A schematic representation of the sol-gel method is shown in Figure 7.2. Numerous experimental parameters can rule the development of the sol-gel processes, hence, influencing the structure and morphology of the resulting oxidic networks: nature of (semi)metal atom and alkyl/alkoxide groups, water/alkoxide ratio, temperature and pH, reaction time, and possible presence of co-solvents like ethanol (Malucelli 2016a).

Concerning the use of sol-gel derived ceramic phases in flame-retarded textiles, it is noteworthy that, as the fabric thickness is limited, the deposited sol-gel ceramic phases can exert only a partial protection on the underlying organic material. Therefore, synergistic or additive effects are frequently needed and thus exploited, combining the sol-gel chemistry with other active flame retardants that usually comprise *P*- and/or *N*-containing species (Brancatelli et al. 2011; Alongi et al. 2012d, 2012e, 2013c, 2013d, 2014a; Alongi and Malucelli 2013; Salmeia et al. 2016).

Besides, quite recently, dual-cure processes, which combine the sol-gel technique with a photo-induced polymerization step, have allowed obtaining hybrid organic-inorganic fire protective coatings (Alongi et al. 2011a; Malucelli 2016b).

7.2.2 (Nano)particle Absorption

This method represents the simplest technique in order to surface modify a textile substrate with micro- to nanoparticles. As for a usual finishing treatment, it requires the fabric dipping into an aqueous (nano)

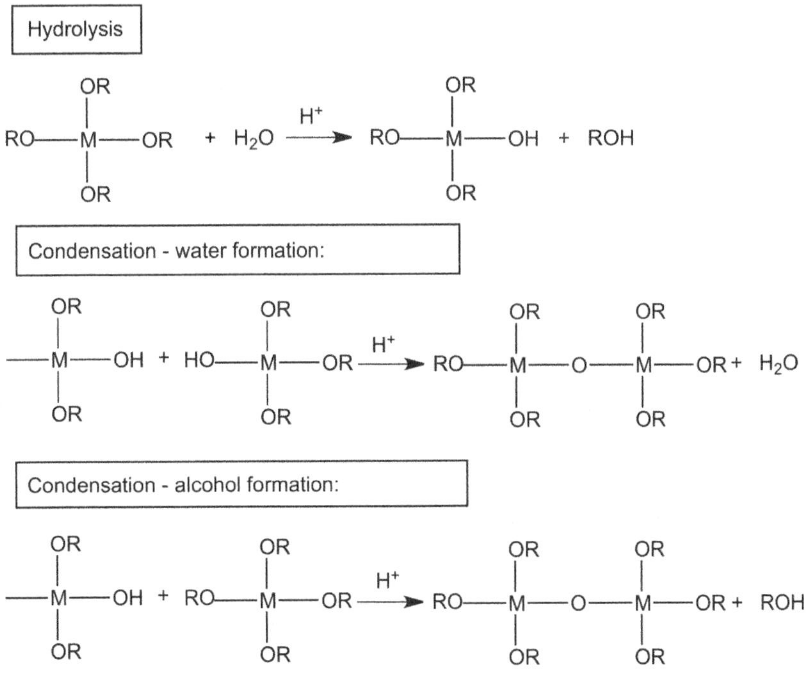

FIGURE 7.2 Scheme of the hydrolysis and condensation reactions occurring during sol-gel processes performed in acidic conditions.

particle suspension: this way, (nano)particles can be absorbed on the fiber surface, giving rise to an inorganic ceramic shield, potentially able to limit the heat and oxygen transfer to the underlying textile substrate. Furthermore, the resulting ceramic coating is able to favor the pyrolysis of the textile, instead of its burning, at the same time entrapping the volatile combustible species that can additionally fuel the combustion process. Unlike the sol-gel approach, nanoparticle absorption represents a bottom-up approach, suitable for both natural and synthetic fabric substrates. So far, several types of (nano)particles have been deposited on different textiles, namely: hydrotalcite, titania, silica, polyhedral oligomeric silsesquioxane (POSS)®, carbon nanotubes, and montmorillonite in the sodic form (Alongi et al. 2011b, 2012a; Tata 2012; Alongi and Malucelli 2015). Besides, specifically referring to polyester substrates, it was found that the nanoparticle adsorption can be remarkably improved by pre-treating the fabrics with cold oxygen plasma: in particular, the selection and fine tuning of different plasma process parameters (namely, power and etching time) allow optimizing the subsequent nanoparticle adsorption, hence, the fire performances. In fact, as demonstrated by cone calorimeter tests, the plasma pre-treatment is able to: (i) increase the density of nanoparticles on the fabric surface, thus increasing the compactness of the resulting ceramic coating and (ii) enhance the interactions of the adsorbed nanoparticles with the fabric surface (Carosio et al. 2011a).

7.2.3 Plasma Treatments

Cold plasma technique is a surface process, useful for grafting small functional groups and macromolecular species as well onto different textile and plastic substrates. One of the main advantages of this method refers to the possibility of changing the surface features of the treated material, without affecting its bulk properties (Yasuda 1985).

Cold plasma technique can be exploited for different purposes. In particular:

- It is suitable for etching the surface structure of the treated material and/or for functionalizing it by means of non-polymerizable gases (such as N_2, O_2, Ar, H_2, ammonia, carbon dioxide, etc.).
- It can be exploited for depositing thick polymeric coatings or inorganic layers on the material surface. This process can be carried out by generating plasma from a volatile organic, organometallic, or organosilicon compound (Farag et al. 2013).
- It is exploitable in two-step polymerization processes, where plasma is firstly utilized for the activation of the material surface, which is then grafted by a selected preformed polymer.
- It can be used for low pressure plasma-induced graft-polymerization processes, which comprise the simultaneous surface activation and the grafting and polymerization of a non-volatile monomer (Tsafack et al. 2004; Tsafack and Levalois-Grutzmacher 2006a, 2006b; Kamlangkla et al. 2011).

In the past, the possibility of carrying out plasma processes for textile applications was somehow limited, since atmospheric pressure plasma treatments, which have only quite recently become commercially available, are needed. The scientific literature is plentiful of very interesting papers that describe the use of atmospheric plasma treatments for conferring flame-retardant features to different textiles (mainly, cotton, polyester, polyacrylonitrile, and polyamide 6 fabrics) (Akovali and Gundogan 1990; Akovali and Takrouri 1991; Bourbigot et al. 1999; Shi 2000a, 2000b; Quede et al. 2002).

7.3 The Layer-by-Layer Approach

7.3.1 Main Characteristics

Quite recently, LbL technique has arisen as a simple, new approach for the molecularly controlled fabrication of nanostructured films. LbL was first demonstrated in 1966 by Iler, through the deposition of oppositely electrically charged particles onto a substrate (Iler 1966). Surprisingly, the potential of the LbL stayed hidden till the early 1990s, when a practical method for assembling polyanions and polycations was designed by Decher and co-workers (Decher and Hong 1991); at present, multi-layer assembly is acquiring popularity, since researchers are exploring endless combinations of components and their respective functionalities (Richardson et al. 2015). In its simplest design, the preparation of the LbL assembly starts by properly preparing the substrate surface, usually implicating a surface charge. Then the charged substrate is exposed, by dipping or spraying, to aqueous solutions or suspensions containing polyelectrolytes or (nano)particles, respectively, with tailored affinities to one other. The duration of each exposure may range between seconds and minutes. Optionally, each deposition step can be followed by rinsing and drying steps, which allow removing any loosely adhered material, preventing, at the same time, possible contamination phenomena taking place between the oppositely charged solutions or suspensions. As a result, every single layer deposited on the substrate surface reverses the actual surface interactions, hence permitting a complementary material to be deposited on the growing assembly. This procedure can be repeated as many times as needed, in order to build up the required assembly.

The exploitation of electrostatic interactions is not a prerequisite for successfully creating the layer-by-layer assembly. In fact, there are other interactions that can be employed for the multi-layer deposition; among them, it is worthy to mention donor/acceptor interactions (Shimazaki et al. 1997, 1998), hydrogen bonding (Stockton and Rubner 1997; Wang et al. 2000), covalent bonds (Fang et al. 1997; Ichinose et al. 1998), stereocomplex formation (Serizawa et al. 2000, 2001) or even specific recognition (Spaeth et al. 1997; Anzai et al. 1998).

Figure 7.3 shows a schematic two-component layer-by-layer process. The multi-layer process is independent from the substrate size and topology: in fact, almost any solvent-accessible surface

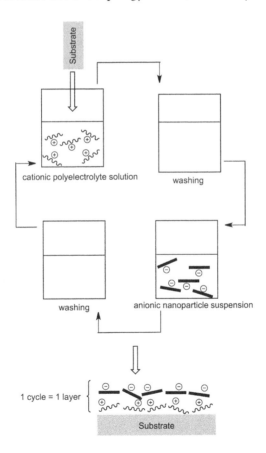

FIGURE 7.3 Scheme of LbL dipping process through electrostatic interaction (simplified concept of the first two adsorption steps): The deposition is illustrated as if it started on a negatively charged substrate; a cationic polyelectrolyte is combined with anionic ceramic platelets (i.e., nanoclay).

can be LbL-treated, from submicron objects (Neu et al. 2001), up to the inside of pipes or objects with a large surface (several square meters).

Cationic or anionic polyelectrolytes (Decher et al. 1992, 1994; Korneev et al. 1995), metallic or oxidic colloids (Kotov et al. 1995; Ariga et al. 1997; Lvov et al. 1997; Koo et al. 2004), and layered silicates (Lvov et al. 1996; Mamedov et al. 2000; Zhou et al. 2002) are a few examples of materials suitable for the LbL deposition. Conversely, the LbL technique is significantly affected by several experimental parameters related to the chosen layer components and deposition conditions, such as chemistry of used polyelectrolytes (Mermut and Barrett 2003), pH of the solutions/suspensions (Shiratori and Rubner 2000; Chang et al. 2008), temperature (Tan et al. 2003), molecular weight (Sui et al. 2003), counterions (Zhang and Ruhe 2003), ionic strength (McAloney et al. 2001), adsorption time (Fujita and Shiratori 2006), and drying step (Lourenco et al. 2007).

During the last 5–10 years, the LbL technique has emerged as a promising tool for providing polymeric substrates (bulk polymers, thin films, fabrics, foams) with flame-retardant features. Compared to more traditional flame retardant (FR) treatments, the LbL approach has many advantages. First, it can be exploited for assembling functional components; furthermore, it can be carried in mild conditions (i.e., at room temperature and atmospheric pressure). Lastly, it possesses environmentally friendly features, as it usually employs water as a solvent for preparing very diluted polyelectrolyte/(nano)particle baths (concentrations below 1 wt%), which can be easily recycled after use.

The concept of flame-retardant LbL coatings dates back to 2006, although the flame-retardant properties were not evaluated using specific tests (namely, flammability or combustion tests) (Srikulkit et al. 2006). However, it was possible to demonstrate that assemblies made of chitosan and polyphosphoric acid bi-layers were capable of enhancing the thermal resistance of silk fabrics.

Since this work, the LbL method has been utilized for depositing nanostructured coatings for different purposes, namely, thermal shielding and intumescent-like coatings. The former, which represent the first type of deposited and properly tested structures, mainly consist of nanoparticles deposited in fully inorganic or hybrid organic-inorganic assemblies. As a consequence of the exposure of the LbL-treated substrate to a heat flux or a flame, thermal shielding coatings give rise to the formation of a ceramic barrier on the underlying material, which exerts a protection effect, limiting the oxygen diffusion and reducing the heat transfer during the combustion process.

Intumescent-like coatings represent the recent trend of flame-retardant LbL treatments. In particular, when exposed to a heat flux or a flame, the designed intumescent formulation develops a foamed carbonaceous shield on its surface (e.g., char), which shows a very high protection of the underlying material. Three components are necessary for the intumescent effect, namely: an acid source (usually phosphoric acid salts), which releases acid, a carbon source (i.e., a material that is able to form a carbonaceous structure), and a blowing agent (such as melamine) which, upon heating, forms non-combustible swelling products (such as ammonia or carbon dioxide) (Bourbigot et al. 2004; Bourbigot and Duquesne 2007). The intumescence is the result of the reactions occurring in between the three components of the intumescent formulation.

In this context, LbL is an efficient method for miniaturizing the intumescent effect in nanostructured coatings comprising an intumescent-like composition that may also include nanoparticles. Next paragraphs will give an overview of the development of flame-retardant LbL coatings over the last 10 years.

7.3.2 Fully Inorganic LbL Assemblies

One of the pioneering works was published by Laufer and co-workers, who assembled positive alumina-coated silica (10 nm) and negative silica nanoparticles (10 or 40 nm) on cotton fabrics, exploiting a dipping method (Laufer et al. 2011). More specifically, 20 silica/silica bi-layers assemblies provided cotton with self-extinction, as assessed by vertical flame spread tests; furthermore, an increase of the time to ignition was observed in cone calorimeter experiments.

Then, cotton fabrics were treated with the same silica/silica assemblies, utilizing a spray-assisted LbL method and comparing the fire performances with those attained by dipping. In particular, using the horizontal spray method allowed to homogeneously cover the fabric surface, providing, at the same time, the highest thermal protection to the underlying substrate during forced combustion tests (Alongi et al. 2013b; Carosio et al. 2013c, 2013d).

Quite recently, the thermal shielding effect offered by LbL silica/silica coatings on polyester fabrics was successfully exploited for polycarbonate (PC) thin and thick films (Carosio et al. 2013b). More specifically, 20 bi-layers (200 nm thickness) were deposited on 200 μm PC samples; then, the treated substrates were subjected to vertical flame spread and cone calorimeter tests performed at 50 kW/m^2. The incandescent melt dripping phenomena during flammability tests were suppressed; conversely, the performance of the same 20 bi-layers assembly were outclassed by the five bi-layers counterpart, for which the highest peak of heat release rate and total heat release reduction (−20% and −30%, respectively) and time to ignition (TTI) increase (+22%) were observed. These findings were ascribed to the coating stability during cone calorimeter tests: in fact, upon exposure to the cone heat flux, the PC films shrank and stretched before melting, hence, compromising the physical stability of the deposited assembly.

Beside silica, alumina nanoparticles have been exploited in order to build all-nanoparticle multifunctional LbL coatings on cotton fabrics; in particular, Ugur and co-workers exploited the amphoteric nature of alumina by tuning the pH in order to have positive (acid) or negative (basic) charges on 50 nm alumina nanoparticles. The obtained LbL-treated fabrics showed enhanced flame retardancy, tensile strength, and UV-transmittance (Ugur et al. 2011).

Lee and co-workers designed LbL assemblies consisting of positive alumina-coated silica nanoparticles (average diameter: 10 nm) coupled with negative silica nanoparticles of different average size (30 and 10 nm). These assemblies were applied to both cotton and PET fabrics (Lee et al. 2006, 2007). Referring to these latter, five and ten bi-layers of 10 nm nanoparticles were capable to reduce the burning time by 63% and 94%, respectively, during vertical flame spread tests.

Several other inorganic nanoparticles have been utilized for fabricating fully inorganic LbL coatings. Among them, POSS˙ salts, positively or negatively charged, have been successfully studied and exploited. Li et al. (2011b) utilized water-soluble OctaAmmonium POSS [(+)POSS] and OctaTetramethylammonium POSS [(−)POSS], respectively, as cationic and anionic components of LbL thin coatings on cotton. In addition, aminopropyl silsesquioxane oligomer was also chosen as possible component for replacing the cationic layer. The obtained LbL coatings promoted the formation of more than 12 wt% char as shown by micro cone calorimeter tests; notwithstanding a remarkable decrease of the afterglow time observed during vertical flame spread tests, the deposited LbL architectures allowed keeping the fabric texture and the shape of the individual fibers.

Carosio et al. (2011b) designed fully inorganic LbL coatings on polyester fabrics, pairing negatively charged α-zirconium phosphate nanoplatelets with either positively charged octapropylammonium-functionalized POSS˙ or alumina-coated silica nanoparticles. The treated PET fabrics showed good FR performances when subjected to forced combustion tests (35 kW/m^2 heat flux): more specifically, silica/α-zirconium phosphate assemblies were able to increase the time to ignition (+42%) and to decrease the peak of heat release rate (−16%). Furthermore, the smoke production rate was reduced (−20%), as well as CO and CO$_2$ yields (−30% and −45%, respectively).

Recently, Carosio et al. deposited LbL assemblies based on ammonium polyphosphate and octapropyl ammonium POSS layers on acrylic fabrics. In particular, four or six bi-layered architectures were found to homogeneously cover each fiber and to protect the underlying substrate from the exposure of a 20 mm methane flame or 35 kW/m^2 heat flux. Besides, the melt dripping phenomena were prevented and the combustion rate significantly reduced. These findings were ascribed to several reasons, namely: the char-former character of ammonium polyphosphate, the thermal insulating protection provided by the ceramic barrier created by POSS, and the intimate contact between these two species, occurring in the LbL coating (Carosio and Alongi 2016).

7.3.3 Intumescent LbL Assemblies

The successful outcomes derived from the fully inorganic LbL assemblies stimulated the scientific community toward the design and exploitation of intumescent LbL coatings: this way, it is possible to combine the barrier effect offered by the inorganic/ceramic layers with the char-forming character provided by the reactive intumescent layers.

The first example of intumescent LbL coating, consisting of polyallylamine layers (carbon source and blowing agent) combined with sodium phosphates layers (acid source), was deposited on cotton by Li and co-workers (2011a). The obtained LbL coating was able to confer self-extinguishment to the treated fabrics during vertical flame spread tests, as well as non-ignitability during forced combustion (cone calorimeter) measurements. The intumescent character of the proposed architectures was proven by SEM analyses performed on the residues after flammability tests.

A similar LbL architecture was then applied to polyamide 6,6 fabrics, which showed a significant decrease of the heat release rate during micro cone calorimeter tests (Apaydin et al. 2013).

Chitosan was exploited as carbon source in combination with ammonium polyphosphate (acid source and blowing agent) in order to design intumescent coatings on cotton-polyester blends (Carosio et al. 2012). The treated fabrics were capable to suppress the afterglow phenomena and to leave a consistent residue after horizontal flame spread tests, reducing, at the same time, the combustion kinetics.

Quad-layer repetitive units consisting of polydiallydimethylammonium chloride/polyacrylic acid/polydiallydimethylammonium chloride/ammonium polyphosphate were designed by Malucelli's group and deposited either on cotton and polyester fabrics (Alongi et al. 2012c; Carosio et al. 2013a). The high char forming character of the obtained LbL coatings was exploited for providing the treated fabrics with self-extinction and for suppressing incandescent melt dripping; in addition, the heat related parameters, regardless of type of LbL-treated substrate, were remarkably reduced.

Quite recently, the concept of intumescent-LbL assemblies was utilized for ramie fibers (Zhang et al. 2013), which were coated with polyethylenimine (carbon source) and ammonium polyphosphate (acid source and blowing agent). The effect of different polyelectrolyte concentrations on the composition of the coating and, subsequently, on the resulting flame-retardant features was thoroughly investigated. In particular, the coatings bearing high ammonium polyphosphate concentration were able to impart self-extinction during vertical flame spread tests.

Pursuing this research, Wang and co-workers treated ramie fabrics with LbL coatings containing branched polyethylenimine (acting as carbon source and blowing agent) and polyvinylphosphonic acid (acid source), in the presence of transition-metal-ions (namely, Cu^{2+} and Zn^{2+} ions) (Wang et al. 2013). The coated fabrics did not self-extinguish, but the presence of the transition-metal-ions in the LbL architectures was found to increase the final residue and to reduce the combustion time. These findings were ascribed to the transition-metal-ions, which are able to promote the release of phosphoric acid from phosphorous-containing compounds at lower temperatures, hence, potentially starting flame-retardant activity earlier.

Huang and co-workers (2012b) polymerized an intumescent flame-retardant polyacrylamide and combined it with graphene oxide (GO) sheets, aiming at exploiting the barrier features of GO. To this aim, acrylamide was polymerized with *N*-(5,5-dimethyl-1,3,2-dioxaphosphinyl-2-yl)-acrylamide (DPAA): this way it was possible to obtain a phosphorus-based charring polymer that was coupled with graphene oxide sheets to get a platelet/char-forming combination coating. The formation of a continuous char during cone calorimeter tests allowed reducing the peak of heat release rate (−50%) and total heat release (−22%), increasing, at the same time, time to ignition (+56%).

The LbL technique has been also exploited for designing and depositing assemblies, where an intumescent composition is coupled with an environmentally friendly nature.

Some nice examples of these peculiar coatings are reported in the open scientific literature. The first describes a LbL assembly of chitosan (carbon source) and phytic acid (acid source) on cotton (Laufer et al. 2012b). By changing the pH of each aqueous solution, the composition of the LbL assembly was tuned in order to reach self-extinction in vertical flame spread tests. In particular, two deposition pH values were selected: accordingly, the coatings deposited at pH=6 were thicker and had 48 wt% phytic acid content, while pH=4 yielded the thinnest deposition with 66 wt% phytic acid.

Malucelli et al. have quite recently proposed the use of deoxyribonucleic acid (DNA) in combination with chitosan in order to deliver novel and environmentally sustainable LbL coatings (Carosio et al. 2013c). In fact, DNA has been clearly proven to be an all-in-one intumescent system: more specifically, the phosphate groups behave as acid source, the deoxyribose units can act as a carbon source, and the nitrogen-containing bases (i.e., adenine, guanine, cytosine, and thymine) may release ammonia as a blowing agent. When applied to cotton, this biomacromolecule was capable of extinguishing the flame during horizontal flammability tests, as well as to strongly reduce the combustion kinetics (Alongi et al. 2013a). The chitosan-DNA LbL assemblies showed an exponential growth as the number of bi-layers increased; when deposited on cotton fabrics, the coatings were able to achieve self-extinction during horizontal flame spread tests. Furthermore, cone calorimeter tests performed at 35 kW/m² showed a reduction of the peak of heat release rate by 40%, when the fabrics were treated with 20 bi-layers.

7.3.4 Hybrid Organic-Inorganic LbL Assemblies

The high FR effectiveness shown by some of the fully inorganic LbL assemblies was the starting point for the design and exploitation of multi-layers combining the nanoparticle barrier effect with the advantages provided by organic components into a single hybrid coating. In fact, this strategy is aimed at improving the protection provided by the charred carbon structures created by the organic component, preventing, at the same time, the destructive outcomes deriving from both flame exposure and mechanical stresses due to air drafts.

One of the first examples of hybrid organic-inorganic coatings refers to the deposition of complex architectures consisting of chitosan, ammonium polyphosphate, and silica nanoparticles bearing positive or negative charge on cotton-PET blends (Alongi et al. 2012b). It was found that the achieved flame-retardant performances are remarkably affected by the morphology and physical stability of the deposited assemblies: in particular, only those that yielded to the formation of a continuous and homogeneous coating were capable of improving both the flammability and combustion properties.

Huang and co-workers (2012a) deposited cotton assemblies of polyacrylic acid coupled with amino-functionalized montmorillonite nanoplatelets on cotton fabrics. Compared with the untreated substrates, the LbL-treated fabrics exhibited reduced combustion kinetics; furthermore, as assessed by SEM microscopy, the residues after combustion tests showed the formation of a blown charred structure embedding nanoplatelets.

Hybrid organic-inorganic LbL assemblies were also deposited on bulk polylactic acid (PLA) films (1 mm) by Laachachi and co-workers (2011); more specifically, a bi-layer structured coating comprising polyallylamine and montmorillonite was applied to the plastic film. The coatings made of 60 bi-layers reached a considerable thickness of 18 μm with nanostructured ordering of the clay nanoplatelets parallel to the surface. After the deposition, the coatings underwent a post-diffusional treatment: to this aim, the LbL-treated PLA film was exposed to a solution of sodium polyphosphate in order to attain the diffusion of the acid source through the whole thickness of the coating. After the post-diffusional process, the flame-retardant assembly acted both as a thermal shielding (due to the presence of montmorillonite nanoparticles) and as intumescent charring system (due to the presence of polyphosphates and polyallylamine). During cone calorimeter tests performed at 35 kW/m², the 60 bi-layers-coated PLA films showed an increase of time to ignition and a decrease of the peak of heat release rate (−37%).

The same LbL assembly was then deposited on polyamide 6 films (500 μm) without performing the post-diffusional treatment (Apaydin et al. 2013). 20 bi-layers assemblies were able to remarkably decrease the peak of heat release rate (−60%, under 25 kW/m² heat).

Another substrate that has being investigated in terms of fire protection is foam. In fact, foamed materials represent, along with fabrics, one of the first ignitable items in domestic fires. In fact, the fuel load most often associated with tragic residential fires can be referred to upholstered furniture, often filled with polyurethane (PU) foams.

From the safety point of view, foams that have not been treated with FR represent a severe threat, since untreated PU foams burn very rapidly, release toxic gases, and show melt dripping phenomena that can easily spread the fire to other ignitable materials (Hirschler 2008).

Referring to the exploitation of the LbL technique, Grunlan and co-workers developed an environmentally friendly coating with renewable composition by combining chitosan with sodium montmorillonite (Laufer et al. 2012a). The thickness of the LbL coating was regulated by adjusting the pH of the chitosan solution: in particular, pH = 6 produced thicker coatings due to the low charge density of chitosan. The deposition of ten bi-layers completely stopped the melt dripping of PU foam exposed to a butane torch flame for 10 s. For the LbL-treated foams, cone calorimeter tests (35 kW/m^2 heat flux) showed a decreased peak of heat release rate (–52%).

A tri-layer assembly made of anionic polyacrylic acid, cationic branched-polyethylenimine, and anionic clay (sodium montmorillonite) was designed by Kim et al. (2012) and deposited on a porous polyurethane foam. The combustion properties of this latter were significantly enhanced in the presence of the LbL assembly: in fact, the peak of heat release rate decreased (–17%), as well as the total burning time (–21%).

7.4 Conclusions and Perspectives

The layer-by-layer assembly has been proven to be a valuable method for nano-structuring different types of material surfaces, providing, at the same time, enhanced flame-retardant features.

One of the main advantages of this surface engineered technique refers to the "unlimited" freedom in the design of the LbL assemblies that can exploit different flame-retardant effects, namely:

- Nanoparticle-based thermal shielding assemblies
- Intumescent assemblies
- Hybrid organic-inorganic assemblies, where the organic layers may behave as intumescent components.

Thus, the LbL strategy also facilitates the incorporation of multiple flame-retardant mechanisms simultaneously, which may allow for the finding of new synergies. However, the final flame-retardant effect provided to the LbL-treated substrate can be achieved by finely tuning the deposition parameters.

Therefore, this nanotechnology strategy seems to represent a possible effective alternative to conventional flame-retardant systems, which are very efficient, but, in some cases, are facing limitations in use, due to their questioned/evaluated toxicity and low eco-sustainability.

On the other hand, LbL is experiencing some limitations, mainly referable to the transfer of this technology from lab-scale to industrial-scale: in this context, the spray LbL method could be applied to roll-to-roll industrial plants, as already demonstrated at a pilot-scale (Chang et al. 2014). Undoubtedly, spray coating is fast and reduces material waste; furthermore, it eliminates the risk of solution/suspension contamination that may take place through repetitive dipping into the solution/suspension baths.

Another disadvantage of the LbL method refers to the durability (i.e., washing fastness) of the most currently employed LbL coatings, since they are usually based on electrostatic interactions. This significantly limits the life time of the flame-retardant LbL coatings: in fact, these latter are waterborne systems that cannot withstand the washing cycles, fabrics are usually subjected to, without losing their flame-retardant effectiveness. In fact, the electrostatic interactions that link the different layers can be degraded when the LbL assemblies are exposed to detergents, specifically exploited for removing non-covalently linked materials from the textile surface. Some research groups have tried to overcome this challenging issue by using the formation of covalent interactions occurring in between the multi-layer assembly. In particular, it could be possible to exploit thermally or UV-curable organic layers (Alongi et al. 2014b; Carosio and Alongi 2015) for the build-up of the LbL coating.

Though it has been rarely discussed, another important issue specifically related to the use of LbL coatings on textiles refers to the evidence of bridging fibers during the LbL deposition: as a consequence, fabric stiffness increases, thus changing the "hand" (comfort) of the treated substrate. In order to overcome this problem, Guin and co-workers (2014) demonstrated that bridging of the fibers can be prevented, without impacting on the flame-retardant features of the coatings, by using ultrasonication during the rinse step.

In conclusion, the suitability of the layer-by-layer approach for the design of new and cost-effective flame-retardant coatings, together with the possibility of using low environmental impact components, foresee its rapid expansion over a multitude of substrates, which may result in commercial LbL-based flame-retardant treatments within the next few years.

References

Akovali, G., and Gundogan, G. 1990. Studies on flame retardancy of polyacrylonitrile fiber treated by flame-retardant monomers in cold plasma. *Journal of Applied Polymer Science* 41: 2011–2019.

Akovali, G., and Takrouri, F. 1991. Studies on modification of some flammability characteristics by plasma. II. Polyester fabric. *Journal of Applied Polymer Science* 42: 2717–2725.

Alongi, J., and Malucelli, G. 2012. State of the art and perspectives on sol–gel derived hybrid architectures for flame retardancy of textiles. *Journal of Materials Chemistry* 22: 21805–21809.

Alongi, J., and Malucelli, G. 2013. Thermal stability, flame retardancy and abrasion resistance of cotton and cotton-linen blends treated by sol–gel silica coatings containing alumina micro- or nanoparticles. *Polymer Degradation and Stability* 98: 1428–1438.

Alongi, J., and Malucelli, G. 2015. Thermal degradation of cellulose and cellulosic substrates. In: *Reactions and Mechanisms in Thermal Analysis of Advanced Materials*, A. Tiwari and B. Raj (Eds.). Beverly, MA: Scrivener Publishing LLC.

Alongi, J., Brancatelli, G., and Rosace, G. 2012a. Thermal properties and combustion; behavior of POSS- and bohemite-finished cotton fabrics. *Journal of Applied Polymer Science* 123: 426–436.

Alongi, J., Carletto, R. A., Di Blasio, A. et al. 2013a. DNA: A novel, green, natural flame retardant and suppressant for cotton. *Journal of Materials Chemistry A* 1: 4779–4785.

Alongi, J., Carosio, F., and Malucelli, G. 2012b. Layer by Layer complex architectures based on ammonium polyphosphate, chitosan and silica on polyester-cotton blends: Flammability and combustion behaviour. *Cellulose* 19: 1041–1050.

Alongi, J., Carosio, F., and Malucelli, G. 2012c. Influence of ammonium polyphosphate-/poly(acrylic acid)-based layer by layer architectures on the char formation in cotton, polyester and their blends. *Polymer Degradation and Stability* 97: 1644–1653.

Alongi, J., Carosio, F., Frache, A. et al. 2013b. Layer by layer coatings assembled through dipping, vertical or horizontal spray for cotton flame retardancy. *Carbohydrate Polymers* 92: 114–119.

Alongi, J., Ciobanu, M., and Malucelli, G. 2011a. Cotton fabrics treated with hybrid organic-inorganic coatings obtained through dual-cure processes. *Cellulose* 18: 1335–1348.

Alongi, J., Colleoni C., Rosace G. et al. 2013d. Phosphorus- and nitrogen-doped silica coatings for enhancing the flame retardancy of cotton: Synergisms or additive effects? *Polymer Degradation and Stability* 98: 579–589.

Alongi, J., Colleoni, C., Malucelli, G. et al. 2012d. Hybrid phosphorus-doped silica architectures derived from a multistep sol-gel process for improving thermal stability and flame retardancy of cotton fabrics. *Polymer Degradation and Stability* 97: 1334–1344.

Alongi, J., Colleoni, C., Rosace, G. et al. 2012e. Thermal stability, flame retardancy and mechanical properties of cotton fabrics treated with inorganic coatings synthesized through sol-gel processes. *Journal of Thermal Analysis and Calorimetry* 110: 1207–1216.

Alongi, J., Colleoni, C., Rosace, G. et al. 2013c. The role of pre-hydrolysis on multistep sol-gel processes for enhancing the flame retardancy of cotton. *Cellulose* 20: 525–535.

Alongi, J., Colleoni, C., Rosace, G. et al. 2014a. Sol-gel derived architectures for enhancing cotton flame retardancy: Effect of pure and phosphorus-doped silica phases. *Polymer Degradation and Stability* 99: 92–98.

Alongi, J., Di Blasio, A., Carosio, F. et al. 2014b. UV-cured hybrid organic-inorganic layer by layer assemblies: Effect on the flame retardancy of polycarbonate films. *Polymer Degradation and Stability* 107: 74–81.

Alongi, J., Tata, J., and Frache, A. 2011b. Hydrotalcite and nanometric silica as finishing additives to enhance the thermal stability and flame retardancy of cotton. *Cellulose* 18(1): 179–190.

Alongi, J., Tata, J., Carosio, F. et al. 2015. A comparative analysis of nanoparticle adsorption as fire-protection approach for fabrics. *Polymers* 7(1): 47–68.

Anzai, J., Kobayashi, Y., Suzuki, Y. et al. 1998. Enzyme sensors prepared by layer-by-layer deposition of enzymes on a platinum electrode through Avidin-Biotin interaction. *Sensors and Actuators B: Chemical* 52: 3–9.

Apaydin, K., Laachachi, A., Ball, V. et al. 2013. Polyallylamine-montmorillonite as super flame retardant coating assemblies by layer-by layer deposition on polyamide. *Polymer Degradation and Stability* 98: 627–634.

Ariga, K., Lvov, Y., Onda, M. et al. 1997. Alternately assembled ultrathin film of silica nanoparticles and linear polycations. *Chemistry Letters* 26: 125–128.

Bourbigot, S., and Duquesne S. 2007. Fire retardant polymers: Recent developments and opportunities. *Journal of Materials Chemistry* 17: 2283–2300.

Bourbigot, S., Jama, C., Le Bras, M. et al. 1999. New approach to flame retardancy using plasma assisted surface polymerization techniques. *Polymer Degradation and Stability* 66(1): 153–155.

Bourbigot, S., Le Bras, M., Duquesne, S. et al. 2004. Recent advances for intumescent polymers. *Macromolecular Materials and Engineering* 289: 499–511.

Brancatelli, G., Colleoni, C., Massafra, M. R. et al. 2011. Effect of hybrid phosphorus-doped silica thin films produced by sol-gel method on the thermal behavior of cotton fabrics. *Polymer Degradation and Stability* 96: 483–490.

Carosio, F., Alongi, J., and Frache, A. 2011a. Influence of surface activation by plasma and nanoparticle adsorption on the morphology, thermal stability and combustion behavior of PET fabrics. *European Polymer Journal* 47(5): 893–902.

Carosio, F., Alongi, J., and Malucelli, G. 2011b. α-Zirconium phosphate-based nanoarchitectures on polyester fabrics through Layer-by-Layer assembly. *Journal of Materials Chemistry* 21: 10370–10376.

Carosio, F., Alongi, J., and Malucelli, G. 2012. Layer by layer ammonium polyphosphate-based coatings for flame retardancy of polyester-cotton blends. *Carbohydrate Polymers* 88: 1460–1469.

Carosio, F., Alongi, J., and Malucelli, G. 2013a. Flammability and combustion properties of ammonium polyphosphate-/poly(acrylic acid)-based layer by layer architectures deposited on cotton, polyester and their blends. *Polymer Degradation and Stability* 98: 1626–1630.

Carosio, F., and Alongi, J. 2015. Few durable layers suppress cotton combustion due to the joint combination of layer by layer assembly and UV-curing. *RSC Advances* 5: 71482–71490.

Carosio, F., and Alongi, J. 2016. Influence of layer by layer coatings containing octapropylammonium polyhedral oligomeric silsesquioxane and ammonium polyphosphate on the thermal stability and flammability of acrylic fabrics. *Journal of Analytical and Applied Pyrolysis* 119: 114–123.

Carosio, F., Di Blasio A., Cuttica, F. et al. 2013d. Flame retardancy of polyester fabrics treated by spray-assisted layer-by-layer silica architectures. *Industrial & Engineering Chemistry Research* 52: 9544–9550.

Carosio, F., Di Blasio, A., Alongi, J. et al. 2013b. Layer by layer nanoarchitectures for the surface protection of polycarbonate. *European Polymer Journal* 49: 397–404.

Carosio, F., Di Blasio, A., Alongi, J. et al. 2013c. Green DNA-based flame retardant coatings assembled through layer by layer. *Polymer* 54: 5148–5153.

Chang, L., Kong, X., Wang, F. et al. 2008. Layer-by-layer assembly of poly (*N*-acryloyl-*N'*-propylpiperazine) and poly (acrylic acid): Effect of pH and temperature. *Thin Solid Films* 516: 2125–2129.

Chang, S., Slopek, R. P., Condon, B. et al. 2014. Surface coating for flame-retardant behavior of cotton fabric using a continuous layer-by-layer process. *Industrial & Engineering Chemistry Research* 53: 3805–3812.

Decher, G., and Hong, J. D. 1991. Buildup of ultrathin multilayer films by a self-assembly process. 1. Consecutive adsorption of anionic and cationic bipolar amphiphiles on charged surfaces. *Makromolekulare Chemie Macromolecular Symposia* 46: 321–327.

Decher, G., Hong, J. D., and Schmitt, J. 1992. Buildup of ultrathin multilayer films by a self-assembly process: III. Consecutively alternating adsorption of anionic and cationic polyelectrolytes on charged surfaces. *Thin Solid Films* 210: 831–835.

Decher, G., Lvov, Y., and Schmitt, J. 1994. Proof of multilayer structural organization in self-assembled polycation-polyanion molecular films. *Thin Solid Films* 244: 772–777.

Fang, M. M., Kaschak, D. M., Sutorik, A. C. et al. 1997. A mix and match ionic-covalent strategy for self-assembly of inorganic multilayer films. *Journal of the American Chemical Society* 119: 12184–12191.

Farag, Z. R., Kruger, S., Hidde, G. et al. 2013. Deposition of thick polymer or inorganic layers with flame-retardant properties by combination of plasma and spray processes. *Surface and Coatings Technology* 228: 266–274.

Fujita, S., and Shiratori, S. 2006. The optical properties of ultra-thin films fabricated by layer-by-layer adsorption process depending on dipping time. *Thin Solid Films* 499: 54–60.

Guin, T., Krecker, M., Milhorn, A. et al. 2014. Maintaining hand and improving fire resistance of cotton fabric through ultrasonication rinsing of multilayer nanocoating. *Cellulose* 21: 3023–3030.

Hirschler, M. M. 2008. Polyurethane foam and fire safety. *Polymers for Advanced Technologies* 19: 521–529.

Huang, G. B., Yang, J. G., Gao, J. R. et al. 2012b. Thin films of intumescent flame retardant-polyacrylamide and exfoliated graphene oxide fabricated via layer-by-layer assembly for improving flame retardant properties of cotton fabric. *Industrial & Engineering Chemistry Research* 51: 12355–12366.

Huang, G., Liang, H., Wang, X. et al. 2012a. Poly (acrylic acid)/clay thin films assembled by layer-by-layer deposition for improving the flame retardancy properties of cotton. *Industrial & Engineering Chemistry Research* 51: 12299–12309.

Ichinose, I., Kawakami, T., and Kunitake, T. 1998. Alternate molecular layers of metal oxides and hydroxyl polymers prepared by the surface sol-gel process. *Advanced Materials* 10: 535–539.

Iler, R. K. 1966. Multilayers of colloidal particles. *Journal of Colloid and Interface Science* 21: 569–594.

Kamlangkla, K., Hodak, S. K., and Levois-Grutzmacher, J. 2011. Multifunctional silk fabrics by means of the plasma induced graft polymerization (PIGP) process. *Surface and Coatings Technology* 205(13–14): 3755–3762.

Kim, Y. S., Harris, R., and Davis, R. 2012. Innovative approach to rapid growth of highly clay-filled coatings on porous polyurethane foam. *ACS Macro Letters* 1: 820–824.

Koo, H. Y., Yi, D. K., Yoo, S. J. et al. 2004. A snowman-like array of colloidal dimers for antireflecting surfaces. *Advanced Materials* 16: 274–277.

Korneev, D., Lvov, Y., Decher, G. et al. 1995. Neutron reflectivity analysis of self-assembled film super-lattices with alternate layers of deuterated and hydrogenated polystyrenesulfonate and polyallyl-amine. *Physica B* 213: 954–956.

Kotov, N. A., Dekany, I., and Fendler, J. H. 1995. Layer-by-layer self-assembly of polyelectrolyte-semiconductor nanoparticle composite films. *The Journal of Chemical Physics* 99: 13065–13069.

Laachachi, A., Ball, V., Apaydin, K. et al. 2011. Diffusion of polyphosphates into (poly(allylamine)-montmorillonite) multilayer films: Flame retardant intumescent films with improved oxygen barrier. *Langmuir* 27: 13879–13887.

Laufer, G., Carosio, F., Martinez, R. et al. 2011. Growth and fire resistance of colloidal silica-polyelectrolyte thin film assemblies. *Journal of Colloid and Interface Science* 356: 35669–35677.

Laufer, G., Kirkland, C., Cain, A. et al. 2012a. Clay–chitosan nanobrick walls: Completely renewable gas barrier and flame-retardant nanocoatings. *ACS Applied Materials & Interfaces* 4: 1643–1649.

Laufer, G., Kirkland, C., Morgan, A. et al. 2012b. Intumescent multilayer nanocoating, made with renewable polyelectrolytes, for flame-retardant cotton. *Biomacromolecules* 13: 2843–2848.

Lee, D., Gemici, Z., Rubner, M. F. et al. 2007. Multilayers of oppositely charged SiO$_2$ nanoparticles: Effect of surface charge on multilayer assembly. *Langmuir* 23: 8833–8837.

Lee, D., Rubner, M. F., and Cohen, R. E. 2006. All-nanoparticle thin-film coatings. *Nano Letters* 6: 2305–2312.

Li, Y. C., Mannen, S., Morgan, A. B. et al. 2011a. Intumescent all-polymer multilayer nanocoating capable of extinguishing flame on fabric. *Advanced Materials* 23: 3926–3930.

Li, Y. C., Mannen, S., Schulz, J. et al. 2011b. Growth and fire protection behavior of POSS-based multilayer thin films. *Journal of Materials Chemistry* 21: 3060–3069.

Lourenco, J. M. C., Ribeiro, P. A., Botelho do Rego, A. M. et al. 2007. Counterions in layer-by-layer films-Influence of the drying process. *Journal of Colloid and Interface Science* 313: 26–33.

Lvov, Y. M., Ariga, K., Onda, M. et al. 1997. Alternate assembly of ordered multilayers of SiO$_2$ and other nanoparticles and polyions. *Langmuir* 13: 6195–6203.

Lvov, Y., Ariga, K., Ichinose, I. et al. 1996. Formation of ultrathin multilayer and hydrated gel from montmorillonite and linear polycations. *Langmuir* 12: 3038–3044.

Malucelli, G. 2016a. Surface engineered fire protective coatings for fabrics through sol-gel and layer-by-layer methods: An overview. *Coatings* 6: 1–23.

Malucelli, G. 2016b. Hybrid organic/inorganic coatings through dual-cure processes: State of the art and perspectives. *Coatings* 6: 1–11.

Malucelli, G., Carosio F., Alongi, J. et al. 2014. Materials engineering for surface-confined flame retardancy. *Materials Science & Engineering R-Reports* 84: 1–20.

Mamedov, A., Ostrander, J., Aliev, F. et al. 2000. Stratified assemblies of magnetite nanoparticles and montmorillonite prepared by the layer-by-layer assembly. *Langmuir* 16: 3941–3949.

McAloney, R. A., Sinyor, M., Dudnik, V. et al. 2001. Atomic force microscopy studies of salt effects on polyelectrolyte multilayer film morphology. *Langmuir* 17: 6655–6663.

Mermut, O., and Barrett, C. J. 2003. Effects of charge density and counterions on the assembly of polyelectrolyte multilayers. *The Journal of Physical Chemistry B* 107: 2525–2530.

Neu, B., Voigt, A., Mitlohner, R. et al. 2001. Biological cells as templates for hollow microcapsules. *Journal of Microencapsulation* 18: 385–395.

Quede, A., Jama, C., Supiot, P. et al. 2002. Elaboration of fire retardant coatings on polyamide-6 using a cold plasma polymerization process. *Surface & Coatings Technology* 151–152: 424–428.

Richardson, J. J., Bjornmalm, M., and Caruso, F. 2015. Technology-driven layer-by-layer assembly of nanofilms. *Science* 348: aaa2491.

Salmeia, K. A., Gaan, S., and Malucelli, G. 2016. Recent advances for flame retardancy of textiles based on phosphorus chemistry. *Polymers* 8: 1–36.

Serizawa, T., Hamada, K., Kitayama, T. et al. 2000. Stepwise assembly of isotactic poly(methyl methacrylate) and syndiotactic poly(methacrylic acid) on a substrate. *Langmuir* 16: 7112–7115.

Serizawa, T., Yamashita, H., Fujiwara, T. et al. 2001. Stepwise assembly of enantiomeric poly(lactide)s on surfaces. *Macromolecules* 34: 1996–2001.

Shi, L. S. 2000a. Characterization of the flame retardancy of EVA copolymer by plasma grafting of acrylic acid. *European Polymer Journal* 36(12): 2611–2615.

Shi, L. S. 2000b. An approach to the flame retardation and smoke suppression of ethylene-vinyl acetate copolymer by plasma grafting of acrylamide. *Reactive and Functional Polymers* 45(2): 85–93.

Shimazaki, Y., Mitsuishi, M., Ito, S. et al. 1997. Preparation of the layer-by-layer deposited ultrathin film based on the charge-transfer interaction. *Langmuir* 13: 1385–1390.

Shimazaki, Y., Mitsuishi, M., Ito, S. et al. 1998. Preparation and characterization of the layer-by-layer deposited ultrathin film based on the charge-transfer interaction in organic solvents. *Langmuir* 14: 2768–2773.

Shiratori, S. S., and Rubner, M. F. 2000. pH-dependent thickness behavior of sequentially adsorbed layers of weak polyelectrolytes. *Macromolecules* 33: 4213–4219.

Spaeth, K., Brecht, A., and Gauglitz, G. 1997. Studies on the biotin-avidin multilayer adsorption by spectroscopic ellipsometry. *Journal of Colloid and Interface Science* 196: 128–135.

Srikulkit, K., Iamsamai, C., and Dubas, S. T. 2006. Development of flame retardant polyphosphoric acid coating based on the polyelectrolyte multilayers technique. *Journal of Metals, Materials and Minerals* 16: 41–45.

Stockton, W. B., and Rubner, M. F. 1997. Molecular-level processing of conjugated polymers. 4. Layer-by-layer manipulation of polyaniline via hydrogen-bonding interactions. *Macromolecules* 30: 2717–2725.

Sui, Z. J., Salloum, D., and Schlenoff, J. B. 2003. Effect of molecular weight on the construction of polyelectrolyte multilayers: Stripping versus sticking. *Langmuir* 19: 2491–2495.

Tan, H. L., McMurdo, M. J., Pan, G. Q. et al. 2003. Temperature dependence of polyelectrolyte multilayer assembly. *Langmuir* 19: 9311–9314.

Tata, J., Alongi, J., and Frache, A. 2012. Optimization of the procedure to burn textile fabrics by cone calorimeter: Part II. Results on nanoparticle-finished polyester. *Fire and Materials* 36(7): 527–536.

Tsafack, M. J., and Levalois-Grutzmacher, J. 2006a. Flame retardancy of cotton textiles by plasma-induced graft-polymerization (PIGP). *Surface & Coatings Technology* 201(6): 2599–2610.

Tsafack, M. J., and Levalois-Grutzmacher, J. 2006b. Plasma-induced graft-polymerization of flame retardant monomers onto PAN fabrics. *Surface & Coatings Technology* 200(11): 3503–3510.

Tsafack, M. J., Hochart, F., and Levalois-Grutzmacher, J. 2004. Polymerization and surface modification by low pressure plasma technique. *European Physical Journal-Applied Physics* 26(3): 215–219.

Ugur, S. S., Sariisik, M., and Aktas, A. H. 2011. Nano-Al_2O_3 multilayer film deposition on cotton fabrics by layer-by-layer deposition method. *Materials Research Bulletin* 46: 1202–1206.

Wang, L. L., Zhang, T., Yan, H. Q. et al. 2013. Modification of ramie fabric with a metal-ion-doped flame-retardant coating. *Journal of Applied Polymer Science* 129: 2986–2997.

Wang, L. Y., Cui, S. X., Wang, Z. Q. et al. 2000. Multilayer assemblies of copolymer PSOH and PVP on the basis of hydrogen bonding. *Langmuir* 16: 10490–10494.

Yasuda, H. 1985. *Plasma Polymerization*. New York: Academic Press.

Zhang, H. N., and Ruhe, J. 2003. Interaction of strong polyelectrolytes with surface-attached polyelectrolyte brushes–polymer brushes as substrates for the layer-by-layer deposition of polyelectrolytes. *Macromolecules* 36: 6593–6598.

Zhang, T., Yan, H., Wang, L. et al. 2013. Controlled formation of self-extinguishing intumescent coating on ramie fabric via layer-by-layer assembly. *Industrial & Engineering Chemistry Research* 52: 6138–6140.

Zhou, Y., Hu, N., Zeng, Y. et al. 2002. Heme protein–clay films: Direct electrochemistry and electrochemical catalysis. *Langmuir* 18: 211–219.

III

Flame Retardant Polymer Nanocomposites

III

8

Use of Carbon Nanotubes and Nanofibers for Multifunctional Flame Retardant Polymer Composites

Alexander B.
Morgan

8.1 Introduction to Carbon Nanotubes and Nanofibers

Before discussing how carbon nanotubes and nanofibers are used in flame retardant (FR) applications, it is helpful to understand what these materials are, and how their chemical structure yields their various mechanical, thermal, and electrical properties. Structure-property relationships are a key part of fundamental material science understanding, and the chemical structure of carbon nanotubes and nanofibers directly relates to their properties.

As the name indicates, carbon nanotubes (CNTs) and carbon nanofibers (CNFs) are carbon-based materials, predominantly composed of sp^2 bonded carbon, which are cylindrical in shape. Lengths and diameters of these cylinders can vary depending upon synthesis method, purification methods, and specific carbon bond geometry. For more details on the breadth of CNT and CNF chemistry, and how they have been put into polymers to date, there are several review papers available (Moniruzzaman and Winey 2006; Bredeau et al. 2008; Al-Saleh and Sundararaj 2009; Bauhofer and Kovacs 2009; Byrne and Gunko 2010; Chou et al. 2010; Qian et al. 2010; Prasek et al. 2011; Rahmat and Hubert 2011; Alig et al. 2012), but for the purposes of this chapter, we will focus on the more commonly used multi-wall

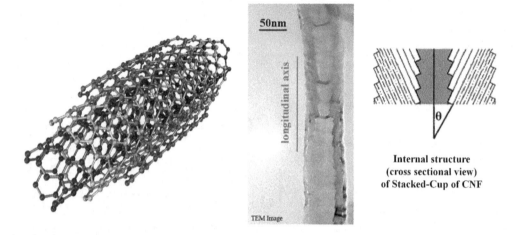

FIGURE 8.1 Idealized chemical structures of (left) MWCNT (From https://en.wikipedia.org/wiki/Carbon_ nanotube#/media/File:Multi-walled_Carbon_Nanotube.png) and (right) VGCNF (Reprinted from Liberata, G. et al. *Nanotechnology*, 24, 305704, 2013. With permission).

carbon nanotubes (MWCNTs) which have a "rolled graphitic sheet" structure (Figure 8.1), and on vapor grown carbon nanofibers (VGCNFs), which are cylindrical, but have more of a "stacked cone" structure (Figure 8.1). The sp^2 carbon present in these chemical structures of MWCNTs and VGCNFs allow these materials to have high thermal and electrical conductivity. Further, these same chemical structures give very high mechanical strength to these tubes and fibers.

Depending upon how the MWCNTs/VGCNFs are made, this can result in changes in structure vs. the idealized forms, including defects around residual metal catalyst being found, oxidation at the ends of the tubes and fibers, and differences in lengths. Therefore, users of MWCNTs and VGCNFs are encouraged to get product data sheets from the vendors, as well as elemental analysis information on residual metal content, prior to end-use of these nanoparticles. While two different vendors may both sell the MWCNTs or VGCNFs, it should not be assumed that the materials from the two vendors (MWCNT compared to MWCNT or VGCNF compared to VGCNF) are of identical purity, tube/fiber length, or ability to be dispersed into a polymer matrix. It is very common to find that each vendor will provide different grades of MWCNTs or VGCNFs, with different levels of purification, residual metal content, and tube/fiber length available. Some notable MWCNT and VGCNF vendors include:

Nanocyl SA (Belgium): http://www.nanocyl.com/
US Research Nanomaterials, Inc. (USA): http://www.us-nano.com/
Thomas Swan & Co., Ltd. (UK): http://www.thomas-swan.co.uk/advanced-materials
Applied Sciences, Inc. (USA): http://pyrografproducts.com/

8.2 Nanotube and Nanofiber Forms in Composites

To impart the thermal, electrical, and mechanical benefits of MWCNTs and VGCNFs to polymers, the nanoparticles will need to be incorporated into the polymer in some way. For traditional carbon-fiber and glass-fiber reinforced polymer composites, the nanoparticles are often incorporated as surface compositions (namely, coatings or papers), but in some cases, the nanotubes can be grown directly onto the fiber prior to polymer infusion to manufacture the final composite (Chou et al. 2010; Qian et al. 2010). These composites where the nanotubes are grown directly on the fiber have not been studied for fire performance to date and so will not be discussed in this chapter. However, for nanotube/nanofiber "papers" and nanotube-nanofiber coatings, these have been studied for fire performance and are discussed below.

8.2.1 Papers

MWCNT or VGCNF loose fibers/particles which can be entangled into a non-woven sheet are often referred to as "papers" due to the similar processes used to convert cellulosic fibers into the paper we know and use for writing and reading (and like the page of this book, if you are reading this chapter in a non-electronic format). These papers are often bonded onto fiber reinforced composites to provide a protective top layer that brings thermal and electrical enhancements to the composite. Specifically, the MWCNT/ VGCNF "paper" on top of the composite has a very high content of nanoparticles and a low polymer content. This means the electrical and thermal properties of the MWCNT or VGCNF dominate the top layer of the composite and potentially bring the maximum effect to that surface, rather than throughout the entire composite. For example, an electrically non-conductive epoxy composite with a VGCNF paper bonded to the surface would have an electrically conductive surface, but the bulk composite would be an insulator. This would enable a multi-functional material that needed to have anti-static surface dissipation properties/electromagnetic shielding properties, with a non-conductive side to the composite for optimized structural/chemical resistance properties. Both VGCNFs and MWCNTs have been combined to produce these papers (sometimes called "buckypapers") which are then bonded/laminated to fiber reinforced polymer composites to provide a top protective layer (Wu et al. 2008, 2010; Zhuge et al. 2012a, 2012b). The benefit of these papers for fire performance was that they lowered both heat release and smoke release, as well as yielded a delay in time to peak heat release rate, as studied by cone calorimeter at a heat flux of 50 kW/m². The mechanism of flame retardancy for these papers is a reduction in mass loss rate/pyrolysis of flammable gases from the polymer composite under the protective "paper," which results in less fuel being available for burning, hence, a lower heat/smoke release. However, the results show that these papers only slow mass loss and delay burning; they do not prevent it. Still, the concentration of carbon nanotubes/ fibers on the surface of the composite shows a flame retardant effect that could be used in combination with additional flame retardants in the polymer composite to provide more robust fire performance, as well as an approach for applying electrical conductive outer layers to a non-conductive composite.

8.2.2 Coatings

Related to top-layer coatings for composites are layer-by-layer (LbL) coatings which are conformal coatings on top of complex and simple shapes, including foams, fabrics, and flat surfaces. LbL coatings are created by building up single polymeric layers one at a time on top of one another, typically through ionic interactions of cationic and anionic layers, although other materials such as clays and ionically functionalized nanoparticles can be used as well (Li et al. 2009; Carosio et al. 2011; Zhou et al. 2011; Laufer et al. 2012; Cain et al. 2013). There are recent examples of MWCNTs or VGCNFs being successfully employed in these LbL coatings for polyurethane foams (Kim et al. 2011; Kim and Davis 2014; Pan et al. 2014; Holder et al. 2016). Similar to the benefit provided by the "papers" described in the previous paragraph, the LbL coatings containing VGCNFs or MWCNTs resulted in reductions of heat release rate in the cone calorimeter (ASTM E1354) test. However, and more importantly for polyurethane foam, the conformal coatings prevented the foams from collapsing into a "melt pool" of decomposing polyurethane. This latter effect indicates that not only do the LbL coatings slow mass loss and pyrolysis rates during burning, but they also generate some localized high viscosity effects that prevent the polyurethane from liquefying during thermal decomposition. Preventing the liquefaction of polyurethane during a fire event will reduce the probability of flashover in a room and will further improve the fire safety performance of what is the largest fire hazard in most modern structures, the polyurethane foam couch/ upholstered furniture item (Hirschler 2008; Krämer et al. 2010; Hall 2015).

8.2.3 Nanocomposites

The most common use of MWCNTs or VGCNFs for flame retardant, and indeed multi-functional applications, is when the MWCNTs or VGCNFs are added directly to polymer matrix to yield a polymer

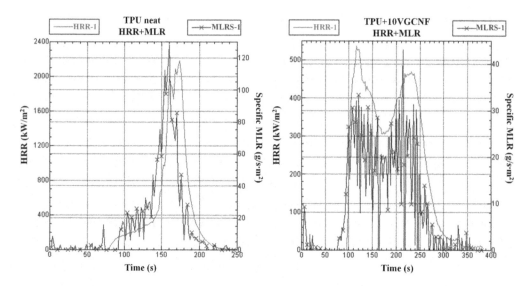

FIGURE 8.2   Heat release rate and specific mass loss rate curves for TPU (left) and TPU + VGCNF (Right). (Reprinted from Morgan, A.B. and Liu, W., *Fire Mater.*, 35, 43–60, 2011. With permission.)

nanocomposite. Polymer nanocomposites are a well-studied material class, with several good review papers available (Sinha Ray and Okamoto 2003; Morgan 2006; Utracki et al. 2007; Paul and Robeson 2008; Kiliaris and Papaspyrides 2010). Further, polymer nanocomposites containing MWCNT or VGCNF have also been well studied to date (Bredeau et al. 2008; Al-Saleh and Sundararaj 2009; Bauhofer and Kovacs 2009; Bose et al. 2010). In general, MWCNT or VGCNF added to polymers results in two benefits of flame retardancy: reduction in mass loss rate and anti-dripping behavior. Specifically, the nanocomposite has a lower rate of fuel release/mass loss during burning, resulting in lower heat release. An example of this behavior for a thermoplastic polyurethane (TPU) combined with 10 wt% VGCNF is shown in Figure 8.2, where the mass loss rates align with the heat release rate (HRR) curves, showing that the VGCNF is reducing the mass loss rate, which in turn lowers heat release rate.

This lower mass loss rate is due to the formation of a network of nanotubes/nanofibers which forms a fire protection barrier and locally increases melt viscosity (Kashiwagi et al. 2004, 2005a, 2005b; Cipiriano et al. 2007; Utracki et al. 2007; Rahatekar et al. 2010). This increase in polymer viscosity is why polymer nanocomposites containing nanotubes or nanofibers tend to char in place rather than melt/deform in a fire, very often maintaining their shape. These nanocomposites with lowered flammability can be combined with additional flame retardants to achieve even greater effect, and examples of these "nanocomposite + FR" materials will be discussed later in this chapter.

8.2.4 Manufacturing Issues and Requirements

In the previous pages the many benefits of polymer nanocomposites have been discussed, but some discussion on the practical issues surrounding manufacturing and end-use requirements are needed, as there are still barriers preventing these materials from being used in commercial use. The barriers can be grouped into three general categories.

1. *Environmental Health & Safety (EH&S)*: MWCNT and VGCNF present some unknown hazards from an EH&S perspective. Toxicity of these materials has not been fully studied (Hurt et al. 2006; Hussain et al. 2009; Wang et al. 2011; Reijnders 2012; Heister et al. 2013), and whether they will remain as discrete nanoparticles, or will agglomerate into a micron sized particle upon release from a polymer is unclear. Issues such as polymer composite sanding, grinding, sawing,

and drilling to make final structures may release MWCNT and VGCNF with unknown consequences. These materials could be released during manufacture and handling of polymer nanocomposites, and further, when the nanocomposites are burned, there appears to be some potential for release of the nanoparticles when the chars containing the MWCNT/VGCNF are disturbed (Chivas-Joly et al. 2010; Nyden et al. 2010). There have been some attempts to chemically functionalize MWCNT and VGCNF so they covalently react into the polymer and therefore present less of a potential EH&S issue (Kim et al. 2006; Wang et al. 2006; Zou et al. 2008; Armstrong et al. 2009; Liu et al. 2009; Yang et al. 2009), but more study is needed to determine if these nanoparticles present a notable hazard or not.

2. *Manufacturing Issues Due to Viscosity Increase:* Many nanoparticles, due to their high surface area and interactions with polymer during processing, typically increase polymer viscosity during processing to high levels. This is true for thermoset polymers where liquid monomers interact with the nanoparticle, as well as for thermoplastic polymers, where molten polymer interacts with the nanoparticle. In both cases, there is typically a large increase in viscosity as the nanoparticles form network structures and slow down the flow of small molecules and polymer chains (Wang et al. 2007; Bredeau et al. 2008; Cipriano et al. 2008; Bangarusampath et al. 2009; Rahmat and Hubert 2011; Alig et al. 2012). This effect can be detrimental for manufacturing and can prevent the final part from being manufactured. For thermoplastic materials, the increase in viscosity can result in injection molding "short shots" where the polymer does not flow to all parts of the mold. For thermoset materials, the monomer "resin" + nanoparticle fluid may not wet out all parts of the mold, and, may not penetrate fully through fiber reinforcement fabrics, resulting in final composites which easily delaminate due to a lack of polymer penetration in between fabric plies and through the fabric. Both issues can be overcome through careful injection mold design and through the use of polymer resin "prepreg," but, there will be cases where the viscosity increase brought by the MWCNT or VGCNF may not be acceptable for some manufacturing processes.

3. *Metal Catalyst Impurities:* Many VGCNF and MWCNT are produced with the use of metal catalysts which "grow" the nanotube or nanofiber off the surface of the metal catalyst. Depending upon how the nanotube or nanofiber is produced, the metal catalyst may be embedded deep in the carbon structure, or, may just be entrapped in the final product as a partly grown MWCNT/VGCNF shrouded by amorphous or other graphitic carbon species. These metal particles can present oxidative thermal stability issues in some applications (where the metal impurities causes the polymer to degrade faster) and may also have an effect on polymer aging properties against UV. The purity of the MWCNT or VGCNF can vary depending upon what degree of post-manufacture "clean-up" has occurred prior to sale of the MWCNT/VGCNF material. Some thermal heat treatment of raw materials will get rid of the metal impurities, but in other cases, acid washing may be required to fully remove the metal catalyst. Users of MWCNT/VGCNF materials are encouraged to inquire about residual catalyst content, and, determine if this residue will or will not present problems in their final product.

8.3 Multi-functional Performance

8.3.1 Electrical and Thermal Conductivity

One of the features that makes use of MWCNT or VGCNF attractive for use as a nanofiller in polymers is the fact that these two carbon-based materials can impart enhanced electrical and thermal conductivity to the polymer they are placed into. Specifically, since MWCNT and VGCNF have high electrical and thermal conductivity, they can impart that functionality to other polymers they are placed into. However, these conductive properties are only provided if the MWCNTs or VGCNFs are well dispersed throughout the polymer and are capable of setting up a conductive network. Depending upon the length of the individual MWCNT/VGCNF nanoparticles and their ability to disperse in the polymer matrix, different loading levels may be required to set up a conductive network when using these nanoparticles. In general, as loading level increases, there tend to be more sites within the polymer for the nanoparticles to come in contact with one another and set up the conductive pathway. Depending upon the number of connection points, the electrical conductivity can be on the level of electrostatic dissipation all the way up to conductive circuit level for highly filled polymer nanocomposites. Good papers on the subject can shed more light on the ranges of electrical performance that can be achieved (Moniruzzaman and Winey 2006; Bredeau et al. 2008; Al-Saleh and Sundararaj 2009; Bauhofer and Kovacs 2009; Byrne and Gunko 2010; Chou et al. 2010; Qian et al. 2010; Cardoso et al. 2012). Once electrical conductivity of some sort is established, these nanocomposites can provide other benefits such as electromagnetic interference (EMI) shielding, since the conductive MWCNT/VGCNF network effectively creates a "Faraday cage" should the nanocomposite be used to encapsulate another electronic device. Anti-static behavior can also be gained, which could help with grounding and preventing electronics from "shorting out" due to touching by static electricity containing bodies.

Related to electrical conductivity is thermal conductivity, which these nanofillers can also provide. However, conduction of heat is different than conduction of electricity, although in general, the more MWCNT/VGCNF nanofiller present, the higher the potential thermal conductivity of the material. The relationship between thermal conductivity of the nanocomposite and flammability is not so clear. In theory, if a material can conduct heat away faster than it can heat up to melting or decomposition temperature, then the material should not degrade or ignite. However, many MWCNT/VGCNF containing nanocomposites are black or dark grey in color and therefore have infrared absorptive properties regarding to heat. Therefore, the benefit of MWCNT/VGCNF thermal conductivity in regards to fire performance may be somewhat negated by the thermal absorptive characteristics of these materials when combined into a polymer. Thermal conductivity of polymer nanocomposites based upon MWCNT/VGCNF is not as well studied to date, and more studies are needed in this area. Based upon observations by Kashiwagi and others (Kashiwagi et al. 2004, 2005a, 2005b; Hurt et al. 2006; Rahatekar et al. 2010), the thermal conductivity of MWCNT and VGCNF do not appear to play a major role in heat release reduction/flame retardancy of these materials.

8.3.2 Enhanced Mechanical Properties

As with many other types of polymer nanocomposites, MWCNT and VGCNF can improve the mechanical properties of polymers they are mixed into. Provided a nanocomposite structure is achieved, there can be notable improvements in flexural and tensile strength. There are several review papers which speak to the range of enhanced mechanical properties which MWCNT and VGCNF impart to polymers (Moniruzzaman and Winey 2006; Bredeau et al. 2008; Byrne and Gunko 2010; Qian et al. 2010; Rahmat and Hubert 2011; Alig et al. 2012). This enhancement of mechanical properties makes them attractive for flame retardant applications (as well as others) because better balance of properties (flammability, thermal, electrical, mechanical) can be achieved with the MWCNT/VGCNF nanocomposite than can be obtained with traditional fillers and additives. When MWCNT/VGCNF are applied as flame retardant coatings, however (such as the above mentioned "buckypapers"), they tend to have minimal impact (positive or negative) on mechanical properties because the carbon nanofiller is localized to the coating, rather than part of the polymer matrix (Krämer et al. 2010; Pan et al. 2014).

8.4 Examples of Flame Retardancy of Carbon Nanotubes and Nanofibers in Polymers and Polymer Composites

There have been several examples of carbon nanotubes or nanofibers combined with other flame retardant materials to achieve a final nanocomposite with enhanced flame retardant properties. As alluded to above, use of nanofiller alone does have a flame retardant effect. Specifically, mass loss rate is reduced which results in lower heat release rate, and the nanotubes/nanofibers provide an anti-drip effect for thermoplastics during burning (Kashiwagi et al. 2004, 2005a, 2005b; Hurt et al. 2006; Bangarusampath et al. 2009; Rahatekar et al. 2010). However, multiple studies have shown that this flame retardant effect by itself is not sufficient to meet most regulatory fire safety tests (Morgan and Wilkie 2007; Bredeau et al. 2008; Paul and Robeson 2008), and therefore the flame retardant effect of carbon nanotubes/nanofibers alone is not sufficient. Still, the lower heat release of the nanocomposite means that less flame retardant additive needs to be added to the polymer to obtain acceptable flame retardant performance in regulatory tests. In the rest of this section, examples will be given for a wide range of flame retardant chemicals, showing how the carbon nanotube/nanofiber works well with many flame retardant chemicals to achieve superior fire safety performance.

8.4.1 Halogenated Flame Retardants

MWCNT has shown effectiveness in reducing heat release when combined with brominated flame retardants, namely, decabromodiphenyl ether with antimony oxide synergist in polystyrene. In a paper with Lu and Wilkie, MWCNT was mixed with flame retardant in polystyrene via melt compounding and notable reductions in peak heat release were observed (Lu and Wilkie 2010). In the same paper, the brominated flame retardants were also combined with an organoclay at the same weight% loading to see which nanoparticle (clay or MWCNT) was more effective at reducing heat release. The results showed that MWCNT was more effective per wt% in reducing heat release.

In another example, a flexible polyurethane foam containing an aliphatic bromine functionalized phosphate was tested for heat release reduction with either VGCNF or organoclay (Zammarano et al. 2008). While both nanoparticles reduced heat release rate alongside the brominated phosphate, only carbon nanofibers fully prevented foam collapse and dripping, yielding a final char that maintained its shape of the original foam. This is noteworthy as polyurethane foam often liquefies during a fire and forms a high intensity "pool fire" which can quickly lead to flashover in a room fire. Lowering heat release *and* preventing the formation of the pool fire will help prevent flashover where the polyurethane foam was the first item ignited.

8.4.2 Phosphorus-Based Flame Retardants

There are several papers available showing how phosphorus-based flame retardants have been combined with MWCNT or VGCNF to lower flammability. Only a few of them are mentioned here to show the range of performance and range of chemistries available.

In one example (Wang et al. 2010), the flame retardant 9,10-dihydro-9-oxy-10-phosphaphenanthrene-10-oxide (DOPO) was reacted with a polysilane which in turn was copolymerized with another phosphate. This polymeric phosphorus/silicon flame retardant was then reacted directly onto a MWCNT which had been functionalized via oxidation and amino-siloxane reactions. This grafted the flame retardant directly onto the MWCNT, which was then melt compounded into polyethylene-co-vinyl acetate. A control sample of flame retardant and MWCNT with no pre-grafting reaction was included in the study. Cone calorimeter testing showed that the all of the samples containing flame retardant and MWCNT lowered heat release, but the flame retardant grafted on to the MWCNT (and therefore intimately mixed throughout the polymer) showed reductions in peak heat release rate as well as delaying time to ignition and total heat release. In effect, the flame retardant grafted into the MWCNT was a more effective solution for flame retardancy.

In a series of two papers (Lao et al. 2009, 2011), the use of VGCNF combined with organophosphinates (Clariant GmbH OP1230 or OP1311/1312) into polyamide 11 or polyamide 12 by melt compounding. These materials were then studied for heat release reduction potential, with the combination of OP1230 and VGCNF showing the greatest reductions in heat release. Further, this combination of additives showed the ability to pass a vertical orientation small flame ignition resistance test (namely, UL-94) with bars having a thickness of 3.2 mm.

In a final example, VGCNF was combined with an organophosphinate (Clariant GmbH OP930) in a bisphenol F, aromatic amine cured epoxy (Morgan and Galaska 2008). Zinc borate was also used in some of the formulations for additional flame retardant enhancement. Via the use of small-scale heat release testing (namely, micro combustion calorimeter, ASTM D7309), the combination of VGCNF, organophosphinate, and zinc borate showed the greatest reduction in heat release when compared to control samples and even samples containing the same flame retardant, but using an organoclay rather than VGCNF as the nanoparticle for flame retardant enhancement.

8.4.3 Mineral Fillers

Mineral fillers, namely, aluminum and magnesium hydroxide, are often used with polyolefins to flame retard materials used in wire and cable jacketing materials. However, very high loading levels of these mineral fillers, typically 50%–80%, may be required to obtain the required fire safety in these products. One study is known where both MWCNT and single-wall CNT were studied in the presence of $Al(OH)_3$ as an additional flame retardant (Beyer 2005). Side by side, the MWCNT reduced the heat release rate of the final material, whereas SWCNT increased heat release. The latter result was unexpected and the authors did not explore it further to see if residual metal catalyst in the SWCNT may have been responsible for the result. The MWCNT system was scaled up and tested in full-scale wire and cable flame spread tests, where it outperformed other materials containing only $Al(OH)_3$. The author of this study also compared MWCNT to clay nanocomposites in combination with $Al(OH)_3$ and found them to be roughly equivalent in performance in heat release reduction (cone calorimeter) and in full-scale wire and cable flame spread tests.

8.4.4 Intumescent Flame Retardants

The one area of flame retardant chemistry which appears to be slightly antagonistic with MWCNT is intumescent flame retardant chemistry. Intumescent flame retardants form a thermally protective "foam" early in the fire growth curve for a material. This thermally protective foam slows mass loss which in turn lowers heat release and also provides thermal insulation which in turn prevents further decomposition of polymer underneath the foam. For a good intumescent foam to form, it must be unimpeded during growth. Therefore, viscosity is very important when considering intumescent foam formation. If the burning polymer is too viscous, the chars will not form well and the flame retardant effect may be diminished. Indeed, this has been observed in a series of three papers (Bourbigot et al. 2010; Gérard et al. 2011, 2012), all of which combined MWCNT with intumescent flame retardants in either thermosetting polymers (epoxy) or thermoplastic materials (thermoplastic polyurethane or polyamide). The results were the same: the combination of intumescent and MWCNT was never as effective as just using the intumescent flame retardant by itself, suggesting that the MWCNT and its increase in melt viscosity during burning prevents the formation of a good intumescent foam.

8.5 Conclusions

Throughout this chapter, the utility of carbon nanotubes/nanofibers to impart fire safety as well as enhanced mechanical, thermal, and electrical properties has been shown. These enhanced properties suggest that these materials could be increasingly used in a wide range of market applications. However, this has not been the case as there are some practical considerations which need to be addressed before they can be used commercially, and some areas of new research needed to further advance the use of flame retardant carbon nanotube/nanofiber containing materials.

8.5.1 Practical Considerations

Like other nanocomposites, MWCNT or VGCNF cannot be simply "added in" to the polymer with subsequent automatic improvements in material properties. Rather, close attention to how the nanotubes or nanofibers are added into the polymer must be undertaken, and, changes in processing of the polymer may be required to achieve good dispersion of the nanoparticle throughout the matrix. As mentioned above when discussing the mechanism of flame retardancy, where the MWCNT or VGCNF causes increases in viscosity during burning, this same increase in viscosity occurs during melt compounding of a thermoplastic polymer or processing of precursor liquid monomers for a thermoset polymer. Unlike clay nanocomposites which can shear thin due to thixotropic behavior (Utracki et al. 2007; Kiliaris and Papaspyrides 2010), MWCNT and VGCNF do not always shear thin and at high loadings can be very difficult to process. For example, resin transfer molding of thermoset monomers into a fiber preform may fail when MWCNT and VGCNF are added. Specifically, the resin may be so viscous that the monomers do not penetrate fully into the mold and do not wet out all the fibers, resulting in a final composite with voids and dry fibers. Additives which lower resin viscosity may be required, along with heating of the resin (assuming the heat does not initiate polymerization), is often required to successfully process a thermoset polymer composite containing MWCNT or VGCNF. For thermoplastic materials, increases in melt temperature and injection molding pressure are often required to get the thermoplastic material to fully fill the mold during injection molding. Care must be taken to not heat the thermoplastic too hot otherwise the polymer may begin to decompose during processing, resulting in loss of mechanical/ thermal properties in the final part. Of final note, MWCNT and VGCNF purity may need to be considered prior to their use in polymers. While many manufacturers of these materials have made great strides in improving the purity of their carbon nanoparticles, there still can be trace residual metal catalyst and amorphous carbon from the process left behind. Likewise, the purification method used on the MWCNT or VGCNF may change some of the inherent properties and surface chemistry of the nanoparticle which may have beneficial or detrimental effects in the final nanocomposite. Heat treatment to remove amorphous carbon, if taken too far, can further graphitize these nanoparticles which may change some of their electrical, mechanical, and thermal properties. Acid treatments to remove metals can oxidize the ends of the nanotubes and nanofibers, which in turn can help with bonding to the polymer resin or cause early initiation of polymerization during processing in cases of select thermoset resin chemistries. The researcher is strongly advised to fully characterize their MWCNT and/ or VGCNF materials prior to use and really understand how those measured properties correlate to observed dispersion in the polymer and final measured properties (mechanical, thermal, electrical, fire) in the polymer nanocomposite.

8.5.2 Further Research Areas

Since it has been shown that MWCNT and VGCNF can be used successfully for flame retardant applications, and that they work with many different types of flame retardants, there are not many remaining areas of research to explore with these materials. The few remaining areas to consider for further research are:

- *Understanding the limits of using MWCNT/VGCNF with other flame retardants*: As discussed above, MWCNT has shown some antagonistic behavior when combined with intumescent flame retardants (Beyer 2005; Morgan and Galaska 2008; Gérard et al. 2012). It is likely that MWCNT/ VGCNF would also show antagonism with flame retardants that promote "dripping away" from flames, such as nitrogen oxy alkyl hindered amine light stabilizer (NOR-HALS) chemistry in polypropylene, melamine cyanurate with aliphatic polyamides, and aliphatic bromides with polystyrene, since the nanoparticles inhibit melt flow
- *Understanding how MWCNT/VGCNF affect long term aging (oxidative, UV, thermal) of polymers*: The high surface area of MWCNT and VGCNF, plus their formation of network structures, may actually slow long term aging effects in the polymer as any cracks or changes in surface chemistry

caused by oxygen, heat, or UV, would have to penetrate through the carbon nanoparticle network to generate damage to the nanocomposite. However, this is just a hypothesis as at the writing of this chapter, the author was unable to find any papers or review articles which fully studied (or even superficially studied) this issue

- *Environmental, Health, & Safety Issues*: Like all new technologies, end-of-life issues for the material containing MWCNT and VGCNF must be considered. Similar to the previous point about long term aging, how the nanocomposite degrades and decomposes will determine if any of these nanoparticles are released into the environment. Some preliminary studies indicate that because these nanoparticles are carbon based, toxicity effects are not always so clear, with studies showing cells interacting well with MWCNT, and other cases showing negative interactions (Hussain et al. 2009; Nyden et al. 2010; Wang et al. 2011; Reijnders 2012; Heister et al. 2013). Further, exposure effects to workers who may be sanding, grinding, or drilling through MWCNT/VGCNF containing materials are unknown as we do not know how those nanoparticles are released. Such studies are needed before these materials are used in commercial applications. Equally important is how to handle the final chars/ashes created from the burning of a MWCNT/VGCNF containing nanocomposite. Several papers have shown that the nanotubes and nanofibers remain behind in the final char, but they can be easily disturbed and released into the air (Kim et al. 2006; Wang et al. 2006), which is a potential real-world scenario for these materials after the fire has been extinguished.

Polymer nanocomposites containing MWCNT or VGCNF have great potential to be multi-functional materials with enhanced mechanical, thermal, electrical, and flammability properties, with potential use in a wide range of applications (aerospace, electronics, mass transport, etc.). This chapter has hopefully shown the reader this potential, as well as what references to read to learn more about this technology. However, at the conclusion, the reader must be aware of the processing difficulties and EH&S issues surrounding these materials. Therefore, the author strongly recommends to those scientists considering future research in this area to not try another permutation of MWCNT/VGCNF in another polymer and measure its properties, but rather take some well-known examples from the literature, reproduce them, and try to answer the long term aging issues, EH&S unknowns, and focus more on process science of nanocomposites.

References

Alig, I., Potschke, P., Lellinger, D. et al. 2012. Establishment, morphology and properties of carbon nanotube networks in polymer melts. *Polymer* 53(1): 4–28.

Al-Saleh, M. H., and Sundararaj, U. 2009. A review of vapor grown carbon nanofiber/polymer conductive composites. *Carbon* 47(1): 2–22.

Armstrong, G., Ruether, M., Blighe, F. et al. 2009. Functionalised multi-walled carbon nanotubes for epoxy nanocomposites with improved performance. *Polymer International* 58(9): 1002–1009.

Bangarusampath, D. S., Ruckdaschel, H., Altstadt, V. et al. 2009. Rheology and properties of melt-processed poly(ether ether ketone)/multi-wall carbon nanotube composites. *Polymer* 50(24): 5803–5811.

Bauhofer, W., and Kovacs, J. Z. 2009. A review and analysis of electrical percolation in carbon nanotube polymer composites. *Composites Science and Technology* 69(10): 1486–1498.

Beyer, G. 2005. Filler blend of carbon nanotubes and organoclays with improved char as a new flame retardant system for polymers and cable applications. *Fire and Materials* 29(2): 61–69.

Bose, S., Khare, R. A., and Moldenaers, P. 2010. Assessing the strengths and weaknesses of various types of pre-treatments of carbon nanotubes on the properties of polymer/carbon nanotubes composites: A critical review. *Polymer* 51(5): 975–993.

Bourbigot, S., Samyn, F., Turf, T. et al. 2010. Nanomorphology and reaction to fire of polyurethane and polyamide nanocomposites containing flame retardants. *Polymer Degradation and Stability* 95(3): 320–326.

Bredeau, S., Peeterbroeck, S., Bonduel, D. et al. 2008. From carbon nanotube coatings to high-performance polymer nanocomposites. *Polymer International* 57(4): 547–553.

Byrne, M. T., and Gunko, Y. K. 2010. Recent advances in research on carbon nanotube–polymer composites. *Advanced Materials* 22(15): 1672–1688.

Cain, A. A., Nolen, C. R., Li, Y. C. et al. 2013. Phosphorous-filled nanobrick wall multilayer thin film eliminates polyurethane melt dripping and reduces heat release associated with fire. *Polymer Degradation and Stability* 98(12): 2645–2652.

Cardoso, P., Silva, J., Klosterman, D. et al. 2012. The role of disorder on the AC and DC electrical conductivity of vapour grown carbon nanofibre/epoxy composites. *Composites Science and Technology* 72(2): 243–247.

Carosio, F., Laufer, G., Alongi, J. et al. 2011. Layer-by-layer assembly of silica-based flame retardant thin film on PET fabric. *Polymer Degradation and Stability* 96(5): 745–750.

Chivas-Joly, C., Guillaume, E., Ducourtieux, S. et al. 2010. Influence of carbon nanotubes on fire behavior and on decomposition products of thermoplastic polymers. *Interflam Proceedings*, July 5–7, Nottingham, UK, pp. 1375–1386.

Chou, T. W., Gao, L., Thostenson, E. T. et al. 2010. An assessment of the science and technology of carbon nanotube-based fibers and composites. *Composites Science and Technology* 70(1): 1–19.

Cipiriano, B. H., Kashiwagi, T., Raghavan, S. R. et al. 2007. Effects of aspect ratio of MWNT on the flammability properties of polymer nanocomposites. *Polymer* 48(20): 6086–6096.

Cipriano, B. H., Kota, A. K., Gershon, A. L. et al. 2008. Conductivity enhancement of carbon nanotube and nanofiber-based polymer nanocomposites by melt annealing. *Polymer* 49(22): 4846–4851.

Gérard, C., Fontaine, G., and Bourbigot, S. 2011. Synergistic and antagonistic effects in flame retardancy of an intumescent epoxy resin. *Polymers for Advanced Technologies* 22(7): 1085–1090.

Gérard, C., Fontaine, G., Bellayer, S. et al. 2012. Reaction to fire of an intumescent epoxy resin: Protection mechanisms and synergy. *Polymer Degradation and Stability* 97(8): 1366–1386.

Hall, J. R. 2015. Estimating fires when a product is the primary fuel but not the first fuel, with an application to upholstered furniture. *Fire Technology* 51(2): 381–391.

Heister, E., Brunner, E. W., Dieckmann, G. R. et al. 2013. Are carbon nanotubes a natural solution? Applications in biology and medicine. *ACS Applied Materials & Interfaces* 5(6): 1870–1891.

Hirschler, M. M. 2008. Polyurethane foam and fire safety. *Polymers for Advanced Technologies* 19(6): 521–529.

Holder, K. M., Cain, A. A., Plummer, M. G. et al. 2016. Carbon nanotube multilayer nanocoatings prevent flame spread on flexible polyurethane foam. *Macromolecular Materials and Engineering* 301(6): 665–673.

Hurt, R. H., Monthioux, M., and Kane, A. 2006. Toxicology of carbon nanomaterials: Status, trends, and perspectives on the special issue. *Carbon* 44(6): 1028–1033.

Hussain, S. M., Braydich-Stolle, L. K., Schrand, A. M. et al. 2009. Toxicity evaluation for safe use of nanomaterials: Recent achievements and technical challenges. *Advanced Materials* 21(16): 1549–1559.

Kashiwagi, T., Du, F., Douglas, J. F. et al. 2005a. Nanoparticle networks reduce the flammability of polymer nanocomposites. *Nature Materials* 4: 928–933.

Kashiwagi, T., Du, F., Winey, K. I. et al. 2005b. Flammability properties of polymer nanocomposites with single-walled carbon nanotubes: Effects of nanotube dispersion and concentration. *Polymer* 46(2): 471–481.

Kashiwagi, T., Grulke, E., Hilding, J. et al. 2004. Thermal and flammability properties of polypropylene/carbon nanotube nanocomposites. *Polymer* 45(12): 4227–4239.

Kiliaris, P., and Papaspyrides, C. D. 2010. Polymer/layered silicate (clay) nanocomposites: An overview of flame retardancy. *Progress in Polymer Science* 35(7): 902–958.

Kim, J. A., Seong, D. G., Kang, T. J. et al. 2006. Effects of surface modification on rheological and mechanical properties of CNT/epoxy composites. *Carbon* 44(10): 1898–1905.

Kim, Y. S., and Davis, R. 2014. Multi-walled carbon nanotube layer-by-layer coatings with a trilayer structure to reduce foam flammability. *Thin Solid Films* 550: 184–189.

Kim, Y. S., Davis, R., Cain, A. A. et al. 2011. Development of layer-by-layer assembled carbon nanofiber-filled coatings to reduce polyurethane foam flammability. *Polymer* 52(13): 2847–2855.

Krämer, R. H., Zammarano, M., Linteris, G. T. et al. 2010. Heat release and structural collapse of flexible polyurethane foam. *Polymer Degradation and Stability* 95(6): 1115–1122.

Lao, S. C., Koo, J. H., Moon, T. J. et al. 2011. Flame-retardant polyamide 11 nanocomposites: Further thermal and flammability studies. *Journal of Fire Sciences* 29(6): 479–498.

Lao, S. C., Wu, C., Moon, T. J. et al. 2009. Flame-retardant polyamide 11 and 12 nanocomposites: Thermal and flammability properties. *Journal of Composite Materials* 43(17): 1803–1818.

Laufer, G., Kirkland, C., Cain, A. A. et al. 2012. Clay–Chitosan nanobrick walls: Completely renewable gas barrier and flame-retardant nanocoatings. *ACS Applied Materials & Interfaces* 4(3): 1643–1649.

Li, Y. C., Schulz, J., and Grunlan, J. C. 2009. Polyelectrolyte/nanosilicate thin-film assemblies: Influence of pH on growth, mechanical behavior, and flammability. *ACS Applied Materials & Interfaces* 1(10): 2338–2347.

Liberata, G., Marialuigia, R., Vittoria, V. et al. 2013. The role of carbon nanofiber defects on the electrical and mechanical properties of CNF-based resins. *Nanotechnology* 24(30): 305704.

Liu, L., Etika, K. C., Liao, K. S. et al. 2009. Comparison of covalently and noncovalently functionalized carbon nanotubes in epoxy. *Macromolecular Rapid Communications* 30(8): 627–632.

Lu, H. D., and Wilkie, C. A. 2010. Synergistic effect of carbon nanotubes and decabromodiphenyl oxide/Sb_2O_3 in improving the flame retardancy of polystyrene. *Polymer Degradation and Stability* 95(4): 564–571.

Moniruzzaman, M., and Winey, K. I. 2006. Polymer nanocomposites containing carbon nanotubes. *Macromolecules* 39(16): 5194–5205.

Morgan, A. B. 2006. Flame retarded polymer layered silicate nanocomposites: A review of commercial and open literature systems. *Polymers for Advanced Technologies* 17(4): 206–217.

Morgan, A. B., and Galaska, M. 2008. Microcombustion calorimetry as a tool for screening flame retardancy in epoxy. *Polymers for Advanced Technologies* 19(6): 530–546.

Morgan, A. B., and Liu, W. 2011. Flammability of thermoplastic carbon nanofiber nanocomposites. *Fire and Materials* 35(1): 43–60.

Morgan, A. B., and Wilkie, C. A. 2007. *Flame Retardant Polymer Nanocomposites*. Hoboken, NJ: John Wiley & Sons.

Nyden, M. R., Zammarano, M., Harris, R. H. et al. 2010. Characterizing particle emissions from burning polymer nanocomposites. *Interflam Proceedings*, July 5–7, Nottingham, UK, pp. 623–628.

Pan, H., Pan, Y., Wang, W. et al. 2014. Synergistic effect of layer-by-layer assembled thin films based on clay and carbon nanotubes to reduce the flammability of flexible polyurethane foam. *Industrial & Engineering Chemistry Research* 53(37): 14315–14321.

Paul, D. R., and Robeson, L. M. 2008. Polymer nanotechnology: Nanocomposites. *Polymer* 49(15): 3187–3204.

Prasek, J., Drbohlavova, J., Chomoucka, J. et al. 2011. Methods for carbon nanotubes synthesis—Review. *Journal of Materials Chemistry* 21(40): 15872–15884.

Qian, H., Greenhalgh, E. S., Shaffer, M. S. P. et al. 2010. Carbon nanotube-based hierarchical composites: A review. *Journal of Materials Chemistry* 20(23): 4751–4762.

Rahatekar, S. S., Zammarano, M., Matko, S. et al. 2010. Effect of carbon nanotubes and montmorillonite on the flammability of epoxy nanocomposites. *Polymer Degradation and Stability* 95(5): 870–879.

Rahmat, M., and Hubert, P. 2011. Carbon nanotube–polymer interactions in nanocomposites: A review. *Composites Science and Technology* 72(1): 72–84.

Reijnders, L. 2012. Human health hazards of persistent inorganic and carbon nanoparticles. *Journal of Materials Science* 47(13): 5061–5073.

Sinha Ray, S., and Okamoto, M. 2003. Polymer/layered silicate nanocomposites: A review from preparation to processing. *Progress in Polymer Science* 28(11): 1539–1641.

Utracki, L. A., Sepehr, M., and Boccaleri, E. 2007. Synthetic, layered nanoparticles for polymeric nanocomposites (PNCs). *Polymers for Advanced Technologies* 18(1): 1–37.

Wang, J., Fang, Z., Gu, A. et al. 2006. Effect of amino-functionalization of multi-walled carbon nanotubes on the dispersion with epoxy resin matrix. *Journal of Applied Polymer Science* 100(1): 97–104.

Wang, K., Tang, C., Zhao, P. et al. 2007. Rheological investigations in understanding shear-enhanced crystallization of isotactic poly(propylene)/multi-walled carbon nanotube composites. *Macromolecular Rapid Communications* 28(11): 1257–1264.

Wang, L., Yu, J., Tang, Z. et al. 2010. Synthesis, characteristic, and flammability of modified carbon nanotube/poly(ethylene-co-vinyl acetate) nanocomposites containing phosphorus and silicon. *Journal of Materials Science* 45(24): 6668–6676.

Wang, X., Xia, T., Addo Ntim, S. et al. 2011. Dispersal state of multiwalled carbon nanotubes elicits pro-fibrogenic cellular responses that correlate with fibrogenesis biomarkers and fibrosis in the murine lung. *ACS Nano* 5(12): 9772–9787.

Wieser, E. https://en.wikipedia.org/wiki/Carbon_nanotube#/media/File:Multi-walled_Carbon_Nanotube.png. Accessed November 7, 2017.

Wu, Q., Zhang, C., Liang, R. et al. 2008. Fire retardancy of a buckypaper membrane. *Carbon* 46(8): 1164–1165.

Wu, Q., Zhu, W., Zhang, C. et al. 2010. Study of fire retardant behavior of carbon nanotube membranes and carbon nanofiber paper in carbon fiber reinforced epoxy composites. *Carbon* 48(6): 1799–1806.

Yang, K., Gu, M., Guo, Y. et al. 2009. Effects of carbon nanotube functionalization on the mechanical and thermal properties of epoxy composites. *Carbon* 47(7): 1723–1737.

Zammarano, M., Krämer, R. H., Harris Jr., R. et al. 2008. Flammability reduction of flexible polyurethane foams via carbon nanofiber network formation. *Polymers for Advanced Technologies* 19(6): 588–595.

Zhou, C. H., Shen, Z. F., Liu, L. H. et al. 2011. Preparation and functionality of clay-containing films. *Journal of Materials Chemistry* 21(39): 15132–15153.

Zhuge, J., Gou, J. and Ibeh, C. 2012b. Flame resistant performance of nanocomposites coated with exfoliated graphite nanoplatelets/carbon nanofiber hybrid nanopapers. *Fire and Materials* 36(4): 241–253.

Zhuge, J., Gou, J., Chen, R. H. et al. 2012a. Fire performance and post-fire mechanical properties of polymer composites coated with hybrid carbon nanofiber paper. *Journal of Applied Polymer Science* 124(1): 37–48.

Zou, W., Du, Z. J., Liu, Y. X. et al. 2008. Functionalization of MWNTs using polyacryloyl chloride and the properties of CNT–epoxy matrix nanocomposites. *Composites Science and Technology* 68(15): 3259–3264.

Graphene/Polymer Composites: A New Class of Fire Retardant

Xin Wang and
Wenwen Guo

9.1 Introduction

Graphene, a single-atom-thick nanosheet of sp^2-hybridized carbon atoms densely packed in a honeycomb crystal lattice, is discovered by the exfoliation of graphite in 2004 (Novoselov et al. 2004). In principle, it can be the basic building block of all sp^2-hybridized carbon allotropes; it can be stacked to form three-dimensional graphite, rolled to be one-dimensional carbon nanotubes (CNTs), and wrapped to form zero-dimensional fullerenes (Allen et al. 2010). There is no doubt that graphene has become one of the emerging two-dimensional exotic nano-materials in the twenty-first century (Kuila et al. 2012; Zhang and Liu 2012). The special long-range π-conjugation monolayer structure in graphene yields its excellent properties. For instance, graphene shows excellent mechanical properties, such as extremely high Young's modulus (~1 TPa) (Lee et al. 2008) and fracture strength (~125 GPa). Graphene exhibits excellent thermal conductivity (5000 W/mK) (Balandin et al. 2008) and electrical conductivity (~106 S/cm), and it also has very high electron mobility at room temperature (250,000 cm^2/V s) (Novoselov et al. 2005) and large surface area (2630 m^2/g) (Zhu et al. 2010). Therefore, the highly conjugated two-dimensional plane monolayer structure and the remarkable aforementioned properties make graphene nanosheets ideal platforms for the fabrication of high performance graphene/polymer composites (Roilo et al. 2017; Saravanan et al. 2014).

Up to now, considerable effort has been expended to prepare graphene/polymer composites (Cai et al. 2017; Du and Cheng 2012; Kim et al. 2010b; Kuilla et al. 2010; Yang et al. 2008). As reported in previous studies, graphene and its derivatives have been incorporated into various polymer matrices, such as epoxy

(Tang et al. 2013; Teng et al. 2011; Wajid et al. 2013), polypropylene (PP) (Huang et al. 2010), polyvinyl alcohol (Liu et al. 2017), polymethyl methacrylate (Pham et al. 2012; Thomassin et al. 2014), polystyrene (PS) (Patole et al. 2010; Yan et al. 2012a), nylon-6 (Xu and Gao 2010), and polyurethane (Wang et al. 2011b), which eventually exhibit remarkably reinforced properties at a very small amount (<5 wt%). Among these property improvements, flame retardancy of polymeric materials is one of the most important issues. With the different morphology from other carbon allotropes such as CNT and fullerene, the two-dimensional layered structure of graphene is very similar to that of montmorillonite (MMT) and layered double hydroxide (LDH), both of which have been widely used to prepare flame retardant polymeric materials over the past 2 decades. Previous investigations have demonstrated that the incorporation of these two-dimensional nanosheets can significantly improve the flame retardancy and thermal stabilities of polymer matrices (Wang et al. 2016). Hence, as a new class of flame retardant, graphene and its derivatives with large aspect ratio have aroused increasing scientific interests in recent years (Wang et al. 2017b). The unique laminar structure makes graphene a good physical barrier which can slow down heat and mass transfer between the gas and condensed phase and protect the underlying polymer matrix from further attack by flame.

The incorporation of graphene or its derivatives has been demonstrated to be effective in reducing the flammability and enhancing the thermal stability of polymer matrices (Feng et al. 2018; Nine et al. 2017b; Pethsangave et al. 2017; Wang et al. 2016, 2017b; Xu et al. 2017). However, due to the high surface area and intrinsic strong van der Waals force between graphene nanosheets, the major challenge is that it is easily restacked, and dispersed with difficulty, within polymer matrix (Fang et al. 2009; Zhu 2008). Therefore, it is necessary to functionalize the graphene nanosheets to improve the dispersion state of graphene and to enhance the compatibility between graphene and polymer matrices. To solve these issues, a large number of methods have been proposed to prepare functionalized graphene via covalent linkage, hydrogen bonding, and electrostatic and π–π interactions (Chen et al. 2012a).

In this chapter, we review the most recent advances regarding the use of graphene in the development of flame retardant polymer composites. The flammability characteristics of the graphene/polymer (nano) composites is the primary area of interest, while a number of other properties including thermal stability and mechanical and electrical properties are also commented. The flame retardant mechanism is also explored for optimizing design of graphene/polymer (nano)composites. Finally, we give a brief comment on challenges and opportunities for future development of this new class of polymer (nano)composites.

9.2 Preparation Methods of Graphene

Usually, single- or few-layered graphene can be fabricated in various exfoliated methods. In this section, we will display the typical synthetic routes of exfoliated graphene from mechanical exfoliation, chemical vapor deposition (CVD), liquid phase exfoliation, electrochemical exfoliation, as well as chemical reduction of graphene oxide.

9.2.1 Mechanical Exfoliation

Mechanical exfoliation is originally known as the "Scotch-tape" method which surprisingly made stand-alone graphene a reality and led to Geim and Novoselov being awarded the Nobel Prize in 2010 (Geim 2011). It is the simplest among the preparation methods. In this technique, graphite was mechanically split into monolayer and few-layer graphene which is best for the study of its fundamental properties. However, this method is unlikely to be suitable for large-scale production due to its obvious disadvantages in terms of yield and throughput (Chen et al. 2012b).

9.2.2 Chemical Vapor Deposition

CVD on transition metal substrates is one of the most promising approaches to prepare graphene of high structural quality, which is inexpensive and produces large-area graphene. During the CVD process, gas species are sent into the reactor and pass through the hot zone with high temperatures, where hydrocarbon precursors will decompose to carbon radicals at the metal substrate surface, and then, shape into single-layer and few-layers graphene. In the reaction process, the metal substrate not only works as a catalyst to decrease the energy barrier for the reaction, but determines the graphene deposition mechanism as well, which eventually affects the quality of graphene (Papageorgiou et al. 2017; Zhang et al. 2013).

There are several types of CVD methods available such as thermal CVD, plasma-enhanced CVD, hot/cold wall CVD, etc. And diverse transition metals can be used to synthesize graphene by CVD method such as nickel, copper, Palladium, iridium, ruthenium, cobalt, etc. (Bhaviripudi et al. 2010). The formation mechanism of graphene depends on the metal substrate, but usually initiates with the growth of carbon atoms which nucleate on the metal after decomposition of the hydrocarbon precursors and the nuclei grow then into large domains. Except for gaseous hydrocarbons such as methane, ethylene, or acetylene, liquid hydrocarbon precursors have also been utilized such as hexane (Srivastava et al. 2010) or pentane (Dong et al. 2011), while the quite different set of materials which can be employed for the CVD production of graphene can even involve food, insects, and waste (Ruan et al. 2011). After synthesis process, the as-grown graphene film would be transferred to other insulator substrates for further characterization and applications. The transfer process can be difficult because of the chemical inertness of graphene and it will bring defects and wrinkles to the material, while the thermal fluctuations can also affect the stability of the grown material (Novoselov et al. 2016). In spite of the complexity of the CVD process, and the high energy demands for the specific method add to the difficulties, CVD method is still one of the most successful methods for the production of high-quality graphene in large-area quantities.

9.2.3 Liquid Phase Exfoliation (LPE)

Liquid phase exfoliation is one of the promising techniques to achieve mass production of graphene. It is becoming more and more interesting because the direct LPE of crystal graphite into graphene assisted by sonication in some certain solvents or solution makes facile, versatile, green, scalable, and low-cost graphene production possible (Yi et al. 2012). The LPE process typically contains three steps: (1) dispersion of pristine graphite in a solvent, (2) exfoliation, and (3) purification in order to remove the non-exfoliated material from the exfoliated one (Ciesielski and Samori 2014). Previous studies used ultrasonication to exfoliate graphite flakes in a suitable organic solvent. During ultrasonication, shear forces and cavitation, namely, the growth and collapse of the micrometer-sized bubbles or voids in liquids due to pressure fluctuations, act upon the bulk material and cause exfoliation. Ultrasonic time is very important, for higher concentrations of graphene can be gained via longer sonication time, at the cost of energy consumption. After exfoliation, the solvent–graphene interaction has to balance the inter-sheet attractive forces. Solvents ideal to disperse graphene are those which can minimize the interfacial tension between liquid and graphene flakes, i.e., the force that can minimizes the area of the surfaces in contact (Ciesielski and Samori 2014). After the ultrasonication step, the obtained dark dispersion was centrifuged to remove any largish flakes and higher centrifugation speeds result in thinner flakes. Due to the existing industrial knowledge and equipment, the direct liquid-phase exfoliation assisted by sonication, which is associated with simplicity, speed, and high throughput, is expected to be the most viable option for up-scaling of graphene production (Papageorgiou et al. 2017).

9.2.4 Electrochemical Exfoliation

Electrochemical exfoliation to produce graphene is a reasonably simple, fast, and low-cost method compared to other methods (Wang et al. 2011a). The typical process includes the use of electrolyte and an electrical current used to consume the electrode consisting of graphite. The exfoliated process is non-spontaneous in which the current flow generates a rearrangement of the charges on the surface of the graphite-based electrode, so that graphite is expanded by intercalation and then heated for expansion into a sudden gaseous growth of intercalation, when the growth is strong enough to overcomes the van der Waals forces joining the layers, graphite will be exfoliated and suspended in the electrolyte. Figure 9.1 schematically illustrates the device and the proposed mechanism of electrochemical exfoliation (Parvez et al. 2013).

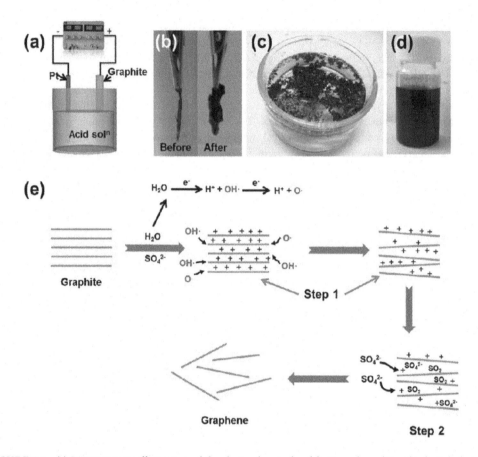

FIGURE 9.1 (a) Diagrammatic illustration of the electrochemical exfoliation of graphite, (b) digital photos of graphite flakes before and after exfoliation, (c) electrochemically exfoliated graphene floating on top of water, (d) dispersed graphene sheets (~1 mg/mL) in DMF, and (e) diagrammatic illustration of the proposed mechanism of electrochemical exfoliation. (Reprinted from Parvez, K. et al., *ACS Nano*, 7, 3598–3606, 2013. With permission.)

The superiority of electrochemical exfoliation method over the other methods is that this route takes place by a very simple single step, it is very easier to operate and very efficient for obtaining graphene oxide in a short space of time. It is a relatively promising approach for large-scale production of graphene and this process is also very easy to be applied in the industrial production of graphene. However, there are several issues that should be additionally considered, such as the difficulty of controlling the process, the expensive ionic liquids used in some routes, and obtaining various types of flakes (monolayer, bilayer, and multi-layers). Apart from the disadvantages above, the obtained exfoliated graphene chemically reacted in the process generate plenty of oxygen functional groups and defects. Hence, there are several factors that should be adjusted to gain flakes in a controlled manner in order to obtain graphene of high quality and high yield.

9.2.5 Chemical Reduction of Graphene Oxide

Among a large deal of methods for graphene preparation, the chemical reduction of graphene oxide (GO) is the most convenient method to date due to its advantages of low-cost and large-scale production of graphene for commercial applications (Chua and Pumera 2014). The preparation process of graphene obtained from chemical reduction of GO contains chemical oxidation, exfoliation, and reduction steps (Figure 9.2). Graphite oxide is a quite important intermediate material for the production of graphene.

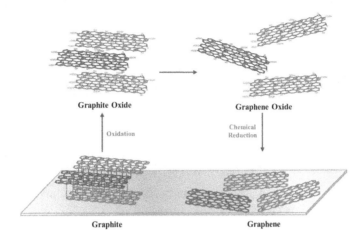

Graphite Oxide　　　　　　　　**Graphene Oxide**

Oxidation　　　　　Chemical Reduction

Graphite　　　　　　　　**Graphene**

FIGURE 9.2 Synthesis of graphene from chemical reduction of GO. (Reprinted from Chua, C.K. and Pumera, M., *Chem. Soc. Rev.*, 43, 291–312, 2014. With permission.)

The oxidation process of graphite helps to enlarge the interlayer distance between graphene sheets in graphite which is beneficial for an easy exfoliation.

Graphite oxide is generally produced via oxidation of natural graphite powder by one of three principal ways developed by Brodie, Staudenmaier, and Hummers, respectively (Stankovich et al. 2007). After the oxidation process, the basal-plane of the graphite oxide is heavily populated with hydroxyls and epoxides, whereas the edge-plane basically consists of carbonyl and carboxyl groups (Gómez-Navarro et al. 2007). These groups give assistance in not only expanding the interlayer distance, but also making the atomic-thick layers hydrophilic. As a consequence, graphite oxide can be exfoliated in water via adequate ultrasonication. If the exfoliated sheets contain only one or few layers of carbon atoms such as graphene with oxygenating groups, these sheets are named as GO (Thakur and Karak 2015).

GO is considered as another key intermediate between graphite and graphene. It is structurally different, but chemically similar to graphite oxide. However, these oxygen-containing groups account for the defective structure which makes it necessary to do further reduction or modification of graphite oxide to get stable and high-quality graphene production. The conversion of graphene oxide to graphene is, by observing experimental phenomena, usually indicated by a color change of the reaction mixture from brown (GO) to black (graphene) (Chua and Pumera 2014; Compton and Nguyen 2010). In order to accomplish the transformation of graphene oxide to graphene, thermal, chemical, and electrochemical reduction methods have been often applied. These three approaches would endow graphene with various performances in terms of electronic, physical, structural, and surface morphological properties. Generally, chemical reduction of GO is more commonly used (Chen et al. 2010b; Sheng et al. 2011). This method is significantly affected by the choice of reducing agent, solvent, and surfactant, which can be combined (Papageorgiou et al. 2017). There is currently a wide range of reducing agents for graphene oxide, such as hydrazine and its derivatives, dimethyl hydrazine, hydroquinone, amino acid, sodium borohydride, vitamin C, etc. Among various reducing agents used for chemical reduction, hydrazine or hydrazine hydrate are widely used reducing agents for their high reduction efficiency (Chen et al. 2010b).

The reduced GO sheets (rGO) have generally been regarded as one kind of chemically derived graphene. rGO is also known by some other names, such as reduced graphene, functionalized graphene, chemically modified graphene, chemically converted graphene, etc (Pei and Cheng 2012). Several important concerns of rGO are in the case of healing the defective graphene oxide, achieving the highest reduction capability, selectively removing a specific type of oxygen moiety, enhancing the dispersion stability of the obtained graphene, as well as applying eco-friendly and economical reducing or modifying agents (Chua and Pumera 2014).

Even though the quality and purity of the material produced by reduction of graphene oxide is imperfect, the reduction of graphene oxide is still applied extensively in the literatures due to its high yield and possibilities for scalability (Papageorgiou et al. 2017). Compared to other methods used for graphene production, chemical reduction of GO is known to be a promising means for the large-scale production of graphene-based materials. As abundant in oxygen-containing functional groups, varying reduction and modification methods can be applied to obtained rGO to meet different requirements. There is no doubting that this method is very important for the applications of graphene in many fields such as composites, coatings, paint/ink, transparent conductive layers, bio-applications, and energy storage (Chua and Pumera 2014).

9.2.6 Comparison of Production Methods

In order to introduce graphene into practical production and applications, intensive research efforts have been made over the years on the developing mass production methods of graphene. Every production method attributes different characteristics to the final products and has different possibilities for up-scaling. Raccichini et al. (2015) have done a nice summary for the advantages and drawbacks of each method in one of their articles. They evaluated the common methods applied for graphene production in terms of the best important aspects of the product graphene (quality and purity) and of every method (scalability, cost, and yield). Figure 9.3 compares the most common methods adopted

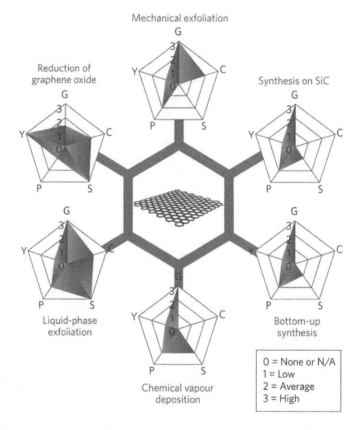

FIGURE 9.3 Schematic comparison of the most common graphene production methods. Each method has been assessed in terms of graphene quality (G), cost aspect (C); a low value corresponds to high cost of production), scalability (S), purity (P), and yield (Y) of the overall production process. (Reprinted from Raccichini, R. et al., *Nat. Mater.*, 14, 271–279, 2015. With permission.)

for graphene production. It is a fact that liquid-phase exfoliation and reduction of graphene oxide are widely employed methods for the mass production of graphene. Nevertheless, each method displays different characteristics, and the methods play a critical role in determining the properties of the final material. So selection of the method should be performed every time, based on the fields the graphene will be used (Papageorgiou et al. 2017).

9.3 Preparation Methods of Graphene/Polymer Composites

In general, there are three major strategies which are widely used to prepare the graphene/polymer nanocomposites: melt compounding, solution blending, and *in situ* polymerization.

9.3.1 Melt Compounding

Melt compounding is a scalable and economically attractive preparation technique favored by industry, which directly disperses fillers into molten polymers under mechanical shear forces. A wide variety of graphene/polymer composites including polystyrene (Qi et al. 2011), polyamide-12 (Yan et al. 2012b), polyethylene terephthalate (Zhang et al. 2010), and natural rubber (Berki et al. 2017) have been prepared via melt compounding. Despite the advantages of this technique, the dispersion state of graphene is usually not satisfactory owing to the high viscosity of the molten polymer as well as the high loading of fillers (Zhan et al. 2011). In contrast to the other preparation techniques for graphene/polymer composites, melt compounding method results in the poorest dispersion state of the fillers, as demonstrated by Kim et al. (2010c) for their thermoplastic polyurethane (TPU) composites filled with graphite and thermally reduced graphene (TRG) (Figure 9.4). It is important to note that even though melt

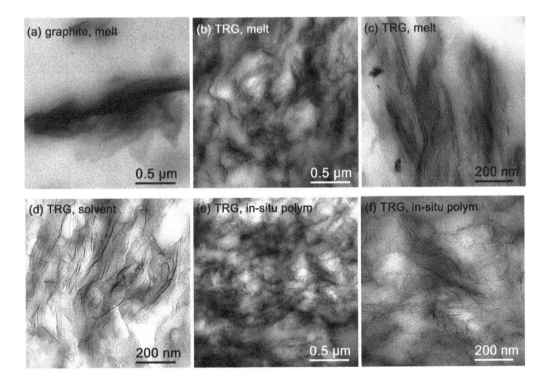

FIGURE 9.4 TEM images of thermoplastic polyurethane nanocomposites with (a) 5 wt% graphite, (b, c) melt-blended, (d) solvent-mixed, (e, f) *in situ* polymerized ~3 wt% TRG. (Reprinted from Kim, H. et al., *Chem. Mater.*, 22, 3441–3450, 2010. With permission.)

compounding does not lead to the satisfactory dispersion state of filler, flame retardancy along with other property improvements have been observed in the previous studies, as will be discussed in the following section.

9.3.2 Solution Blending

Solution blending is a technique that is widely used in academic investigations for the fabrication of graphene/polymer nanocomposites (Al-Attabi et al. 2017; Alireza 2014; Cao et al. 2010; Chang et al. 2015; Chen et al. 2010a; Ionita et al. 2013; Lago et al. 2016; Liang et al. 2009; Mo et al. 2015; Ramanathan et al. 2008; Stankovich et al. 2006a; Vilcinskas et al. 2015; Xu et al. 2009; Yang et al. 2010; Zhao et al. 2010). Generally, this technique involves the blending of colloidal suspensions of graphene nanoplatelets with the host polymer, followed by removal of solvent. Since the precursor, graphene oxide, could be stably dispersed in the aqueous suspensions, solution blending strategy is particularly appealing for water-soluble polymers including polyethylene oxide (PEO) (Chang et al. 2015), polyvinyl alcohol (PVA) (Liang et al. 2009; Mo et al. 2015; Xu et al. 2009; Yang et al. 2010; Zhao et al. 2010), alginate (Ionita et al. 2013; Vilcinskas et al. 2015), and chitosan (Alireza 2014). If water-insoluble polymers are used as the host matrices, such as polyurethane (Al-Attabi et al. 2017), polystyrene (Stankovich et al. 2006a), poly-methyl methacrylate (Ramanathan et al. 2008), polycarbonate (Lago et al. 2016), polylactic acid (PLA) (Cao et al. 2010), and polyimides (Chen et al. 2010a) composites could be fabricated in organic solvents. Graphene nanoplatelets could be easily exfoliated in the suspensions under the assistance of ultrasoni-cation or high-speed shear mixing, which ensures good dispersion of the filler in the polymeric matrix. Despite many advantages of solution blending, the re-stacking of well exfoliated graphene nanoplatelets might occur during the slow procedure of solvent evaporation. Furthermore, the high expense of sol-vents and their disposal limits the scalable production of graphene/polymer nanocomposites using this technique by industrial community. Finally, this technique is strongly dependent upon the solubility of the host polymer. For example, poor solubility of most conjugated polymers restricts the utilization of solution blending method in fabricating their nanocomposites.

9.3.3 *In Situ* Polymerization

In situ polymerization is an effective approach that has been previously used to fabricate layered-silicate/polymer nanocomposites with intercalated and/or exfoliated structure (Sinha Ray and Okamoto 2003). This method generally involves the blending of nano-fillers in neat monomer or a monomer solution, followed by *in situ* polymerization in the presence of well dispersed nano-fillers. Monomers could be covalently grafted onto graphene, followed by *in situ* polymerization to form graphene/polymer nanocomposites. For example, graphene/polyurethane (PU) nanocomposite has been successfully prepared using this strategy (Wang et al. 2011b). Also, *in situ* polymerization has been employed for the preparation of non-covalent graphene/polymer nanocomposites. Monomer and initiator are intercalated into the interlayer spacing of graphene or its precursor as guest species, and then the initiators trigger the *in situ* formation of polymers, like graphene/polymethyl methac-rylate (PMMA) (Thomassin et al. 2014) and graphene/polyethylene (Zhang et al. 2017a). Despite the highest homogeneity of nano-fillers within polymer matrix, this method has difficulty controlling the molecular weight of the polymers in the resultant nanocomposites.

9.4 Application of Graphene in Flame Retardant Polymers

As aforementioned, the fascinating properties of graphene enable itself one promising nano-filler to be employed in fabrication of polymer nanocomposites. In this section, the performances of graphene-based polymer nanocomposites will be commented with special emphasis on the flame retardant property.

9.4.1 Utilization of Pristine Graphene

Graphite oxide has been widely used as a halogen-free flame retardant additive in polymers over the past few decades (Higginbotham et al. 2009; Lee et al. 2011; Uhl and Wilkie 2004). Graphite oxide can be easily exfoliated into GO; however, GO is not thermally stable and apt to be consumed when exposed to heat sources according to the previous report (Figure 9.5a) (Shi and Li 2011). Interestingly, GO can be thermally reduced into a highly thermal stable form, graphene. Graphene is quite stable against a natural gas flame for a few seconds. The burned part turns red, but does not propagate, and is quenched after the flame removal. No burning occurs after the thermal reduction of GO, clearly demonstrating the high intrinsic flame retardancy of graphene (Figure 9.5b) (Shi and Li 2011). Besides thermally reduced graphene, chemically reduced graphene by flame retardants is another method to obtain highly flame retardant materials. For example, *in situ* polymerization of dopamine occurred on the surface of GO nanoplatelets, accompanying by the chemical reduction of GO (Luo et al. 2017a). The resulting graphene-polydopamine film shows high flexibility, anisotropic thermal conductivity, and excellent flame resistance without ignition as well as very peak low heat release rate (40 W/g). Furthermore, chemically reduced GO can be self-assembled into graphene aerogels with light-weight, good compressibility, high porosity, and outstanding fire resistance (Li et al. 2014a). The highly porous structure allows graphene aerogels as efficient absorbent for organic liquids, while the exceptional fire resistance enables it to be recyclably absorbent since the absorbed organic liquids can be easily removed by combustion. Doped graphene with flame retardant elements such as phosphorus (Kim et al. 2014) and aluminum (Jeon et al. 2017) has also been demonstrated as an effective flame retardant for polymer materials.

Owing to its excellent intrinsic fire resistant property, pristine graphene has been used directly to fabricate flame retardant polymer nanocomposites through solution blending or melt compounding approaches. Graphene/waterborne polyurethane (WPU) nanocomposites were fabricated by an *in situ* emulsion blending method (Hu and Zhang 2014). When the graphene fraction is 1 wt%, the peak heat release rate (pHRR) and total smoke release (TSR) are reduced by 18% and 25%, respectively, compared to those of WPU. Pristine graphene has also been introduced into flame retardant epoxy resin (Liu et al. 2014b). Although the increased limiting oxygen index (LOI) value and the decreased total heat release (THR) are observed with the addition of graphene, it is noted that the pHRR exhibits a first increase and

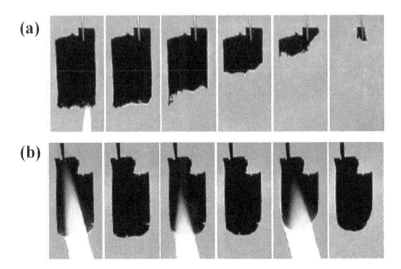

FIGURE 9.5 Snapshots showing (a) flame treatments on rGO film with intervals of 3–4 seconds and (b) flame treatment on a rGO contaminated with KOH salts (1 wt%). (Adapted from Kim, F. et al., *Adv. Funct. Mater.*, 20, 2867–2873, 2010. With permission.)

subsequent decrease trend with the increase of graphene fraction, in comparison to that of neat epoxy resins. This phenomenon is primarily attributed to the balance between the thermal conductivity and the barrier effect of graphene (Liu et al. 2014b).

The flame retardant efficiency of graphene in polymer composites is closely dependent on its oxidation degree. Han et al. (2013) found that both the thermal stability and the pHRR reduction of the resulting polystyrene nanocomposites increase with the decreased oxidation degree of GO. It has been reported that high oxidation degree was detrimental to the flame retardant effect of GO owing to a weak intumescent ability (Lee et al. 2011). Compared to GO, graphene imparted better flame retardancy as well as thermal stability to polystyrene nanocomposites due to the elimination of thermally liable oxygen-containing groups on the graphene sheets (Han et al. 2013). Similar phenomenon is also observed in another study (Han et al. 2016). The reduced graphene at higher reduction temperature displays better thermal stability and flame retardant performance for polystyrene nanocomposites. The optimal flame retardancy is achieved with the reduced graphene at 800°C (5 wt%), in which case the pHRR reduction is about 47.5% as compared to pure polystyrene.

9.4.2 Organic Flame Retardants Functionalized Graphene

GO, as a precursor of graphene, contains a great amount of functional groups including epoxy, hydroxyl, and carboxyl groups at the basal planes and edge areas (Figure 9.6) (Stankovich et al. 2006b), which facilitate the functionalization of graphene via covalent bonding. The utilization of covalently grafted organic flame retardants is of particular interest for the fabrication of functionalized graphene. There are two major strategies to synthesize organic flame retardant functionalized graphene: *grafting to* and *grafting from*.

The *grafting to* method has been widely used to prepare organic flame retardants functionalized graphene. This method generally involves the synthesis of a small molecule or polymer with a reactive end group which is covalently linked to the surface of graphene or GO. 9,10-Dihydro-9-oxa-10-phosphaphenanthrene-10-oxide (DOPO) (Liao et al. 2012; Luo et al. 2016, 2017b) and its derivatives (Chen et al. 2017; Feng et al. 2017a; Guo et al. 2017; Shi et al. 2018a; Sun et al. 2016) are the most

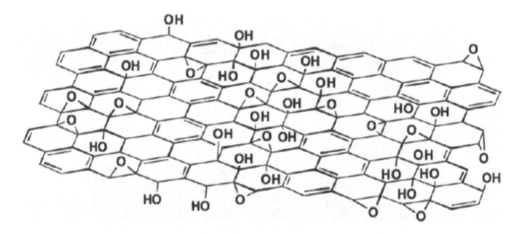

FIGURE 9.6 Schematic illustration of a proposed model of graphene oxide structure. (Reprinted from He, H. et al., *Chem. Phys. Lett.*, 287, 53–56, 1998. With permission.)

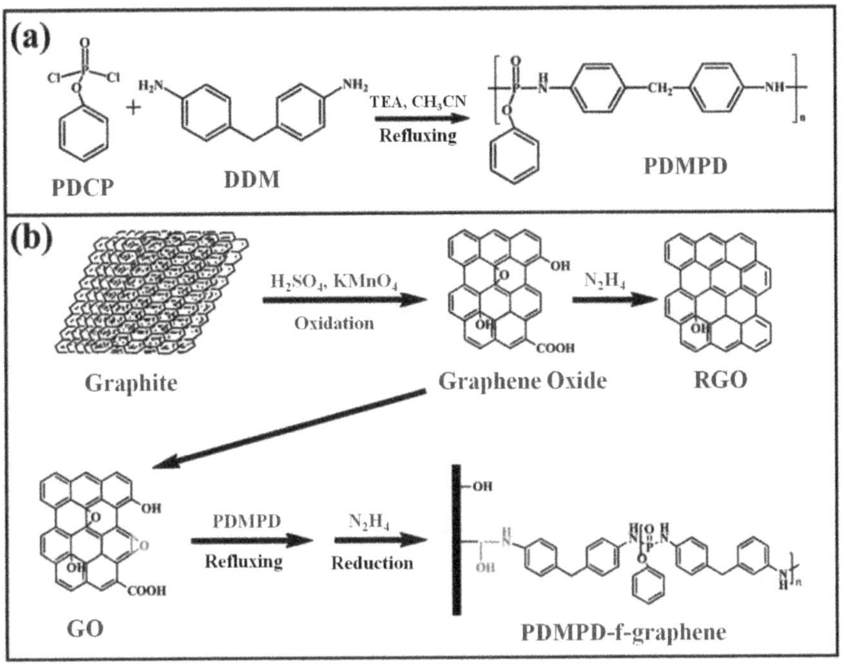

FIGURE 9.7 Schematic illustrations of synthetic routes of (a) PDMPD and (b) PDMPD-f-graphene. (Adapted from Feng, Y. et al., *J. Mater. Chem. A*, 5, 13544–13556, 2017. With permission.)

common kind of small molecular flame retardants used for functionalized graphene. Alternatively, various polymeric flame retardants, including polyethyleneimine (PEI)-based phosphonamidate (Shi et al. 2018b), polydimethylsiloxane (PDMS) (Luo et al. 2017c), polyphosphamide (PPA) (Wang et al. 2014b; Yu et al. 2014), poly(piperazine spirocyclic pentaerythritol bisphosphonate) (PPSPB) (Huang et al. 2012a), poly(4,4′-diaminodiphenyl methane phenyl dichlorophosphate) (PDMPD) (Feng et al. 2017b), poly(methylphosphonyl bis(hydroxymethyl) hyposphate) (PMPPD) (Qiu et al. 2015), and hyperbranched flame retardants (Hu et al. 2014, 2015) have also been grafted to the surface of graphene or GO. The typical preparation procedure of organic flame retardant functionalized graphene via *grafting to* method is illustrated in Figure 9.7.

The *grafting from* method is another emerging strategy to prepare functionalized graphene with flame retardants. This strategy generally involves *in situ* polymerization of the monomer directly from the graphene or GO. Grafting from method has been employed to grow polyaniline (PANI) (Yuan et al. 2016) and different polymeric flame retardants from condensation between phosphorus-nitrogen (Yu et al. 2015) and phosphorus-silicon (Zhang et al. 2017b) compounds. However, it is difficult for this strategy to explicitly control the molecular weight and polydispersity index of the polymeric flame retardants grown on graphene or GO. The typical preparation procedure of organic flame retardant functionalized graphene via *grafting from* method is illustrated in Figure 9.8.

Table 9.1 summarizes several organic flame retardant functionalized graphene and their flame retardant performance in polymer composites. As expected, organic flame retardant functionalized graphene displays more enhanced flame retardant performance than either organic flame retardants or pristine graphene in terms of LOI, UL-94, rating and pHRR. For instance, epoxy composites with

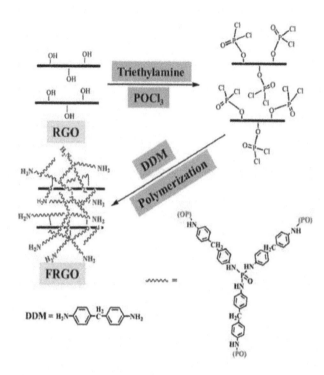

FIGURE 9.8 Schematic illustrations of the preparation of flame retardant functionalized GO (FRGO). (Reprinted from Yu, B. et al., *J. Mater. Chem. A*, 3, 8034–8044, 2015. With permission.)

TABLE 9.1 Flammability of Organic Flame Retardant Functionalized Graphene/Polymer Nanocomposites

Type of Polymer	Type and Content of Functionalized Graphene	Highlights	References
Epoxy	4 wt% of DOPO-based phosphonamidate functionalized graphene	A UL-94 V-0 rating was attained compared to pure epoxy (no rating [NR]*) and epoxy with 4 wt% DOPO-based phosphonamidate (V-1); 43% and 30% reduction in pHRR and THR, respectively, measured by microscale combustion calorimeter (MCC)	Yu et al. 2015
Epoxy	10 wt% of DDPES-graphene	Improvement in LOI to 36% and UL-94 V-0 grade, respectively, in comparison to pure epoxy (20%, NR), DPPES-10%/epoxy (26%, NR), and graphene-10%/epoxy (26%, NR)	Li et al. 2014b
Epoxy	4 wt% of DPP-graphene	LOI value attained 25.2% along with UL-94 V-0 rating (1.6 mm); for DPP-4%/epoxy composite, LOI value of 22.6%, fail in UL-94 test (1.6 mm)	Chen et al. 2017
Epoxy	1 wt% of PEI-DOPO phosphonamidate functionalized graphene	22% reduction in pHRR observed from cone calorimeter, better than that of GO-1%/epoxy (12%)	Shi et al. 2018c
Epoxy	5 wt% of DOPO-graphene	Improvement in char yield to 25.6% from 21% (DOPO-5%/epoxy) and 20.1% (GO-5%/epoxy); increase in LOI to 25% from 23% (DOPO-5%/epoxy) and 23% (GO-5%/epoxy)	Liao et al. 2012

(Continued)

TABLE 9.1 (*Continued*) Flammability of Organic Flame Retardant Functionalized Graphene/Polymer Nanocomposites

Type of Polymer	Type and Content of Functionalized Graphene	Highlights	References
Epoxy	8 wt% of PPA-graphene	41% and 50% reduction in pHRR and THR were observed in cone calorimeter, respectively, while those are 35% and 46% for PPA-8%/epoxy composites	Wang et al. 2014b
PP	2 wt% of melamine-graphene	29% reduction in pHRR measured by cone calorimeter, while that are 6% for epoxy/2 wt% GO	Yuan et al. 2015
PP	20 wt% of PPA-graphene	67% and 24% reduction in pHRR and THR were observed in cone calorimeter, respectively, while those were 48% and 20% for PPA-20%/PP composite	Yu et al. 2014
Ethylene-vinyl acetate copolymer (EVA)	1 wt% of PPSPB-graphene	46% reduction in pHRR and time to ignition (TTI) of 75 seconds were observed from cone calorimeter, whereas those are 31% and 61 seconds in the case of graphene-1%/EVA	Huang et al. 2012a
PS	5 wt% of PMPPD-graphene	38% and 21% reduction in pHRR and THR, respectively, observed from MCC, much better than those of GO-5%/epoxy (13% and 7%)	Qiu et al. 2015
Cross-linked polyethylene (XLPE)	1 wt% of hyper-branched flame retardant- graphene	28% reduction in pHRR observed from cone calorimeter, while it is only 8% for GO-1%/XLPE.	Hu et al. 2014

10 wt% of 2-(diphenylphosphino) ethyltriethoxy silane (DPPES) functionalized graphene could pass UL-94 V-0 rating, while its counterparts with the equivalent fraction of either graphene or DPPES cannot (Li et al. 2014b).

9.4.3 Inorganic/Graphene Hybrids

Besides covalent functionalization of graphene by organic flame retardant, a wide variety of inorganic/graphene hybrids have been synthesized through non-covalent modification. Although organic flame retardant functionalized graphene is effective in lowering the heat release rate during combustion, their smoke suppression effect is not so notable in most cases. Therefore, to combine the inorganic nanomaterials with catalysis and/or absorption effect with graphene to obtain hybrid flame retardants is expected. These inorganic/graphene hybrids usually can be prepared by co-precipitation (Wang et al. 2013b) or hydrothermal method (Zhou et al. 2016b). Taking LDH/graphene hybrid as an example, the preparation procedure is schematically presented in Figure 9.9. The metal cations are firstly anchored to the negatively charged GO by electrostatic attraction, and then GO is chemically reduced to graphene. Finally, graphene nanoplatelets decorated by metal cations are self-assembled into LDH/graphene hybrid nanostructure. Similar preparation procedure has been employed for metal oxide/graphene hybrids.

   One of the pioneering investigations on inorganic/graphene hybrids as a flame retardant additive was reported by Wang et al. (2012), where the addition of 2 wt% Co_3O_4/graphene hybrid led to a 31% and 40% reduction in pHRR for the resulting polybutylene succinate (PBS) and PLA composites, respectively. In another study (Wang et al. 2014a), metal oxide/graphene hybrids were demonstrated to be effective in improving the thermal stability of polymers: the temperature at 5% weight loss is increased by 37°C and 27°C for epoxy composites with 2 wt% SnO_2/graphene and 2 wt% Co_3O_4/graphene, respectively,

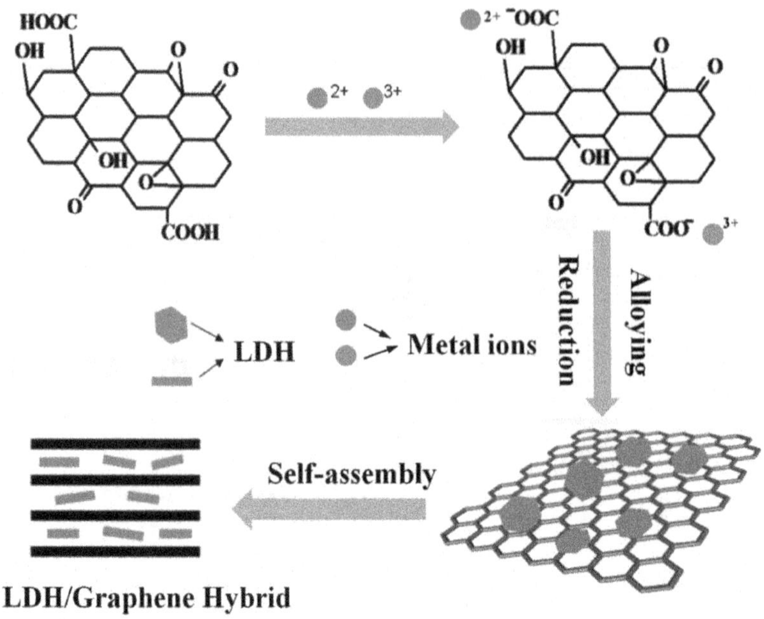

FIGURE 9.9 Typical preparation procedure of layered double hydroxide (LDH)/graphene hybrid. (Reprinted from Wang, X. et al., *Prog. Polym. Sci.*, 69, 22–46, 2017. With permission.)

in comparison to that of pure epoxy resin. Table 9.2 lists several inorganic/graphene hybrids and their flame retardant performance in polymer composites. In most cases, the pHRR of the polymer composites is reduced by at least 30% with a relatively low loading of inorganic/graphene hybrids (usually <3 wt%) (Hong et al. 2013; Jiang et al. 2014b; Wang et al. 2012, 2013a, 2013b; Xu et al. 2016; Yang et al. 2017). The notable reduction in heat release rate might be assigned to the formation of a char layer with high graphitized carbons (as demonstrated by Raman spectra) and thermal resistance (as demonstrated by X-ray photoelectron spectroscopy [XPS]) (Wang et al. 2013b), which serves as a good barrier to inhibit energy and mass transfer and shield the underlying matrix from heat irradiation. In addition to LDH/graphene and metal oxide/graphene hybrids, other inorganic/graphene hybrids including halloysite/graphene (Li et al. 2014c), molybdenum disulfide/graphene (Wang et al. 2013a), ZnS/graphene (Jiang et al. 2014a), aluminum hypophosphite/graphene (Qi et al. 2017), metal-organic frameworks/graphene (Zhang et al. 2018), zirconium phosphate/graphene (Xu et al. 2018), ferrocene/graphene (Zhou et al. 2016a), zinc hydroxystannate/graphene (Liu et al. 2016b), and zinc ferrite/graphene (Yang et al. 2017), have also been demonstrated to be effective in reducing the heat release rate of various polymers such as epoxy (Jiang et al. 2014a; Liu et al. 2016b; Wang et al. 2013a; Xu et al. 2018; Yang et al. 2017), polystyrene (Zhou et al. 2016a), polybutylene terephthalate (Qi et al. 2017), polylactic acid (Zhang et al. 2018), and polyamide-6 (Li et al. 2014c).

Besides the significant suppression effect on the heat release rate, inorganic/graphene hybrids also exhibit good suppression effect on smoke production and gas toxicity during combustion in contrast to pristine graphene (Wang et al. 2013b). The smoke suppression might be originated from the decreased amount of the volatile products evolved from polymer decomposition, because the volatile products are the main source of smoke particles (Wang et al. 2014a). Moreover, in some cases, the gas toxicity like CO is reduced due to the catalysis and/or absorption effect of inorganic species (Wang et al. 2012, 2014a), which is favorable to reduce the casualties in fire incidents.

TABLE 9.2 Flammability of Polymer Nanocomposites with Inorganic/Graphene Hybrids

Polymer	Type and Content of Inorganic/ Graphene Hybrids	Highlights	References
Epoxy	2 wt% of NiFe-LDH/graphene	61% and 60% reduction in pHRR and THR, respectively, which are superior over either NiFe-LDH or pristine graphene	Wang et al. 2013b
Epoxy	2 wt% of Ce-doped MnO_2/ graphene	54% and 41% reduction in pHRR and TSR, respectively, observed from cone calorimeter	Jiang et al. 2014b
PBS	2 wt% of Co_3O_4/graphene	31% reduction in pHRR of PBS composite as well as a significant decrease in the carbon monoxide release	Wang et al. 2012
PLA	2 wt% of Co_3O_4/graphene	40% reduction in pHRR of PLA composite as well as a significant decrease in the carbon monoxide release	Wang et al. 2012
PU	2 wt% of MoO_3/graphene	69% reduction in pHRR observed from cone calorimeter	Xu et al. 2016
PP	2.5 wt% of Ni–Ce mixed oxide/ graphene	37% and 36% reduction in pHRR and TSR, respectively, obtained from cone calorimeter, which is much better than graphene/PP composite	Hong et al. 2013
Epoxy	3 wt% of zinc hydroxystannate/ graphene	50%, 39%, and 10% reduction in pHRR, THR and average CO yield, respectively, observed from cone calorimeter, better than those containing either zinc hydroxystannate or graphene	Liu et al. 2016b
Epoxy	3 wt% of zinc ferrite/graphene	The pHRR, TSR, and peak CO productive rate are reduced by 39.6%, 32.6%, and 58.8%, respectively, in the cone calorimeter	Yang et al. 2017
Epoxy	2 wt% of MoS_2/graphene	46% and 30% reduction in pHRR and TSR, respectively, accompanying with a 53°C increment in the onset thermal degradation temperature (T_{onset}) under air atmosphere	Wang et al. 2013a

9.4.4 Synergism between Graphene and Other Flame Retardants

Despite the high efficiency in reducing the heat release rate of polymers by the utilization of organic flame retardants functionalized graphene and inorganic/graphene hybrids, most of these composites still cannot meet the industrial standard tests such as the LOI and UL-94 vertical burning measurements. Therefore, graphene has been combined with various conventional flame retardants for flame retarding polymers in order to create a synergism. Table 9.3 gives several synergistic systems between graphene and other flame retardants. Notable synergism created by combining CNTs and graphene has been observed: with 0.5 wt% of CNT and 0.5 wt% of graphene, the pHRR of the resulting PP composite is reduced by 73%, which is much better than its counterparts containing 1.0 wt% of graphene or CNT (Song et al. 2013). Striking multiple synergistic effects on the electrical conductivity (+32.3%), tensile strength (+14.3%), Young's modulus (+27.1%), and thermal conductivity (+34.6%) are also observed (Song et al. 2013). In addition to the significant reduction in heat release rate, the linear increment trend of LOI values with increasing addition of graphene nanoplatelets is also reported (Idumah et al. 2017). The LOI value increases from 18% for neat PP to 31% for kenaf flour polypropylene composite with 5 phr graphene. With regards to UL-94 vertical burning tests, kenaf flour polypropylene composites with 3 phr or 5 phr graphene attain V-0 rating (Idumah et al. 2017). However, in some cases, the high graphene loading results in an increased trend in pHRR of the resulting flame retardant polymer composites

TABLE 9.3 Combination of Graphene with Other Flame Retardant Additives in Polymer Composites

Polymer	Content of Graphene	Type and Content of Synergist	Highlights	References
PBS	2 wt%	APP (12 wt%) and MA (6 wt%)	An UL-94 V-0 rating material was attained with a high LOI value of 33.0% compared to pure PBS (23.0% in LOI and no rating in UL-94 vertical burning test)	Wang et al. 2011c
PVA	1 wt%	MPP (10 wt%)	Compared to pure PVA, the pHRR was decreased by about 60%. As well, the LOI value of 29.6% and UL-94 V0 grade was achieved	Huang et al. 2012d
Acrylonitrile–butadiene–styrene copolymer (ABS)	2 wt%	Cobalt hydroxide (4 wt%)	The pHRR exhibited a 30.5% reduction, and the tensile strength was increased by 45.1%	Hong et al. 2014
PLA	0.6 wt%	Polyphosphonate (2.4 wt%)	UL-94 V0 grade and a LOI value of 36.0% can be achieved	Jing et al. 2018
Epoxy	7 wt%	Alumina (68%) and magnesium hydroxide (5%)	The epoxy composite showed a high LOI of 39% and UL-94 V-0 rating as well as a thermal conductivity of 2.2 W/(mK), 11 times higher than that of neat epoxy	Guan et al. 2016
Epoxy	2.5 wt%	DOPO (2.5 wt%)	The pHRR was reduced by 67%, which was more significant than that of the epoxy composites with 5 wt% graphene or DOPO	Liu et al. 2016a
PP	0.5 wt%	CNT (0.5 wt%)	73% and 38% reduction in pHRR and average mass loss rate (AMLR), respectively, observed from cone calorimeter	Song et al. 2013
Cellulose	10 wt%	Sepiolite nanorods (10 wt%) and boric acid (3 wt%)	The flame did not self-propagate in vertical burning test and the LOI was as high as 34%	Wicklein et al. 2015
PU	2 wt%	microencapsulated ammonium polyphosphate (MCAPP) (12 wt%) and MA (6 wt%)	LOI value increased to 34.0% accompanying with good anti-dripping properties as well as UL-94 V-0 rating	Gavgani et al. 2014

(Hong et al. 2014). As aforementioned, the optimal flame resistance may be a balance between the barrier effect and the thermal conductivity of graphene which is similar to that of CNTs.

Recently, bioinspired PVA nanocomposites containing graphene and MMT nanosheets have been prepared via a vacuum-assisted filtration self-assembly process (Ming et al. 2015). The ternary graphene–MMT–PVA nanocomposites display exceptional integration of strength and toughness of 356.0 MPa and 7.5 MJ/m^3, respectively, which are much higher than those of pure PVA and graphene films.

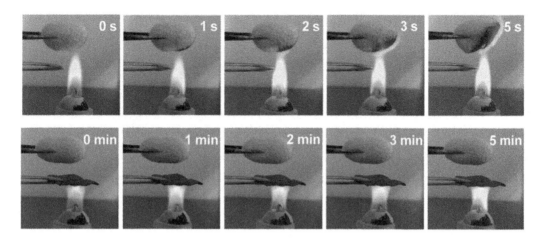

FIGURE 9.10 Snapshots showing that the ternary graphene–MMT–PVA nanocomposites act as a fire shield to protect a silk cocoon: (top row) without nanocomposites and (bottom row) with nanocomposites. (Reprinted from Ming, P. et al., *J. Mater. Chem. A*, 3, 21194–21200, 2015. With permission.)

As well, the ternary graphene–MMT–PVA nanocomposites show outstanding fire resistance to open flame, which could provide an effective insulator for combustible materials (Figure 9.10) (Ming et al. 2015).

In another study, lightweight foamy materials have been fabricated by freeze casting suspension of cellulose nanofibers, GO, and sepiolite nanorods (SEP) (Wicklein et al. 2015). The optimal flame retardancy is attained in the formulation of nanocellulose-based composite foam containing 10 wt% of GO and 10 wt% of SEP, which exhibits a self-extinguishing behavior in the vertical burning tests. Nanocellulose-based composite foams with no GO or low SEP content display inferior fire retardancy, demonstrating the existence of synergism between GO and SEP. Along with the excellent fire resistance, the composite foam shows a thermal conductivity as low as 15 mW/m·K, making it a promising candidate as thermally insulating material for construction and buildings.

The synergism between graphene with other flame retardants also imparts excellent anti-dripping property to polymers during the combustion. For example, PBS composite with 18 wt% intumescent flame retardant (APP and melamine [MA] by the weight ratio of 2:1) and 2 wt% graphene attains a UL-94 V-0 rating along with excellent anti-dripping behavior (Wang et al. 2011c). The enhanced melt viscosity by graphene has been proposed as anti-dripping mechanism. Similar anti-dripping behavior has also been reported in other synergistically flame retardant polymer systems such as polyurethane (Gavgani et al. 2014), polypropylene (Idumah et al. 2017), and epoxy resin (Liu et al. 2014a).

9.4.5 Flame Retardant Coatings Made from Graphene

In recent years, flame retardant coatings made from graphene have been deposited onto the surface of bulk polymer materials for fire-proof protection. The most common used techniques for fabricating graphene-based coatings include layer-by-layer (LbL) assembly and dipping or spraying. Layer-by-layer assembly of negatively charged GO and positively charged intumescent flame retardant-polyacrylamide (IFR-PAM) coating has been investigated to improve the flame retardant properties of cotton fabric (Huang et al. 2012c). With the 20 bilayers (BLs) of IFR-PAM/GO coating (add-on content of 3.04%), the pHRR of the treated cotton was decreased by 50% and the TTI was prolonged by 23 seconds in contrast to those of the control fabric. Wang and co-workers have compared GO and phosphorylated graphene oxide (PGO) in the flame retardant coating via LbL deposition onto cotton fabric (Wang et al. 2017a). The pHRR and THR values of the treated cotton with PGO/PEI coating are found to be lower than those of the samples with GO/PEI coating. However, these coatings still cannot stop flame propagation of cotton fabrics in the vertical burning tests. Chen et al. have deposited functionalized graphene oxide/PEI coating onto PVA film, and it is found that the pHRR and THR of all the coated PVA samples decreases gradually with increasing BLs number of functionalized graphene oxide (FGO)-based coating (Chen et al. 2018). Furthermore, compared to PVA composite blended with the equivalent content of FGO, the coated PVA shows much lower pHRR and THR along with better fire resistance against open flame (Chen et al. 2018). Fire retardant properties are also imparted to flexible polyurethane foams by constructing a ternary coating of chitosan/GO/alginate via LBL deposition (Zhang et al. 2016). In particular, the treated flexible polyurethane foam (FPUF) sample with ten trilayers of chitosan/GO/alginate (8.31 wt% in weight gain) shows a significant reduction in the pHRR (59.9%), peak smoke production rate (SPR) (45.6%), TSR (30.5%), and peak CO production (54.0%).

As aforementioned, the LbL assembly method has been adopted to construct effective fire protective coatings for polymeric materials. Unfortunately, this method generally requires many immersion and rinse steps. Alternatively, graphene-based flame retardant coatings have been applied onto polymeric materials by a simple dipping or spraying step. Very recently, GO hydrosol has been deposited onto silk fabrics by a dry-coating method and subsequently reduced by L-ascorbic acid (Ji et al. 2017). Through this method, the graphene sheets (up to 19.5 wt% relative to the original fabric) are deposited uniformly on the silk fabric and attached firmly to silk fibers. With increasing the deposited graphene amount, the LOI shows a significantly increasing trend, and the damage length and the time to self-extinguish reveals a decreasing trend, as shown in Table 9.4. The graphene-coated silk fabric (19.5 wt% graphene) also displays good electrical conductivity even if exposed to an ethanol flame for 60 seconds (Figure 9.11), which enables graphene-coated silk fabric to be applied as a fire-resistant conductor.

Nine et al. (2017a) have treated the pine wood slat by spraying a thin coating layer of graphene/sodium metaborate (SMB). The treated wood displays strong intumescent effect and self-extinguishing ability when exposed to the butane flame, whereas the untreated wood slat completely burns out within 40 seconds (Figure 9.12). Apart from the outstanding fire resistance, the graphene/SMB coating also exhibits an excellent mechanical robustness (ASTM-class 4B) and a suppression rate of bacterial colonization up to ~99.92%. Graphene-based coatings have also been used to impart excellent flame retardant property to cotton fabric (Liu et al. 2018) and polyurethane sponges (Huijuan et al. 2017; Pan et al. 2016). However, the deposition of graphene-based coatings changes the color of polymer matrix to dark brown or black, which will limit its applications.

TABLE 9.4 LOI Value of Pristine Silk Fabric and Coated Silk Fabrics with Different Amounts of Graphene Coating and Their Damage Length, Time to Self-Extinguish, and Percentage of Residual Char after Vertical Flame Testing

Samples	LOI (%)	Damaged Length (cm)	Time to Self-Extinguish (s)	Residual Char (%)
Pristine silk fabric	24.0 ± 0.3	27.1 ± 2	12 ± 1	$35.5 \pm 5\%$
With 3.9 wt% graphene	28.8 ± 0.5	17 ± 0.5	8 ± 0.5	$85.3 \pm 1\%$
With 7.8 wt% graphene	29.6 ± 0.5	14.7 ± 0.5	7.5 ± 0.5	$86.1 \pm 1\%$
With 15.6 wt% graphene	37.5 ± 1.0	13.2 ± 0.5	6 ± 0.5	$88.7 \pm 1\%$
With 19.5 wt% graphene	43.5 ± 2.0	12.8 ± 0.5	5 ± 0.5	$90.1 \pm 1\%$

Source: Ji, Y. et al., *Mater. Des.*, 133, 528–535, 2017.

FIGURE 9.11 LED connected with a burning graphene-coated silk fabric (19.5 wt% rGO). (Adapted from Ji, Y. et al., *Mater. Des.*, 133, 528–535, 2017.)

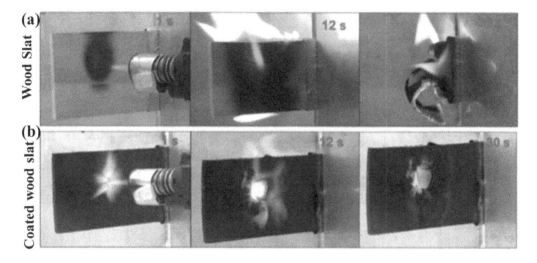

FIGURE 9.12 (a) Butane flame introduced for 12 seconds to pine wood board and (b) flame retardant ability of pine wood board treated with graphene/SMB coating. (Reprinted from Nine, J. et al., *Carbon*, 117, 252–262, 2017. With permission.)

9.4.6 Comparison of Graphene with Other Carbon Materials

It is well known that flame retardant property of polymer composites depends strongly upon particle sizes and shapes of additives. Schartel and co-workers have investigated the influence of different carbon additives including thermally reduced graphite oxide (TRGO), multi-layer graphene (MLG250), carbon black (CB), multi-wall nanotubes (MWNTs), and expanded graphite (EG) on the thermal stability and flame retardant property of polypropylene (Dittrich et al. 2013). The additional amount of these carbon additives is kept at 5.0 wt% concentration. The reduction in pHRR becomes more pronounced in the following sequence of materials: PP/EG40 < PP/MWNT < PP/CB < PP/TRGO (Figure 9.13a), indicating that a well-exfoliated layered morphology as in TRGO shows superior flame retardant efficiency over spherical particles (CB), tubes (MWNT), and platelets (EG). Furthermore, the reduction in pHRR becomes more pronounced with increasing the quality of dispersion of the layered particles, following the order of EG40 ≈ EG60 << MLG250 ≈ TRGO (Figure 9.13b). The reduction in pHRR becomes gradually pronounced with increasing Brunauer–Emmett–Teller (BET) surface area as shown in Figure 9.13c. Similar trend in the onset decomposition temperature is also observed (Figure 9.13d).

Synergism between halogenated flame retardants and the different carbon materials has been further investigated (Daniel et al. 2013). The pronounced dependency between the BET surface area and flame retardant property is analogous to that for PP composites containing single carbon material. By comparison to EG, nano-scaled CB, and MWNTs, only graphene including MLG250 and TRGO affords uniform dispersion within PP matrix accompanied with a significantly reduced pHRR by 76%. The char residues of the halogenated flame retardant PP containing MLG250 (5 wt%) and TRGO (5 wt%) seem to be more compact and with a uniform structure (Figure 9.14), which is responsible for the strong reduction in pHRR.

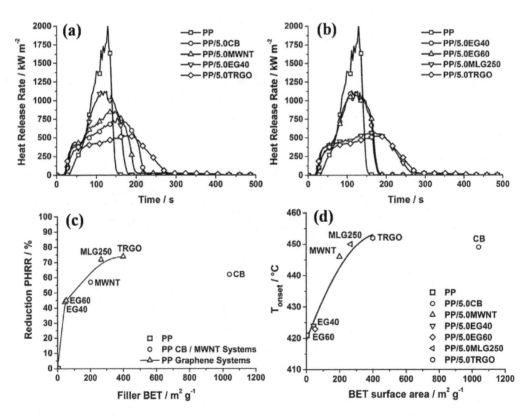

FIGURE 9.13 Heat release rate curves of PP and its carbon nanocomposites with nanofillers of (a) different morphologies and (b) various surface areas. (c) Reduction in pHRR and (d) onset degradation temperature (T_{onset}) plotted against the BET surface area of carbon nanofillers. (Reprinted from Dittrich, B. et al., *Polym. Degrad. Stab.*, 98, 1495–1505, 2013. With permission.)

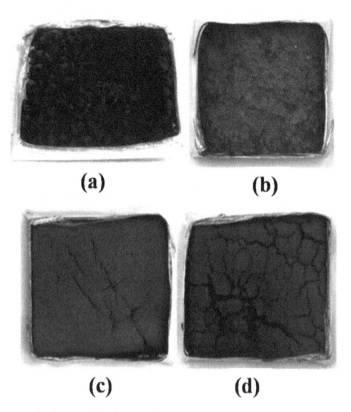

FIGURE 9.14 Photographs showing that char residues after cone calorimeter tests of flame retardant PP containing 5.0 wt% of (a) CB, (b) EG40, (c) MLG250, and (d) TRGO. (Adapted from Hofmann, D. et al., *Macromol. Mater. Eng.*, 298, 1322–1334, 2013.)

9.5 Flame Retardant Mechanism of Graphene

As a kind of typical layered nano-material, the flame retardant action of graphene remains similar to that observed in other layered nano-materials such as nano-clay. The excellent barrier property has been already reported for nano-clay (Messersmith and Giannelis 1995). Since defect-free graphene platelets are impermeable to all gas molecules (Bunch et al. 2008), it is anticipated that graphene could be used as an alternative to nano-clay that can enhance the barrier property of polymers during the thermal decomposition process. The barrier property of graphene can lead to a so-called "tortuous path" effect for polymers, which alters the diffusion path of decomposition volatiles (Figure 9.15) (Wang et al. 2011b). Therefore, the onset thermal decomposition temperature is improved as well as the mass loss rate is reduced. Well dispersed graphene platelets with high aspect ratio may lead to further improvement in barrier property of polymers (Paul and Robeson 2008). The incorporation of graphene in polymers involves a flame retardant mechanism mainly in the condensed phase resulting in the formation of a compact and uniform carbonaceous layer during combustion. Such a carbonaceous char layer effectively retards the inner thermal decomposition products to feed the flame as well as the permeation of outer oxygen and heat into the underlying matrix (Huang et al. 2012b). In the gaseous phase, the presence of graphene does not alter the composition of the evolved pyrolysis products, as measured by thermogravimetric analysis coupled with online Fourier-transformed infrared technique (Bettina et al. 2013). However, the accurate identification of the evolved pyrolysis products in the presence of graphene needs to be determined by thermogravimetric analysis combined with gas chromatography-mass spectrometer.

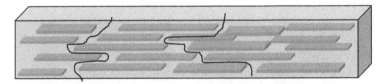

Permeation path imposed by nanoplatelet modification of polymer composites

FIGURE 9.15 Schematic illustration of the "tortuous path" effect caused by graphene nanoplatelet modification of polymer composites. (Reprinted from Wang, X. et al., *Prog. Polym. Sci.*, 69, 22–46, 2017. With permission.)

9.6 Conclusions and Perspective

Since its discovery in 2004 (Novoselov et al. 2004), graphene has spawned huge research interests in preparation of flame retardant polymer composites over the last decade. By comparison to conventional flame retardants, the addition of graphene could enhance the fire retardant property of polymers as well as the thermal stability, mechanical property, thermal conductivity, and electrical conductivity, which is a fascinating strategy to develop multi-functional polymer composites. Usually, utilization of a relatively low graphene content (≤5 wt%) leads to significant reduction in heat release rate and smoke production. Among different carbon-based materials, graphene displays the best flame retardant efficiency as compared to carbon black, expanded graphite, and multi-walled nanotubes (Bettina et al. 2013; Daniel et al. 2013; Dittrich et al. 2013). The general principles of the flame retardant action of graphene do not significantly differ from other layered nano-materials. The incorporation of graphene in polymers involves a flame retardant mechanism mainly in the condensed phase resulting in the formation of a compact and uniform carbonaceous layer during combustion, which can protect the underlying matrix against flame and meanwhile retard the "fuel flow" (flammable volatiles from the polymer decomposition) to feed the flame.

However, one great challenge for graphene/polymer composites is the poor compatibility between pristine graphene and most of organic polymers, which restrains the maximum reinforcement of graphene. In order to overcome this obstacle, functionalization and hybridization of graphene has been adopted to enhance the dispersion state of graphene in polymer matrices as well as their interfacial interaction. A wide variety of functionalized graphene and hybrids has been prepared via covalent linkage, hydrogen bonding, and electrostatic and π–π interactions, which show improved dispersion state of graphene within polymers. In the future, functionalization and hybridization of graphene are still two important methods to further improve the fire retardancy as well as other properties of polymer composites due to the advantages of multi-function, enhanced interfacial interaction, and synergism between various components.

Up to now, the application of graphene in the field of flame retardant polymers is still at the level of laboratory experiments or pilot batches. However, such situation might be changed with the scalable production of graphene in a foreseen future, since several new techniques have been developed to prepare graphene from low-cost materials. Compared to solution blending and *in situ* polymerization, melt compounding is a more suitable approach for industrial applications. Furthermore, the combination of graphene with the commercial flame retardants has imparted good comprehensive properties to polymers, which is anticipated to be commercialized in the future. Despite the existence of these challenges, the multi-functional property enhancements already demonstrated in graphene/polymer composites may expedite the applications of this new class of material in future.

References

Al-Attabi, N. Y., Kaur, G., Adhikari, R. et al. 2017. Preparation and characterization of highly conductive polyurethane composites containing graphene and gold nanoparticles. *Journal of Materials Science* 52: 11774–11784.

Alireza, A. 2014. Effects of graphene on the behavior of chitosan and starch nanocomposite films. *Polymer Engineering & Science* 54: 2258–2263.

Allen, M. J., Tung, V. C., and Kaner, R. B. 2010. Honeycomb carbon: A review of graphene. *Chemical Reviews* 110: 132–145.

Balandin, A. A., Ghosh, S., Bao, W. Z. et al. 2008. Superior thermal conductivity of single-layer graphene. *Nano Letters* 8: 902–907.

Berki, P., Laszlo, K., Tung, N. T. et al. 2017. Natural rubber/graphene oxide nanocomposites via melt and latex compounding: Comparison at very low graphene oxide content. *Journal of Reinforced Plastics and Composites* 36: 808–817.

Bettina, D., Karen-Alessa, W., Daniel, H. et al. 2013. Carbon black, multiwall carbon nanotubes, expanded graphite and functionalized graphene flame retarded polypropylene nanocomposites. *Polymers for Advanced Technologies* 24: 916–926.

Bhaviripudi, S., Jia, X. T., Dresselhaus, M. S. et al. 2010. Role of kinetic factors in chemical vapor deposition synthesis of uniform large area graphene using copper catalyst. *Nano Letters* 10: 4128–4133.

Bunch, J. S., Verbridge, S. S., Alden, J. S. et al. 2008. Impermeable atomic membranes from graphene sheets. *Nano Letters* 8: 2458–2462.

Cai, X. Y., Lai, L. F., Shen, Z. X. et al. 2017. Graphene and graphene-based composites as Li-ion battery electrode materials and their application in full cells. *Journal of Materials Chemistry A* 5: 15423–15446.

Cao, Y., Feng, J., and Wu, P. 2010. Preparation of organically dispersible graphene nanosheet powders through a lyophilization method and their poly(lactic acid) composites. *Carbon* 48: 3834–3839.

Chang, Y. W., Lee, K. S., Lee, Y. W. et al. 2015. Poly(ethylene oxide)/graphene oxide nanocomposites: Structure, properties and shape memory behavior. *Polymer Bulletin* 72: 1937–1948.

Chen, D., Feng, H., and Li, J. 2012a. Graphene oxide: Preparation, functionalization, and electrochemical applications. *Chemical Reviews* 112: 6027–6053.

Chen, D., Zhu, H., and Liu, T. 2010a. In situ thermal preparation of polyimide nanocomposite films containing functionalized graphene sheets. *ACS Applied Materials & Interfaces* 2: 3702–3708.

Chen, J., Duan, M., and Chen, G. 2012b. Continuous mechanical exfoliation of graphene sheets via three-roll mill. *Journal of Materials Chemistry* 22: 19625–19628.

Chen, W., Liu, P., Min, L. et al. 2018. Non-covalently functionalized graphene oxide-based coating to enhance thermal stability and flame retardancy of PVA film. *Nano-Micro Letters* 10: 39.

Chen, W., Liu, Y., Liu, P. et al. 2017. The preparation and application of a graphene-based hybrid flame retardant containing a long-chain phosphaphenanthrene. *Scientific Reports* 7: 8759.

Chen, W., Yan, L., and Bangal, P. R. 2010b. Chemical reduction of graphene oxide to graphene by sulfur-containing compounds. *The Journal of Physical Chemistry C* 114: 19885–19890.

Chua, C. K., and Pumera, M. 2014. Chemical reduction of graphene oxide: A synthetic chemistry viewpoint. *Chemical Society Reviews* 43: 291–312.

Ciesielski, A., and Samori, P. 2014. Graphene via sonication assisted liquid-phase exfoliation. *Chemical Society Reviews* 43: 381–398.

Compton, O. C., and Nguyen, S. T. 2010. Graphene oxide, highly reduced graphene oxide, and graphene: Versatile building blocks for carbon-based materials. *Small* 6: 711–723.

Daniel, H., Karen-Alessa, W., Ralf, T. et al. 2013. Functionalized graphene and carbon materials as additives for melt-extruded flame retardant polypropylene. *Macromolecular Materials and Engineering* 298: 1322–1334.

Dittrich, B., Wartig, K. A., Hofmann, D. et al. 2013. Flame retardancy through carbon nanomaterials: Carbon black, multiwall nanotubes, expanded graphite, multi-layer graphene and graphene in polypropylene. *Polymer Degradation and Stability* 98: 1495–1505.

Dong, X., Wang, P., Fang, W. et al. 2011. Growth of large-sized graphene thin-films by liquid precursor-based chemical vapor deposition under atmospheric pressure. *Carbon* 49: 3672–3678.

Du, J. H., and Cheng, H. M. 2012. The fabrication, properties, and uses of graphene/polymer composites. *Macromolecular Chemistry and Physics* 213: 1060–1077.

Fang, M., Wang, K. G., Lu, H. B. et al. 2009. Covalent polymer functionalization of graphene nanosheets and mechanical properties of composites. *Journal of Materials Chemistry* 19: 7098–7105.

Feng, Y., He, C., Wen, Y. et al. 2017a. Improving thermal and flame retardant properties of epoxy resin by functionalized graphene containing phosphorous, nitrogen and silicon elements. *Composites Part A: Applied Science and Manufacturing* 103: 74–83.

Feng, Y., He, C., Wen, Y. et al. 2018. Superior flame retardancy and smoke suppression of epoxy-based composites with phosphorus/nitrogen co-doped graphene. *Journal of Hazardous Materials* 346: 140–151.

Feng, Y., Hu, J., Xue, Y. et al. 2017b. Simultaneous improvement in the flame resistance and thermal conductivity of epoxy/Al_2O_3 composites by incorporating polymeric flame retardant-functionalized graphene. *Journal of Materials Chemistry A* 5: 13544–13556.

Gavgani, J. N., Adelnia, H., and Gudarzi, M. M. 2014. Intumescent flame retardant polyurethane/reduced graphene oxide composites with improved mechanical, thermal, and barrier properties. *Journal of Materials Science* 49: 243–254.

Geim, A. K. 2011. Nobel lecture: Random walk to graphene. *Reviews of Modern Physics* 83: 851–862.

Gómez-Navarro, C., Weitz, R. T., Bittner, A. M. et al. 2007. Electronic transport properties of individual chemically reduced graphene oxide sheets. *Nano Letters* 7: 3499–3503.

Guan, F. L., Gui, C. X., Zhang, H. B. et al. 2016. Enhanced thermal conductivity and satisfactory flame retardancy of epoxy/alumina composites by combination with graphene nanoplatelets and magnesium hydroxide. *Composites Part B-Engineering* 98: 134–140.

Guo, W., Yu, B., Yuan, Y. et al. 2017. In situ preparation of reduced graphene oxide/DOPO-based phosphonamidate hybrids towards high-performance epoxy nanocomposites. *Composites Part B:Engineering* 123: 154–164.

Han, Y., Wu, Y., Shen, M. et al. 2013. Preparation and properties of polystyrene nanocomposites with graphite oxide and graphene as flame retardants. *Journal of Materials Science* 48: 4214–4222.

Han, Y. Q., Wang, T. Q., Gao, X. X. et al. 2016. Preparation of thermally reduced graphene oxide and the influence of its reduction temperature on the thermal, mechanical, flame retardant performances of PS nanocomposites. *Composites Part A-Applied Science and Manufacturing* 84: 336–343.

He, H., Klinowski, J., Forster, M. et al. 1998. A new structural model for graphite oxide. *Chemical Physics Letters* 287: 53–56.

Higginbotham, A. L., Lomeda, J. R., Morgan, A. B. et al. 2009. Graphite oxide flame-retardant polymer nanocomposites. *ACS Applied Materials & Interfaces* 1: 2256–2261.

Hong, N., Zhan, J., Wang, X. et al. 2014. Enhanced mechanical, thermal and flame retardant properties by combining graphene nanosheets and metal hydroxide nanorods for Acrylonitrile–Butadiene–Styrene copolymer composite. *Composites Part A: Applied Science and Manufacturing* 64: 203–210.

Hong, N. N., Pan, Y., Zhan, J. et al. 2013. Fabrication of graphene/Ni-Ce mixed oxide with excellent performance for reducing fire hazard of polypropylene. *RSC Advances* 3: 16440–16448.

Hu, J., and Zhang, F. 2014. Self-assembled fabrication and flame-retardant properties of reduced graphene oxide/waterborne polyurethane nanocomposites. *Journal of Thermal Analysis and Calorimetry* 118: 1561–1568.

Hu, W., Yu, B., Jiang, S. D. et al. 2015. Hyper-branched polymer grafting graphene oxide as an effective flame retardant and smoke suppressant for polystyrene. *Journal of Hazardous Materials* 300: 58–66.

Hu, W., Zhan, J., Wang, X. et al. 2014. Effect of functionalized graphene oxide with hyper-branched flame retardant on flammability and thermal stability of cross-linked polyethylene. *Industrial & Engineering Chemistry Research* 53: 3073–3083.

Huang, G., Chen, S., Tang, S. et al. 2012a. A novel intumescent flame retardant-functionalized graphene: Nanocomposite synthesis, characterization, and flammability properties. *Materials Chemistry and Physics* 135: 938–947.

Huang, G., Gao, J., Wang, X. et al. 2012b. How can graphene reduce the flammability of polymer nano-composites? *Materials Letters* 66: 187–189.

Huang, G., Yang, J., Gao, J. et al. 2012c. Thin films of intumescent flame retardant-polyacrylamide and exfoliated graphene oxide fabricated via layer-by-layer assembly for improving flame retardant properties of cotton fabric. *Industrial & Engineering Chemistry Research* 51: 12355–12366.

Huang, G. B., Liang, H. D., Wang, Y. et al. 2012d. Combination effect of melamine polyphosphate and graphene on flame retardant properties of poly(vinyl alcohol). *Materials Chemistry and Physics* 132: 520–528.

Huang, Y., Qin, Y., Zhou, Y. et al. 2010. Polypropylene/graphene oxide nanocomposites prepared by in situ Ziegler–Natta polymerization. *Chemistry of Materials* 22: 4096–4102.

Huijuan, W., Zhaoqi, Z., Hanxue, S. et al. 2017. Graphene and poly(ionic liquid) modified polyurethane sponges with enhanced flame-retardant properties. *Journal of Applied Polymer Science* 134: 45477.

Idumah, C. I., Hassan, A., and Bourbigot, S. 2017. Influence of exfoliated graphene nanoplatelets on flame retardancy of kenaf flour polypropylene hybrid nanocomposites. *Journal of Analytical and Applied Pyrolysis* 123: 65–72.

Ionita, M., Pandele, M. A., and Iovu, H. 2013. Sodium alginate/graphene oxide composite films with enhanced thermal and mechanical properties. *Carbohydrate Polymers* 94: 339–344.

Jeon, I. Y., Shin, S. H., Choi, H. J. et al. 2017. Heavily aluminated graphene nanoplatelets as an efficient flame-retardant. *Carbon* 116: 77–83.

Ji, Y., Li, Y., Chen, G. et al. 2017. Fire-resistant and highly electrically conductive silk fabrics fabricated with reduced graphene oxide via dry-coating. *Materials & Design* 133: 528–535.

Jiang, S. D., Bai, Z. M., Tang, G. et al. 2014a. Synthesis of ZnS decorated graphene sheets for reducing fire hazards of epoxy composites. *Industrial & Engineering Chemistry Research* 53: 6708–6717.

Jiang, S. D., Bai, Z. M., Tang, G. et al. 2014b. Fabrication of Ce-doped MnO_2 decorated graphene sheets for fire safety applications of epoxy composites: Flame retardancy, smoke suppression and mechanism. *Journal of Materials Chemistry A* 2: 17341–17351.

Jing, J., Zhang, Y., Tang, X. L. et al. 2018. Combination of a bio-based polyphosphonate and modified graphene oxide toward superior flame retardant polylactic acid. *RSC Advances* 8: 4304–4313.

Kim, F., Luo, J., Cruz-Silva, R. et al. 2010a. Self-propagating domino-like reactions in oxidized graphite. *Advanced Functional Materials* 20: 2867–2873.

Kim, H., Abdala, A. A., and Macosko, C. W. 2010b. Graphene/polymer nanocomposites. *Macromolecules* 43: 6515–6530.

Kim, H., Miura, Y., and Macosko, C. W. 2010c. Graphene/polyurethane nanocomposites for improved gas barrier and electrical conductivity. *Chemistry of Materials* 22: 3441–3450.

Kim, M. J., Jeon, I. Y., Seo, J. M. et al. 2014. Graphene phosphonic acid as an efficient flame retardant. *ACS Nano* 8: 2820–2825.

Kuila, T., Bose, S., Mishra, A. K. et al. 2012. Chemical functionalization of graphene and its applications. *Progress in Materials Science* 57: 1061–1105.

Kuilla, T., Bhadra, S., Yao, D. H. et al. 2010. Recent advances in graphene based polymer composites. *Progress in Polymer Science* 35: 1350–1375.

Lago, E., Toth, P. S., Pugliese, G. et al. 2016. Solution blending preparation of polycarbonate/graphene composite: Boosting the mechanical and electrical properties. *RSC Advances* 6: 97931–97940.

Lee, C., Wei, X. D., Kysar, J. W. et al. 2008. Measurement of the elastic properties and intrinsic strength of monolayer graphene. *Science* 321: 385–388.

Lee, Y. R., Kim, S. C., Lee, H. I. et al. 2011. Graphite oxides as effective fire retardants of epoxy resin. *Macromolecular Research* 19: 66–71.

Li, J., Li, J., Meng, H. et al. 2014a. Ultra-light, compressible and fire-resistant graphene aerogel as a highly efficient and recyclable absorbent for organic liquids. *Journal of Materials Chemistry A* 2: 2934–2941.

Li, K. Y., Kuan, C. F., Kuan, H. C. et al. 2014b. Preparation and properties of novel epoxy/graphene oxide nanosheets (GON) composites functionalized with flame retardant containing phosphorus and silicon. *Materials Chemistry and Physics* 146: 354–362.

Li, L. L., Chen, S. H., Ma, W. J. et al. 2014c. A novel reduced graphene oxide decorated with halloysite nanotubes (HNTs-d-rGO) hybrid composite and its flame-retardant application for polyamide 6. *Express Polymer Letters* 8: 450–457.

Liang, J. J., Huang, Y., Zhang, L. et al. 2009. Molecular-level dispersion of graphene into poly(vinyl alcohol) and effective reinforcement of their nanocomposites. *Advanced Functional Materials* 19: 2297–2302.

Liao, S. H., Liu, P. L., Hsiao, M. C. et al. 2012. One-step reduction and functionalization of graphene oxide with phosphorus-based compound to produce flame-retardant epoxy nanocomposite. *Industrial & Engineering Chemistry Research* 51: 4573–4581.

Liu, H., Du, Y., Yang, G. et al. 2018. Flame retardance of modified graphene to pure cotton fabric. *Journal of Fire Sciences* 36: 111–128.

Liu, P., Chen, W., Jia, Y. et al. 2017. Fabrication of poly (vinyl alcohol)/graphene nanocomposite foam based on solid state shearing milling and supercritical fluid technology. *Materials & Design* 134: 121–131.

Liu, S., Fang, Z. P., Yan, H. Q. et al. 2016a. Superior flame retardancy of epoxy resin by the combined addition of graphene nanosheets and DOPO. *RSC Advances* 6: 5288–5295.

Liu, S., Yan, H., Fang, Z. et al. 2014a. Effect of graphene nanosheets and layered double hydroxides on the flame retardancy and thermal degradation of epoxy resin. *RSC Advances* 4: 18652–18659.

Liu, S., Yan, H., Fang, Z. et al. 2014b. Effect of graphene nanosheets on morphology, thermal stability and flame retardancy of epoxy resin. *Composites Science and Technology* 90: 40–47.

Liu, X., Wu, W., Qi, Y. et al. 2016b. Synthesis of a hybrid zinc hydroxystannate/reduction graphene oxide as a flame retardant and smoke suppressant of epoxy resin. *Journal of Thermal Analysis and Calorimetry* 126: 553–559.

Luo, F., Wu, K., Guo, H. et al. 2016. Anisotropic thermal conductivity and flame retardancy of nano-composite based on mesogenic epoxy and reduced graphene oxide bulk. *Composites Science and Technology* 132: 1–8.

Luo, F., Wu, K., Shi, J. et al. 2017a. Green reduction of graphene oxide by polydopamine to a construct flexible film: Superior flame retardancy and high thermal conductivity. *Journal of Materials Chemistry A* 5: 18542–18550.

Luo, F. B., Wu, K., Guo, H. L. et al. 2017b. Simultaneous reduction and surface functionalization of graphene oxide for enhancing flame retardancy and thermal conductivity of mesogenic epoxy composites. *Polymer International* 66: 98–107.

Luo, J., Yang, S., Lei, L. et al. 2017c. Toughening, synergistic fire retardation and water resistance of polydimethylsiloxane grafted graphene oxide to epoxy nanocomposites with trace phosphorus. *Composites Part A: Applied Science and Manufacturing* 100: 275–284.

Messersmith, P. B., and Giannelis, E. P. 1995. Synthesis and barrier properties of poly(ε-caprolactone)-layered silicate nanocomposites. *Journal of Polymer Science Part A: Polymer Chemistry* 33: 1047–1057.

Ming, P., Song, Z., Gong, S. et al. 2015. Nacre-inspired integrated nanocomposites with fire retardant properties by graphene oxide and montmorillonite. *Journal of Materials Chemistry A* 3: 21194–21200.

Mo, S., Peng, L., Yuan, C. et al. 2015. Enhanced properties of poly(vinyl alcohol) composite films with functionalized graphene. *RSC Advances* 5: 97738–97745.

Nine, M. J., Tran, D. N. H., ElMekawy, A. et al. 2017a. Interlayer growth of borates for highly adhesive graphene coatings with enhanced abrasion resistance, fire-retardant and antibacterial ability. *Carbon* 117: 252–262.

Nine, M. J., Tran, D. N. H., Tung, T. T. et al. 2017b. Graphene-borate as an efficient fire retardant for cellulosic materials with multiple and synergetic modes of action. *ACS Applied Materials & Interfaces* 9: 10160–10168.

Novoselov, K. S., Geim, A. K., Morozov, S. V. et al. 2004. Electric field effect in atomically thin carbon films. *Science* 306: 666–669.

Novoselov, K. S., Geim, A. K., Morozov, S. V. et al. 2005. Two-dimensional gas of massless Dirac fermions in graphene. *Nature* 438: 197–200.

Novoselov, K. S., Mishchenko, A., Carvalho, A. et al. 2016. 2D materials and van der Waals heterostructures. *Science* 353: aac9439.

Pan, H., Yu, B., Wang, W. et al. 2016. Comparative study of layer by layer assembled multilayer films based on graphene oxide and reduced graphene oxide on flexible polyurethane foam: Flame retardant and smoke suppression properties. *RSC Advances* 6: 114304–114312.

Papageorgiou, D. G., Kinloch, I. A., and Young, R. J. 2017. Mechanical properties of graphene and graphene-based nanocomposites. *Progress in Materials Science* 90: 75–127.

Parvez, K., Li, R., Puniredd, S. R. et al. 2013. Electrochemically exfoliated graphene as solution-processable, highly conductive electrodes for organic electronics. *ACS Nano* 7: 3598–3606.

Patole, A. S., Patole, S. P., Kang, H. et al. 2010. A facile approach to the fabrication of graphene/polystyrene nanocomposite by in situ microemulsion polymerization. *Journal of Colloid and Interface Science* 350: 530–537.

Paul, D. R., and Robeson, L. M. 2008. Polymer nanotechnology: Nanocomposites. *Polymer* 49: 3187–3204.

Pei, S., and Cheng, H. M. 2012. The reduction of graphene oxide. *Carbon* 50: 3210–3228.

Pethsangave, D. A., Khose, R. V., Wadekar, P. H. et al. 2017. Deep eutectic solvent functionalized graphene composite as an extremely high potency flame retardant. *ACS Applied Materials & Interfaces* 9: 35319–35324.

Pham, V. H., Dang, T. T., Hur, S. H. et al. 2012. Highly conductive poly(methyl methacrylate) (pmma)-reduced graphene oxide composite prepared by self-assembly of PMMA latex and graphene oxide through electrostatic interaction. *ACS Applied Materials & Interfaces* 4: 2630–2636.

Qi, X. Y., Yan, D., Jiang, Z. et al. 2011. Enhanced electrical conductivity in polystyrene nanocomposites at ultra-low graphene content. *ACS Applied Materials & Interfaces* 3: 3130–3133.

Qi, Y. X., Wu, W. H., Liu, X. W. et al. 2017. Preparation and characterization of aluminum hypophosphite/reduced graphene oxide hybrid material as a flame retardant additive for PBT. *Fire and Materials* 41: 195–208.

Qiu, S., Hu, W., Yu, B. et al. 2015. Effect of functionalized graphene oxide with organophosphorus oligomer on the thermal and mechanical properties and fire safety of polystyrene. *Industrial & Engineering Chemistry Research* 54: 3309–3319.

Raccichini, R., Varzi, A., Passerini, S. et al. 2015. The role of graphene for electrochemical energy storage. *Nature Materials* 14: 271–279.

Ramanathan, T., Abdala, A. A., Stankovich, S. et al. 2008. Functionalized graphene sheets for polymer nanocomposites. *Nature Nanotechnology* 3: 327.

Roilo, D., Patil, P. N., Brusa, R. S. et al. 2017. Polymer rigidification in graphene based nanocomposites: Gas barrier effects and free volume reduction. *Polymer* 121: 17–25.

Ruan, G., Sun, Z., Peng, Z. et al. 2011. Growth of graphene from food, insects, and waste. *ACS Nano* 5: 7601–7607.

Saravanan, N., Rajasekar, R., Mahalakshmi, S. et al. 2014. Graphene and modified graphene-based polymer nanocomposites-A review. *Journal of Reinforced Plastics and Composites* 33: 1158–1180.

Sheng, K. X., Xu, Y. X., Li, C. et al. 2011. High-performance self-assembled graphene hydrogels prepared by chemical reduction of graphene oxide. *New Carbon Materials* 26: 9–15.

Shi, X., Peng, X., Zhu, J. et al. 2018a. Synthesis of DOPO-HQ-functionalized graphene oxide as a novel and efficient flame retardant and its application on polylactic acid: Thermal property, flame retardancy, and mechanical performance. *Journal of Colloid and Interface Science* 524: 267–278.

Shi, Y., and Li, L. J. 2011. Chemically modified graphene: Flame retardant or fuel for combustion? *Journal of Materials Chemistry* 21: 3277–3279.

Shi, Y., Yu, B., Zheng, Y. et al. 2018b. Design of reduced graphene oxide decorated with DOPO-phosphanomidate for enhanced fire safety of epoxy resin. *Journal of Colloid and Interface Science* 521: 160–171.

Shi, Y. Q., Yu, B., Zheng, Y. Y. et al. 2018c. Design of reduced graphene oxide decorated with DOPO-phosphanomidate for enhanced fire safety of epoxy resin. *Journal of Colloid and Interface Science* 521: 160–171.

Sinha Ray, S., and Okamoto, M. 2003. Polymer/layered silicate nanocomposites: A review from preparation to processing. *Progress in Polymer Science* 28: 1539–1641.

Song, P. A., Liu, L. N., Fu, S. Y. et al. 2013. Striking multiple synergies created by combining reduced graphene oxides and carbon nanotubes for polymer nanocomposites. *Nanotechnology* 24: 125704.

Srivastava, A., Galande, C., Ci, L. et al. 2010. Novel liquid precursor-based facile synthesis of large-area continuous, single, and few-layer graphene films. *Chemistry of Materials* 22: 3457–3461.

Stankovich, S., Dikin, D. A., Dommett, G. H. B. et al. 2006a. Graphene-based composite materials. *Nature* 442: 282–286.

Stankovich, S., Dikin, D. A., Piner, R. D. et al. 2007. Synthesis of graphene-based nanosheets via chemical reduction of exfoliated graphite oxide. *Carbon* 45: 1558–1565.

Stankovich, S., Piner, R. D., Nguyen, S. T. et al. 2006b. Synthesis and exfoliation of isocyanate-treated graphene oxide nanoplatelets. *Carbon* 44: 3342–3347.

Sun, F., Yu, T., Hu, C. et al. 2016. Influence of functionalized graphene by grafted phosphorus containing flame retardant on the flammability of carbon fiber/epoxy resin (CF/ER) composite. *Composites Science and Technology* 136: 76–84.

Tang, L. C., Wan, Y. J., Yan, D. et al. 2013. The effect of graphene dispersion on the mechanical properties of graphene/epoxy composites. *Carbon* 60: 16–27.

Teng, C. C., Ma, C. C. M., Lu, C. H. et al. 2011. Thermal conductivity and structure of non-covalent functionalized graphene/epoxy composites. *Carbon* 49: 5107–5116.

Thakur, S., and Karak, N. 2015. Alternative methods and nature-based reagents for the reduction of graphene oxide: A review. *Carbon* 94: 224–242.

Thomassin, J. M., Trifkovic, M., Alkarmo, W. et al. 2014. Poly(methyl methacrylate)/graphene oxide nanocomposites by a precipitation polymerization process and their dielectric and rheological characterization. *Macromolecules* 47: 2149–2155.

Uhl, F. M., and Wilkie, C. A. 2004. Preparation of nanocomposites from styrene and modified graphite oxides. *Polymer Degradation and Stability* 84: 215–226.

Vilcinskas, K., Norder, B., Goubitz, K. et al. 2015. Tunable order in alginate/graphene biopolymer nanocomposites. *Macromolecules* 48: 8323–8330.

Wajid, A. S., Ahmed, H. S. T., Das, S. et al. 2013. High-performance pristine graphene/epoxy composites with enhanced mechanical and electrical properties. *Macromolecular Materials and Engineering* 298: 339–347.

Wang, D., Zhou, K., Yang, W. et al. 2013a. Surface modification of graphene with layered molybdenum disulfide and their synergistic reinforcement on reducing fire hazards of epoxy resins. *Industrial & Engineering Chemistry Research* 52: 17882–17890.

Wang, J., Manga, K. K., Bao, Q. et al. 2011a. High-yield synthesis of few-layer graphene flakes through electrochemical expansion of graphite in propylene carbonate electrolyte. *Journal of the American Chemical Society* 133: 8888–8891.

Wang, W., Wang, X., Pan, Y. et al. 2017a. Synthesis of phosphorylated graphene oxide based multi-layer coating: Self-assembly method and application for improving the fire safety of cotton fabrics. *Industrial & Engineering Chemistry Research* 56: 6664–6670.

Wang, X., Hu, Y., Song, L. et al. 2011b. In situ polymerization of graphene nanosheets and polyurethane with enhanced mechanical and thermal properties. *Journal of Materials Chemistry* 21: 4222–4227.

Wang, X., Kalali, E. N., Wan, J. T. et al. 2017b. Carbon-family materials for flame retardant polymeric materials. *Progress in Polymer Science* 69: 22–46.

Wang, X., Kalali, E. N., and Wang, D. Y. 2016. Two-dimensional inorganic nanomaterials: A solution to flame retardant polymers. *Nano Advances* 1: 1–16.

Wang, X., Song, L., Yang, H. et al. 2011c. Synergistic effect of graphene on antidripping and fire resistance of intumescent flame retardant poly(butylene succinate) composites. *Industrial & Engineering Chemistry Research* 50: 5376–5383.

Wang, X., Song, L., Yang, H. et al. 2012. Cobalt oxide/graphene composite for highly efficient CO oxidation and its application in reducing the fire hazards of aliphatic polyesters. *Journal of Materials Chemistry* 22: 3426–3431.

Wang, X., Xing, W., Feng, X. et al. 2014a. The effect of metal oxide decorated graphene hybrids on the improved thermal stability and the reduced smoke toxicity in epoxy resins. *Chemical Engineering Journal* 250: 214–221.

Wang, X., Xing, W., Feng, X. et al. 2014b. Functionalization of graphene with grafted polyphospha-mide for flame retardant epoxy composites: Synthesis, flammability and mechanism. *Polymer Chemistry* 5: 1145–1154.

Wang, X., Zhou, S., Xing, W. et al. 2013b. Self-assembly of Ni-Fe layered double hydroxide/graphene hybrids for reducing fire hazard in epoxy composites. *Journal of Materials Chemistry A* 1: 4383–4390.

Wicklein, B., Kocjan, A., Salazar-Alvarez, G. et al. 2015. Thermally insulating and fire-retardant light-weight anisotropic foams based on nanocellulose and graphene oxide. *Nature Nanotechnology* 10: 277–283.

Xu, L., Xiao, L., Jia, P. et al. 2017. Lightweight and ultrastrong polymer foams with unusually superior flame retardancy. *ACS Applied Materials & Interfaces* 9: 26392–26399.

Xu, W. Z., Liu, L., Zhang, B. L. et al. 2016. Effect of molybdenum trioxide-loaded graphene and cuprous oxide-loaded graphene on flame retardancy and smoke suppression of polyurethane elastomer. *Industrial & Engineering Chemistry Research* 55: 4930–4941.

Xu, W. Z., Wang, X. L., Wang, G. S. et al. 2018. A novel graphene hybrid for reducing fire hazard of epoxy resin. *Polymers for Advanced Technologies* 29: 1194–1205.

Xu, Y., Hong, W., Bai, H. et al. 2009. Strong and ductile poly(vinyl alcohol)/graphene oxide composite films with a layered structure. *Carbon* 47: 3538–3543.

Xu, Z., and Gao, C. 2010. In situ polymerization approach to graphene-reinforced nylon-6 composites. *Macromolecules* 43: 6716–6723.

Yan, D., Zhang, H. B., Jia, Y. et al. 2012b. Improved electrical conductivity of polyamide 12/graphene nanocomposites with maleated polyethylene-octene rubber prepared by melt compounding. *ACS Applied Materials & Interfaces* 4: 4740–4745.

Yan, D. X., Ren, P. G., Pang, H. et al. 2012a. Efficient electromagnetic interference shielding of light-weight graphene/polystyrene composite. *Journal of Materials Chemistry* 22: 18772–18774.

Yang, C., Li, Z., Yu, L. et al. 2017. Mesoporous zinc ferrite microsphere-decorated graphene oxide as a flame retardant additive: Preparation, characterization, and flame retardance evaluation. *Industrial & Engineering Chemistry Research* 56: 7720–7729.

Yang, X., Li, L., Shang, S. et al. 2010. Synthesis and characterization of layer-aligned poly(vinyl alcohol)/graphene nanocomposites. *Polymer* 51: 3431–3435.

Yang, Y. G., Chen, C. M., Wen, Y. F. et al. 2008. Oxidized graphene and graphene based polymer composites. *New Carbon Materials* 23: 193–200.

Yi, M., Shen, Z., Ma, S. et al. 2012. A mixed-solvent strategy for facile and green preparation of graphene by liquid-phase exfoliation of graphite. *Journal of Nanoparticle Research* 14: 1003.

Yu, B., Shi, Y. Q., Yuan, B. H. et al. 2015. Enhanced thermal and flame retardant properties of flame-retardant-wrapped graphene/epoxy resin nanocomposites. *Journal of Materials Chemistry A* 3: 8034–8044.

Yu, B., Wang, X., Qian, X. et al. 2014. Functionalized graphene oxide/phosphoramide oligomer hybrids flame retardant prepared via in situ polymerization for improving the fire safety of polypropylene. *RSC Advances* 4: 31782–31794.

Yuan, B., Wang, B., Hu, Y. et al. 2016. Electrical conductive and graphitizable polymer nanofibers grafted on graphene nanosheets: Improving electrical conductivity and flame retardancy of polypropylene. *Composites Part A: Applied Science and Manufacturing* 84: 76–86.

Yuan, B. H., Sheng, H. B., Mu, X. W. et al. 2015. Enhanced flame retardancy of polypropylene by melamine-modified graphene oxide. *Journal of Materials Science* 50: 5389–5401.

Zhan, Y. H., Wu, J. K., Xia, H. S. et al. 2011. Dispersion and exfoliation of graphene in rubber by an ultrasonically-assisted latex mixing and in situ reduction process. *Macromolecular Materials and Engineering* 296: 590–602.

Zhang, C., and Liu, T. X. 2012. A review on hybridization modification of graphene and its polymer nanocomposites. *Chinese Science Bulletin* 57: 3010–3021.

Zhang, H. B., Zheng, W. G., Yan, Q. et al. 2010. Electrically conductive polyethylene terephthalate/graphene nanocomposites prepared by melt compounding. *Polymer* 51: 1191–1196.

Zhang, H. X., Z., Park, J. H., Moon, Y. K. et al. 2017a. Preparation of polyethylene/graphene nanocomposites with octadecylamine-modified graphene oxide-MgCl-supported Ziegler–Natta catalyst. *Journal of Polymer Science Part A: Polymer Chemistry* 55: 855–860.

Zhang, M., Shi, X., Dai, X. et al. 2018. Improving the crystallization and fire resistance of poly(lactic acid) with nano-ZIF-8@GO. *Journal of Materials Science* 53: 7083–7093.

Zhang, X., Shen, Q., Zhang, X. et al. 2016. Graphene oxide-filled multilayer coating to improve flame-retardant and smoke suppression properties of flexible polyurethane foam. *Journal of Materials Science* 51: 10361–10374.

Zhang, Y., Zhang, L., and Zhou, C. 2013. Review of chemical vapor deposition of graphene and related applications. *Accounts of Chemical Research* 46: 2329–2339.

Zhang, Z., Yuan, L., Guan, Q. et al. 2017b. Synergistically building flame retarding thermosetting composites with high toughness and thermal stability through unique phosphorus and silicone hybridized graphene oxide. *Composites Part A: Applied Science and Manufacturing* 98: 174–183.

Zhao, X., Zhang, Q., Chen, D. et al. 2010. Enhanced mechanical properties of graphene-based poly(vinyl alcohol) composites. *Macromolecules* 43: 2357–2363.

Zhou, K., Gui, Z., and Hu, Y. 2016a. The influence of graphene based smoke suppression agents on reduced fire hazards of polystyrene composites. *Composites Part A: Applied Science and Manufacturing* 80: 217–227.

Zhou, K., Gui, Z., Hu, Y. et al. 2016b. The influence of cobalt oxide–graphene hybrids on thermal degradation, fire hazards and mechanical properties of thermoplastic polyurethane composites. *Composites Part A: Applied Science and Manufacturing* 88: 10–18.

Zhu, J. 2008. Graphene production: New solutions to a new problem. *Nature Nanotechnology* 3: 528–529.

Zhu, Y. W., Murali, S., Cai, W. W. et al. 2010. Graphene and graphene oxide: Synthesis, properties, and applications. *Advanced Materials* 22: 3906–3924.

Functionalized Layered Nanomaterials towards Flame Retardant Polymer Nanocomposites

Ye-Tang Pan and
De-Yi Wang

10.1 Introduction

Polymeric materials commonly have a fatal drawback of high flammability, which has severely restricted their application in the fields requiring remarkable flame retardancy (Pan et al. 2017b). Over the past decades, multiple flame retardants (FRs) have been incorporated into polymers to improve their flame retardancy. Although a high level fire resistance has been achieved, a major problem encountered with the conventional flame retardants was the high loading required. The emergence of nanocomposite technology has provided a revolutionary new solution to flame retardant polymer materials, and it was accepted that a significant improvement in the fire retardancy has been achieved by introducing a low loading of nanofillers (Zhang et al. 2015). Polymer nanocomposites containing inorganic additives with layered structures have been an area of both significant academic interest and commercial importance due to their enhanced features compared to conventional composite materials (Zhu et al. 2012). Layered nanomaterials, a series of versatile nanofill-ers, involving graphene, layered molybdenum disulfide (MoS$_2$), and layered double hydroxides (LDHs), have been widely employed in polymer nanocomposites. It is worth noting that one of the key dominating factors for polymer nanocomposites is the compatibility between the nanofiller and the polymeric matrix, which is directly relative to the dispersion and interface interaction of constitutive phases and further influences the macroscopic property (Bao et al. 2011). However, the homogeneous dispersion of layered nanomaterials in a polymer matrix is usually a challenge for the preparation of nanocomposites (Xie et al. 2004). Therefore, many efforts have been focused on achieving a homogeneous dispersion state by developing either covalent or non-covalent functionalization of the nanofiller surface (Bartholome et al. 2008). The intercalation of large molecules into the gallery between layers is also an effective method to enlarge the inter-layer space so that the layered nanomaterials are able to exfoliate in the polymer matrix to avoid agglomeration. On the other hand, since the inorganic layered nanomaterials have weak interaction with polymer matrix due to their hydrophilic feature, the organo-modifications preferentially enhance the compatibility within the interfacial section. Therefore, the controlling factors for polymer nanocomposites are the dispersion of the nanoscale inorganics and interfacial interaction of constituents which influence the properties of the composite materials. In this chapter, the modification of representative layered nanomaterials, graphene, MoS_2, and LDHs, and their performance in flame retardant polymer nanocomposites will be introduced, respectively.

Graphene is a carbon material which has good compatibility with the formed amorphous carbon. It could act as another char layer during thermal degradation. The huge challenge for the applications of graphene/polymer nanocomposites at present is the dispersion of graphene in the polymer matrix due to the intrinsic van der Waals force between graphene sheets, which results in the re-aggregating and restacking of graphene (Dappe et al. 2006). The bare graphene is easily burnt out upon exposure to a heat flow under air atmosphere, which could not act as an effective barrier to prevent the escape of organic volatiles. Due to the ease of thermal decomposition of graphene oxide (GO) nanosheets, its nanocomposites merely demonstrated slightly enhanced or even deteriorate fire retardancy. Graphene oxide features a variety of oxygen-containing functionalities such as hydroxyl and carboxyl on the basal plane and along the edges, providing a platform for the surface modification (Gao et al. 2009; Salavagione et al. 2009).

Layered MoS_2 is one of the more commercially available members, which is composed of three stacked atom layers (S-Mo-S) held together by van der Waals forces (Lukowski et al. 2013). Like some other members of the family of 2D layered materials, MoS_2 has many comprehensive applications due to their unmatched properties. As a typical inorganic analogue of graphene, MoS_2 nanosheets are expected to be well dispersed in polymer matrices. Nonetheless, bulk MoS_2 is not appropriate for intercalation by large species such as polymer chains because MoS_2 has a remarkable tendency to agglomerate and even restack in polymer matrices owing to their large specific surface area and van der Waals interactions, which limits the well dispersion of MoS_2 nanosheets in polymer nanocomposites. Therefore, it is of significance to enhance the dispersion of MoS_2 nanosheets in a polymer matrix.

LDHs have attracted considerable interest in the field of flame retardant polymers in the recent years (Evans and Duan 2006; Wang et al. 2009). LDH has a layered structure that can be represented by the chemical formula of $[M^{2+}_{1-x}M^{3+}_x (OH)_2](A^{n-})_{x/n} \cdot mH_2O$, where M^{2+} is a divalent cation, M^{3+} is a trivalent cation, and A^{n-} denotes an inorganic or organic anion with negative charge n (Duceac et al. 2017; Han et al. 2007). When used as flame retardants for polymeric materials, it is well recognized that the flame retardant mechanism of LDH is the "barrier effect" of the nanolayers and/or the formation of ceramic-like materials. During combustion, LDHs lose the inter-layer water, intercalated anions, and dehydroxylate to a mixed metal oxide (Nyambo et al. 2009; Pereira et al. 2009). These processes adsorb huge amounts of heat, dilute the concentration of oxygen, promote the formation of an expanded carbonaceous coating or char on the polymer, protecting the bulk polymer from being exposed to air, and suppress smoke production due to suffocation. However, LDHs have much higher charge density in the inter-layer and stronger interaction among the hydroxide sheets than layered silicates, which makes the exfoliation of LDHs much more difficult. In order to make enough inter-layer spacing to interact with most hydrophobic polymers, LDHs have to be modified with various anionic surfactants (Basu et al. 2014). In most cases, organically modified LDHs used to prepare nanocomposites improve the solubility and dispersion, but simultaneously deteriorate the flame retardancy of materials due to surfactants' inherent flammability.

As layered nanomaterials are widely employed in various areas, this chapter mainly emphasizes on the functionalization of layered nanomaterials utilized in the application to flame retardant polymer nanocomposites. Herein, the functionalization methods of layered nanomaterials aiming to improve their flame retardant effects in polymers will be reviewed and also the characterization for the modified layered nanomaterials and corresponding polymer nanocomposites. The distinctive and typical results reported recently will be selected to illustrate the role of modification on layered nanomaterials in improving the flame retardancy of polymer nanocomposites.

10.2 Functionalization of Layered Nanomaterials

The dispersion and interface problems are key factors for preparation of the polymer nanocomposites. Functionalization of the layered nanomaterials can generally enhance their dispersion and interaction within the polymer matrix.

10.2.1 Functionalization of Graphene

To date, there have been mainly two strategies, namely, covalent and non-covalent functionalization, to improve the dispersion and compatibility of graphene or graphene-based materials with polymers.

10.2.1.1 Covalent Modification

The covalent bonds form on the surface of graphene with the aid of oxygen-containing functional groups including hydroxyl, carboxyl, and epoxy via the chemical reaction with the organic modifiers. Direct covalent functionalization not only significantly improves the dispersion, but also forms strong interfacial interactions with polymer materials via covalent linkages. A phenyl glycidol chlorophosphate (PGC) was synthesized and connected on the graphene oxide by covalent modification according to the reaction between P–Cl and –OH (Guo et al. 2011). Thus, graphene oxide was functionalized with flame retardant elemental phosphorous and reactive epoxy groups to obtain better dispersion in epoxy resin. Another example related to the covalent bonding between P–Cl and –OH ascribed to the functionalization of hexachlorocyclotriphosphazene (HCCP, molecular formula: $P_3N_3Cl_6$) onto graphene oxide. Meanwhile, hydroxyethyl acrylate (HEA) was bonded to HCCP introducing C=C groups to functionalized GO which can be polymerized with monomers, so strong covalent interfacial interactions are available among the functionalized GO and polymers. Phenyl-bis-(triethoxysilylpropyl) phosphamide (PBTP) was synthesized, and functionalized GO nanosheets were obtained through the reaction between the trialkoxy groups on PBTP and the hydroxyl groups on the GO surface (Wang et al. 2014). Phosphorous oxychloride ($POCl_3$) and excess triethylamine as the catalyst were modified on graphene oxide followed by further modification with 4,4-diaminodiphenyl methane (DDM). The functionalized graphene oxide with many active amine groups easily formed covalent bonds with the epoxy resin (EP) matrix (Yu et al. 2015).

Besides hydroxyl, epoxy groups and carboxyl groups also give rise to the covalent modification with suitable organic flame retardants. Among non-reactive compounds containing phosphorous, 9,10-dihydro-9-oxa-10-phosphaphenanthrene-10-oxide (DOPO) shows outstanding flame retarding properties in polymeric materials especially epoxy-based because of its high gas-phase activity and condensed-phase activity (through char formation) at the same time. DOPO-functionalized reduced graphene oxide (DOPO-rGO) was synthesized by the direct reaction between the active hydrogen of DOPO and the epoxy groups on GO (Liao et al. 2012). The grafting of the DOPO molecules onto the surface of the GO simultaneously reduced to graphene partially. The same group also prepared an organic-inorganic flame retardant (9,10-dihydro-9-oxa-10-phosphaphenanthrene-10-oxide–vinyltrimethoxysilane [DOPO-VTS]) containing both silicon and organophosphorus. Then graphene was introduced into the organic-inorganic flame retardants through firstly reacting with (3-isocyanatopropyl) triethoxysilane by using carboxyl groups and through the sol-gel process (Qian et al. 2013). The polymeric flame retardant, poly(4,4′-diaminodiphenyl methane phenyl dichlorophophate) (PDMPD) was synthesized by Feng et al. (2017) and covalently bonded with graphene through epoxy groups on the surface. The long chains and the phosphamide groups of polymeric flame retardant not only improved the dispersion of graphene in the matrix, but also increased its flame retarding efficiency. An intumescent flame retardant, poly (piperazine spirocyclic pentaerythritol bisphosphonate) (PPSPB) has been covalently grafted onto the surface of graphene oxide with exploiting both epoxy and carboxyl groups through esterification of GO and PPSPB (Huang et al. 2012a). In another research, the carboxyl groups on graphene oxide was firstly modified by $SOCl_2$ and then connected with a polymeric flame retardant, 10-dihydro-9-oxa-10-phosphaphenanthrene-10-oxide-g-(2,3-epoxypropoxy) propyltrimethoxysilane (DPP) (Chen et al. 2017). The long-chain and bulky polymeric flame retardant grafted onto GO weaken the van der Waal forces and expand the space between neighboring GO sheets due to "steric propping-open effects," which promotes the dispersion and exfoliation of the GO nanosheets. In addition, the extended chains on the graphene sheets easily entangle with the polymer chains and thus enhance the interfacial combination between graphene and the polymer matrix. Likewise, a novel organophosphorus oligomer (PMPPD) was covalently grafted on the surface of GO with $SOCl_2$ as the bridge as well (Qiu et al. 2015).

The previous studies have demonstrated that the amount of oxygen-containing groups on the surface of graphene was limited (oxygen atom percentage of approximately 30% for graphene oxide), and thus only limited flame retardants could be grafted onto graphene (Khanra et al. 2012). Masterbatch-flame retardants are prepared by incorporating abundant graphene into flame retardants *via in situ* reaction. As compared with solvent blending, the use of organic solvents is significantly decreased; as compared with traditional melt blending, the dispersion and exfoliation of graphene is visibly improved due to the previous exfoliation of graphene in the flame retardants; as compared with the *in situ* polymerization, the masterbatch-flame retardants are fit for industrial applications. Therefore, Hu's group developed a novel strategy to overcome the challenges by modifying GO with DDM and then *in situ* incorporating into phosphoramide oligomer, resulting in a nanocomposite flame retardant containing exfoliated graphene (Yu et al. 2014).

Apart from the organic molecules, the nano- or micro-particles are also available to modify graphene by covalent bonds. A facile approach was demonstrated to functionalize and *in situ* reduce GO with octa-aminophenyl polyhedral oligomeric silsesquioxanes (OapPOSS) without the addition of conventional chemical reducing agents. The amino groups of OapPOSS covalently bonded with epoxy groups of GO (Wang et al. 2012b).

Recently, our group prepared MCM-41 nanospheres containing –NH_2 groups with the help of silane coupling agents as shown in Figure 10.1. The NH_2-MCM-41 covalently assembled on GO nanosheets by the reaction with epoxy groups (Li et al. 2017). Large-size reduced graphene oxide could also attach on ammonium polyphosphate (APP) in the medium of (3-aminopropyl) triethoxysilane (Zhang et al. 2017). In this sense, (3-aminopropyl) triethoxysilane functioning as surface modifier cannot only reduce the polarity of APP, but also attach the graphene sheets onto its surface.

10.2.1.2 Non-covalent Modification

Non-covalent modification of graphene normally consists of π–π interaction, hydrogen bonding, and electrostatic interaction between the modifiers and graphene substrate.

Graphene is non-covalently modified by hex(4-carboxylphenoxy)cyclotriphosphazene (HCPCP) by π–π interaction to form flame retardant-functionalized graphene sheets during exfoliation process (Cai et al. 2017a). The cyclophosphazene structure containing phosphorus and nitrogen elements can improve the flame retardant efficiency of functionalized graphene. The graphene was non-covalently modified by ligninsulfonate and iron ion (Fe-lignin) to form the flame retardant-functionalized graphene sheets (Cai et al. 2016). The iron lignosulfonate with high char yield was employed to surface modify the graphene in a non-covalent way and protect the lamellar structure from falling apart. The sodium lignosulfonate combined with graphene nanosheets *via* formation of hydrogen-bond action as shown in Figure 10.2. The mechanism of sodium ligninsulfonate-functionalized graphene is based on

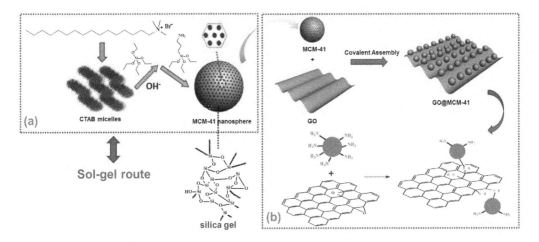

FIGURE 10.1 (a) Preparation of amino grafting MCM-41 nanospheres; (b) preparation of GO decorated with MCM-41 nanospheres. (Reprinted from Li, Z. et al., *Compos. B Eng.*, 138, 101–112, 2017. With permission.)

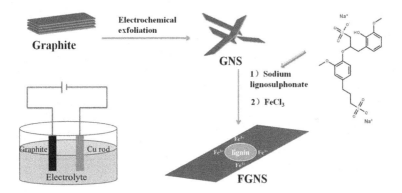

FIGURE 10.2 Illustration for electrochemical exfoliation of graphite and the functionalization process. (Reprinted from Cai, W. et al., *Ind. Eng. Chem. Res.*, 55, 10681–10689, 2016. With permission.)

π–π interaction and hydrogen bonding. The non-covalent modification of graphene with biopolymers has shown that the homogeneous dispersion of graphene can be acquired without damaging its pristine properties. Chitosan (CS) is very promising to reduce and non-covalently modify the graphene oxide simultaneously with environmentally friendly advantage and abundant resources. The reduction of GO is presumably due to a direct redox reduction between GO and the –NH_2 groups from CS. CS has –OH and –NH_2 or O=C–NH_2 groups on its macromolecular chains, and these functional groups can form hydrogen bonding and electrostatic interactions with the residual oxygen-containing groups of GO (Feng et al. 2013).

Graphene oxide has a negative charge nature due to the presence of a small amount of carboxyl, epoxy, and hydroxyl groups. As aforementioned, nanoparticles are always modified with NH_2 groups donated by silane coupling agents to covalently bond with epoxy groups of graphene oxide. Differently, a report mentioned that Ce-doped MnO_2 nanoparticles were reacted with aminopropyltriethoxysilane to modify their surface with NH_2 just to impart positive charge nature to the nanoparticles. Then the modified Ce-MnO_2 non-covalently bonded with graphene oxide *via* electrostatic interactions (Jiang et al. 2014a). Molybdenum trioxide (MoO_3)-loaded graphene hybrids and cuprous oxide (Cu_2O)-loaded graphene hybrids were successfully prepared *via* the hydrothermal method and one-pot co-precipitation method, respectively. Metal salts as precursors provide metal cation with positive charge easily adsorbed by GO nanosheets. In this case, the Cu^{2+} located onto GO surface by electrostatic interaction, and subsequently transformed into Cu_2O. However, if the metal precursors exist in the form of metal oxyanions, such as $(NH_4)_6Mo_7O_{24} \cdot 4H_2O$, the acid aqueous solution should be added to alter the charge nature of graphene oxide. Herein, 10 mL of HNO_3 was utilized to render the successful residing of molybdenum oxyanion onto graphene (Xu et al. 2016a). Similarly, Cu_2O and TiO_2 both grafted on graphene through a one-pot hydrothermal reaction (Wang et al. 2016). The oxygen-containing groups on the plane of graphene oxide would provide anchoring sites for the *in-situ* growth of Cu_2O and TiO_2. Besides metal oxides, metal dicyclopentadienyl such as ferrocene could also decorate on graphene nanosheets by a solvothermal method (Zhou et al. 2016a). Aluminum hypophosphite/reduced graphene oxide hybrid flame retardant was also successfully prepared by a one-step method consisting of the simultaneous reduction of graphene oxide and the deposition of aluminum hypophosphite on graphene (Qi et al. 2017).

10.2.2 Functionalization of Layered MoS_2

Analogous to graphene, the functionalization of layered MoS_2 also includes the covalent modification and non-covalent modification. However, since the functional groups on the surface of layered MoS_2 are insufficient compared with those of graphene, non-covalent modification is dominant during the functionalized process of layered MoS_2.

10.2.2.1 Non-covalent Modification

In fact, the preparation of polymer/layered MoS_2 is difficult to achieve. The amphiphobic nature of the MoS_2 nanosheets causes poor dispersion in a variety of organic solvents and are not compatible with the organic polymer matrix. Most monomers and polymers cannot be easily intercalated between MoS_2 nanosheets in their pristine state which is mainly due to no reactive groups on the surface of MoS_2, and parent MoS_2 lacks both the space and affinity for polymer molecules (or monomers) to be intercalated into its galleries. Furthermore, the individual sheets tend to restack owing to their large specific surface area and van der Waals interactions between the inter-layers of MoS_2 nanosheets, which make the inter-layer distance between MoS_2 nanoplatelets very narrow. Unlike graphene oxide, MoS_2 nanosheets have no abundant oxygen-containing groups, which worsen the dispersion and interface interaction with polymer matrix, and restrict organic functionalization. An effective way to overcome the agglomeration and enhance the compatibility between MoS_2 nanosheets and polymer matrix is the surface modification of MoS_2 nanosheets *via* non-covalent and covalent interaction. But unfortunately, it is rather difficult to modify the MoS_2 nanosheets through covalent functionalization method because there are no function groups on the surface of MoS_2 nanosheets such as GO with hydroxyl, carboxyl, and so forth. However, it is well recognized that the bulk MoS_2 can be exfoliated into single or few layers with negative charge by intercalating with strong reducing agents. Fortunately, the layered structure of MoS_2 enables easy intercalation of metal ions, such as Li^+ and Mg^{2+} (Chang et al. 2011). So it is convenient to prepare polymer nanocomposites by the intercalation of metal ions (Li^+) and then exfoliated to single or few layers through the hydrolysis of the Li^+. Therefore, flocculation of the MoS_2 single-layers in the presence of lipophilic cationic surfactants results in the formation of intercalation compounds.

Hu's group has carried out a large amount of systematic work about the functionalization of MoS_2. Exfoliated MoS_2 nanosheets are firstly obtained from the hydrolysis of Li-intercalated MoS_2 in the alkaline suspension. Then a cationic surfactant, cetyl trimethyl ammonium bromide (CTAB) non-covalently modified on MoS_2 nanosheets by electrostatic interaction (Zhou et al. 2014a, 2015). The fabrication of organic-modified MoS_2 nanosheets can facilitate the dispersion and increase the compatibility of MoS_2 nanosheets in the polymer matrix and then improve the performances of the nanocomposites. However, the intercalating agents such as the organic quaternary ammonium salts (C_{12}–C_{18}) are combustible and have poor thermal stability, resulting in negative effect on flame retardancy and thermal stability of the polymer materials. Therefore, Zhou et al. (2016b) adopted a flame retardant, melamine phosphate as intercalating agent to modify MoS_2 by cation-exchange reaction. They also used non-electron donors such as ferrocene which has a high ionization potential (6.88 eV) to be intercalated into MoS_2 by restacking of exfoliating MoS_2 (Zhou et al. 2014b). The novel intercalation compounds containing ferrocene and MoS_2 were used as flame retardant and smoke suppression agents. In a recent report, Wang et al. (2015a) exfoliated ultrathin MoS_2 nanosheets and non-covalently modified by ultrasonication in an aqueous solution of chitosan, a natural polymer cationic surfactant with multi-hydroxyl and multi-amino groups (Figure 10.3). Chitosan-modified MoS_2 nanosheets could be homogeneously dispersed in water because the attached chitosan donates a partial positive charge to the system, which provide an electrostatic repulsion force against the van der Waals force of the sheets. In addition, the liquid sonication accompanied by natural polymeric surfactants can stabilize exfoliated MoS_2 nanosheets against re-aggregation and provide hydrogen-bond interaction possibility between MoS_2 nanosheets and a polar polymer matrix. It is widely accepted that MoS_2 wrapped by polymer chains can result in stronger interface interactions with the polymer matrix. However, previous works mostly adopted small molecules to decorate the surface of MoS_2 nanosheets. The polyethylenimine (PEI) is positively charged in aqueous solutions. Thus, Cai et al. first wrapped chemically exfoliated MoS_2 nanosheets with PEI by electrostatic interaction. Afterwards, DOPO was also used to modify PEI-MoS_2 and was expected to improve the interfacial interaction in terms of enhanced π–π interaction (Cai et al. 2017b).

10.2.2.2 Covalent Modification

It has been suggested that edge sites in basal plane of MoS_2 nanosheet possess excellent affinity for thiol molecules. Because sulfur-containing functional groups can form covalent bonds with unsaturated Mo edges or sulfur vacancies in MoS_2, thiol derivatives with various functional groups can easily be absorbed and tightly bound to edge sites of MoS_2, providing the pathway to organic functionalization of MoS_2. It is well known that chemical exfoliation of bulk MoS_2 using lithium intercalation is the most common method to produce single-layer MoS_2 nanosheets in scalable quantities. The reaction of inserted lithium and water can produce hydrogen gas at the interface, highly yielding monolayer MoS_2 nanosheets with the aid of ultrasonication. Due to the violent nature of this reaction, MoS_2 crystal structure becomes deformed and internal edge sites become visible.

Up to now, the functionalization of MoS_2 nanosheets with coupling agents has been rarely reported. Jiang et al. (2014b) functionalized MoS_2 nanosheets with dithioglycol (DITG), and then DITG acted as a bridge to connect octa-vinyl polyhedral oligomeric silsesquioxanes. It can remarkably enhance the flame retardancy of polymer and significantly decrease the gaseous products, including hydrocarbons, carbonyl compounds, and carbon monoxide, which is attributed to the synergistic effect of OvlPOSS and MoS_2. Recently, Wang et al. (2017) reported functionalized MoS_2 nanosheets containing phosphorous, nitrogen, and silicon elements synthesized by a "thiol-ene" click reaction between MoS_2 nanosheets with

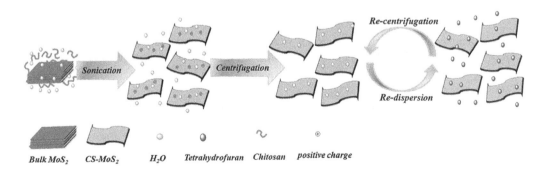

FIGURE 10.3 Schematic illustration for the exfoliation of MoS_2 nanosheets by ultrasonication in an aqueous solution of chitosan and re-dispersion of CS-MoS_2 nanosheets in tetrahydrofuran by the solvent exchange method. (Reprinted from Wang, D. et al., *J. Mater. Chem. A*, 3, 14307–14317, 2015. With permission.)

sulfydryl groups and ene-terminated hyperbranched polyphosphate acrylate (HPA) by covalent modification through C-S bond. First, defects into MoS$_2$ nanosheets to expose edge sites were engineered by designing the reaction with sodium molybdate and excess amount of thiourea. Then dithioglycol molecules are chemisorbed onto defect-rich MoS$_2$ to obtain sulfydryl-terminated MoS$_2$. HPA was synthesized by Michael addition polymerization of tri(acryloyloxyethyl)phosphate and (3-aminopropyl) triethoxysilane, which could react with sulfydryl-terminated MoS$_2$ by thiol-ene reaction.

10.2.3 Functionalization of Layered Double Hydroxides

Unlike graphene and MoS$_2$, the unique structure of layered double hydroxides results in the modification mostly happened in the inter-layer instead of on the surface. The intercalation modification of LDHs by organic long-chain molecules not only endows hydrophobic properties to the inorganic filler, but also enlarges the inter-layer distance of LDHs beneficial to the exfoliation in order to improve the dispersion in the polymer matrix. Some of the intercalated molecules could also impart extra flame retardant effects to functionalized LDHs. The functionalized LDHs are generally synthesized via co-precipitation, ion exchange, and reconstruction methods.

10.2.3.1 Co-precipitation

Co-precipitation method is the most facile and simple way to prepare intercalated LDHs since it only contains one step in most cases. All the precursors including the intercalating agents are mixed together and, during the precipitation process initiated by the alkaline environment, the intercalating agents enter into the gallery of LDHs inter-layers. An organo-modified MgAl LDH intercalated with glycinate was prepared through a co-precipitation process (Becker et al. 2011). The magnesium/aluminum nitrates along with glycine were added at the same time into an aqueous alkaline solution. It should be noted that the pH value of the solution had to be kept in a steady state to assure the successful fabrication. Similar, sodium dodecyl sulfonate (SDS) and sodium dodecylbenzenesulfonate (SDBS) are the most commonly seen as organic intercalating agents to modify LDHs by co-precipitation method with a pH value of the solution in a steady state (Gao et al. 2014; Shabanian et al. 2014; Wang et al. 2011, 2012a).

However, SDBS is a flammable surfactant like most modifiers for LDH. Thus, it is risky to decrease the flame retardancy of LDH/polymer nanocomposites by using SDBS for LDH's modification. In comparison to the structure of SDBS, most of the acid dyes contain non-fused aromatic, naphthalene, or anthracene rings, possessing many rigid structures that may lead to the formation of rich char residues and less volatile fuels during combustion or decomposition. Thus, our group adopted two kinds of anionic dyes to modify LDH by one-step co-precipitation method (Kang et al. 2013). Afterwards, our group reported a novel silicon- and phosphorus-containing intercalant (SIEPDP) which is derived from bio-based eugenol for MgAl LDH. SIEPDP possesses phosphine in its molecular structure which would yield high flame retardancy of the composites and the siloxane linked aromatic moieties would increase the inter-layered distance of the SIEPDP-LDH inter-layer, and in turn promote the char formation and reinforce the strength of the char layer formed during combustion (Li et al. 2015). Another bio-based modifier (cardanol-BS) was also successfully synthesized from renewable resource cardanol *via* the ring-opening of 1,4-butane sultone (BS). Cardanol-BS-modified LDH was developed through a one-step co-precipitation method (Wang et al. 2015b).

Our group also developed a novel modification method firstly for LDHs named after a multi-modifier system. Hydroxypropyl-sulfobutyl-beta-cyclodextrin sodium (sCD), SDBS, and taurine (T) were simultaneously intercalated into inter-layers of LDHs (Figure 10.4). In this multi-modifier system, sCD was synthesized as a first modifier for LDHs since sCD contains glucopyranose units possessing a large number of hydroxyl groups that may lead to the formation of rich char residues; SDBS was used as a second co-modifier to further enlarge the inter-layer distance of LDHs to reach a better dispersion state in the polymer matrix; meanwhile, T was used as the third co-modifier for LDHs that acted as a

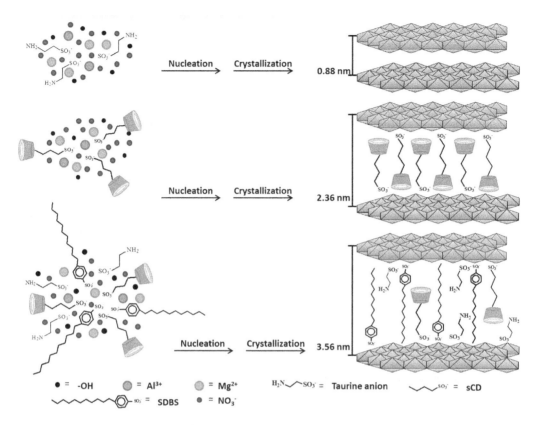

FIGURE 10.4 Schematic diagram of the anionic structure intercalated in the functionalized LDHs by a one-step synthesis method. (Reprinted from Kalali, E.N. et al., *J. Mater. Chem. A*, 3, 6819–6826, 2015. With permission.)

bio-based cross-linking agent to improve the interaction between the functionalized LDHs and polymer matrix (Kalali et al. 2015). Soon afterwards, multi-functional intercalation in LDHs has been developed *via* the design of multi-modifiers with varied functions in order to transfer these functions to polymeric materials by using nanocarriers. The functions of the multi-modifier system include: (i) functional-ized sCD and phytic acid (Ph) aiming at enhancing flame retardancy; (ii) SDBS enlarges the inter-layer space of LDHs for facilitating LDH's dispersion in the polymer matrix; and (iii) chalcone imparts the anti-UV property to the epoxy matrix. In contrast to conventional LDH-based polymer composites, the functionalized LDH-based polymer nanocomposites show a significant improvement in both flame retardancy and anti-UV property (Kalali et al. 2016a).

Although organic anion-modified LDHs facilitate the miscibility in the hydrophobic polymer, the method has serious limitations including unwanted change in the polymer characteristics, a limited selection of anions that can be intercalated, and the fact that organo-LDHs have low decomposition temperatures. Among the many inorganic anions-intercalated LDHs, borate-intercalated LDHs have been regarded as the best candidate. LDHs intercalated with borate anions may combine the advantages of both $Mg(OH)_2$, $Al(OH)_3$, and zinc borate. Wang et al. reported an aqueous miscible organic solvent treatment method for the synthesis of stable, transparent dispersions of the hydrophilic LDH-borate in non-polar solvents with HBO_3 as the borate resource (Wang et al. 2013b). This technique is capable of tuning the surface of LDHs to be hydrophobic, rending them to be highly dispersible in non-polar solvents such as xylene, and enabling to prepare LDHs/polymer nanocomposites using solvent mixing method for the first time ever.

10.2.3.2 Ion Exchange

The ion exchange intercalation usually involves two steps: preparation of LDHs with small inorganic anions such as NO_3^-, CO_3^{2-}, and Cl^- in the inter-layers and subsequently mixing the LDHs with intercalating agents accompanied by the occurrence of ion exchange. Co-precipitation allows direct access to, for instance, NO_3^--LDH or Cl^--LDH that can easily be functionalized by ion-exchange, but co-precipitation delivers rather small aspect ratios. Edenharter employed urea hydrolysis method to synthesize MgAl LDH (Cl as anion) with large aspect ratio which was then ion-exchanged with a functional molecule, phenyl phosphate, a model compound for P-containing flame retardant (Edenharter and Breu 2015). The mechanism of this anion exchange could be shown to be topotactic rather than of dissolution/reprecipitation type. An effective reactive phosphorus-containing flame retardant for polyesters, 2-carboxylethyl phenyl-phosphinic acid (CEPPA) was used to intercalate LDH through hydrothermal reaction followed by ion exchange method. CEPPA molecule contains carboxylic acid groups which can exchange the anions in the pristine LDH layers, which assists CEPPA to be molecularly dissolved in water and facilitate the intercalation of CEPPA (Ding et al. 2015).

Besides organic molecules, inorganic anion could also intercalate into LDHs by ion exchange method. Xu et al. (2016b) employed $Mo_7O_{24}^{6-}$ to intercalate MgAl LDH by the ion exchange method since molybdenum compounds are effective flame retardant and smoke suppression agent.

10.2.3.3 Reconstruction

The reconstruction method avails oneself of the memory effect, i.e., the ability to recover the original layered structure when the mixed M^{II} (M^{III})O oxide, obtained by a mild calcination (500°C) of the LDH precursor, is immersed in a solution of the anion to be intercalated. The recovery of the original layered structure by reconstruction method, based on the memory effect, appears unconvincing, and recently it is reported that such method is simply a direct synthesis of these attractive solids (Mascolo and Mascolo 2015).

A novel phosphorus-nitrogen-containing compound, N-(2-(5, 5-dimethyl-1, 3, 2-dioxaphosphinyl-2-ylamino)hexylic)-acetamide-2-propyl acid (PAHPA) was synthesized by Huang et al. Then LDHs-carbonate underwent heat treatment at high temperature followed by mixing with PAHPA in water solution to obtain PAHPA-intercalated LDHs. Phosphorus-nitrogen-containing compound treatment for LDHs improved both the dispersion of LDHs in polymer matrix and flame retardancy of the nano-composites (Huang et al. 2012b). The authors assumed that the PAHPA-LDHs were prepared through ion exchange of LDHs with PAHPA. But this process should be ascribed to reconstruction method strictly since due to the pronounced thermodynamic stability of CO_3^{2-}-LDH, it is unfortunately trouble-some to functionalize the material by ion-exchange.

Our group developed a Ph and (hydroxypropyl)sulfobutyl-β-cyclodextrin sodium (CDBS)-intercalated LDH decorated by iron oxide nanoparticle on the surface by reconstruction method as shown in Figure 10.5 (Kalali et al. 2016b). The MgAl LDH first calcined and milled with Fe_3O_4 nanoparticles. Next, the Fe_3O_4-doped-calcined LDH was reconstructed into LDH with the intercalation by Ph and CDBS simultaneously. Phytic acid, a natural resource compound with abundant phosphorous, can catalyze polymer matrix to form intumescent chars, while a cyclodextrin derivative with many rigid structures and abundant hydroxyl groups may lead to high yields of char residues. Furthermore, transition-metal oxides are usually used to improve the thermal resistance of char residues.

10.2.4 Hybrids of Multiple Layered Nanomaterials

The engineering of nanocomposites made up by multiple layered nanomaterials combines the unique properties attributing to each component and stimulates unexpected new performance in the polymer matrices.

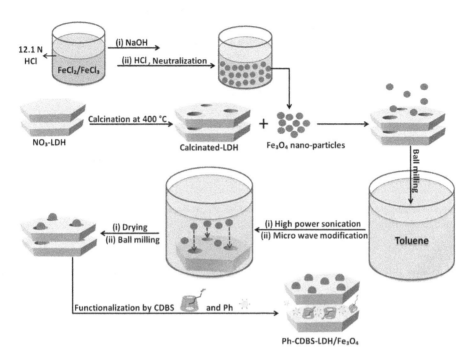

FIGURE 10.5 Preparation process of a Fe_3O_4@Ph-CDBS-LDH hybrid. (Reprinted from Kalali, E.N. et al., *Ind. Eng. Chem. Res.*, 55, 6634–6642, 2016. With permission.)

The oxidation of graphene nanosheets at high temperature would fade the barrier effect. Thus, it is a serious subject to improve thermal oxidative resistance of graphene nanosheets (Wang et al. 2013a). Surface modification of graphene with layered compounds is a valid way to inhibit oxygen from accessing graphene. Layered MoS_2 is a good choice, therefore layered MoS_2/graphene hybrids were synthesized by the hydrothermal method. $Na_2MoO_4 \cdot 2H_2O$ and thiourea were employed as the precursors for layered MoS_2 and mixed with GO to prepare the hybrid in Teflon-lined autoclave.

Inspired by the synergistic effect between LDH and carbon nanomaterials, Hu's group initiated hybrids of GO and LDH containing transitional metals, in an attempt to design a more efficient flame retardant (Wang et al. 2013c). In addition, the distribution of LDH on the surface of RGO could hinder the restacking of inter-layer nanosheets, thus increasing the degree of an exfoliation. Wang et al. synthesized NiFe LDH/graphene hybrids by a one-pot *in situ* solvothermal route to study the synergistic and catalytic effects of the hybrid in flame retardant polymer nanocomposites. The precursors for LDH were mixed together with GO in an autoclave to prepare the hybrid. The presence of the hybrid in the polymer acted as a physical barrier and resulted in the formation of a compact char layer. However, the dispersion of LDH on GO and their cohesion are poor according to the morphological study (Wang et al. 2013c). Hong et al. (2014) also fabricated a hybrid consisting of GO and NiAl LDH by a one-pot co-precipitation method to improve the flame retardancy of the polymer. But the dispersion of LDH on graphene was also non-uniform.

Recently, our group delicately designed a "3D fabrication method" to uniformly distribute LDH nanoplatelets on graphene slab (Pan et al. 2017a). The sandwich-type three-dimensional rGO@LDH hybrid showed LDH standing vertically or lying horizontally on both sides of graphene nanosheets. Unlike the one-step co-precipitation method, we firstly arranged acid-sensitive metal organic framework (MOF) polymorph on the graphene and then etched the MOF into tiny LDH plates by adding

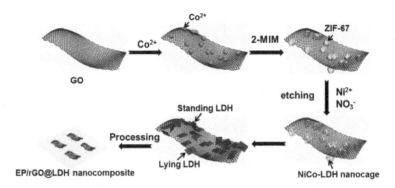

FIGURE 10.6   Schematic illustration of synthesis of rGO@LDH and preparation of EP/rGO@LDH nanocomposites. (Reprinted from Pan, Y.T. et al., *Chem. Eng. J.*, 330, 1222–1231, 2017. With permission.)

metal salt as shown in Figure 10.6. By this novel method, LDH plates located uniformly on graphene nanosheets. In addition, the whole process was carried out at room temperature thereby avoiding the energy consuming. The subtle hybrid presented high efficiency in improving the flame retardancy of the polymer nanocomposites while imparting extra properties to the system.

10.3 Characterization for Functionalized Layered Nanomaterials and Functionalized Layered Nanomaterials/Polymer Nanocomposites

10.3.1 Characterization for Functionalized Layered Nanomaterials

The characterization by different measurement techniques for functionalization layered nanomaterials is essential to have knowledge of the structures, components, and properties for the synthetic materials after proper modification. In the section, several typical models in the previous report will be given as example to help researchers to realize the characterization methods for functionalized layered nanomaterials and their application.

10.3.1.1 Functionalized Graphene

As a typical example, a phosphorous containing flame retardant combined with glycidol, phenyl glycidol chlorophosphate is modified on graphite oxide (Guo et al. 2011). All-sided characterization techniques are carried out to monitor the grafting behavior. The functionalized graphite oxide has an X-ray diffraction (XRD) pattern similar to graphite oxide, indicating that the modification does not destroy the layered structure of graphite oxide. In other report, the typical (001) diffraction of GO shifted to a lower value after modification, implying that the d-spacing was increased due to the binding of modifiers. And sometimes no signal for the phases of graphene (002) or GO (001) could be detected in the functionalized graphene, probably due to the relatively high content and good crystallinity of modifiers in the hybrid which performed strong diffraction peaks, covering the diffraction of the GO. Another reason ascribed to the broken face-to-face stacking due to the introduction of the modifiers on both sides of graphene sheets. Raman scattering spectroscopy is an effective method to examine the microstructure of carbonaceous materials because of its sensitivity toward these materials. The Raman spectra of functionalized graphite oxide is similar to that of graphite oxide, with a slight increase of vibration frequency of D and G band and a decrease of D/G intensity ratio which may be caused by reaction of modifier with graphite oxide (Figure 10.7). In Fourier transform infrared (FTIR) spectra of functionalized graphite

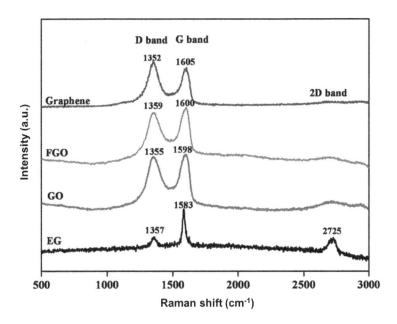

FIGURE 10.7 Evolution of the Raman spectra during the oxidation of expandable graphite (EG), and the reduction and functionalization processes of GO. (Reprinted from Guo, Y. et al., *Ind. Eng. Chem. Res.*, 50, 7772–7783, 2011. With permission.)

oxide, the peak assigned to P–Cl disappears due to the consumption by the –OH group from graphite oxide. Meanwhile, the peak attributed to –OH becomes very weak. The new emerged bands correspond to the O=P– and –P–O–C groups both deriving from the phosphorous functionality. X-ray photoelectron spectra (XPS) analysis was used to investigate the surface composition and oxidation state of functionalized graphene. The new peaks for C–O–P group in C_{1s} XPS and the O=P–O–C group in P_{2p} XPS spectra caused by the phosphorus functionality are highly consistent with the FTIR results. In the thermogravimetric analysis (TGA) of graphitic oxide, the char yield obtained at 800°C is 45.9 wt%, while the modifier on the edge of graphitic oxide sheets improves the thermal stability during the whole thermal decomposition process with an increase of char residue of 50.8 wt%. Transmission electron microscopy (TEM) evaluation of the functionalized graphitic oxide describes that its layer structure is preserved, but some cracks, perhaps induced by the chemical modification process, producing defects are observed. Atomic force microscopy (AFM) observations show that graphitic oxide after modification maintains layer structure with the same average thickness relative to the unmodified one, as a complementary measurement for TEM technique. All these results confirm the successful grafting on graphite oxide sheets.

The restoration of conjugated C=C bonds in functionalized graphene during chemical reduction can be analyzed by UV-vis spectroscopy. Liao et al. (2012) grafted DOPO onto the surface of graphene oxide by reacting epoxy ring groups which also partly restored the conjugate structure of graphene oxide. The UV-vis spectrum of graphene oxide exhibits a strong adsorption peak at 228 nm, ascribing to π–π* transitions of aromatic C–C bonds, and a shoulder at ~300 nm that is assigned to n–π* transitions of C=O bonds. After chemical reduction, the adsorption peak of DOPO-modified graphene oxide is red-shifted to 263 nm, which indicates that the electronic conjugation has been restored through the removal of oxygenic groups on the modified graphene sheets. It is noteworthy that the strong adsorption peak at around 282 nm that belongs to DOPO appears in the spectrum of modified graphene, confirming the strong adsorption between aromatic DOPO and graphene.

10.3.1.2 Functionalized Layered MoS$_2$

The intense reflection attributed to the (002) plane of MoS$_2$ in XRD pattern always shifts to lower value after modification. It has been generally accepted that the expansion of the inter-layer space is due to the successful functionalization of MoS$_2$ nanosheets with functional molecules (Jiang et al. 2014b). The FTIR spectra of pristine MoS$_2$ and modified MoS$_2$ show additional evidence for the functionalization of MoS$_2$. Zhou et al. modified MoS$_2$ with CTAB, and compared with pristine MoS$_2$, the spectra of CTAB-MoS$_2$ exhibiting additional absorption peaks at 1020, 1108, and 1385 cm^{-1} are ascribed to aliphatic C–H bonding, the peak at 1470 cm^{-1} is due to ammonium salt, and the peaks at 2920 and 2850 cm^{-1} are attributed to CH$_2$ anti-symmetric and symmetric stretching vibration, respectively. The FTIR results can be considered to be indicative of successful modification (Zhou et al. 2014a). In addition, TGA was carried out to provide further evidence for successful modification onto MoS$_2$. The total weight loss of MoS$_2$ is 11 wt% when it is heated to 700°C. This value is close to the weight loss of the sulfur dioxide molecules (10 wt%), it means that the MoS$_2$ transforms to molybdenum oxide and sulfur oxide under air atmosphere in heating process and reaches the maximum rate at about 510°C. Comparing the TGA curve of MoS$_2$ with that of ferrocene-modified MoS$_2$ samples, it can be found that Fe-MoS$_2$ hybrid exhibits a similar TGA profile, and the mass loss of the modified MoS$_2$ sample in the temperature range of 100°C–700°C is much larger than that of bulk MoS$_2$. The temperature of the maximum mass loss rate is in advance to 450°C and the total weight loss reaches 23 wt% at 700°C. Based on the information above, it is reasonable that ferrocene has been successfully intercalated into MoS$_2$ (Zhou et al. 2014b).

The morphologies of MoS$_2$ nanosheets before and after the CS-assisted modification could be investigated by TEM. The TEM image of bulk MoS$_2$ (Figure 10.8a) showed large-area stacked nanosheets with lengths of up to 1 mm and widths of about 500 nm are clearly visible. After sonication, the seriously stacked nanosheets of bulk MoS$_2$ were exfoliated to few- and mono-layered structures. Compared with bulk MoS$_2$, CS-MoS$_2$ (Figure 10.8b) exhibits a thinner and smaller nanoplatelet shape. Tapping-mode of AFM (Figure 10.8c and d) was further used to determine the size and thickness of CS-MoS$_2$ nanosheets along with atomic thickness. The exfoliated CS-MoS$_2$ nanosheets are tens or hundreds nanometers wide, with thicknesses varying from 0.8 to 1.3 nm, which is fairly consistent with the single-layer MoS$_2$ after exfoliation (Wang et al. 2015a).

Wang et al. (2017) synthesized HPA to modify MoS$_2$, and Raman spectroscopy was carried out to garner structural information about the modification. Pure MoS$_2$ has two dominant peaks at 380.2 and 405.5 cm^{-1}, ascribed to in-plane E$^1_{2g}$ and out of-plane A^1_g vibrational modes, respectively. It is observed that E$^1_{2g}$ and A^1_g have different degrees of shift to lower frequencies, and the shift of E$^1_{2g}$ is larger than that of A^1_g, suggesting that surface HPA molecules introduce tensile strain into MoS$_2$ layers. To further confirm the existence of C–S bond in MoS$_2$-HPA, XPS spectra were carried out to garner the composition and chemical status of various elements in MoS$_2$-HPA hybrids. The calculated atomic ratio of S to Mo (2.65) is much larger than theoretical values of MoS$_2$, due to ligand conjugation of MoS$_2$ with thiol-terminated derivative. As observed in the high resolution C1s spectrum, the peaks centered at 282.7, 284.6, 286.1, 287.0, 287.7, 288.4, and 289.3 eV are assigned to C–Si, C=C/C–C, C–S, C–O, C–N, C=O, and COO– (ester) species. The peak at 164.2 eV also indicates the formation of C–S bond demonstrating the covalent interaction between defect-rich MoS$_2$ nanosheet and HPA molecules.

10.3.1.3 Functionalized Layered Double Hydroxides

The typical XRD pattern for LDHs exhibits the hydrotalcite-like characteristic reflections of a hexagonal unit cell with a three-layer polytype of rhombohedral symmetry (Jayanthi et al. 2015). The characteristic peaks at the angles in a geometric progression can be ascribed to the (003), (006), and (012) plain diffraction peaks (Long et al. 2015). And in some cases, high energy facet such as (110) and (113) plane diffractions may appear in the high angle. From these parameters, the basal spacing d$_{(003)}$ of LDH can be calculated according to Bragg's Equation (Wang et al. 2015c). After functionalization by the intercalants, the diffraction peaks move to the direction of low angle as evidenced in Figure 10.9, responding to the increase of basal spacing of LDH caused by the successful interaction of the modifiers which enlarge the inter-layer spacing (Kang et al. 2013).

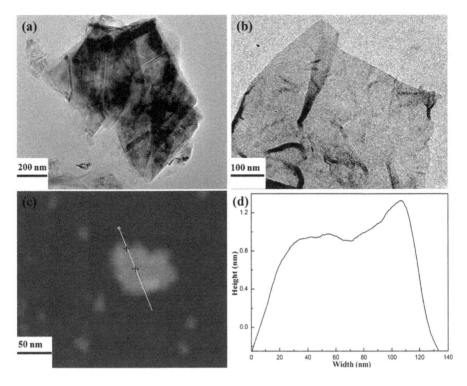

FIGURE 10.8 TEM images of bulk MoS_2 (a) and the ultrathin $CS-MoS_2$ nanosheets (b); AFM images (c); and thickness profiles (d) of $CS-MoS_2$ nanosheets. (Reprinted from Wang, D. et al., *J. Mater. Chem. A*, 3, 14307–14317, 2015. With permission.)

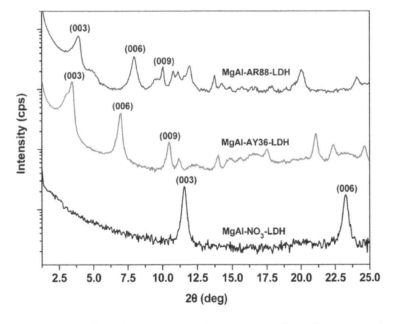

FIGURE 10.9 XRD patterns of MgAl-NO_3-LDH, MgAl-AY36-LDH, and MgAl-AR88-LDH. (Reprinted from Kang, N.J. et al., *ACS Appl. Mater. Interfaces*, 5, 8991–8997, 2013. With permission.)

The electron microscopy analysis is used to study the morphologies and sizes of the LDH before and after modification. After intercalated by CEPPA, the LDH sheets maintained the lamellar morphology, but their thickness was increased. This indicates that the CEPPA molecule successfully entered into the LDH interlayers and did not destroy the lamellar morphology of the LDH nanosheets. The inductively coupled plasma (ICP) spectrometer test was employed to determine the compositions of pristine LDH and CEPPA-modified LDH samples. The molar ratio of Ni^{2+} and Al^{3+} in pure LDH was about 2:1. The value is in good agreement with the feeding ratio. After intercalation, the molar ratio of Ni^{2+} and Al^{3+} in modified LDH was decreased to 1.4:1, indicating the leaching out of Ni^{2+} ions during the long intercalation process (Ding et al. 2015).

In the FTIR spectrum of pristine MgAl LDH (Figure 10.10a), several characteristic peaks are observed: the broad peak around $3490 \ cm^{-1}$ can be ascribed to the stretching of OH groups attached to Al and Mg ions in the layers; the small peak at $1620 \ cm^{-1}$ is assigned to the bending vibration of the inter-layer water; and the strong band at $1385 \ cm^{-1}$ is owing to the asymmetric stretching of the carbonate anion. In contrast to the pristine LDH, some new peaks appear in the FTIR spectrum (Figure 10.10b) of cardanol-BS-modified LDH. The appearance of the $-CH_3$ and $-CH_2-$ stretching peaks (2930 and $2860 \ cm^{-1}$) together with the sulfonate stretching bands ($1185 \ cm^{-1}$) confirms that

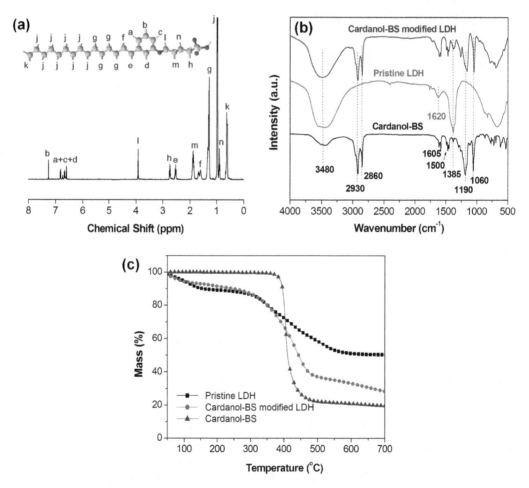

FIGURE 10.10 (a) ¹H NMR spectrum of 4-(3-pentadecylphenoxy)butane-1-sulfonic salt; (b) FTIR spectra of cardanol-BS, pristine LDH, and cardanol-BS-modified LDH; and (c) thermogravimetric analysis profiles of pristine LDH, cardanol-BS, and cardanol-BS-modified LDH. (Reprinted from Wang, X. et al., *ACS Sustain. Chem. Eng.*, 3, 3281–3290, 2015. With permission.)

cardanol-BS molecules have been intercalated into the inter-layer space of MgAl LDH. In this research, TGA was carried out to provide further evidence for successful modification of LDH, and TGA profiles were depicted in Figure 10.10c. Pristine LDH starts to lose weight upon heating even below 200°C, which is due to the removal of absorbed water, and the major mass loss occurs from 300°C to 550°C, presumably attributable to the dehydroxylation and the removal of inter-layer anions. In contrast to pristine LDH, cardanol-BS-modified LDH shows the different thermal degradation behaviors with much more mass loss at high temperature, which is probably attributed to the surface modification of LDH by cardanol-BS (Wang et al. 2015b).

Elemental analysis is a useful tool to determine the chemical formula of LDHs. The LDHs could be represented by the chemical formula $Mg_2Al \cdot (OH)_6 \cdot (dye\ anion)_x \cdot (NO_3)_{1-x} \cdot 0.4H_2O$ in one of our research (Kang et al. 2013), where the dye anion is AY36 or AR88 and x is the degree of intercalation of the dye anion. On the basis of elemental analysis data of sulfur [S (wt%)], the degree of intercalation of AY36 or AR88 (x) can be calculated from the equation:

$$S(wt\%) = \frac{32x}{246.2 + x(M_w - 62)} \times 100\%,$$

where M_w is the molecular weight of dye anion, 352.39 for AY36 and 377.39 for AR88. The results show the MgAl-AR88-LDH shows a degree of intercalation of 97%, which means almost all the nitrate ions (NO_3^-) in the inter-layer region have been exchanged for AR88 anions. MgAl-AY36-LDH has also achieved 88% successful substitution.

Raman spectroscopy can be used to characterize the species of the inter-layer anions. In the Raman spectrum of the NO_3-MgAl LDH sample, the peak at 557 cm^{-1} is attributed to the lattice vibrations of the brucite octahedral layers (Al-O-Al and Al-O-Mg), and the peaks at 710 and 1055 cm^{-1} are attributed to the nitrate vibrations. In the Raman spectrum of the Mo-MgAl LDH sample, the peaks corresponding to nitrate vibrations disappear with the emergence of some new peaks assigned to $Mo_7O_{24}^{6-}$. Specifically, the most intense band 947 cm^{-1} characteristic for $Mo_7O_{24}^{6-}$ can be clearly discerned, and the peaks at 219 and 560 cm^{-1} belong to Mo-O-Mo stretching and blending vibrations, and the peak at 358 cm^{-1} could be assigned to the Mo=O deformation vibration. It can be confirmed that $Mo_7O_{24}^{6-}$-intercalated LDHs are synthesized successfully through the ion exchange method by the Raman spectra results (Xu et al. 2016b).

10.3.2 Morphology and Structure of Functionalized Layered Nanomaterials/Polymer Nanocomposites

The morphology and structure of polymer/layered inorganic nanocomposites are generally characterized by XRD and TEM analysis. The XRD technique was used to characterize the crystallographic structure of the functionalized layered nanomaterials and their dispersed homogeneity in the polymer. The disappearance of characteristic peaks for layered nanomaterials in the XRD pattern of polymer nanocomposites is one of the important symbols that indicate the highly exfoliated and/or well intercalated LDH platelets within the polymer matrices, but the conjecture needs to be confirmed by other powerful techniques also, such as TEM. In Figure 10.11, the unmodified LDH/epoxy composite showed an intense diffraction peak at 2 theta = 10.1° corresponding to an inter-gallery spacing of 0.88 nm, indicating the presence of the ordered structure of unmodified LDH in the polymer matrix. Taurine-modified LDH/epoxy composite showed a similar XRD pattern as that of the unmodified LDH/epoxy composite because the inter-gallery spacing of them was not large enough to allow the diffusion of the resin/curing agent mixture into the inter-layers. For the sCD-modified LDH/epoxy composite, a weak diffraction peak at 2 theta = 3.8° was observed corresponding to an inter-gallery spacing of 2.32 nm, suggesting the swelling of LDH platelets upon polymerization. In the case of the sCD-SDBS-T-modified

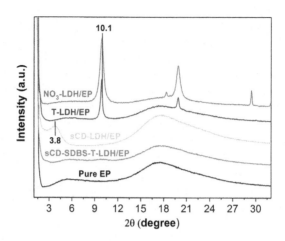

FIGURE 10.11 XRD patterns of pure EP, NO_3-LDH/EP, T-LDH/EP, sCD-LDH/EP, and sCD-DBS-T-LDH/EP composites. (Reprinted from Kalali, E.N. et al., *J. Mater. Chem. A*, 3, 6819–6826, 2015. With permission.)

LDH/epoxy composite, no visible reflection peaks were observed below 10°, indicating highly exfoliated and/or well intercalated LDH platelets within the epoxy resin (Kalali et al. 2015).

The microstructures of the nanocomposites were characterized by means of TEM employed to provide more information in order to investigate the results obtained from the XRD analysis as well as to directly observe the dispersion state of LDH within the polymer matrix. Low and high magnification TEM images (Figure 10.12a–c) of the unmodified LDH/epoxy composite (NO_3-LDH/EP) show that large agglomerations of LDH plates with the average size of more than 500 nm were formed. This is the reason why the (003) reflection is observed in the XRD patterns of NO_3-LDH/EP. The dispersion state of NO_3-LDH/EP composite was quite different from that of multi-functionally intercalated LDH/ epoxy (fCD-SDBS-Ph-LDH/EP). The low and medium magnification TEM image of fCD-SDBS-Ph-LDH/EP (Figure 10.12d and e) reveals a much better dispersion of LDH in the epoxy matrix, and the higher magnification image of fCD-SDBS-Ph-LDH/EP (Figure 10.12f) shows a homogeneously disordered nanostructure. Selected area electron diffraction (SAED) patterns were also utilized by TEM in order to investigate the crystalline parameter structures of the specimens. In the composite, SAED patterns of NO_3-LDH/EP indicated a clear corona around the pointer (inset of Figure 10.12c), implying the existence of large poly-crystalline structures. In contrast, in the selected area electron diffraction image of fCD-SDBS-Ph-LDH/EP (inset of Figure 10.12f), there was no corona observed around the pointer, which indicated the existence of scant crystallites or excellent exfoliated nano-particles (Kalali et al. 2016a).

The direct measurement of interfacial interactions is not easy, and TEM studies offer circumstantial evidence. In several works, the ultrathin sections of graphene-polymer composites are usually smooth, but the ultrathin sections in Bao's work are obviously rougher, especially around the graphene layers which are modified by hexachlorocyclotriphosphazene and hydroxyethyl acrylate, implying that functionalized graphene has formed strong interfacial interactions with polymer matrix due to the functionalization (Bao et al. 2012a).

The attenuated total reflectance (ATR)-FTIR can also be used to evaluate the interface interaction between nanofiller and polymer matrix. ATR-FTIR spectra of pure thermoplastic polyurethane (TPU) and TPU blended with DOPO-functionalized MoS_2 have two different bands: a main peak located at 1732 and a relatively small shoulder around 1700 cm^{-1} (Cai et al. 2017b). The peak at 1732 cm^{-1} is attributed to –C=O groups that are "free" (non-hydrogen bonded), and the shift to 1700 cm^{-1} results from hydrogen bonding with functionalized MoS_2. For TPU polymer nanocomposite, the intensity ratio

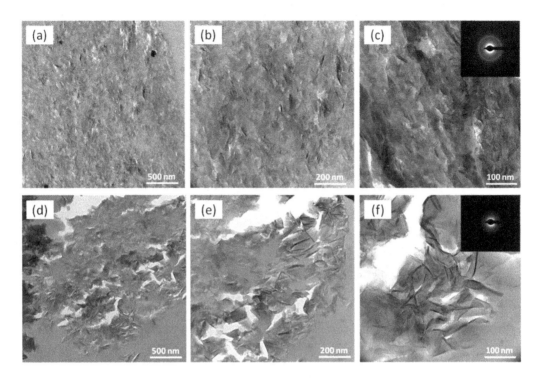

FIGURE 10.12 TEM images of NO$_3$-LDH/EP (a–c) and fCD-SDBS-Ph-LDH/EP (d–f) nanocomposites. (Reprinted from Kalali, E.N. et al., *J. Mater. Chem. A*, 4, 2147–2157, 2016. With permission.)

between hydrogen-bonded and "free" carbonyl domain is visibly changed and hydrogen-bonded carbonyl is dominant. This phenomenon expounds that hydrogen bonding enhances the interfacial interaction between functionalized MoS$_2$ and the polymer matrix.

The compatibility of modified layered nanomaterials and bulk one within polymer matrix can be compared facilely by their performance in organic solvents. The compatibility of CTAB-MoS$_2$ in dimethyl formamide (DMF) is compared with that of an unmodified one (Zhou et al. 2014a). Virgin MoS$_2$ and CTAB-MoS$_2$ are dispersed in DMF, undergo 20 minutes of ultrasonication, and are then kept standing for 24 hours. Unmodified MoS$_2$ precipitates, while CTAB-MoS$_2$ remains well dispersed in DMF. Thus, it can be concluded that the compatibility of CTAB-MoS$_2$ is improved as compared with that of bulk MoS$_2$. The compatibility between the layered nanomaterials and polymer matrix not only affects the dispersion of nanofillers, but also influences the properties of the polymer nanocomposites.

10.4 Structure-Property Relationship of Functionalized Layered Nanomaterial-Based Flame Retardant Polymer Nanocomposites

The functionalized layered nanomaterials exhibit enhanced flame retardant effect, uniform dispersion, and/or strong interaction within the polymer matrices definitely improving the flame retardancy of the polymer nanocomposites compared with the unmodified counterparts. The structure-property relationship of functionalized layered nanomaterial-based flame retardant polymer nanocomposites reported in recent years will be reviewed in this section, and we only select several representative examples for a brief illustration.

10.4.1 Flame Retardancy of Functionalized Graphene-Based Polymer Nanocomposites

10.4.1.1 Epoxy Resin Nanocomposites

Few reports are carried out that nanocomposites can reach up to the flame-retardant-rating via the test of limited oxygen index and vertical burning method (Zhang et al. 2017). The limiting oxygen index (LOI) value of DOPO/epoxy and unmodified GO/epoxy nanocomposites showed a slight increase, but these values are still higher than that of the neat epoxy system. In contrast, the flame retardancy of DOPO-modified GO (DOPO-rGO)/EP nanocomposites improved with increasing the modified GO content. The LOI values of DOPO-rGO/EP nanocomposites improved significantly, namely, from 21% to 26% when the DOPO-rGO content was increased from 0 to 10 wt%. The dramatic enhancements with DOPO-rGO can be attributed to the following reasons: (1) Faster decomposition of phosphate group at lower temperature forms a phosphorus-rich residue to prevent further decomposition of the resin matrix and promote the formation of char residue. (2) The origin of the flame-retarding behavior of GO is thought to be its ability to form a continuous, protective char layer that acts as a thermal insulator and a mass transport barrier (Liao et al. 2012). With the incorporation of 5 wt% of DOPO-VTS-modified GO (FRs-rGO) into EP, satisfied flame retardant grade (V-0) and the LOI as high as 29.5 were obtained. Moreover, the peak heat release rate (PHRR) value of FRs-rGO/EP was significantly reduced by 35% compared to those of neat EP. The flame retardancy strategy of FRs-rGO combines condensed phase and gas phase flame retardant strategies such as the nanocomposites technique, phosphorus-silicon synergism systems in the condensed phase, and DOPO flame retardant systems in the gas phase (Qian et al. 2013).

For the sample of EP/OapPOSS-modified GO (OapPOSS-rGO), the PHRR exhibited reduction compared with both unmodified EP/GO and EP/OapPOSS as shown in Figure 10.13 (Wang et al. 2012b). The best fire retardant properties of EP/OapPOSS-rGO could be attributed to two aspects: firstly, the reduction of GO by OapPOSS occurs to convert GO into a more stable form, rGO; secondly, POSS can create a stable silica layer on the char surface of EP, which reinforces the barrier effect of graphene. Total heat release (THR) values of EP/GO, EP/OapPOSS, and EP/OapPOSS-rGO are significantly reduced by 16%, 23%, and 36%, respectively, compared to that of pure EP. In addition, CO production for EP/OapPOSS-rGO is much lower than that of pure epoxy resins, which is probably due to the adsorption and barrier effect of graphene.

In another report, neat EP exhibits a LOI value of 25.0% and no classification in the UL-94 vertical burning test with dripping phenomenon. Incorporating RGO and polymeric flame retardant-modified reduced graphene oxide (PFR-fRGO) induces an opposite change trend for the LOI value for the EP composites: The LOI values for EP/RGO and EP/PFR-fRGO are 24.3% and 26.3%, respectively. In addition, the EP/PFR-fRGO composite shows the V-2 rating, while EP/RGO has no classification in UL-94 test. In the cone calorimeter test, the introduction of 1 wt% RGO or PFR-*f*RGO decreases the PHRR from 1137.6 kW/m² (neat EP) to 972.5 kW/m² (EP/RGO) and 853.3 kW/m² (EP/PFR-*f*RGO), the THR and total smoke production (TSP) also show a <10% reduction compared to neat EP. The weak improvement in flame retardancy may be ascribed to the low graphene loading (1 wt%), which is lower than the threshold value needed to form a network structure in EP matrix (Ma et al. 2008). In contrast, the flame retardancy of EP/PFR-*f*RGO is better than that of EP/RGO, due to the wrapping of PDMPD chains, which not only improves the dispersion and interfacial compatibility of PFR-*f*RGO in resulting composites, but also catalyzes charring to form a protective layer on the surface of polymer. These results suggest that using PFR-fRGO is a more effective way to improve the flame retardancy than using unmodified RGO (Feng et al. 2017).

10.4.1.2 Polystyrene (PS) Nanocomposites

The production of the chemical char results from a catalysis reaction in the presence of phosphorous-containing acids at high temperature, accordingly, a certain amount of time is needed for the release of acids, as well as the formation of the barrier chars when flame occurs. This means that

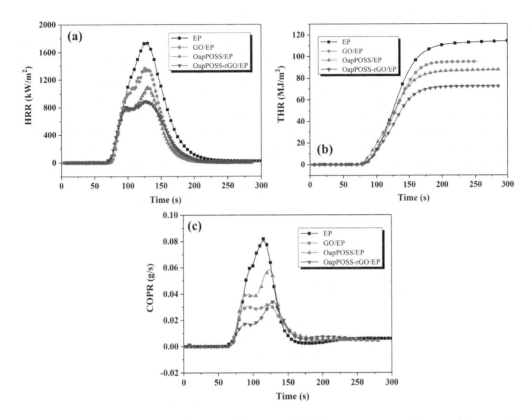

FIGURE 10.13 (a) HRR, (b) THR and (c) CO production rate (COPR) versus time curves of epoxy and its composites. Reprinted with permission from Wang, X., et al. "Simultaneous reduction and surface functionalization of graphene oxide with POSS for reducing fire hazards in epoxy composites", Journal of Materials Chemistry 22, (2012): 22037–22043.

phosphorous-containing flame retardants play ineffective roles during the initial combustion of materials. In comparison, after anchoring on the graphene sheets, the two-dimensional graphene sheets can provide barrier shields in the beginning of flame. Such an advantage overcomes the lagging effect of chemical charring to a degree (Chen et al. 2017).

HCCP and HEA were used to covalently functionalize GO (fGO) (Bao et al. 2012a). When fGO-polystyrene composites are degraded or burned, fGO catalyzes the char formation from polystyrene (Char A). Char A protects fGO from burning out and then fGO acts as a graphitic char (Char B). Thanks to the coordination of Char A, Char B, the physical barrier effects of fGO, and the strong interfacial interactions of fGO and polymers, the fire safety properties of the fGO–polystyrene composites are improved, resulting in much better performance in improving the fire safety properties of PS compared with the unmodified GO. As reported in many works, graphene-based materials usually facilitate the ignition of graphene-based polymer composites (Bao et al. 2012b, 2012c). In this work, time to ignition (TTI) is also decreased with increasing the fGO loading level. Such a decrease can be ascribed to the high thermal conductivity of graphene-based materials, which speed up the diffusion of heat from the surface into the body of composites. When samples are heated in a cone calorimeter, graphene-based composites are more easily able to increase their temperature as compared with neat polymer samples because of the high thermal conductivity of graphene-based materials, and thus, the ignition of graphene-based composites is facilitated. Such facilitation is still a challenge to be overcome because it makes burning easy.

In another work, GO was functionalized by an organophosphorous oligomer (PMPPD) (Qiu et al. 2015). As expected, introducing 5 wt% GO into PS matrix (PS-GO5.0) makes the PHRR decrease to 806.3 kW/m² by 24.4%, compared to that of pure PS. Moreover, the PHRR of the

PS/5 wt% PMPPD-modified GO sample (PS-fGO5.0) exhibits further reduction relative to that of the PS-GO5.0 sample, the PS-fGO5.0 sample presents the lowest PHRR value of 733.9 kW/m^2 (31.2% reduction). The better flame retardant properties of fGO/PS nanocomposites are attributed to two factors: first, the functionalization of GO by PMPPD reinforces the physical barrier effect of graphene; second, PMPPD facilitates the charring process of PS nanocomposites. The THR value of the PS-GO5.0 sample is slightly reduced by 2.6%. For fGO/PS nanocomposites, the THR value decreases as the additive amount of fGO increases. The lowest THR value of 97.3 MJ/m^2 (4.9% reduction) is observed for the PS-fGO5.0 sample. The results suggest that fGO enhances fire safety of PS nanocomposites.

10.4.1.3 Thermoplastic Polyurethane Nanocomposites

Cai et al. (2016) first employed electrochemistry exfoliated graphene to improve the flame retardancy of TPU. The iron lignosulfonate (Fe-lignin) was modified on the graphene in a non-covalent way. It could be found that the neat TPU burned violently with a sharp HRR peak (1408 kW/m^2). Evidently, compared to pure TPU, the PHRR value of TPU nanocomposites gradually decreased with increasing the contents of Fe-lignin-modified GO nanosheets (functionalized GO nanosheets [fGNS]). Through adding 2.0 wt% fGNS (TPU-2.0), PHRR decreased by more than half of the original level (from 1408 to 523 kW/m^2). Moreover, the THR of TPU-2.0 (78.6 MJ/m^2) was considerably reduced compared to pure TPU (63.9 MJ/m^2). Benefiting from the catalytic carbonization of Fe-lignin, *in situ*-formed char on the surface of graphene prevents the lamella structure of FGNS from over burning and thus forms the protective char layer. This stable char layer effectively hinders the delivery of degradation production with a shield effect, thus imparting the excellent flame retardancy to TPU materials.

Low oxidation graphene can prevent the lamellar structure from over burning. In order to reduce the oxidation degree of graphene, the same group of the above work masterly designed a multi-functional flame retardant HCPCP to simultaneously exfoliate and modify graphene functionalized GO nanosheets [fGNS] (Cai et al. 2017a). The benzene ring structure of the flame retardant is used to reduce the oxidation degree of graphene by the reaction between radicals and benzene rings, thus improving the layer quality and decreasing the defect (Xu et al. 2014). The enhanced interfacial adhesion is due to three reasons: the π–π action exists between the benzene rings of the absorbed HCPCP and the hard segment of TPU; the enhanced density of hydrogen bonds by the graphene; and high specific surface area. Compared with pure TPU, the composites contained in fGNS all present a gradually decreasing PHRR value with increasing the contents. With 4.0 wt% fGNS in TPU matrix (TPU-4.0), a remarkable decline in PHRR value (768 kW/m^2) can be obtained. Moreover, similar phenomenon can be observed in THR curve, which demonstrates additional fGNS can more effectively enhances fire safety of TPU than unmodified graphene or HCPCP. With an increasing fGNS content, the THR values of TPU composites display stepwise diminishing trend. The THR value of TPU-4.0 (64.0 kJ/g) is considerably reduced compared to that of pure TPU (79.1 kJ/g). It is reasonable to assume that HCPCP containing phosphorus and nitrogen elements promotes the formation of char residue during combustion, and graphene sheets play a barrier effect to inhibit the escape of gradation production. Benefiting from the combination of the high-quality layer structure and the coated flame retardant, graphene-based char layer is remarkably promoted. This stable char layer effectively inhibits the delivery of heat and degradation products, thus enhancing the flame retardancy to TPU. Therefore, fGNS imparts higher fire safety to TPU matrix than unmodified graphene and HCPCP.

The LOI test cannot detect the difference among the flame retardant TPU composites. That is because nanoparticles themselves are often not efficient in reducing the overall flammability of polymers (Pan and Wang 2015). The UL-94 classification and oxygen index usually remain similar or are even worsened by the addition of nanoparticles (Yi et al. 2014). However, the vertical burning test highlights the differences. The neat TPU cannot reach any ratings with melt dripping during the vertical burning test. With the addition of 5% APP into TPU composites, the LOI value can improve obviously, while the sample still exhibits serious dripping during the vertical burning test and reaches only V-2 rating. However, the APP particles encapsulated by only 0.25% or 0.5% of large-size reduced graphene oxide (LRGO) can depress the droppings in TPU/large-size graphene wrapped APP (LRAPP) composites, reaching a V-0 rating. When the content of LRGO increases to 1%, the LRAPP 1 cannot achieve such an effect. As reported, the synergistic effect of LRGO increases the melt viscosity of TPU composites which hinders the polymer material from dripping and improves the char residues (Dittrich et al. 2014). However, too much excess LRGO can cause the opposite effect, which might be because more LRGO on the surface of APP cannot only result in excessive accumulation, but also restrains the reaction between APP and TPU resin during burning (Zhang et al. 2017).

10.4.2 Flame Retardancy of Functionalized Layered MoS$_2$-Based Polymer Nanocomposites

10.4.2.1 Epoxy Resin Nanocomposites

Wang et al. (2015a) prepared chitosan-modified MoS$_2$ (CS-MoS$_2$) to study the flame retardancy of EP/CS-MoS$_2$ nanocomposites. The PHRR of the neat EP reaches up to 1592 kW/m^2, exhibiting high inflammability. On incorporating CS-MoS$_2$ into the EP matrix, the PHRR values of EP nanocomposites are gradually decreased with gradually increasing content. The addition of 2 wt% CS-MoS$_2$ significantly reduces the PHRR of EP nanocomposite to 902 kW/m^2, corresponding to a 43.3% reduction compared to neat EP. On the contrary, the PHRR of 2 wt% EP/MoS$_2$ nanocomposite is decreased by only 26.0%. The THR values of the EP/CS-MoS$_2$ nanocomposites decrease progressively until the additive amount is increased to 1 wt%. The decreased THR could be explained by the large contact area of CS-MoS$_2$ nanosheets to suppress the release of combustible gas and then promote charring. When the CS-MoS$_2$ further increases to 2 wt%, the THR value is increased to 33.9 MJ/m^2, probably because of the stronger suppression of additional CS-MoS$_2$ nanosheets that further reduces the effusion rate of combustible gas, causing the more complete oxidative combustion of the combustible volatiles such as hydrocarbons to produce more heat. In the case of EP/MoS$_2$, the PHRR value is higher than that of the neat EP instead (Table 10.1).

The flame retardant mechanism for the modified MoS$_2$ in the EP nanocomposites can be explained as follows: Inside the EP nanocomposites, each independent CS-MoS$_2$ nanosheet serves as a nano-barrier to prevent the permeation of external heat as well as oxygen and retard the escape of pyrolytic products such as hydrocarbons, CO, and aromatic compounds. The nano-barrier effects of CS-MoS$_2$ nanosheets result in a "tortuous path" to lower the release rate of combustible gas including hydrocarbons to support combustion, thus decreasing the heat-release rate. Furthermore, the nano-barrier effects of CS-MoS$_2$ nanosheets also promote the aggregation of hydrocarbons and aromatic compounds to produce smoke particles and then accumulate into char residues, thus reducing total heat release. Unfortunately, the low release rate of combustible gas may facilitate its complete oxidation-combustion to produce more THR.

TABLE 10.1 The Data of Cone Calorimeter Tests of Neat EP and Its Composites

Sample	PHRR (kW/m^2)	THR (MJ/m^2)
EP	1592	39.7
EP/0.5 wt% CS-MoS$_2$	1243	35.9
EP/1 wt% CS-MoS$_2$	1107	28.6
EP/2 wt% CS-MoS$_2$	902	33.9
EP/2 wt% MoS$_2$	1178	40.1

Source: Wang, D. et al., *J. Mater. Chem. A*, 3, 14307–14317, 2015.

10.4.2.2 Polystyrene Nanocomposites

The introduction of CTAB-modified MoS_2 (CTAB-MoS_2) led to a decline of the PHRR of PS nanocomposites and showed better performance than that of PS/unmodified MoS_2 (Zhou et al. 2014a). The incorporation of 3 wt% CTAB-MoS_2 nanosheets resulted in a reduction of PHRR by 20%. There were no obvious changes for the THR of pure PS and its composites, indicating that the PS resin was burnt out. The results of the fire safety test demonstrated that the fire resistance of PS/CTAB-MoS_2 nanocomposites is better than that of pure PS and PS/MoS_2 composites, due to the homogeneous dispersion of CTAB-MoS_2 in PS and an improved physical barrier effect. Moreover, the combustion property of the composites is closely related to the formation of carbonaceous char during thermal degradation. The carbonaceous char can reduce the spread of combustible volatile fragments generated. The TGA results indicated that the carbonaceous char residues of the PS/CTAB-MoS_2 nanocomposites are larger than pure PS and PS/MoS_2 composites, which is important for the shielding effect that the composites provide. For PS/MoS_2 composites, a cracked and fragile crust was preserved after combustion in the cone calorimeter. However, the PS/CTAB-MoS_2 system gave rise to the formation of a uniform and cohesive carbonaceous char residue, which can be explained by the homogeneous dispersion of the modified nanosheets forming a compact char layer. The dense and continuous char layer is a good barrier to protect the underlying materials and inhibit the exchange of combustible gases, degradation products, and oxygen. Therefore, the homogeneous dispersion or excellent compatibility, physical barrier effect of nanosheets, and the promotion of the charring effect are the main reasons for the excellent fire resistance properties of the PS/CTAB-MoS_2 nanocomposites.

Then the same author modified MoS_2 with ferrocene (Fe-MoS_2) and then studied the synergistic effect of ferrocene and MoS_2 on the fire resistance properties of the PS composites (Zhou et al. 2014b). The HRR, in particular the PHRR value, proves to be the most important parameter to evaluate fire safety. The reduction in PHRR is important for fire safety, as PHRR represents the point in a fire where heat is likely to propagate further, or ignite adjacent objects (Manzi-Nshuti et al. 2009). PS/ferrocene composites have a higher PHRR value than the pure PS, it may be attributed to the combustion-supporting effect of the ferrocene which had been usually used for rocket fuel additives. As for the PS/MoS_2 composites, the PHRR value of PS/MoS_2 composites are 13.0% lower than that of virgin PS. But for the PS/Fe-MoS_2 composites, the PHRR value decreased 21% comparing with the virgin PS, it is evident that the addition of Fe-MoS_2 can improve the fire resistance of the composites more obviously than PS/MoS_2 and PS/ferrocene composites, it means the presence of ferrocene and MoS_2 also have a significant synergistic effect in improving the flame retardancy of the PS composites. The better flame retardancy of the PS/Fe-MoS_2 composites is mainly due to the better dispersion of Fe-MoS_2 nanosheets in polymer matrix which can improve the physical barrier effect of the layered nanosheets. In addition, the fire resistance of the composites is also strongly dependent on the formation of carbonaceous char during the thermal degradation which reduces the diffusion of volatile combustible fragments generated by polymer degradation which diffuse toward the surface of the burning polymer to evaporate to feed the flame.

10.4.3 Flame Retardancy of Functionalized Layered Double Hydroxides-Based Polymer Nanocomposites

10.4.3.1 Epoxy Resin Nanocomposites

Our group has developed a series of modified LDHs to improve the flame retardancy of EP nanocomposites. LDHs show an advantage over layered silicates because the latter remain chemically inactive during the combustion of the nanocomposite and simply act as a physical barrier between the flame-front and the burning surface. Li et al. (2015) synthesized a bio-based eugenol derivative (SIEPDP) containing silicon and phosphorous and used it to modify MgAl LDH (SIEPDP-LDH). In this work, the LOI value of both the EP/SIEPDP-LDH and EP/NO_3-LDH gradually increased with addition of SIEPDP-LDH and NO_3-LDH as shown in Figure 10.14. It was noted that 4%, 6%, and 8% EP/SIEPDP-LDH nanocomposites expressed higher LOI values than 4%, 6%, and 8% EP/NO_3-LDH nanocomposites. This was caused by the functionalization of LDH. On one side, SIEPDP-LDH in the epoxy matrix was partially exfoliated

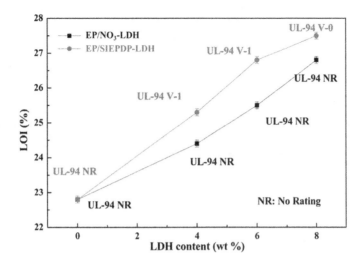

FIGURE 10.14 The LOI values and UL-94 ratings of EP/SIEPDP-LDH and EP/NO3-LDH (NLDH). (Reprinted from Li, C. et al., *J. Mater. Chem. A*, 3, 3471–3479, 2015. With permission.)

and intercalated, while the NO_3-LDH in epoxy resin exhibited severe aggregation; on the other side, the functional modifier, SIEPDP, contained Si and P which provided the flame retardancy to epoxy as well. All of the EP/NO_3-LDH nanocomposites had no classification in the UL-94 tests. In contrast, the epoxy with 8% SIEPDP-LDH was classified as V-0 rating in the UL-94 tests.

The cone calorimeter test results were summarized in Table 10.2. The PHRR of EP/4% NO_3-LDH and EP/4% SIEPDP-LDH were 835 and 658 kW/m^2, respectively. Compared with pure EP (920 kW/m^2), both of the modified and unmodified LDH-based EP showed a reduced PHRR. The pure epoxy only exhibited one broad peak with PHRR, it appeared at about 140 seconds. In comparison, the EP/4% SIEPDP-LDH had two HRR peaks: the first narrow peak (around 300 kW/m^2) appeared at around 70–80 seconds. Then, the HRR value sharply increased to 680 kW/m^2 (at 160–170 seconds, second peak). The second peak was higher and broader than the first one. This difference was attributed to the presence of SIEPDP-LDH in char residue at the initial stage, which hampered the flame by limiting the volatiles release, but not strong enough to stop the fire development. Comparatively, for the EP/4% NO_3-LDH, there was only one broad HRR peak, it appeared at around 120–130 seconds, and the PHRR value was much higher than that of EP/4% SIEPDP-LDH. The PHRR value of EP/4% NO_3-LDH was 835 kW/m^2, while the PHRR value of EP/4% SIEPDP-LDH was only 658 kW/m^2. Besides, at 55–60 seconds there appeared a very small peak. The reason was in the beginning the surface formed a char in the surface rapidly. However, the char was not strong enough and easily broken by both the constant heat flux and a continuous heat feedback. In addition, the EP/4% SIEPDP-LDH also expressed a lower THR value.

TABLE 10.2 Cone Calorimeter Test Results of EP Composites

Samples	EP	EP/4% NO_3-LDH	EP/4% SIEPDP-LDH
PHRR (KW/m^2)	920	835	658
THR (MJ/m^2)	90.5	89.6	86.9
Char residues (%)	9.7	16.4	21.6

Source: Li, C. et al., *J. Mater. Chem. A*, 3, 3471–3479, 2015.

Then Kalali et al. (2015) developed a multi-modifiers-intercalated LDH system, namely, sCD-SDBS-T-modified LDH (sCD-SDBS-T-LDH). In this work, pure epoxy exhibited a LOI value of 23.0% and no classification in the UL-94 vertical burning test. When NO_3-LDH was added, the LOI value rises to 25.2%, but still maintained no rating in the UL-94 test. Incorporating T-LDH or sCD-LDH improved the LOI value slightly, but neither of them passed the UL-94 V-0 rating. However, it was noted that adding sCD-SDBS-T-LDH into epoxy resulted in a great improvement in fire resistance and the UL-94 V-0 rating was achieved in the vertical burning test. In the cone calorimeter test, it was found that pure epoxy burnt very rapidly after ignition and the PHRR value is 931 kW/m². Compared with pure EP, EP/NO_3-LDH and EP/T-LDH composites burnt relatively slowly and the PHRR decreased from 931 to 621 kW/m² and 491 kW/m², respectively, and the reduction in PHRR was 33% and 47%. However, in the case of EP/sCD-SDBS-T-LDH composites, the PHRR exhibited a further decrease to 318 kW/m². In comparison to that of pure epoxy, it had 66% reduction of PHRR.

The improved fire retardancy of EP/sCD-SDBS-T-LDH was attributed to two aspects: firstly, the largest inter-layer distance of sCD-SDBS-T-LDH led to good dispersion of LDH in the epoxy matrix; secondly, sCD species improved the char yield of epoxy composites during combustion. At the end of burning, pure EP had released a THR of 81 MJ/m², the EP/NO_3-LDH had released almost the same amount of heat (80 MJ/m²), whereas only 53 MJ/m² had been released by the EP/sCD-SDBS-T-LDH nanocomposite. The significant reduction in THR meant more organic structures in the epoxy resin participated in the carbonization process and kept in the condensed phase, rather than converted to "fuel" in the gas phase. Based on the cone calorimeter results, the EP/sCD-SDBS-T-LDH composite showed a significantly lower mass loss and higher char yield compared to other samples, indicating a condensed phase mechanism for its fire resistance. The higher char yield meant that less epoxy resin degraded into flammable gases, and also a thick char layer was formed on the surface of the matrix. This thick char layer served as a thermal insulating barrier that stimulated the extinguishing of the flame and prevented combustible gases from feeding the flame zone, and also separated oxygen from burning materials. This was also in accordance with the reduced heat release rate and total heat release of the EP/sCD-SDBS-T-LDH composite in the cone calorimeter test.

The same author also adopted functionalized sCD (fCD), SDBS, and Ph to multi-intercalated LDH (fCD-SDBS-Ph-LDH) and studied the flame retardancy of EP/fCD-SDBS-Ph-LDH nanocomposites (Kalali et al. 2016a). In this study, pure epoxy showed a LOI value of 23.0% and no classification in the UL-94 vertical burning test. By adding NO_3-LDH, the LOI value increased to 25.2%, but still indicated no rating in the UL-94 test. Incorporation of Ph-LDH and sCD-Ph-LDH improved the LOI value slightly, but neither of them can pass the UL-94 V-0 rating. However, it was noted that adding sCD-SDBS-Ph-LDH or fCD-SDBS-Ph-LDH into epoxy provided a great improvement in LOI, 26% and 26.5%, respectively, and the UL-94 V-0 rating was achieved in the vertical burning test. In the cone calorimeter test (curves were shown in Figure 10.15), compared with pure EP, EP/NO_3-LDH, EP/sCD-Ph-LDH, and EP/sCD-SDBS-Ph-LDH composites, the peak of HRR decreased from 835 to 679 kW/m² (−19%), 766 kW/m² (−8%), and 349 kW/m² (−58%), respectively. In the case of EP/fCD-SDBS-Ph-LDH composites, it can be observed that EP/fCD-SDBS-Ph-LDH ignited rapidly due to the early initial degradation stage after ignition. The PHRR of EP/fCD-SDBS-Ph-LDH dramatically decreased to 232 kW/m², corresponding to a 72% reduction compared to that of pure epoxy. The good dispersion state of the LDH layers into the epoxy matrix was due to the enlarged inter-layer distance of fCD-SDBS-Ph-LDH and the improved char yield of epoxy nanocomposites during combustion due to the existence of fCD and Ph species together are the major aspects of the improved fire retardancy of fCD-SDBS-Ph-LDH. At 350 seconds after the ignition, pure EP released a total heat of 96.7 MJ/m², the EP/NO_3-LDH, EP/sCD-Ph-LDH, and sCD-SDBS-Ph-LDH/EP released 87.8, 76.6, and 66.4 MJ/m², respectively, whereas the EP/fCD-SDBS-Ph-LDH nanocomposite only released a total heat of 47.2 MJ/m². Participation of the organic species in the carbonization process and their presence in the condensed phase instead of going into the gas phase as "fuel" are the main reasons for the significant reduction of THR in the epoxy resin.

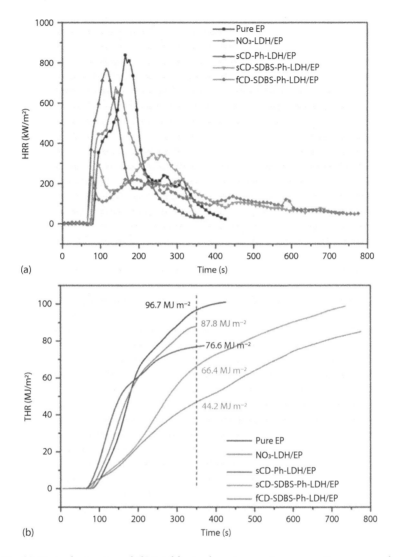

FIGURE 10.15 (a) Heat release rate and (b) total heat release temperature versus time curves of epoxy and its nanocomposites from cone calorimeter tests. (Reprinted from Kalali, E.N. et al., *J. Mater. Chem. A*, 4, 2147–2157, 2016. With permission.)

10.5 Conclusions

In this chapter, three prevailing layered nanomaterials widely employed in polymer nanocomposites to improve their flame retardancy, graphene, layered MoS_2, and layered double hydroxides are introduced mainly based on the functionalization, characterization, and flame retardant properties. With proper surface modification, the compatibility between the inorganic nanofillers and hydrophobic polymer matrices can be improved, leading to a much stronger interaction within the polymer nanocomposites. Meanwhile, the modification decreases the possibility of agglomeration caused by the strong electrostatic interaction or van der Walls force between the layers of the layered nanomaterials. The suitable intercalants enlarge the inter-layer distances of layered nanomaterials, facilitating the exfoliation and dispersion of the nanofillers in the polymer systems. By means of the modifiers with flame retardant properties, the modified layered nanomaterials impart reinforced flame retardant properties to the polymer nanocomposites. Multiple characterization techniques could be exploited to determine the successful functionalization of layered nanomaterials and the dispersion combined with interaction of modified layered nanomaterials in the polymer nanocomposites. Numerous reports have confirmed excellent flame retardancy of functionalized layered nanomaterials in various polymeric material systems. In the future, we do believe multi-function and smartness of the nano-flame retardant would be a main direction in developing new generation flame retardants.

References

Bao, C., Guo, Y., Song, L. et al. 2011. In situ preparation of functionalized graphene oxide/epoxy nanocomposites with effective reinforcements. *Journal of Materials Chemistry* 21: 13290–13298.

Bao, C., Guo, Y., Yuan, B. et al. 2012a. Functionalized graphene oxide for fire safety applications of polymers: A combination of condensed phase flame retardant strategies. *Journal of Materials Chemistry* 22: 23057–23063.

Bao, C., Song, L., Wilkie, C. A. et al. 2012b. Graphite oxide, graphene, and metal-loaded graphene for fire safety applications of polystyrene. *Journal of Materials Chemistry* 22: 16399–16406.

Bao, C., Song, L., Xing, W. et al. 2012c. Preparation of graphene by pressurized oxidation and multiplex reduction and its polymer nanocomposites by masterbatch-based melt blending. *Journal of Materials Chemistry* 22: 6088–6096.

Bartholome, C., Miaudet, P., Derré, A. et al. 2008. Influence of surface functionalization on the thermal and electrical properties of nanotube–PVA composites. *Composites Science and Technology* 68: 2568–2573.

Basu, D., Das, A., Stöckelhuber, K. W. et al. 2014. Advances in layered double hydroxide (LDH)-based elastomer composites. *Progress in Polymer Science* 39: 594–626.

Becker, C. M., Gabbardo, A. D., Wypych, F. et al. 2011. Mechanical and flame-retardant properties of epoxy/Mg–Al LDH composites. *Composites Part A: Applied Science and Manufacturing* 42: 196–202.

Cai, W., Feng, X., Hu, W. et al. 2016. Functionalized graphene from electrochemical exfoliation for thermoplastic polyurethane: Thermal stability, mechanical properties, and flame retardancy. *Industrial & Engineering Chemistry Research* 55: 10681–10689.

Cai, W., Feng, X., Wang, B. et al. 2017a. A novel strategy to simultaneously electrochemically prepare and functionalize graphene with a multifunctional flame retardant. *Chemical Engineering Journal* 316: 514–524.

Cai, W., Zhan, J., Feng, X. et al. 2017b. Facile construction of flame-retardant-wrapped molybdenum disulfide nanosheets for properties enhancement of thermoplastic polyurethane. *Industrial & Engineering Chemistry Research* 56: 7229–7238.

Chang, K., Chen, W., Ma, L. et al. 2011. Graphene-like MoS_2/amorphous carbon composites with high capacity and excellent stability as anode materials for lithium ion batteries. *Journal of Materials Chemistry* 21: 6251–6257.

Chen, W., Liu, Y., Liu, P. et al. 2017. The preparation and application of a graphene-based hybrid flame retardant containing a long-chain phosphaphenanthrene. *Scientific Reports* 7: 8759.

Dappe, Y. J., Basanta, M. A., Flores, F. et al. 2006. Weak chemical interaction and van der Waals forces between graphene layers: A combined density functional and intermolecular perturbation theory approach. *Physical Review B* 74: 205434.

Ding, P., Kang, B., Zhang, J. et al. 2015. Phosphorus-containing flame retardant modified layered double hydroxides and their applications on polylactide film with good transparency. *Journal of Colloid and Interface Science* 440: 46–52.

Dittrich, B., Wartig, K.-A., Mülhaupt, R. et al. 2014. Flame-retardancy properties of intumescent ammonium poly(phosphate) and mineral filler magnesium hydroxide in combination with graphene. *Polymers* 6: 2875–2895.

Duceac, L. D., Dobre, C. E., Pavaleanu, I. et al. 2017. Diseases prevention by water defluoridation using hydrotalcites as decontaminant materials. *Small* 1: 2.

Edenharter, A., and Breu, J. 2015. Applying the flame retardant LDH as a Trojan horse for molecular flame retardants. *Applied Clay Science* 114: 603–608.

Evans, D. G., and Duan, X. 2006. Preparation of layered double hydroxides and their applications as additives in polymers, as precursors to magnetic materials and in biology and medicine. *Chemical Communications* 5: 485–496.

Feng, X., Wang, X., Xing, W. et al. 2013. Simultaneous reduction and surface functionalization of graphene oxide by chitosan and their synergistic reinforcing effects in PVA films. *Industrial & Engineering Chemistry Research* 52: 12906–12914.

Feng, Y., Hu, J., Xue, Y. et al. 2017. Simultaneous improvement in flame resistance and thermal conductivity of epoxy/Al_2O_3 composites by incorporating polymeric flame retardant-functionalized graphene. *Journal of Materials Chemistry A* 5: 13544–13556.

Gao, L., Zheng, G., Zhou, Y. et al. 2014. Synergistic effect of expandable graphite, diethyl ethylphosphonate and organically-modified layered double hydroxide on flame retardancy and fire behavior of polyisocyanurate-polyurethane foam nanocomposite. *Polymer Degradation and Stability* 101: 92–101.

Gao, W., Alemany, L. B., Ci, L. et al. 2009. New insights into the structure and reduction of graphite oxide. *Nature Chemistry* 1: 403–408.

Guo, Y., Bao, C., Song, L. et al. 2011. In situ polymerization of graphene, graphite oxide, and functionalized graphite oxide into epoxy resin and comparison study of on-the-flame behavior. *Industrial & Engineering Chemistry Research* 50: 7772–7783.

Han, Y., Liu, Z. H., Yang, Z. et al. 2007. Preparation of Ni^{2+}-Fe^{3+} layered double hydroxide material with high crystallinity and well-defined hexagonal shapes. *Chemistry of Materials* 20: 360–363.

Hong, N., Song, L., Wang, B. et al. 2014. Co-precipitation synthesis of reduced graphene oxide/NiAl-layered double hydroxide hybrid and its application in flame retarding poly (methyl methacrylate). *Materials Research Bulletin* 49: 657–664.

Huang, G., Chen, S., Tang, S. et al. 2012a. A novel intumescent flame retardant-functionalized graphene: Nanocomposite synthesis, characterization, and flammability properties. *Materials Chemistry and Physics* 135: 938–947.

Huang, G., Fei, Z., Chen, X. et al. 2012b. Functionalization of layered double hydroxides by intumescent flame retardant: Preparation, characterization, and application in ethylene vinyl acetate copolymer. *Applied Surface Science* 258: 10115–10122.

Jayanthi, K., Nagendran, S., and Kamath, P. V. 2015. Layered double hydroxides: Proposal of a one-layer cation-ordered structure model of monoclinic symmetry. *Inorganic Chemistry* 54: 8388–8395.

Jiang, S. D., Bai, Z. M., Tang, G. et al. 2014a. Fabrication of Ce-doped MnO_2 decorated graphene sheets for fire safety applications of epoxy composites: Flame retardancy, smoke suppression and mechanism. *Journal of Materials Chemistry A* 2: 17341–17351.

Jiang, S. D., Tang, G., Bai, Z. M. et al. 2014b. Surface functionalization of MoS_2 with POSS for enhancing thermal, flame-retardant and mechanical properties in PVA composites. *RSC Advances* 4: 3253–3262.

Kalali, E. N., Wang, X., and Wang, D. Y. 2015. Functionalized layered double hydroxide-based epoxy nanocomposites with improved flame retardancy and mechanical properties. *Journal of Materials Chemistry A* 3: 6819–6826.

Kalali, E. N., Wang, X., and Wang, D. Y. 2016a. Multifunctional intercalation in layered double hydroxide: Toward multifunctional nanohybrids for epoxy resin. *Journal of Materials Chemistry A* 4: 2147–2157.

Kalali, E. N., Wang, X., and Wang, D. Y. 2016b. Synthesis of a Fe_3O_4 nanosphere@Mg–Al layered-double-hydroxide hybrid and application in the fabrication of multifunctional epoxy nanocomposites. *Industrial & Engineering Chemistry Research* 55: 6634–6642.

Kang, N. J., Wang, D. Y., Kutlu, B. et al. 2013. A new approach to reducing the flammability of layered double hydroxide (LDH)-based polymer composites: Preparation and characterization of dye structure-intercalated LDH and its effect on the flammability of polypropylene-grafted maleic anhydride/d-LDH composites. *ACS Applied Materials & Interfaces* 5: 8991–8997.

Khanra, P., Kuila, T., Kim, N. H. et al. 2012. Simultaneous bio-functionalization and reduction of graphene oxide by baker's yeast. *Chemical Engineering Journal* 183: 526–533.

Li, C., Wan, J., Kalali, E. N. et al. 2015. Synthesis and characterization of functional eugenol derivative based layered double hydroxide and its use as a nanoflame-retardant in epoxy resin. *Journal of Materials Chemistry A* 3: 3471–3479.

Li, Z., González, A. J., Heeralal, V. B. et al. 2017. Covalent assembly of MCM-41 nanospheres on graphene oxide for improving fire retardancy and mechanical property of epoxy resin. *Composites Part B: Engineering* 138: 101–112.

Liao, S. H., Liu, P. L., Hsiao, M. C. et al. 2012. One-step reduction and functionalization of graphene oxide with phosphorus-based compound to produce flame-retardant epoxy nanocomposite. *Industrial & Engineering Chemistry Research* 51: 4573–4581.

Long, X., Xiao, S., Wang, Z. et al. 2015. Co intake mediated formation of ultrathin nanosheets of transition metal LDH—An advanced electrocatalyst for oxygen evolution reaction. *Chemical Communications* 51: 1120–1123.

Lukowski, M. A., Daniel, A. S., Meng, F. et al. 2013. Enhanced hydrogen evolution catalysis from chemically exfoliated metallic MoS_2 nanosheets. *Journal of the American Chemical Society* 135: 10274–10277.

Ma, H. Y., Tong, L. F., Xu, Z. B. et al. 2008. Functionalizing carbon nanotubes by grafting on intumescent flame retardant: Nanocomposite synthesis, morphology, rheology, and flammability. *Advanced Functional Materials* 18: 414–421.

Manzi-Nshuti, C., Chen, D., Su, S. et al. 2009. Structure–property relationships of new polystyrene nanocomposites prepared from initiator-containing layered double hydroxides of zinc aluminum and magnesium aluminum. *Polymer Degradation and Stability* 94: 1290–1297.

Mascolo, G., and Mascolo, M. C. 2015. On the synthesis of layered double hydroxides (LDHs) by reconstruction method based on the "memory effect." *Microporous and Mesoporous Materials* 214: 246–248.

Nyambo, C., Kandare, E., and Wilkie, C. A. 2009. Thermal stability and flammability characteristics of ethylene vinyl acetate (EVA) composites blended with a phenyl phosphonate-intercalated layered double hydroxide (LDH), melamine polyphosphate and/or boric acid. *Polymer Degradation and Stability* 94: 513–520.

Pan, Y. T., Wan, J., Zhao, X. et al. 2017a. Interfacial growth of MOF-derived layered double hydroxide nanosheets on graphene slab towards fabrication of multifunctional epoxy nanocomposites. *Chemical Engineering Journal* 330: 1222–1231.

Pan, Y. T., and Wang, D. Y. 2015. One-step hydrothermal synthesis of nano zinc carbonate and its use as a promising substitute for antimony trioxide in flame retardant flexible poly(vinyl chloride). *RSC Advances* 5: 27837–27843.

Pan, Y. T., Zhang, L., Zhao, X. et al. 2017b. Interfacial engineering of renewable metal organic framework derived honeycomb-like nanoporous aluminum hydroxide with tunable porosity. *Chemical Science* 8: 3399–3409.

Pereira, C. M. C., Herrero, M., Labajos, F. M. et al. 2009. Preparation and properties of new flame retardant unsaturated polyester nanocomposites based on layered double hydroxides. *Polymer Degradation and Stability* 94: 939–946.

Qi, Y., Wu, W., Liu, X. et al. 2017. Preparation and characterization of aluminum hypophosphite/reduced graphene oxide hybrid material as a flame retardant additive for PBT. *Fire and Materials* 41: 195–208.

Qian, X., Song, L., Yu, B. et al. 2013. Novel organic–inorganic flame retardants containing exfoliated graphene: Preparation and their performance on the flame retardancy of epoxy resins. *Journal of Materials Chemistry A* 1: 6822–6830.

Qiu, S., Hu, W., Yu, B. et al. 2015. Effect of functionalized graphene oxide with organophosphorus oligomer on the thermal and mechanical properties and fire safety of polystyrene. *Industrial & Engineering Chemistry Research* 54: 3309–3319.

Salavagione, H. J., Gomez, M. A., and Martínez, G. 2009. Polymeric modification of graphene through esterification of graphite oxide and poly(vinyl alcohol). *Macromolecules* 42: 6331–6334.

Shabanian, M., Basaki, N., Khonakdar, H. A. et al. 2014. Novel nanocomposites consisting of a semi-crystalline polyamide and Mg–Al LDH: Morphology, thermal properties and flame retardancy. *Applied Clay Science* 90: 101–108.

Wang, D., Kan, Y., Yu, X. et al. 2016. In situ loading ultra-small Cu_2O nanoparticles on 2D hierarchical TiO_2-graphene oxide dual-nanosheets: Towards reducing fire hazards of unsaturated polyester resin. *Journal of Hazardous Materials* 320: 504–512.

Wang, D., Song, L., Zhou, K. et al. 2015a. Anomalous nano-barrier effects of ultrathin molybdenum disulfide nanosheets for improving the flame retardance of polymer nanocomposites. *Journal of Materials Chemistry A* 3: 14307–14317.

Wang, D., Wen, P., Wang, J. et al. 2017. The effect of defect-rich molybdenum disulfide nanosheets with phosphorus, nitrogen and silicon elements on mechanical, thermal, and fire behaviors of unsaturated polyester composites. *Chemical Engineering Journal* 313: 238–249.

Wang, D., Zhou, K., Yang, W. et al. 2013a. Surface modification of graphene with layered molybdenum disulfide and their synergistic reinforcement on reducing fire hazards of epoxy resins. *Industrial & Engineering Chemistry Research* 52: 17882–17890.

Wang, D. Y., Das, A., Leuteritz, A. et al. 2012a. Structural characteristics and flammability of fire retarding EPDM/layered double hydroxide (LDH) nanocomposites. *RSC Advances* 2: 3927–3933.

Wang, D. Y., Leuteritz, A., Kutlu, B. et al. 2011. Preparation and investigation of the combustion behavior of polypropylene/organo-modified MgAl-LDH micro-nanocomposite. *Journal of Alloys and Compounds* 509: 3497–3501.

Wang, L., Su, S., Chen, D. et al. 2009. Variation of anions in layered double hydroxides: Effects on dispersion and fire properties. *Polymer Degradation and Stability* 94: 770–781.

Wang, Q., Undrell, J. P., Gao, Y. et al. 2013b. Synthesis of flame-retardant polypropylene/LDH-borate nanocomposites. *Macromolecules* 46: 6145–6150.

Wang, X., Kalali, E. N., and Wang, D. Y. 2015b. Renewable cardanol-based surfactant modified layered double hydroxide as a flame retardant for epoxy resin. *ACS Sustainable Chemistry & Engineering* 3: 3281–3290.

Wang, X., Song, L., Yang, H. et al. 2012b. Simultaneous reduction and surface functionalization of graphene oxide with POSS for reducing fire hazards in epoxy composites. *Journal of Materials Chemistry* 22: 22037–22043.

Wang, X., Spörer, Y., Leuteritz, A. et al. 2015c. Comparative study of the synergistic effect of binary and ternary LDH with intumescent flame retardant on the properties of polypropylene composites. *RSC Advances* 5: 78979–78985.

Wang, X., Zhou, S., Xing, W. et al. 2013c. Self-assembly of Ni–Fe layered double hydroxide/graphene hybrids for reducing fire hazard in epoxy composites. *Journal of Materials Chemistry A* 1: 4383–4390.

Wang, Z., Wei, P., Qian, Y. et al. 2014. The synthesis of a novel graphene-based inorganic–organic hybrid flame retardant and its application in epoxy resin. *Composites Part B: Engineering* 60: 341–349.

Xie, X. L., Liu, Q. X., Li, R. K. Y. et al. 2004. Rheological and mechanical properties of $PVC/CaCO_3$ nanocomposites prepared by in situ polymerization. *Polymer* 45: 6665–6673.

Xu, G. R., Xu, M. J., and Li, B. 2014. Synthesis and characterization of a novel epoxy resin based on cyclotriphosphazene and its thermal degradation and flammability performance. *Polymer Degradation and Stability* 109: 240–248.

Xu, W., Liu, L., Zhang, B. et al. 2016a. Effect of molybdenum trioxide-loaded graphene and cuprous oxide-loaded graphene on flame retardancy and smoke suppression of polyurethane elastomer. *Industrial & Engineering Chemistry Research* 55: 4930–4941.

Xu, W. Z., Wang, S. Q., Liu, L. et al. 2016b. Synthesis of heptamolybdate-intercalated MgAl LDHs and its application in polyurethane elastomer. *Polymers for Advanced Technologies* 27: 250–257.

Yi, D., Yang, R., and Wilkie, C. A. 2014. Full scale nanocomposites: Clay in fire retardant and polymer. *Polymer Degradation and Stability* 105: 31–41.

Yu, B., Shi, Y., Yuan, B. et al. 2015. Enhanced thermal and flame retardant properties of flame-retardant-wrapped graphene/epoxy resin nanocomposites. *Journal of Materials Chemistry A* 3: 8034–8044.

Yu, B., Wang, X., Qian, X. et al. 2014. Functionalized graphene oxide/phosphoramide oligomer hybrids flame retardant prepared via in situ polymerization for improving the fire safety of polypropylene. *RSC Advances* 4: 31782–31794.

Zhang, J., de Juan, S., Esteban-Cubillo, A. et al. 2015. Effect of organo-modified nanosepiolite on fire behaviors and mechanical performance of polypropylene composites. *Chinese Journal of Chemistry* 33: 285–291.

Zhang, Y., Wang, B., Yuan, B. et al. 2017. Preparation of large size reduced graphene oxide wrapped ammonium polyphosphate and its enhancement on the mechanical and flame retardant properties of thermoplastic polyurethane. *Industrial & Engineering Chemistry Research* 56: 7468–7477.

Zhou, K., Gui, Z., and Hu, Y. 2016a. The influence of graphene based smoke suppression agents on reduced fire hazards of polystyrene composites. *Composites Part A: Applied Science and Manufacturing* 80: 217–227.

Zhou, K., Jiang, S., Shi, Y. et al. 2014a. Multigram-scale fabrication of organic modified MoS_2 nanosheets dispersed in polystyrene with improved thermal stability, fire resistance, and smoke suppression properties. *RSC Advances* 4: 40170–40180.

Zhou, K., Liu, J., Gui, Z. et al. 2016b. The influence of melamine phosphate modified MoS_2 on the thermal and flammability of poly (butylene succinate) composites. *Polymers for Advanced Technologies* 27: 1397–1400.

Zhou, K., Liu, J., Zeng, W. et al. 2015. In situ synthesis, morphology, and fundamental properties of polymer/MoS_2 nanocomposites. *Composites Science and Technology* 107: 120–128.

Zhou, K., Zhang, Q., Liu, J. et al. 2014b. Synergetic effect of ferrocene and MoS_2 in polystyrene composites with enhanced thermal stability, flame retardant and smoke suppression properties. *RSC Advances* 4: 13205–13214.

Zhu, J., Zhang, X., Haldolaarachchige, N. et al. 2012. Polypyrrole metacomposites with different carbon nanostructures. *Journal of Materials Chemistry* 22: 4996–5005.

The Use of Polyhedral Oligomeric Silsesquioxane in Flame Retardant Polymer Composites

Lei Song and
Wei Cai

11.1 Introduction

Polyhedral oligomeric silsesquioxane (POSS) compounds have been around for some time, but only recently have they been sufficiently functionalized and developed for incorporation into traditional synthetic polymer systems. As an important component of the silicon-oxygen family, POSS has a high dielectric constant and nanoscale structure generated from unique molecular structure (Leu et al. 2003a). The molecular structure of POSS is a steady hexahedral cage composed of a silicon-oxygen framework, with a molecule size of 1 ~ 3 nm. The special structure also brings superior rigidity and stability for POSS (Kannan et al. 2005). Besides, the general formula is designed as $RSiO_{3/2}$ (Figure 11.1), where R is H, alkyl, alkylene, aryl, aromatic alkylene, or their derivative groups (Liu et al. 2007). These different functional groups on the exterior of POSS are capable of enhancing its solubility and compatibility in a polymer matrix, thus promoting the application of POSS in polymer composites. More importantly, the presence of reactive functional groups also provides availability for stem grafting and copolymerization to pristine polymer materials (Liao et al. 2014). Except for the external functional groups, the internal cavity structure allows the insertion of a different volume of groups, including ion, atom, and radical (He et al. 2013).

So far, the promising potential of POSS on enhancing the fire safety and thermal stability has been represented and investigated because of their environmental neutrality, excellent heat resistance, as well as thermoxidative stability (Hartmann-Thompson 2011). Though the research works about the influence of POSS on polymer is not limited in the above two aspects, the focus of this chapter was placed on fire performance and thermal stability of polymer/POSS composites. In the first part, brief information was demonstrated for the dispersity and interfacial compatibility of POSS within a polymeric matrix. The second part is based on the kinds of functionalized groups to illustrate the enhancement of

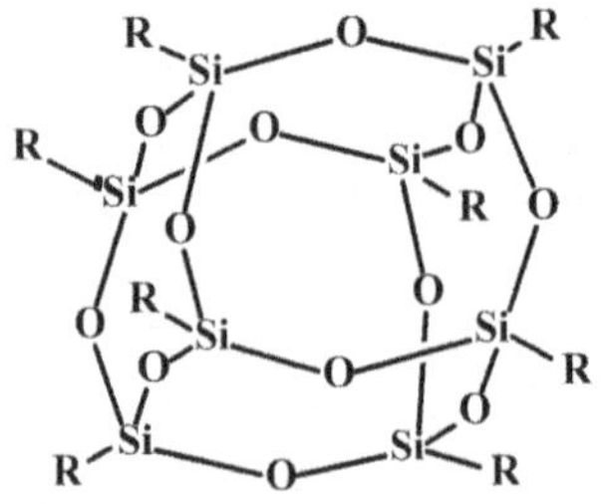

FIGURE 11.1   Molecular structure of POSS.

modified POSS on flame retardant and thermal stability. In the final part, different kinds of polymeric material that are enhanced by functionalized POSS were represented. The examples reported in this chapter are mostly drawn from the open literature. Patent literature is not included since the reported claims regarding the fire retardant mechanism are more explicitly analyzed in the authors' research articles, found in the scientific literature.

11.2 Dispersity and Interfacial Compatibility of POSS

The uniform dispersion of POSS in a host polymer matrix is responsible for the enhancement of mechanical properties, reduction in flammability, and also of heat release on burning. For the present, the main incorporation way is either through chemical addition or physical mixture. In general, the physical mixture is effortless and being used for the industrial application of POSS, due to an inexpensive, fast, and versatile technology. Both melt blending and solution mixing make up the main way of physical mixture. The solution mixing usually achieves good dispersion of POSS in a polymeric matrix, while the vast consumption of solution that generates higher cost was adverse to practical application. For the melt mixing, aggregation of POSS during the extrusion process is difficult to control and that will influence the final dispersion of POSS nanoparticles. In the chemical approach, POSS nanoparticles are bonded covalently with a polymer through grafting reaction or polymerization. Even though the dispersion state was significantly improved, the fabrication process is complex and at the cost of vast solution and precious catalyst. Although POSS nanoparticles could be incorporated into almost all the polymers by chemical reaction or physical blending, the control of nanostructure and dispersion of nanoparticles in polymer/POSS nanocomposites remain open challenges.

11.3 Functionalization Approach of POSS

11.3.1 Amine Functionalized POSS

In previous literatures, Meenakshi and coworkers have reported the fabricated method of amino-POSS, which mixed a certain amount of POSS-triol and 3-aminopropyltriethoxysilane (APTES) in toluene solution (Meenakshi et al. 2011, 2012). The mixed solution was then refluxed for 8 hours at 90°C, and the resultant products were collected with solvent evaporation. The experimental results indicate that, with incorporation of amino-POSS, the limit oxygen index (LOI) and char yield of phosphorus tetraglycidyl epoxy nanocomposites were increased to 50% and 73 wt% from 43% and 58 wt%, respectively. The greatly enhanced flame retardant was attributed to the low surface energy of Si-O-Si present in amino-POSS, which migrates to the surface of char residue and protects the underlying matrix. In other literatures, the synthesis of amino-terminal POSS had been widely reported (Leu et al. 2003a, 2003b).

The incorporation of amino-POSS into a polymer matrix mainly uses two approaches, i.e., solution blend and melt mixing. In view of the well dispersed state, the additional method of amino-POSS usually adopted a solution blend. For example, Wang et al. (2012b) successfully achieved simultaneous reduction and surface functionalization of graphene oxide by simple refluxing of graphene oxide (GO) with amino-POSS without the use of any reducing agents. Then, the resultant product was blended uniformly within epoxy resin, with the assistance of acetone solution. With the incorporation of 2.0 wt% of POSS-reduced graphene oxide (rGO), the onset thermal degradation temperature of the epoxy composite was significantly increased by 43°C. Moreover, the peak heat release rate, total heat release, and CO production rate values of POSS-rGO/epoxy resins (EP) were significantly reduced by 49%, 37%, and 58%, respectively, compared to those of neat epoxy. In addition, the employment of amino-POSS was also processed during the polymerization of a polymer for enhancing the thermal stability and flame retardancy. For instance, Sethuraman et al. (2014) reinforced the thermal stability of the epoxy resin by the reaction between amino-POSS and the epoxy group of diglycidyl ether of bisphenol A (DGEBA) epoxy resin. The resultant product with 5 wt% amino-POSS presents an increase of 53.1°C in the temperature of 20 wt% mass loss, and meanwhile the char residue at 700°C was improved to 10.7 wt% from zero.

The enhancement of amino-POSS in other polymer composites was also reported. As an example, Bouza et al. (2014) studied the crystallization, morphology, and flame retardancy of polypropylene (PP) composites with aminopropylisobutyl-POSS. With assistance of a thermogravimetric analysis (TGA) test, it was found that the thermal degradation process of PP was significantly hindered, conforming by an increase in the onset temperature of up to 30°C. In this work, the flame retardancy of PP composites was evaluated with LOI. The enhancement in fire safety of PP composites only with amino-POSS was indistinctive. And, LOI values of the blends PP/amino-POSS remain constant (about 17.6), regardless of the amount of amino-POSS (2%–10%). Similarly, Du et al. (2011) also reported the synergetic effect between amino-POSS and intumescent flame retardant (IFR) for enhancing the fire safety of PP composites. It was found that the effect of amino-POSS on thermal stability of PP composites was not obvious at a low temperature zone, but became effective at a high temperature zone due to the two-type thermal degradation behaviors for POSS. Although no remarkable increase of peak of rate of heat release (PHRR) values was found, the use of amino-POSS leads to the increase of total smoke release. The improvement of amino-POSS in polyimide composites (PI) was investigated by Devaraju et al. With thermal imidization, the polycondensation between amino-POSS and polyamic acid was successfully performed (Devaraju et al. 2012). With the content of 0, 5, 10, and 15 wt%, the residual char yields at 900°C were 43.4%, 47.6%, 48.8%, and 50.4%, respectively, indicating that these hybrid materials were thermally stable. Fan et al. also studied the reinforcement of amino-POSS in PI composites (Fan and Yang 2013). As presented in this work, amino-POSS was employed as cross-linked agent for in situ formation of PI composites, with thermal solidification. With XRD patterns of PI composites, the dispersed state of amino-POSS in PI matrix was desirable, indicating the good compatibility. For the practical application, melt blend may be a better choice. Wang et al. (2012a) fabricated flame retardant polybutylene succinate (PBS) composites by melt blending PBS with melamine phosphate (MP), using amino-POSS as synergist. The incorporation of amino-POSS increased the char residue at 600°C of PBS to 9.4 wt% from 0.4 wt% for pure PBS and 7.8 wt% for PBS with 20 wt% MP.

11.3.2 Vinyl-Functionalized POSS

The influences of octavinyl polyhedral oligomeric silsesquioxane (vinyl-POSS) on thermal degradation and flame retardancy of polymer composites have been widely studied. After preparing phosphorus-containing epoxy resin, Wing et al. added the vinyl-POSS into EP matrix with solution blend (Wang et al. 2010). For investigating the dispersed state of vinyl-POSS, TEM was employed to directly observe the morphology of the phosphorous-containing epoxy resin (PCEP)/vinyl-POSS hybrids. It was seen that the dark area (the portion of the hybrid) was homogenous and no localized domains were detected at this scale, implying that the POSS component was homogenously dispersed in the continuous epoxy matrix at the nanoscale. With inclusion of 3 wt% vinyl-POSS, char residue at 700°C and $T_{50}\%$ was increased to 23.23 and 417.5 from 15.06 wt% and 404.1°C, compared to phosphorus-containing epoxy resin. Moreover, thermal stabilities of experimental curves were all better than the calculated ones, which indicated the occurrence of an interaction between the phosphorus and silicon element on the flame retardant. In the case of

PCEP/3.0 wt% vinyl-POSS, the peak HRR and total heat evolved (THR) were decreased obviously to 156.0 and 16.6 from 227.0 W/g and 19.3 kJ/g. The reason was mainly attributed to the presence of vinyl-POSS that can promote the formation of a protective char and the char could protect the underlying materials from further burning. The effect of vinyl-POSS on the fire safety of UV-cured epoxy acrylate had been studied (Zhang et al. 2010). The incorporated vinyl-POSS presented a synergistic effect with dimethyl methylphosphonate (DMMP), which greatly improved the flame retardancy of the UV-cured epoxy acrylate (EA)/micro-PCM composite. When 2 wt% vinyl-POSS replaced the equiponderant DMMP, the synergistic effect between octavinyl polyhedral oligomeric silsesquioxane (OVPOSS) and DMMP showed the best power, which could dramatically catalyze the char formation. The obtained char layer exhibited a protective shield effect for the bottom polymer matrix from fire zone.

Due to the existence of vinyl group, the copolymerization between vinyl-POSS and polyethylene was able to be performed. Waddon et al. (2002) added vinyl-POSS into the molecule chains of polyethylene and found the thermal stability of polyethylene composites was significantly increased. Besides polyethylene, Vahabi et al. (2012) successfully processed the physical addition and reactive incorporation of POSS into polymethyl methacrylate (PMMA). It had been highlighted that the copolymerizable introduction of the vinyl-POSS caused an enhanced thermal stability, which presented 30 wt% char residue for copolymer (MMA-phosphonated modified PMMA [MAPCl])/trisilanolphenyl (Tr). POSS. In addition, chemical incorporation of POSS into PMMA offered better fire safety (lower sum of heat release capacity [sumHRC] and THR) than physical incorporation. A comparative study on physical blending and reactive blending for POSS-filled polypropylene was performed by Zhou et al (2008). Random copolymers of syndiotactic polystyrene (sPS) and POSS had been synthesized (Figure 11.2) (Zheng et al. 2002). It was observed that the incorporation caused the melting-point depression of semicrystalline sPS, with a reduced melting point. Compared to atactic poly(4-methylstyrene)–POSS, the T_{intial} of sPS–POSS copolymers was improved. This increase may be attributed to the semicrystalline nature of the polymer matrix influencing the aggregation of the inorganic POSS component of the copolymers. Under nitrogen, the char yield of sPS–POSS-4 was increased to 17.2 from 1.1 wt%. This improvement of thermal oxidative stability was attributed to the formation of a silica layer on the surface of the polymer melt, which served as a barrier preventing further degradation of the underlying polymer. The effect of vinyl-POSS on enhancing the dispersed state and interfacial interaction of MoS_2 within the polyvinyl

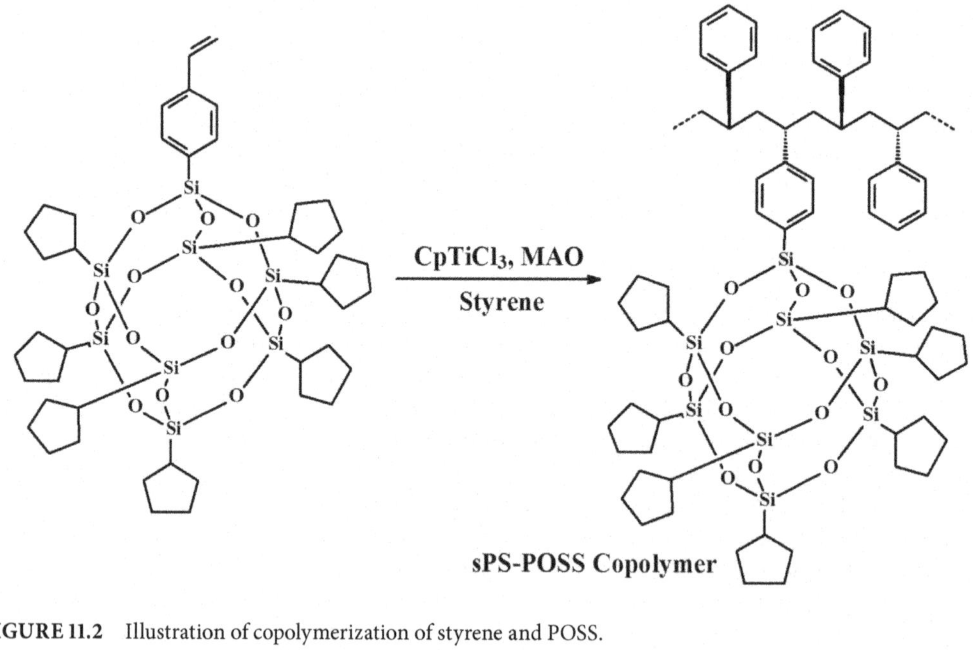

FIGURE 11.2   Illustration of copolymerization of styrene and POSS.

alcohol (PVA) matrix was investigated (Jiang et al. 2014). With assistance of dithioglycol, vinyl-POSS was successfully grafted on the surface of MoS_2 nanosheets. Combined with the XRD, TEM, and SEM results, it can be concluded that vinyl-POSS-modified MoS_2 achieved a homogeneous dispersion state within the polyvinyl alcohol (PVA) matrix. In comparison with neat PVA, the maximum decomposition temperature (T_{max}) of the PVA/MoS_2 nanocomposite was increased due to the physical barrier effect of MoS_2. When vinyl-POSS-modified MoS_2 was added, PVA/vinyl-POSS/MoS_2 depicted the best thermal stability, indicating an increment of 23°C in T_{max}. In addition, after vinyl-POSS/MoS_2 was incorporated, the glass transition temperature (Tg) was found to be increased by 10.2°C, compared with that of pure PVA. The flame retardancy of PVA nanocomposites was further determined. Neat PVA burned extremely rapidly after ignition and the PHRR value reached to 375 kW/m². After incorporating MoS_2 nanosheets, the PHRR was decreased to 297 kW/m². It was consistent with the TGA test, PHRR of PVA/vinyl-POSS/MoS_2 was farthest reduced to 268 kW/m², presenting a 29% decrease compared to that of pure PVA. In addition, Jiang et al. also found the addition of vinyl-POSS/MoS_2 significantly decreased the gaseous products, including hydrocarbons, carbonyl compounds, and carbon monoxide, which was attributed to the synergistic effect of vinyl-POSS and MoS_2.

11.3.3 Epoxy-Functionalized POSS

The octaepoxy polyhedral oligomeric silsesquioxane was also copolymerized and physically incorporated into polymer matrix. For example, Lu et al. (2006) prepared a multiepoxy cubic silsesquioxane by the hydrolysis and polycondensation of trifunctional monomer (γ-glycidoxypropyl)trimethoxysilane in a solvent mixture of methyl isobutyl ketone and anhydrous ethanol with a tetraethyl ammonium hydroxide aqueous solution acting as the catalyst. With mutual cure reaction, as-prepared epoxy-POSS was successfully added into the cyanate matrix and together formed a highly crosslinked organic-inorganic hybrid composites. The structure and properties of the composites were characterized and largely depended on the epoxy-POSS concentration. The presence of epoxy-POSS changed the local structure of the molecule, made the chain more rigid, restricted the chain mobility, and eventually improved the thermal stability and flame retardancy of the resin. As TGA results presented, $T_{initial}$ of the cyanate composite with 50 mol% epoxy-POSS was significantly increased to 510°C from 411°C (pure cyanate composite). Moreover, char residue at 800°C of 65.80 wt% was left for the cyanate composite with 50 mol% epoxy-POSS. The LOI of the cyanate composite containing 50 mol% epoxy-POSS was dramatically improved than that of pure cyanate composite (from 32% to 61%). It was deemed that the high epoxy-POSS concentration led to hybrid composites with a higher crosslinked hybrid network, bulk size, and mass cubic silsesquioxane structure and, as a result, excellent flame-retardant properties and good thermal stability. Franchini et al. (2009) introduced two kinds of POSS as co-monomer into an epoxy-amine formulation in order to obtain hybrid organic/inorganic epoxy networks. It was shown that POSS-bearing phenyl groups were far more effective than POSS with isobutyl groups, and that the presence of a chemical linkage between the phenyl-based POSS clusters and the matrix favored the dispersion of the nanoclusters, resulting in enhanced fire retardancy. With an addition of 3.7 wt% phenyl-containing POSS, the time to ignition of epoxy composite was increased to 153 seconds from 132 seconds for neat epoxy. The fire behavior was obviously modified by the presence of the POSS clusters: for both types of POSS cages, a significant decrease was observed in the peak of the heat release rate. When 3.7 wt% phenyl-containing POSS was incorporated, HRR was decreased to 622 kW/m², corresponding to a 40% reduction. The flame retardant property of epoxy composites was further investigated with UL-94 vertical burning tests. Neat epoxy network and the network containing POSS-bearing isobutyl groups exhibited the worst results in terms of UL-94 vertical burning performances: all these materials burned up to the clamps with the release of incandescent drops (Figure 11.3a and b). By contrast, the epoxy network with phenyl-containing POSS (Figure 11.3c) showed self-extinguishing behaviors in UL-94 tests. In addition to commercial polymer materials, POSS was also utilized to impart flame retardancy to the nano-fibrillated cellulose (Fox et al. 2012).

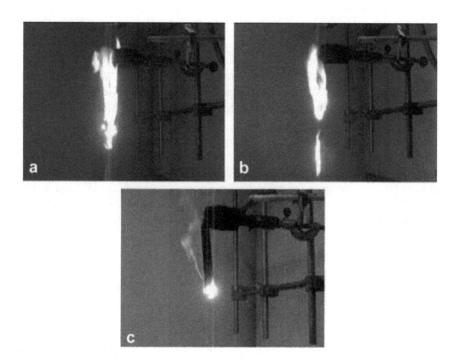

FIGURE 11.3 UL-94 vertical burning tests pictures: (a) neat epoxy matrix, (b) epoxy with 3.7 wt% isobutyl-containing POSS, and (c) epoxy with 3.7 wt% phenyl-containing POSS. (Reprinted from Franchini, E. et al., *Polym. Degrad. Stab.*, 94, 1728–1736, 2009. With permission.)

11.3.4 Phenyl-Functionalized POSS

The incorporation of phenyl-bearing POSS for enhancing the fire safety of polymeric materials had been intensively studied, due to the presence of phenyl groups that improved the interfacial interaction. In the research work of Laik et al. (2016), POSS containing phenyl groups and silanol functions were investigated in two different epoxy systems. The first epoxy system was the mixture of tetraglycidyl(diaminodiphenyl) methane (TGDDM) and 4,4'-methylenebis(2-isopropyl-6-methylaniline) (MMIPA) which was called MVR, while the other was the mixture of TGDDM and 4,4'-methylene bis(2,6-diethylaniline (MDEA) which was called TM. Besides, an aluminum salt was also introduced in the presence of POSS in order to enhance the reaction of the additive with the epoxy monomer. With aid of SEM photographs, angular aggregates of a few microns and some small nodules outside the aggregates were found in the rough fracture surfaces of the TM-based epoxy matrix. The phenomenon was similar to MVR-based networks. After the TGA test, the same profiles of thermal degradation were obtained for both types of epoxy networks. Indeed, the composition of the TM- and the MVR-based networks was close, and the slight differences between the two formulations—additional epoxy prepolymer and different crosslinking agents in the MVR network did not cause significant modifications in the degradation paths. It was notable that the one-step degradation in nitrogen atmosphere of the two epoxy systems were in the same range of temperature (320°C–330°C) and were unaffected whether the POSS and the aluminum-based additive were present or not. In air atmosphere, the addition of POSS led to a slight enhancement in the stability of intermediate char, which prevents the thermal degradation of epoxy matrix. However, Laik et al. found that the experimental residual weights were generally in good accordance with the theoretical residual weights, thus indicating that POSS did not play any synergistic contribution as concerns the residual yield. The reaction to fire of the neat and hybrid networks was also assessed by cone calorimeter tests (CCT). The time to ignition (TTI) in TM-based networks was hardly influenced by the existence of POSS and/or aluminum acetonate. The PHRR was reduced by the addition of POSS and even more in the Al(acac)3-assisting network, which was decreased to 1221 and 1061 from 1535 kW/m^2. Except for the chemical addition, Zhang et al. (2012) directly incorporated phenyl-bearing POSS into the EP matrix, evaluating the variation of thermal stability and fire safety. Results from the TGA test presented an obvious enhancement in $T_{initial}$, T_{maxI}, and T_{max2}, which were increased to 364°C, 379°C, and 576°C from 345°C, 370°C, 559°C. Due to the silsesquioxane creating a stable char layer of silicon dioxide during combustion, values of peak HRR and total smoke release (TSR) of EP containing 4.1 wt% POSS were respectively decreased by 26.8% and 10.8%. The formation of droplets was not being observed during combustion. The synergistic effect of this system containing POSS and 9,10-dihydro-9-oxa-10- phosphaphenanthrene-10-oxide (DOPO) was surveyed in the EP matrix (Li and Yang 2014; Zhang et al. 2012). Li et al. employed a mixture of POSS and DOPO to reinforce the thermal stability and fire safety of EP composites (Li and Yang 2014). The content of P and Si was precisely controlled with the ratio of POSS and DOPO. From the relevant TGA test, similar T_{intial} and T_{max} were noticed in EP-1 containing 5 wt% POSS by comparing to pure EP. The introduction of POSS has barely impacted on the decomposition temperature of the EP composites. However, with the increase of the DOPO load, both T_{intial} and T_{max} of EP composites were obviously decreased. Compared with the pure EP, TTI of flame-retarded EP composites had been slightly decreased, which may be attributed to the flame retardant promoting the resin matrix to degrade at lower temperature. Moreover, the DOPO promoted the EP composites to degrade more quickly than the POSS. It could be observed that the pure EP burned rapidly after ignition and PHRR of 821 kW/m^2. All samples containing POSS and/or DOPO remarkably reduced the heat release of the EP composites during the combustion. The PHRR of EP-1, EP-2, and EP-4 were 417, 461, and 438 kW/m^2, which waere reduced nearly a half to pure EP. However, the PHRR of EP-1 was lower than EP-2 and EP-4, which implied that the char layer promoted by POSS was much more stable and compact. Unfortunately, no synergistic effect of POSS and DOPO was observed in the curves of HRR versus time.

The application of POSS containing phenyl groups is not only confined in EP materials. The role of POSS incorporated as additional filler in polyamide 6/clay composite was studied by Dasari et al. during the combustion. It was speculated that POSS was capable of transforming to a glassy material upon the heat of fire and enhancing the coupling of silicate layers to each other. Thus, the enhanced fire safety was able to be observed. In addition, POSS as a novel synergist with glass-fiber had reinforced the flame retardancy of polybutylene terephthalate (PBT) (Louisy et al. 2013). The individual use of POSS had been investigated by Lu et al. (Zhu et al. 2017). A novel phosphorus–nitrogen-containing POSS (F-POSS) (Figure 11.4) was

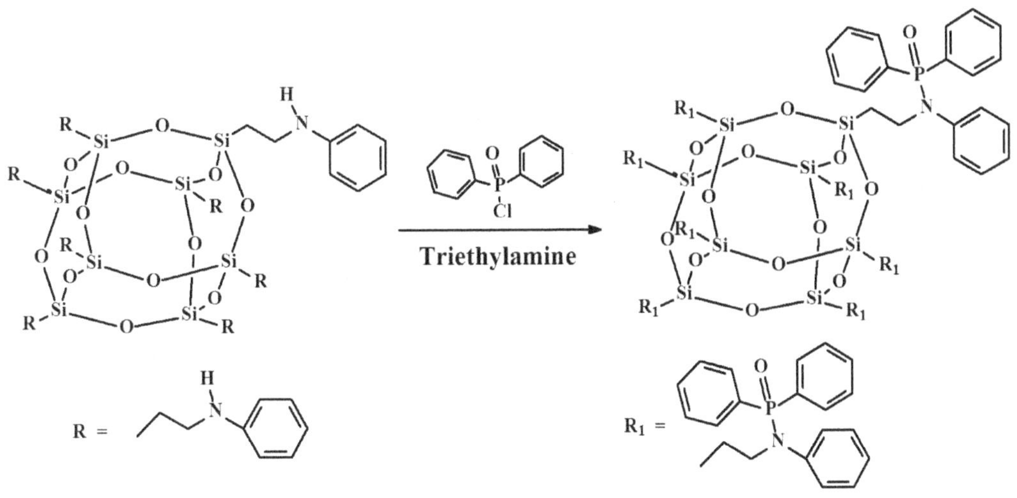

FIGURE 11.4 Schematic representation of the synthesis route for a novel phosphorus–nitrogen-containing POSS (F-POSS).

incorporated into PBT in order to improve properties, including thermal stability and flame retardant properties. The mixture of F-POSS and PBT was processed by the melt blending method. After the incorporation of POSS and F-POSS, the thermal stability and char yield of PBT composites were improved. Compared to the pure PBT, the $T_5\%$ values were increased from 367°C for virgin PBT to 380°C and 382°C, respectively, for PBT/POSS and PBT/F-POSS. The residual char of PBT/F-POSS was higher than that of pure PBT, implying that the F-POSS with phosphinic groups effectively promotes the char formation. The results from the CCT of the PBT composites indicated that the incorporation of POSS did not produce a desirable effect on hindering the heat release process. However, the inclusion of F-POSS brings lower HRR, reduced THR, as well as extended burn time. PHRR was decreased from 1104 kW/m² for pure PBT to 556 kW/m² for PBT/F-POSS with a significant reduction of 50%. The simultaneous presence of resorcinol bis(diphenyl phosphate) (RDP) and POSS in enhancing fire behavior of polycarbonate nanocomposites prepared by melt blending has been studied by Vahabi et al. (2013). The synergistic system resulted in a significant reduction for the flammability of polycarbonate (PC), confirmed by an approximately 21.9% in THR. Except the heat release, the smoke release process of PBT composites was also monitored. Pure PC presented an extremely high TSR value at 3602 m²/m². In sharp contrast, a postponed and reduced TSR curve was observed in PC containing a mixture of POSS and RDP. Didane et al. (2012a) performed the comparative study of three kinds of POSS as synergists with zinc phosphinates for poly(ethylene terephthalate) (PET) fire retardant. The presence of zinc phosphinates in PET does not seem to have relevant effects on delaying combustion (only 10 seconds), while adding 1 wt% of POSS decreases the time to ignition of the loaded materials to approximately 250 seconds. Besides, the incorporation of 10 wt% zinc phosphinates also decreased the PHRR of PET composites, reducing to 365 from 500 kW/m². It was notable that the substitute of 1 wt% phosphinates by POSS containing phenyl group further decreased the HRR of PET composite, as low as 226 kW/m². Correspondingly, THR of PET composites containing 1 wt% phenyl-POSS and 9 wt% zinc phosphinate was reduced by about 53.8% compared to pure PET. Concerning gas emissions, it was found that the cumulative CO in PET composites containing zinc phosphinates was increased. However, the addition of POSS reduced again the release of CO, leading to the same CO quantities released by pure PET. The residual condensed char structures suggested that mechanisms leading to combustion rate reductions were of a physical nature rather than a chemical one. Complete burn of PET containing 10 wt% zinc phosphinates gave a thick and dense carbonaceous layer without outstanding swelling and volume gain. The system containing 10 wt% POSS-phenyl presented the most expanded chars, revealing that the chars produced when POSS-phenyl was used have more effective thermal insulating properties. Apart from zinc phosphinates, the synergistic effect of aluminum phosphinate and POSS was also found in retarding the fire hazards of PET textiles (Didane et al. 2012b). The blend way of aluminum phosphinate and POSS was based on the melt spinning process. Improved performances of the PET textiles have been noticed such as a decrease in the dripping effect, in the peak of heat release rate (reduced by 19.6%), and in the total heat (11.5%) evolved during combustion. Compared to the addition of POSS containing phenyl, more efficient enhancement in fire safety was presented in PET textile containing methyl-POSS. It did not conflict with their research work (Didane et al. 2012a). By the solution blend, PMMA composites containing two different types of POSS were successfully prepared by Vahabi et al. (2012). Through the direct observation of an SEM photograph of the fracture surface of PMMA, spherically agglomerated morphologies could be observed by using both types of POSS. Of course, there was some obvious distinction. In the case of phenyl-POSS, the size of agglomerated particles is between 1 and 2 mm in PMMA. The size of ethyl-POSS particles in PMMA was between 500 and 1 mm. The results illustrated that phenyl-POSS produced a worse dispersion state in the PMMA matrix than ethyl-POSS. In addition, both types of POSS presented poorly dispersed morphologies in a phosphonated modified PMMA compared to those of PMMA blends. Results from the CCT confirmed that the incorporated phenyl-POSS slightly decreased sumHRC, from 438 to 414 J/g·K, meanwhile the reduction of THR was small.

The enhancement of phenyl-POSS in thermal stability and flammability of polypropylene nanocomposites was revealed (Barczewski et al. 2014). The addition of as low as 2.0 wt% POSS did not increase the $T_{initial}$ under nitrogen, on the contrary, reducing it to 362.6°C from 371.1°C. However, $T_{initial}$ was obviously enhanced when the content of nanofillers was 10 wt%. Meanwhile, the residual mass was improved to 5.33 wt% from 0 wt%. The dipping phenomenon in a UL-94 test was effectively improved. When the additional content of POSS was 10 wt%, the melt dripping of PP no longer happened. Liu et al. (2010) surveyed properties of polystyrene composites by adding phenyl-POSS. The dispersion state of nanofillers in host composites was characterized by TEM, which indicated that POSS nanoparticles were homogeneously dispersed in the base polymer and densities of fillers were gradually increased with the increase of phenyl-POSS content. The peak value of HRR in PS with 30 wt% POSS was significantly reduced by 73.1%. And the bromide derivative of phenyl-POSS was applied to enhance the flame retardancy of PS. After coupling with Sb_2O_3, the mixture system presented a most significant enhancement in HRR (decreased by 84.2%).

11.3.5 Methyl-Functionalized POSS

The combination of methyl-POSS in with ammonium polyphosphate (APP) produced an intumescence phenomenon during the combustion of the EP resin (Gérard et al. 2012). Gérard et al. found that the interactions between these fillers modified the viscosity of the degraded matrix, which hindered the thermal degradation process. The evidence from the CCT showed that PHRR in EP/APP-4/POSS-1 was approximately decreased by 50%. Figure 11.5 shows vertical sections of the residues from the CCT of epoxy containing APP alone or combined with POSS observed by digital microscopy. Areas containing big bubbles (red circles in web version) were next to areas where small bubbles were mostly found (blue rectangles). When APP was incorporated solely into the epoxy, foaming was produced. Large bubbles were distinguished in the whole samples. In stark contrast, the combination between APP and POSS leads to two different morphologies. Big bubbles were still produced, but mostly in the lower part of the sample. A large area with thinner ones was located on top of the residue. An evaluation of the size of the bubbles in the red circles revealed that the mean diameter was about 1 mm in both cases. The size of the big bubbles was therefore similar in the two formulations. In contrast, the percentage corresponding to the area with almost only thin bubbles compared to the whole sample was 25% for Epoxy/APP-5 and 70% for Epoxy/APP-4/POSS-1. Therefore, when the matrix did not resist the internal pressure (presence of gases), big bubbles were created in both cases. However, these bubbles were mostly located at the bottom of the sample in the case of Epoxy/APP-4/POSS-1, whereas they were found everywhere in Epoxy/APP-5. Subsequently, the percentage of the whole surface corresponding to denser areas (thin bubbles, blue rectangles on Figure 11.5), which were supposed to provide better protection against heat, was lower in the case of Epoxy/APP-5.

In some work, methyl-POSS was regarded as a novel synergist to reinforce the properties of polybutylene terephthalate/glass fiber (PBT/GF) composites (Louisy et al. 2013). Though the stiffness, strength, and the dimensional stability of PBT were effectively improved, applications of PBT/GF had been limited due to its high flammability. Thus, flame retardants must be incorporated into the PBT/GF matrix.

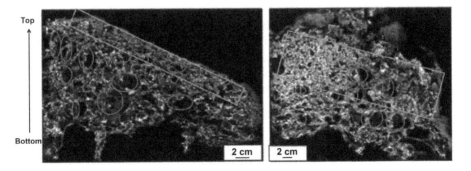

FIGURE 11.5 Cross-sections of the whole residue from Epoxy/APP-5 (left) and Epoxy/APP-4/POSS-1 (right). (Reprinted from Gérard, C. et al., *Polym. Degrad. Stab.*, 97, 1366–1386, 2012.)

Louisy et al. (2013) evaluated the fire performance of PBT/GF composites containing the methyl-POSS. An obvious decrease of 31.2% was achieved in the system that contained aluminum diethylphosphinate (18 wt%) and methyl-POSS (2 wt%). The suppression effect of methyl-POSS was superior to the butyl-POSS. In addition, the substitution of aluminum diethylphosphinate by methyl-POSS (2 wt%) still kept V-0 classification and decrease in maximum total burning time. Karlsson et al. (2011) compared the suppression effect of MMT, sepiolite, and POSS on fire hazards of ethylene-acrylate copolymer. The mass loss curve was shifted toward lower temperatures after POSS was added into the material. Thus, this view that POSS was not able to influence the degradation temperature of the polymer was presented. In addition, POSS was likely to lubricate the polymer at low concentrations, which results in lower melt viscosities and faster release of volatile degradation products. Earlier ignition was observed when POSS was incorporated into the polymer matrix, attributing to the lower viscosity. The protective silica layer derived from POSS could also be expected to reduce heat release in the cone calorimeter, but the results showed the opposite. This could possibly be explained by the fact that the polymer-POSS swell more during burning in the cone calorimeter than their reference as a result of the lower melt viscosity. This, combined with the earlier ignition was likely to prevent formation of a homogeneous silica layer on the surface.

Fire retarded poly(ethylene terephthalate) PET had been obtained by the incorporation of methyl-POSS and Exolit OP950 (zinc diethylphosphinate), in recycled PET (Vannier et al. 2008). PET alone exhibited an LOI value of 26 vol%. When 20 wt% of OP950 was added, the LOI value had gone up to 35 vol%. When OP950 was substituted by POSS, LOI values were higher (37 and 38 vol%) than those with OP950 or POSS alone (in that case LOI was equal to 24 vol%) up to 5 wt% substitution. Specially, the intumescence carbonaceous layer was more developed when OP950 was substituted by 2.0 wt% methyl-POSS (Figure 11.6a). The addition of 20.0 wt% of OP950 led to an important decrease in the PHRR (-30%) from 708 kW/m^2 for virgin PET to 496 kW/m^2 for the fire retardant formulation. When 2 wt% of OP950 was substituted with POSS, the PHRR was decreased by 50% compared to the one with 20 wt% OP950 (PHRR = 247 kW/m^2), which made a decrease of 65% compared to PET alone (Figure 11.6b). The introduction of OP950 led the material to intumesce, but there was no plateau in the HRR curve, whereas when POSS was incorporated with phosphinates, an expanded carbonaceous structure that was more stable was formed.

Besides, Didane prepared fire resistant PET fibrous structures by the mixture of aluminum phosphinate and POSS (Didane et al. 2012b). All samples showed similar degradation pathways with a degradation step followed by a char formation and a progressive mass loss at high temperatures. PET containing 9.0 wt% aluminum phosphinate and 1.0 wt% methyl-POSS presented a lower initial thermal decomposition temperature confirmed by an approximately 9°C reduction compared with PET/ aluminum phosphinate/phenyl-POSS. However, the incorporation of methyl-POSS made more char residue being reserved at 480°C (18.0 wt%). In the UL-94 test, PET specimens showed bad fire properties, the dripping was almost instantaneous with a release of inflamed drops that burned the cotton within the first seconds of flame application and all the samples were totally consumed. Then, addition of aluminum phosphinate and methyl-POSS remarkably improved the reaction to fire of the textiles, the afterflame times never exceeded 1 second and the textiles did not burn completely. Thus, the material classification upgraded to V-2. It was found that the presence of aluminum phosphinate facilitated the degradation of the PET textile during the CCT, and the effective heat of combustion was reduced from 1.85 to 1.11 MJ/m^2. Another phenomenon was noticed in that the ignition time of the textiles was delayed to 586 seconds, while virgin PET fabrics burned at 363 seconds. Similar results were also reported by Alongi et al. (2011). Replacing phosphorus fillers by phenyl-POSS at 1 wt% had no significant effect on the ignition time, neither on the total heat evolved, but it reduced the PHRR to 258 kW/m^2, achieving a reduction of 20% compared to the PHRR of PET. Using methyl-POSS instead of phenyl-POSS completely modifies the fire behavior of the fibrous materials. Combustion occurs sooner and a considerable decrease of PHRR was noted with a reduction of 50% compared to the reference material.

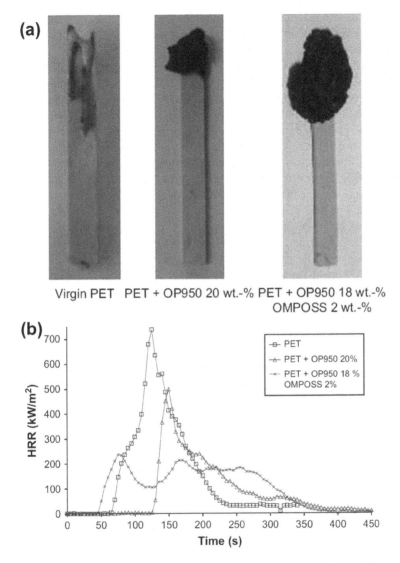

FIGURE 11.6 (a) LOI barrels after lighting (at LOI-1 vol.%) and (b) HRR of PET/OP950/POSS formulations containing 20 wt% additives. (Reprinted from Vannier, A. et al., *Polym. Degrad. Stab.*, 93, 818–826, 2008.)

The total heat evolved was also lowered to 16 MJ/m², and the fire resistant material protected about 68% of its initial weight.

Song et al. (2012) investigated the effect of methyl-POSS on the thermal degradation behaviors of intumescent flame retardant polylactide composites (incorporated by pentaerythritol phosphate, melamine phosphate) under nitrogen or air atmosphere. Under nitrogen atmosphere, polylactic acid (PLA) begun to degrade at 334°C, and the maximum mass loss rate occurred at about 370°C. The residue percentage for pure PLA at 600°C left nearly nothing. For IFRPLA/POSS samples, the initial degradation temperature was lower than that of pure PLA due to the earlier degradation of IFR. The IFR starts to decompose at the temperature range of 200°C–300°C, which catalyzed the degradation of the PLA matrix. However, the char yields were increased with the increase of methyl-POSS addition, indicating the enhanced thermal stability effect of methyl-POSS. This behavior was in accordance with the mechanism of improved fire performance via synergistic effect of phosphorous and silicon: phosphorous catalyzed the char formation and silicon protected the char from further thermal degradation. This char layer could reduce the weight loss rate, increase the thermal stability at higher temperature, and improve the fire retardant.

In addition, the dispersion state of POSS within the PLA matrix was also assessed by Wang et al. (2012c) with aid of TEM observing the morphology of the PLA nanocomposites. The TEM image clearly showed many ellipsoidal methyl-POSS particles with a wide size distribution in the range of 20–200 nm, suggesting that the individual methyl-POSS molecules tended to aggregate together at a nanoscale level rather than becoming uniformly dispersed at a molecular level in the matrix. The formation of methyl-POSS aggregates was probably dependent on the formation of intermolecular hydrogen bonds of the hydroxyl groups between the methyl-POSS molecules. The fire safety of PLA composites was assessed by LOI and UL-94 tests. Pure PLA was highly combustible, and its LOI value was only 20.0%. When the IFR load was 30 wt%, the LOI of the treated PLA increased from 20.0% to 33.0%, but it still could not pass the UL-94 V-0 rating test. When POSS and IFR were combined with PLA, its flame retardant properties were greatly improved. With IFR (22.5 wt%) and POSS (7.5 wt%), the sample PLA-5 achieved the V-0 rating in a UL-94 test, and its LOI value was 40.0%.

The influence of POSS with methyl, vinyl, and phenyl groups on mechanical properties, thermal stability, and combustion properties of PP composites was also investigated (Fina et al. 2010). By the melt blending, different POSSs with cage/ladder mixed structures were successfully incorporated into the PP. With the observation of the SEM photographs of the fractured surface, both at 1.5 wt% and 5 wt% loading, some methyl-POSS micron-sized solid aggregates were observable on the fracture surface, as confirmed by energy-dispersive X-ray spectroscopy (EDS) analyses. This suggested a limited compatibility between methyl-POSS and PP, which was in analogy with the results previously reported on the system PP/methyl-POSS (Fina et al. 2005). It was worth noting that methyl-POSS is a liquid at room temperature: therefore, it was likely that the observed aggregates were formed through POSS crosslinking.

From the TGA results, it was found that POSS did not influence PP degradation under nitrogen, whereas their presence clearly affected the polymer stability under air. In particular, the weight loss curves for PP 1.5 wt% POSS show only small improvements in thermoxidative stability with respect to neat PP, in terms of slightly delayed weight loss, mainly with vinyl-POSS and phenyl-POSS. Increasing the POSS loading to 5 wt%, the thermoxidative stability of PP/POSS was much improved as shown by the delay in the weight loss curves toward higher temperatures (Figure 11.7). Indeed, while the onset temperature for thermoxidative degradation was not significantly modified, the weight loss rate observed for PP/POSS in the 240°C–280°C temperature range was considerably lower as compared to reference PP. After incorporating methyl-POSS, a rapid increase in the weight loss rate above 280°C was shown, namely, related to rupture of the protective layer, leading to a limited delay of the maximum weight loss rate temperature (10°C). As far as fire hazards are concerned, the loading of 5 wt% methyl-POSS produced a reduction of 18.8% and 14.0% in PHRR and THR.

Aiming to obtain well dispersed POSS/PS systems, dumbbell-shaped POSS had been incorporated into the PS nanocomposites by in situ polymerization of styrene in the presence of POSS which has non-polymerizable groups (Blanco et al. 2013). With the aid of FTIR spectra, the POSS-PS interaction was also investigated by Blanco et al. Compared with FTIR spectra of neat PS and POSS, the shift of the sharp band at 1089 cm^{-1} presented in the POSS spectra and attributable to Si-O bonds, was observed in the nanocomposites spectra (1096 cm^{-1} for 3.6 wt% POSS), thus indicating the presence of filler-polymer interactions. As for the thermal stability, it was found that PS and PS-1 and PS-2 degraded, in every case, up to complete mass loss in temperature intervals ranging from 300°C to 700°C about in inert environment and 250°C–550°C about in oxidative atmosphere. Different from the TG curves of PS and composites at lower filler percentage, degradation curves of PS-3 containing 11.2 wt% POSS evidenced the formation of a stable residue, 12.5 wt% in inert environment and 7.3 wt% in oxidative atmosphere, respectively, up to 700°C. Another important difference was also exhibited from the TG curves: while the PS degraded completely in a single step, all composites showed a first very sharp degradation stage, followed by a second one at a low degradation rate in the last piece of the curves. This second stage was more evident on increasing the filler content in the nanocomposite and less evident for the low filler content sample.

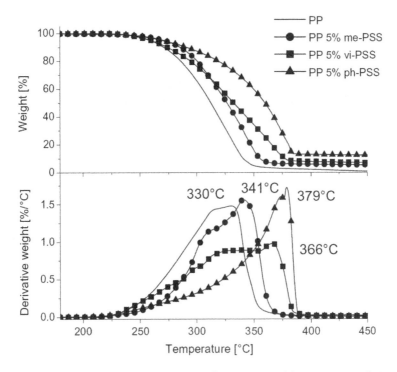

FIGURE 11.7 TGA plots of PP/5% POSS under air atmosphere. (Reprinted from Fina, A. et al., *Eur. Polym. J.*, 46, 14–23, 2010. With permission.)

Devaux et al. (2002) respectively prepared polyurethane/clay and polyurethane/POSS (methyl-POSS and vinyl-POSS) nanocomposites as a flame retarded coating for polyester and cotton fabrics. As demonstrated in this literature, the PU is widely used as coatings for textile fabrics in order to improve some properties, including mechanical behaviour, water repellency, and air impermeability. As shown by the CCT, although it was noticed that the slight shoulder observed after 50 seconds for the virgin PU had disappeared, the PU/methyl-POSS composite coating did not impart an enhancement for the fire safety of knitted fabrics more than the pure PU coating. However, the incorporation of either clay or vinyl-POSS all improved the fire safety of knitted fabrics confirmed by decreased PHRR (55%) in the PU/vinyl-POSS and increased ignition time in the PU/clay.

11.3.6 Other POSS

11.3.6.1 Thiol-POSS

The thiol-POSS was prepared by hydrolysis of (3-Mercaptopropyl) trimethoxysilane and reacted with tri(acryloyloxyethyl) phosphate to obtain a novel branched phosphonate acrylate monomer (Yu et al. 2015b). Then, the resultant was incorporated into EA resins in different ratios using ultraviolet curing technology. As can be seen from the TGA results, the incorporation of functionalized POSS promoted the initial decomposition of the EA composites, confirmed by a lower temperature at 10 wt% mass loss and max mass loss rate. Under nitrogen, T_{max} of the EA containing 30 wt% POSS was reduced by 96°C, while the char at 700°C was increased to 34.8 from 12.9 wt%. The same phenomenon was also observed in the TGA result under air. To estimate the flame retardancy of composite films, microscale combustion calorimeter (MCC) was performed on a series of cured films. Neat EA film was highly flammable with high PHRR and THR values, which were 296 W/g and 20.5 kJ/g, respectively. Incorporating branched phosphonate acrylate monomer (BPA) into the EA remarkably reduced the PHRR and THR values. With the addition of 30 wt% BPA, the PHRR and THR of the cured film were decreased by 50.8% and 29.8%, respectively, relative to those of pure EA. The systemic analyses based on the TGA, TG-IR and real-time (RT)-IR results revealed that the incorporation of the POSS accelerated the degradation of the EA matrix, reduced the flammable and toxic gases release, and promoted char formation. Moreover, the formation of silicon dioxide originating from the degradation of BPA on the surface of the residual chars reinforced the char layer.

11.3.6.2 DOPO-POSS

DOPO-containing POSS (DOPO-POSS) was successfully synthesized by Zhang et al. by the hydrolytic condensation of a modified silane (Zhang and Yang 2011). The starting material was a phosphorus-containing triethoxy silane (DOPO-VTES), which was synthesized by addition reaction between DOPO and vinyl triethoxy silane (VTES). This product was subjected to hydrolytic condensation using an HCl catalyst in methanol. A series of flame retarded EPs was prepared with DOPO-POSS (Zhang et al. 2011). The flame retardancy of these EPs was tested by the LOI, UL-94, which indicated that DOPO-POSS has meaningful effects on the flame retardancy of EP composites. When 2.5 wt% DOPO-POSS was added into the EP matrix, the LOI value of epoxy resin was leveled up from 25.0% to 30.2%. However, the LOI values reduced obviously with 10 wt% DOPO-POSS content. These results were different from the traditional notion. As far as UL-94 was concerned, the drippings were no longer observed for all the samples and the self-extinguishing was observed with 1.5–5.0 wt% DOPO-POSS loading. The best UL-94 rating was very close to a V-0 rating can be obtained when the DOPO-POSS content was 2.5 wt%. All the TGA and differential thermal gravity (DTG) results showed that DOPO-POSS has a positive effect on the char yield and T_{max2} of the DOPO-POSS/EP composites; on the contrary, the introduction of 2.5 wt% DOPO-POSS increased the heat release rate of the EP matrix. In general, flame retardancy of epoxy resin were improved by adding DOPO-POSS, but in different ranges of DOPO-POSS content in the resins, the composites showed different flame-retardant behaviors.

11.4 POSS Nanohybrid with Other Nanomaterials

11.4.1 Graphene and Graphene Oxide

Zhang et al. (2016) masterly utilized hexamethylenediamine and chloropropyl silsesquioxane to prepare a hyperbranched silsesquioxane polymer that grafted on the surface of the graphene oxide. The graphene-based resultant was signed as hyperbranched silsesquioxane polymer (HPP)-graphene oxide (GO) and incorporated into dicyclopentadiene bisphenol dicyanate ester (DCPDCE) to prepare composites. As presented by the TGA test, $T_{initial}$ value of the HPP-GO/DCPDCE system was increased by about 24°C in comparison with that of pure DCPDCE resin, suggesting that 0.6 wt% HPP-GO/DCPDCE composite has better thermal stability than pure DCPDCE resin. The thiol-POSS was also used to fabricate functionalized graphene through a thiol–ene click approach, then the graphene reinforced polyurethane acrylate (PUA) nanocomposite was fabricated by UV irradiation technology (Yu et al. 2015a). The $T_{initial}$ was gradually increased with increasing of content of functionalized graphene and FRGO/PUA1.0 was increased from 299°C for neat PUA to 311°C, while the char residues are increased from 0.5% to 2.3%. Without presence of thiol-POSS, RGO/PUA1.0 exhibited a lower $T_{initial}$ of 304°C and a char residue of 1.5%, due to the poor dispersion and weak interfacial interactions of RGO in the PUA matrix.

Simultaneous reduction and surface functionalization of GO was realized by simple refluxing of GO with amino-POSS without the use of any reducing agents (Wang et al. 2012b). The thermal stability and flame retardancy of the epoxy composite was significantly enhanced. In the case of the GO–EP composite, its $T_{initial}$ was lower than pure EP, since GO is thermally unstable and its major mass loss occurs below 250°C due to the decomposition of the oxygen-contained functional moieties. The addition of amino-POSS exhibited a reverse trend in the $T_{initial}$ compared to GO, and the $T_{initial}$ of amino-POSS/EP was increased by 24°C compared to that of pure EP. Owing to the multifunctional groups of amino-POSS, the crosslinking density of amino-POSS/EP was increased, resulting in the improvement of thermal stability. When GO was reduced by amino-POSS to convert into amino-POSS/rGO, amino-POSS/rGO/EP showed the best thermal stability, a 43°C increment compared to that of pure EP (Xu and Gao 2010). As the MCC results present, the addition of GO and amino-POSS gave rise to a 36% and 31% reduction in peak HRR, respectively, compared to that of pure EP. Furthermore, incorporating amino-POSS/rGO into the EP exhibited the lowest peak HRR value (nearly 58% reduction). For further investigation of fire hazards, Wang et al. (2012) also performed the cone calorimeter test. It was found that pure epoxy resin burned very rapidly after ignition, and the peak heat release rate value was 1730 kW/m². As expected, incorporating GO into the epoxy resins made the peak heat release rate decrease to 1345 kW/m². For the sample of amino-POSS/EP, the PHRR also decreases by 38% compared to that of pure EP. Moreover, the PHRR of amino-POSS-rGO/EP exhibited further reduction compared to both GO/EP and amino-POSS/EP.

11.4.2 Carbon Nanotubes

Nanocomposites of polyamide 12 with multiwall CNT and POSS were studied to assess their flame retardancy properties (Gentiluomo et al. 2016). Figure 11.8 displays results obtained for the heat release rate of the PA 12 composites containing 4 wt% of CNT, 3.3 wt% of POSS, and both CNT and POSS, as compared to PA 12, with an external heat flux of 50 kW/m² applied in the cone calorimeter. PA 12 showed a HRR curve which was typical of many thermoplastic materials, where the heat release showed a sudden acceleration that causes a peak in the heat release rate. After adding 3.3 wt% of POSS, the fire behavior of the PA 12 did not change with the exception of the time to ignition that was decreased to

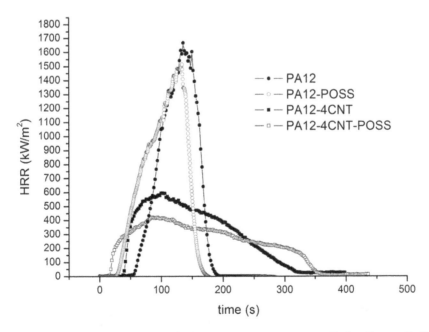

FIGURE 11.8 Heat release rate of the PA 12 and PA 12 composites containing 4 wt% of CNT, 3.3 wt% of POSS, and both CNT and POSS, with an external heat flux of 50 kW/m². (Reprinted from Gentiluomo, S. et al., *Polym. Degrad. Stab.*, 134, 151–156, 2016. With permission.)

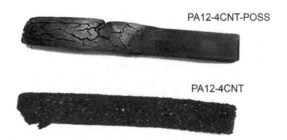

FIGURE 11.9 specimens after ul-94 test. (reprinted from gentiluomo, s. et al., *polym. degrad. stab.*, 134, 151–156, 2016. with permission.)

36 seconds. The butyl-POSS showed evaporation or sublimation in inert atmosphere, over a range of 250°C–300°C (Fina et al. 2005). This means that probably in the cone calorimeter experiment, POSS would evaporate earlier than the thermal degradation products of the polymer, increasing the ignitability. No significant change in the THR was found. The presence of 4 wt% of CNT changed the fire behavior and the heat release rate curve typical for PA 12 was changed into a flattered curve shape similar to those observed in layered silicate polymer nanocomposites (Kashiwagi et al. 2004). As observed before by other authors, the fire behavior of CNT composites was influenced by the consolidation of an interconnected network structure causing viscosity effects that hinder the release of pyrolysis products (Schartel et al. 2005). The simultaneous presence of POSS and CNT in the composites led to a further diminishing in the HRR curve, with a peak of heat released of 425 kW/m^2, and a longer time to stop flaming.

As expected, using at the same time POSS and CNT, the LOI of PA 12 was increased to 27.0% for PA 12-2CNT-POSS and 27.5% for PA 12-4CNT-POSS. During the combustion, the samples burned leaving a solid black residue with a specimen-like shape. In the UL-94 test, the PA 12 resulted as non-classifiable, because the flame arrived to the sample end, producing a remarkable amount of flaming dripping material. Even being not classifiable, PA 12-4CNT showed a totally different behavior compared to the unfilled PA 12: there was no dripping and the flame transit induced the formation of a self-supporting residue, whose shape and size corresponded to those of the specimen, as can be observed in Figure 11.9. The fire behavior of PA 12-POSS was similar to those observed in LOI test. The sample produced a thin carbon layer not sufficient to hinder the polymer dripping and left no residue. On the other hand, the composites containing both POSS and CNT (PA 12-4CNT-POSS) cease to burn before the 100 mm mark, reaching the UL-94 HB classification. Indeed, the flame has extinguished at 27 mm from the sample's tip, with an average time of burning of 108 seconds. The flame front transit induced the formation of a self-supporting residue, whose shape and size correspond to that of the specimen.

These data indicated that both POSS and CNT used alone were not sufficient to achieve a flame retardant effect, even worsening the fire performance in the LOI test. However, the simultaneous use of the two nanofillers showed the possibility of obtaining a self-extinguishing composite, due to a synergistic effect between the fillers. CNTs were able to produce a network during the combustion, but were not able to produce an effective protective shield. The POSS alone was not able to produce a layer of silica to make the polymer effectively flame retardant. In these conditions, POSS undergoes volatilization. On the other hand, the action of the POSS supported by the network of CNT was able to produce an effective shield capable to reduce the flammability.

The combination effect of CNT and POSS on the EP resin was also deeply investigated (Wang et al. 2016). The experimental results showed that the introduction of PB-POSS or MWCNTs further improved the LOI values of the epoxy resin, while the highest LOI value (32.8%) was obtained for the formulation containing 14.6 wt% PB-POSS and 0.4 wt% MWCNTs.

11.4.3 MoS$_2$

Jiang et al. (2014) had analyzed the integrated effect of MoS$_2$ and POSS on the thermal stability and flame retardancy of PVA composites. Functionalized MoS$_2$ was prepared by a simple reflux reaction between DITG-MoS$_2$ and vinyl-POSS. Vinyl-POSS/MoS$_2$ was dispersed well in the matrix due to the good interfacial interaction between the functionalized MoS$_2$ and PVA. The dispersed state and interfacial interaction of MoS$_2$ with the PVA matrix were characterized by XRD and TEM analyses. The neat PVA exhibited a broad crystalline peak at 19.5°, corresponding to the (101) and (101) crystalline reflections of PVA. For the PVA/vinyl POSS-MoS$_2$ system, it displayed similar characteristic diffraction peaks to the PVA matrix. The diffraction peaks observed in the XRD patterns of the MoS$_2$ disappeared in the patterns of the PVA/vinyl-POSS MoS$_2$ composites. The composites exhibited none of the MoS$_2$ diffraction peaks, implying the possible exfoliation of the MoS$_2$ layered structure. In contrast, the strongest diffraction peak (002) of the MoS$_2$ was still observed in the XRD patterns of the PVA/MoS$_2$ composite, indicating the presence of the stacked structure of MoS$_2$. Moreover, the fractured surface of neat PVA was quite smooth, while that of the nanocomposite sample was very rough. It was also worthy to note that a uniform dispersion of the MoS$_2$ sheets within the polymer matrix was achieved, with no visible aggregation of MoS$_2$. This phenomenon was in good agreement with the previous literatures (Alzari et al. 2011; Zhao et al. 2011).

The hindering effect of vinyl-POSS/MoS$_2$ on the thermal decomposition and fire hazards of the PVA were investigated by TGA test and MCC. When MoS$_2$ was modified by vinyl-POSS to convert into vinyl-POSS/MoS$_2$, PVA/vinyl-POSS/MoS$_2$ depicted the best thermal stability, a 23°C increment in T$_{max}$ compares to that of pure PVA. By comparing MCC results with that of pure PVA, the PVA nanocomposites achieved significant improvements in flame retardancy. The PHRR and THR of all the samples were lower than pure PVA. Both PHRR and THR showed the lowest value when vinyl-POSS/MoS$_2$ was added. Compared with the pure PVA, the PHRR and THR decreased apparently, falling by 41% and 16%, respectively.

11.5 Conclusions

In general, the excellent compatibility produced by intimate bond of the exterior organic groups of POSS cage with the polymeric matrix fully develop the reinforcement effect of POSS on the thermal stability and flame retardancy of polymer nanocomposites. Certainly, the reinforcement mechanisms of POSS highly depend on the instinct structure and properties. The production of highly stable char layer that protects the bottom polymeric is considered as the principal contribution for the improvement of flame retardancy of polymer/POSS nanocomposites. During the combustion, the organic groups on POSS cages firstly undergo the decomposition process and produce free-radicals that accelerate the degradation of polymers, thus resulting in a reduction of TTI. On the other hand, the decomposition of POSS timely promotes the char formation and improves the robustness and stability of the char layer, which retards the further combustion of the bottom polymer. In addition, POSS presents a significant synergistic effect with traditional flame retardants even at a very low dosage. Especially when char-forming flame retardant was added, the quality of the produced char layer was significantly enhanced due to the presence of silicon.

Even though the promising prospect of POSS in the field of flame retarded composites had been confirmed, the high cost only allows its application in numbered occasions where cost is unconcerned, including the military field, aviation, and aerospace fields. Certainly, look at it from different perspectives, high efficiency of POSS-based fire retardant decreases the consumption of current flame retardant, thus reducing the cost of practical application. Likewise, reinforcement effect of POSS is not limited in thermal stability and flame retardant, unexpected properties are also obtained. Finally, the widespread use of polymer/POSS nanocomposites in the flame retardant field is still at the initial stage, and it is necessary to continue to exploit the vast potential of POSS in industries application.

References

Alongi, J., Ciobanu, M., and Malucelli, G. 2011. Novel flame retardant finishing systems for cotton fabrics based on phosphorus-containing compounds and silica derived from sol–gel processes. *Carbohydrate Polymers* 85: 599–608.

Alzari, V., Nuvoli, D., Sanna, R. et al. 2011. In situ production of high filler content graphene-based polymer nanocomposites by reactive processing. *Journal of Materials Chemistry* 21: 16544–16549.

Barczewski, M., Chmielewska, D., Dobrzyńska-Mizera, M. et al. 2014. Thermal stability and flammability of polypropylene-silsesquioxane nanocomposites. *International Journal of Polymer Analysis and Characterization* 19: 500–509.

Blanco, I., Bottino, F. A., Cicala, G. et al. 2013. A kinetic study of the thermal and thermal oxidative degradations of new bridged POSS/PS nanocomposites. *Polymer Degradation and Stability* 98: 2564–2570.

Bouza, R., Barral, L., Díez, F. J. et al. 2014. Study of thermal and morphological properties of a hybrid system, iPP/POSS. Effect of flame retardance. *Composites Part B: Engineering* 58: 566–572.

Devaraju, S., Vengatesan, M. R., Selvi, M. et al. 2012. Synthesis and characterization of bisphenol-A ether diamine-based polyimide POSS nanocomposites for low K dielectric and flame-retardant applications. *High Performance Polymers* 24: 85–96.

Devaux, E., Rochery, M., and Bourbigot, S. 2002. Polyurethane/clay and polyurethane/POSS nanocomposites as flame retarded coating for polyester and cotton fabrics. *Fire and Materials* 26: 149–154.

Didane, N., Giraud, S., Devaux, E. et al. 2012a. A comparative study of POSS as synergists with zinc phosphinates for PET fire retardancy. *Polymer Degradation and Stability* 97: 383–391.

Didane, N., Giraud, S., Devaux, E. et al. 2012b. Development of fire resistant PET fibrous structures based on phosphinate-POSS blends. *Polymer Degradation and Stability* 97: 879–885.

Du, B., Ma, H., and Fang, Z. 2011. How nano-fillers affect thermal stability and flame retardancy of intumescent flame retarded polypropylene. *Polymers for Advanced Technologies* 22: 1139–1146.

Fan, H., and Yang, R. 2013. Flame-retardant polyimide cross-linked with polyhedral oligomeric octa (aminophenyl) silsesquioxane. *Industrial & Engineering Chemistry Research* 52: 2493–2500.

Fina, A., Tabuani, D., and Camino, G. 2010. Polypropylene–polysilsesquioxane blends. *European Polymer Journal* 46: 14–23.

Fina, A., Tabuani, D., Frache, A. et al. 2005. Polypropylene–polyhedral oligomeric silsesquioxanes (POSS) nanocomposites. *Polymer* 46: 7855–7866.

Fox, D. M., Lee, J., Zammarano, M. et al. 2012. Char-forming behavior of nanofibrillated cellulose treated with glycidyl phenyl POSS. *Carbohydrate Polymers* 88: 847–858.

Franchini, E., Galy, J., Gérard, J.-F. et al. 2009. Influence of POSS structure on the fire retardant properties of epoxy hybrid networks. *Polymer Degradation and Stability* 94: 1728–1736.

Gentiluomo, S., Veca, A. D., Monti, M. et al. 2016. Fire behavior of polyamide 12 nanocomposites containing POSS and CNT. *Polymer Degradation and Stability* 134: 151–156.

Gérard, C., Fontaine, G., Bellayer, S. et al. 2012. Reaction to fire of an intumescent epoxy resin: Protection mechanisms and synergy. *Polymer Degradation and Stability* 97: 1366–1386.

Hartmann-Thompson, C. 2011. *Applications of Polyhedral Oligomeric Silsesquioxanes*. Dordrecht, the Netherlands: Springer Science & Business Media.

He, H.-B., Li, B., Dong, J.-P. et al. 2013. Mesostructured nanomagnetic polyhedral oligomeric silsesquioxanes (POSS) incorporated with dithiol organic anchors for multiple pollutants capturing in wastewater. *ACS Applied Materials & Interfaces* 5: 8058–8066.

Jiang, S.-D., Tang, G., Bai, Z.-M. et al. 2014. Surface functionalization of MoS 2 with POSS for enhancing thermal, flame-retardant and mechanical properties in PVA composites. *RSC Advances* 4: 3253–3262.

Kannan, R. Y., Salacinski, H. J., Butler, P. E. et al. 2005. Polyhedral oligomeric silsesquioxane nanocomposites: The next generation material for biomedical applications. *Accounts of Chemical Research* 38: 879–884.

Karlsson, L., Lundgren, A., Jungqvist, J. et al. 2011. Effect of nanofillers on the flame retardant properties of a polyethylene–calcium carbonate–silicone elastomer system. *Fire and Materials* 35: 443–452.

Kashiwagi, T., Harris, R. H., Zhang, X. et al. 2004. Flame retardant mechanism of polyamide 6–clay nanocomposites. *Polymer* 45: 881–891.

Laik, S., Galy, J., Gerard, J. F. et al. 2016. Fire behaviour and morphology of epoxy matrices designed for composite materials processed by infusion. *Polymer Degradation and Stability* 127: 44–55.

Leu, C. M., Chang, Y. T., and Wei, K. H. 2003a. Synthesis and dielectric properties of polyimide-tethered polyhedral oligomeric silsesquioxane (POSS) nanocomposites via POSS-diamine. *Macromolecules* 36: 9122–9127.

Leu, C. M., Reddy, G. M., Wei, K. H. et al. 2003b. Synthesis and dielectric properties of polyimide-chain-end tethered polyhedral oligomeric silsesquioxane nanocomposites. *Chemistry of Materials* 15: 2261–2265.

Li, Z., and Yang, R. 2014. Study of the synergistic effect of polyhedral oligomeric octadiphenylsulfonylsilsesquioxane and 9, 10-dihydro-9-oxa-10- phosphaphenanthrene-10-oxide on flame-retarded epoxy resins. *Polymer Degradation and Stability* 109: 233–239.

Liao, W. H., Yang, S. Y., Hsiao, S. T. et al. 2014. Effect of octa (aminophenyl) polyhedral oligomeric silsesquioxane functionalized graphene oxide on the mechanical and dielectric properties of polyimide composites. *ACS Applied Materials & Interfaces* 6: 15802–15812.

Liu, L., Hu, Y., Song, L. et al. 2010. Preparation, characterization and properties of polystyrene composites using octaphenyl polyhedral oligomeric silsesquioxane and its bromide derivative. *Iranian Polymer Journal* 19: 937–948.

Liu, Y., Huang, Y., and Liu, L. 2007. Thermal stability of POSS/methylsilicone nanocomposites. *Composites Science and Technology* 67: 2864–2876.

Louisy, J., Bourbigot, S., Duquesne, S. et al. 2013. Novel synergists for flame retarded glass-fiber reinforced poly(1,4-butylene terephthalate). *Polimery* 58: 403–412.

Lu, T., Liang, G., and Guo, Z. 2006. Preparation and characterization of organic–inorganic hybrid composites based on multiepoxy silsesquioxane and cyanate resin. *Journal of Applied Polymer Science* 101: 3652–3658.

Meenakshi, K. S., Sudhan, E. P. J., and Kumar, S. A. 2012. Development and characterization of new phosphorus based flame retardant tetraglycidyl epoxy nanocomposites for aerospace application. *Bulletin of Materials Science* 35: 129–136.

Meenakshi, K. S., Sudhan, E. P. J., Kumar, S. A. et al. 2011. Development and characterization of novel DOPO based phosphorus tetraglycidyl epoxy nanocomposites for aerospace applications. *Progress in Organic Coatings* 72: 402–409.

Schartel, B., Potschke, P., Knoll, U. et al. 2005. Fire behaviour of polyamide 6/multiwall carbon nanotube nanocomposites. *European Polymer Journal* 41: 1061–1070.

Sethuraman, K., Prabunathan, P., and Alagar, M. 2014. Thermo-mechanical and surface properties of POSS reinforced structurally different diamine cured epoxy nanocomposites. *RSC Advances* 4: 45433–45441.

Song, L., Xuan, S., Wang, X. et al. 2012. Flame retardancy and thermal degradation behaviors of phosphate in combination with POSS in polylactide composites. *Thermochimica Acta* 527: 1–7.

Vahabi, H., Eterradossi, O., Ferry, L. et al. 2013. Polycarbonate nanocomposite with improved fire behavior, physical and psychophysical transparency. *European Polymer Journal* 49: 319–327.

Vahabi, H., Ferry, L., Longuet, C. et al. 2012. Combination effect of polyhedral oligomeric silsesquioxane (POSS) and a phosphorus modified PMMA, flammability and thermal stability properties. *Materials Chemistry and Physics* 136: 762–770.

Vannier, A., Duquesne, S., Bourbigot, S. et al. 2008. The use of POSS as synergist in intumescent recycled poly (ethylene terephthalate). *Polymer Degradation and Stability* 93: 818–826.

Waddon, A., Zheng, L., Farris, R. et al. 2002. Nanostructured polyethylene-POSS copolymers: Control of crystallization and aggregation. *Nano Letters* 2: 1149–1155.

Wang, Q., Xiong, L., Liang, H. et al. 2016. Synergistic effect of polyhedral oligomeric silsesquioxane and multiwalled carbon nanotubes on the flame retardancy and the mechanical and thermal properties of epoxy resin. *Journal of Macromolecular Science, Part B* 55: 1146–1158.

Wang, X., Hu, Y., Song, L. et al. 2010. Thermal degradation behaviors of epoxy resin/POSS hybrids and phosphorus–silicon synergism of flame retardancy. *Journal of Polymer Science Part B: Polymer Physics* 48: 693–705.

Wang, X., Hu, Y., Song, L. et al. 2012a. Comparative study on the synergistic effect of POSS and graphene with melamine phosphate on the flame retardance of poly (butylene succinate). *Thermochimica Acta* 543: 156–164.

Wang, X., Song, L., Yang, H. et al. 2012b. Simultaneous reduction and surface functionalization of graphene oxide with POSS for reducing fire hazards in epoxy composites. *Journal of Materials Chemistry* 22: 22037–22043.

Wang, X., Xuan, S., Song, L. et al. 2012c. Synergistic effect of POSS on mechanical properties, flammability, and thermal degradation of intumescent flame retardant polylactide composites. *Journal of Macromolecular Science, Part B* 51: 255–268.

Xu, Z., and Gao, C. 2010. In situ polymerization approach to graphene-reinforced nylon-6 composites. *Macromolecules* 43: 6716–6723.

Yu, B., Shi, Y., Yuan, B. et al. 2015a. Click-chemistry approach for graphene modification: Effective reinforcement of UV-curable functionalized graphene/polyurethane acrylate nanocomposites. *RSC Advances* 5: 13502–13506.

Yu, B., Tao, Y., Liu, L. et al. 2015b. Thermal and flame retardant properties of transparent UV-curing epoxy acrylate coatings with POSS-based phosphonate acrylate. *RSC Advances* 5: 75254–75262.

Zhang, M., Yan, H., Yuan, L. et al. 2016. Effect of functionalized graphene oxide with hyperbranched POSS polymer on mechanical and dielectric properties of cyanate ester composites. *RSC Advances* 6: 38887–38896.

Zhang, P., Song, L., Dai, K. et al. 2010. Preparation and thermal properties of the UV-cured epoxy acrylate/microencapsulated phase-change material. *Industrial & Engineering Chemistry Research* 50: 785–790.

Zhang, W., Li, X., Fan, H. et al. 2012. Study on mechanism of phosphorus–silicon synergistic flame retardancy on epoxy resins. *Polymer Degradation and Stability* 97: 2241–2248.

Zhang, W., Li, X., and Yang, R. 2011. Novel flame retardancy effects of DOPO-POSS on epoxy resins. *Polymer Degradation and Stability* 96: 2167–2173.

Zhang, W. C., and Yang, R. J. 2011. Synthesis of phosphorus-containing polyhedral oligomeric silsesquioxanes via hydrolytic condensation of a modified silane. *Journal of Applied Polymer Science* 122: 3383–3389.

Zhao, X., Zhang, Q., Chen, D. et al. 2011. Enhanced mechanical properties of graphene-based poly (vinyl alcohol) composites. *Macromolecules* 44: 2392.

Zheng, L., Kasi, R. M., Farris, R. J. et al. 2002. Synthesis and thermal properties of hybrid copolymers of syndiotactic polystyrene and polyhedral oligomeric silsesquioxane. *Journal of Polymer Science Part A: Polymer Chemistry* 40: 885–891.

Zhou, Z., Zhang, Y., Zeng, Z. et al. 2008. Properties of POSS-filled polypropylene: Comparison of physical blending and reactive blending. *Journal of Applied Polymer Science* 110: 3745–3751.

Zhu, S. E., Wang, L. L., Wang, M. Z. et al. 2017. Simultaneous enhancements in the mechanical, thermal stability, and flame retardant properties of poly (1, 4-butylene terephthalate) nanocomposites with a novel phosphorus–nitrogen-containing polyhedral oligomeric silsesquioxane. *RSC Advances* 7: 54021–54030.

IV

Applications

12

Flame Retarded Polymer Foams for Construction Insulating Materials

Zhengzhou Wang,
Xiaoyan Li, and
Lei Liu

12.1 Introduction

Polymer foams are composed of a polymer matrix incorporated with either gas bubbles or gas tunnels, with either a closed-cell or open-cell structure. Polymer foams have many advantages, for example, lightless, low thermal conductivity, easy process, and affordable cost, etc. and are widely used in many sectors of industry, e.g., building, transportation, furniture, chemical industry, and so on. Because the combustible volatile hydrocarbon liquids are often used as blowing agents and most polymer matrices themselves are also flammable, polymer foams, such as polystyrene (PS) foam and polyurethane (PU) foam, are easily ignited. Ignition of these polymer foamy materials is one of the most common reasons accounting for residential home fires and a large amount of fire civilian fatalities (Wang et al. 2014), such as the 2010 Shanghai high-rise apartment building fire (58 deaths) and the 2017 London Grenfell Tower fire (72 deaths). Therefore, flame retardation of PU and PS foams is generally needed in the building and transportation industry. The commonly used flame retardants for polymer foams include two categories: halogen-containing and halogen-free flame retardants. Halogen-containing compounds usually allow considerable improvements of flame retardancy of the foams. However, these kinds of flame retardants tend to release some toxic smoke and gases during fire. Due to environmental concerns, people pay more attention to halogen-free flame retardation of polymer foams.

Compared with PU and PS foams, phenolic foams (PF) have many advantages, such as good heat resistance, excellent flame retardancy, and low smoke and toxic gas release on burning. However, PF foams also have some disadvantages, such as brittleness and friability, which greatly limit their applications. The commonly used toughening agents for PF foams including ethylene glycol, polyethylene glycol (PEG), polyurethane pre-polymer, some elastomers, etc., however, may deteriorate the excellent flame retardancy of PF foams because those agents are highly flammable. Therefore, it is ideal for toughening agents to toughen PF foams without the cost of their excellent flame retardancy. Moreover,

236

Flame Retardant Polymeric Materials

some solid fillers, i.e., fibers (glass and aramid fibers) and nano-materials (clays, carbon nanotubes, graphene) can increase the toughness of PF foams, generally without decreasing the flame retardancy of PF foams. For those solid fillers, especially nano-fillers, their uniform dispersion in the PF resin is a challenging task.

In this chapter, we will review the recent advances of flame retardation of rigid PU foams and PS foams, as well as toughening and flame retardation of PF foams.

12.2 Flame Retardation of Rigid Polyurethane Foams

PU foams are mainly divided into flexible polyurethane (FPU) foams and rigid polyurethane (RPU) foams, their market share of PU foams is about 48% and 28%, respectively (Singh and Jain 2009). FPU foams have been widely used in furniture cushioning and mattresses, and RPU foams have widespread applications in building, transportation, refrigeration industry, and so on (Al-Homoud 2005; Wu et al. 1999). PU foams are highly combustible; therefore, flame retardation of the foams is needed in most applications. A comprehensive review of the flame retardancy of polyurethanes including PU foams with emphasis on flame retardants in commercial use was published by Weil and Levchik (2004). Due to the environmental concerns of halogen-containing flame retardants, the development and application of halogen-free flame retardants for RPU foams has become a subject of extensive investigation.

The approaches for improving the fire retardancy of RPU foams consist of two categories: (1) reactive-type flame retardation, e.g., reactive flame retardant diols or polyols containing phosphorus and/or halogen, or containing phosphorus and/or nitrogen are generally used to react with isocyanates to prepare the flame retarded (FR) PU resin; (2) additive-type flame retardation, e.g., flame retardants are directly added by simple mechanical mixing at the compounding stage of PU foams (Levchik et al. 2004; Singh and Jain 2009). The additive-type flame retardants usually do not participate in the foaming reaction. These kinds of flame retardants can act as plasticizers if they are compatible with the polymer; otherwise, they are considered as fillers (Meng et al. 2009). As for the reactive-type fire retardants, they can build chemically into the PU molecule chain during the foaming reactions, and it is generally thought that greater permanence of flame retardancy can be maintained by this method. This part aims to review the recent developments regarding the use of halogen-free flame retardants for RPU foams.

12.2.1 Reactive-Type Flame Retardants

The non-halogenated reactive flame retardants are generally diols or polyols containing P, N, or $P-N$ groups. The increased flame retardancy of phosphorus-containing RPU foams is attributed to the fact that phosphorus both promotes carbonization in the condensed phase and inhibits combustion in the gas phase. For instance, Wu et al. (2013) synthesized ethanolamine spirocyclic pentaerythritol bisphosphonate (EMSPB) and found that the RPU foam with 25 wt% EMSPB can pass the UL-94 V-0 rating with a limiting oxygen index (LOI) of 27.5%, however, the compressive strength and initial thermal stability of the FR RPU foams decrease compared to the pure RPU foam. Two reactive flame retardants, e.g., hexa-(phosphite-hydroxyl-methyl-phenoxyl)-cyclotriphosphazene (HPHPCP) (Figure 12.1) containing phosphazene and phosphate, and hexa-(5,5-dimethyl-1,3,2-dioxaphosphinane-hydroxyl-methyl-phenoxyl)-cyclotriphosphazene (HDPCP) (Figure 12.1) containing phosphazene and cyclophosphonate were synthesized, and it was observed that the LOI values of the RPU foams containing 20 wt% HPHPCP and 25% HDPCP are 26% and 25%, respectively (Yang et al. 2015, 2017). HPHPCP has a positive effect on compressive strength, while HDPCP has a negative influence on the strength of the RPU foams which may be because powdered HDPCP cannot react completely during the fast foaming process. Introduction of both HPHPCP and HDPCP resulted in an increase in the initial thermal decomposition temperature ($T_5\%$) and the residue of the FR RPU foams at high temperatures, which was attributed to the extent of enhancement of higher crosslinking due to the multifunctional reactive groups of the reactive flame retardants. Nevertheless, the incorporation of the two flame retardants,

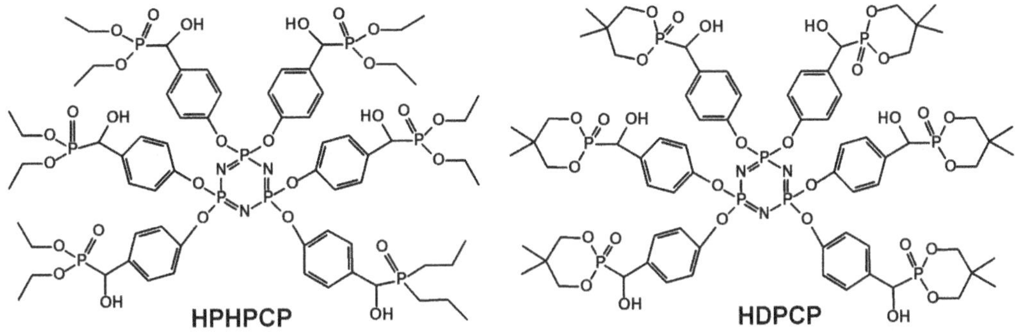

FIGURE 12.1 Chemical structures of hexa-(phosphite-hydroxyl-methyl-phenoxyl)-cyclotriphosphazene (HPHPCP) and hexa-(5,5-dimethyl-1,3,2-dioxaphosphinane-hydroxyl-methyl-phenoxyl)-cyclotriphosphazene (HDPCP).

especially at their high loadings showed a negative influence on thermal insulation of the RPU foams. Liu et al. (2016) synthesized a melamine-based polyether polyol (HMMM–PG), and found that the physical, mechanical, and fire-retardant properties of RPU foams were improved due to the formation of a continuous and dense char layer.

Recently, many researchers have paid attention to the modification of bio-based polyols to improve the mechanical and flame retarded properties of RPU foams. Cardanol has many applications in polymer chemistry, for example, in epoxy resins, phenol–formaldehyde resins, non-ionic surfactants, and polyurethanes. Ionescu et al. (2012) synthesized new bio-based Mannich polyols using cardanol and introduced the new Mannich polyols into RPU foams. It was found that both physico-mechanical and fire retardant properties of RPU foams were improved. Hydroxylated vegetable oils, whether synthesized or natural ones (castor oil), have been used to prepare polyurethanes through the reaction of hydroxyl groups with isocyanates. A phosphorylated polyol (Polyol-P, Figure 12.2) derived from vegetable oils was successfully synthesized, and its effect on the flame retardancy of RPU foams was investigated. RPU foams with densities between 30 and 39 kg/m^3 demonstrated elliptical shaped closed-cells and homogeneous size distribution, due to the fact that phosphorylated polyols are made up of phosphate mono and diesters (Heinen et al. 2014). It was reported that the LOI of the RPU foams prepared with a phosphorus-containing castor oil-based flame-retardant polyol (COFPL, Figure 12.2) reached to 24.3% without any other flame retardant, and the compression strength of the modified RPU foams was improved (Zhang et al. 2014). The SEM images showed that the overall cell structure of the COFPL modified foams became more uniform, the cell walls became thinner, and the amount of broken cells decreased.

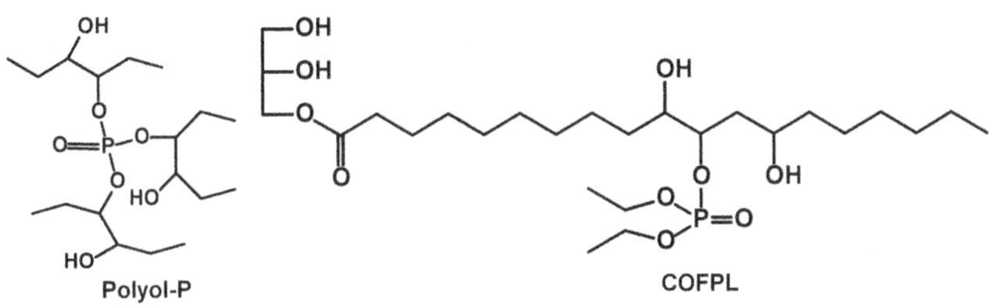

FIGURE 12.2 Structure of phosphorylated polyols (Polyol-P) and castor oil based flame-retardant polyols (COFPL).

In addition, there are some publications on RPU foams modified by boron-containing diols or polyols. Different boron-containing triols were synthesized from the reaction of boric acid with diols such as 1,2-propanediol, 1,3-butanediol or with aminoalcohols such as monoethanol-amine (Czupryński and Paciorek 1999; Czupryński et al. 2002). The incorporation of tri(hydroxyetylamine) borate obviously improved the flame retardancy of RPU foams, but a decrease in the brittleness of the foams was observed (Czupryński et al. 2002). Paciorek-Sadowska synthesized organic polyols based on boric acid and di(hydroxymethyl)urea derivatives, and their results showed that the boron- and nitrogen-containing polyols effectively increased fire resistance of RPU foams (Paciorek-Sadowska et al. 2015a, 2015b).

The reactive-type flame retardant diols or polyols aforementioned are usually applied in combination with regular diols or polyols to prepare FR RPU foams. How the FR diols or polyols participate in the foaming process of RPU foams is still not quite clear. Moreover, the fire retardancy stringent requirements of FR RPU foams prepared by this method are generally difficult to meet (e.g., Fire Retardant Grade B1) in the building and construction industry.

To improve further flame retardancy of RPU foams, reactive-type flame retardants are often combined with other additive-type flame retardants like expandable graphite (EG). A phosphorus-containing polyol (BHPP) and a nitrogen-containing polyol (MADP) as shown in Figure 12.3 were synthesized through dehydrochlorination and Mannich reaction, respectively, and the synergistic effect among BHPP, MADP, and EG in RPU foams was investigated (Yuan et al. 2016b). Based on LOI and cone calorimeter test results, it was concluded that the combination of BHPP and MADP in RPU foams dramatically increased the LOI value and promotes the formation of a compact and continuous char which was attributed to the presence of O=P–O– and triazine ring groups, as well as NH_3 emission. The incorporation of EG into the RPU/BHPP/MADP foams further improved the flame retardant properties of the RPU foams. The LOI of the RPU/BHPP/MADP foams was 33.5%, while the value of the RPU/EG foam was only 27.0%. It was also reported that the RPU foams with a reactive-type flame retardant named diethyl-*N*,*N*-bis(2-hydroxyethyl) phosphoramide (DEPA, Figure 12.3) and EG were prepared through free foaming technique (Zhao et al. 2017). Compared to DEPA or EG alone, the combination (EG/DEPA) endowed RPU foams enhanced LOI values, higher protective char yields, and lower heat release and smoke production, which show a synergistic flame retardant effect between them. The results showed that DEPA decomposed to produce potential free radical scavenger (PO•), this promoted generating more H_2O and CO_2 in gaseous phase and facilitated the formation of phosphorus-rich residue.

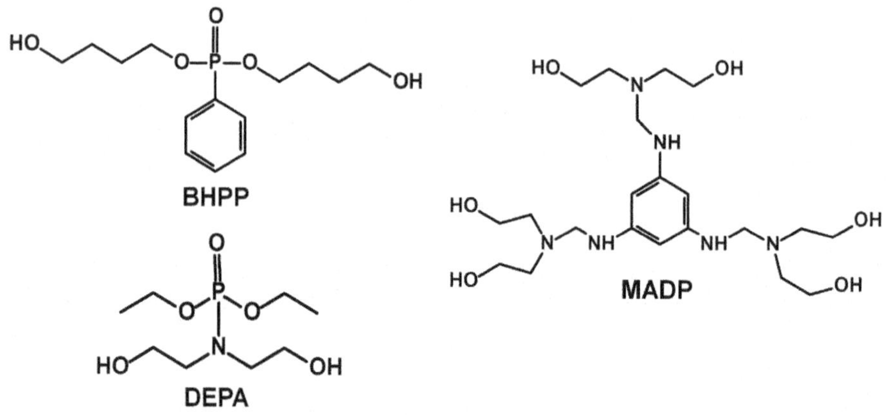

FIGURE 12.3 Structures of phosphorus-containing polyol (BHPP), nitrogen-containing polyol (MADP), and diethyl-*N*,*N*-bis(2-hydroxyethyl) phosphoramide (DEPA).

12.2.2 Additive-Type Flame Retardants

12.2.2.1 Expandable Graphite

EG is usually prepared by the treatment of graphite with sulfuric acid, nitric acid, or acetic acid, which is intercalated into the crystal structure of graphite (Modesti and Lorenzetti 2003). EG on heating expands, and its expansion ratio can reach up to several 100 times. EG exerts its flame retardant role mainly in the condensed phase by means of the formation of intumescent char, which works as an insulating layer to reduce the heat and mass transfer between the flame zone and the burning substrate (Modesti and Lorenzetti 2002a). Many researchers have done extensive investigation on the use of EG in RPU foams (Modesti and Lorenzetti 2002b; Ye et al. 2009). The factors influencing the flame retardant efficiency of EG in RPU foams include its size, expansion ratio, density, and content as well. It was reported by many researchers that the lower the size of the EG particles, the better the flame retardancy for RPU foams (Shi et al. 2005, 2006). Although EG is an efficient flame retardant for RPU foams, the addition of EG usually caused considerable deterioration of mechanical properties of the foams (Modesti et al. 2002; Ye et al. 2009). The deterioration of the mechanical properties of the EG/RPU foams can be explained by the collapse of RPU's cells, which is due to the large size of EG particles and the poor interfacial adhesion between EG particles and the PU matrix. Thereby, the main challenges encountered in developing desirable EG/RPU foams are: (i) to achieve uniform dispersion of EG particles in the polymer matrix without destroying the integrity of the RPU cell system, and (ii) to improve interfacial adhesion between EG and RPU foams. In general, the smaller the filler, the better the composite properties (Ye et al. 2010). However, they do not give good flame retardancy when EG particles were pulverized to very fine particles (PEG) (Shi et al. 2006). As EG particles get smaller, it is difficult for them to hold enough amounts of oxidant (e.g., H_2SO_4) inside the particles upon heating, the blowing gases originating from the reaction between oxidant and graphite easily escape from the edges of the flakes resulting in less expandability, and thus they exert less flame retardancy (Modesti et al. 2002).

Except for the reduction of the mechanical properties caused by larger EG particles, the compatibility between EG and PU resin is poor. To improve the interfacial adhesion between EG particles and RPU foams, one possible approach is to modify the EG particles' surface by introducing some functional groups such as –OH and –NH$_2$, which can react with –NCO of isocyanurate monomer for RPU foams. For instance, the encapsulation or grafting of polymers onto the EG surface can improve the dispersion and compatibility of EG in RPU foams, and thus enhance the mechanical properties of EG-filled RPU foams (Ye et al. 2009). Zhang's group prepared EG-polymethyl methacrylate-acrylic acid copolymer (PMA) and EG-polyglycidyl methacrylate (PGMA) composite particles and found that the RPU foam with 10 wt% of the treated PMA-EG particles maintained good flame retardancy (LOI, 26 vol%; V-1 rating in UL-94), and the PGMA significantly improved the expandability of EG particles from 42 to 70 mL/g (Zhang et al. 2011). EG particles can also be encapsulated with a layer of melamine-formaldehyde (MF) resin via in situ polycondensation. It was found that the expandability of the coated EG particles with the MF resin was significantly enhanced, and thus the coated particles showed good flame-retardant performance in RPU foams (Duan et al. 2014).

EG is an efficient flame retardant in RPU foams, however, the mechanical properties of the foams filled with EG particles were deteriorated. Many efforts have been paid to investigate the synergistic flame retardant effect between EG and other flame retardants in order to reduce EG loadings in RPU foams. It was reported there is a synergistic effect between EG and aluminum hypophosphite (AHP) in RPU foams (Xu et al. 2015d; Yang et al. 2014b). At the weight ratio of EG and AHP 3/1, the LOI value of the RPU/15EG/5AHP foam is higher than that of the RPU/20EG foam or the RPU/20AHP foam. The cone calorimeter results indicated that the peak heat release rate and total heat release of the RPU/15EG/5AHP foam were the lowest among the three. The improved flame retardancy was caused by the combined effect of physical barrier of EG and the catalyzing charring formation of AHP. When EG was partly replaced by AHP, the mechanical strength of the RPU foams was improved compared with the foams filled with EG alone.

Some researchers used reinforcing fillers to increase the mechanical properties of the EG-filled RPU foams. The incorporation of whisker silicon (WSi) and hollow glass microsphere (HGM) particles increased the compressive strength and modulus of RPU foams, and the addition of an appropriate loading of WSi or HGM improved the flame retardancy of EG/RPU foams as well. The dynamical mechanical analysis showed that both particles led to up-shift of the glass transition temperature and storage modulus of RPU foam (Bian et al. 2008a, 2008b). It was also found that appropriate additive contents of HGM and glass fibers (GF) improved the compressive strength and maximum torque of the EG/RPU foams (Cheng et al. 2014). The compressive strength and maximum torque of the RPU/16EG foam were 2.54 MPa and 2.64 N·m, respectively. In contrast, the values reached up to 4.42 MPa and 5.03 N·m, respectively, after appropriate additions of HGM and GF into the above foam. However, HGM and GF had little effect on flame retardancy.

12.2.2.2 Melamine-Based Compounds

Melamine (MEL) and its derivatives are widely used as flame retardant additives in polymers, including RPU foams. MEL and its derivatives usually exert their flame retardant action in both the gas phase and the condensed phase (Xu et al. 2015b). The effect of melamine polyphosphate (MPP) and melamine cyanurate (MC) (Figure 12.4) on the mechanical and flame retardant properties of RPU foams were studied, and it was found that the flame retardancy of MPP-filled RPU foams was better than that of the MC-filled foams (Thirumal et al. 2010). The flame retardancy of MC-filled RPU foams is mainly due to its endothermic decomposition that leads to the evolution of non-combustible gases (e.g., ammonia, N_2) which dilute concentration of combustible gases in the gas phase. Despite the gas phase mechanism, MPP decomposes to form polyphosphoric acid, which promotes the formation of a carbonaceous char layer on the sample surface. In addition, the density and mechanical properties of MPP-filled RPU foams decrease, whereas the properties of MC-filled RPU foams improve at higher loadings which is possibly caused by different cellular structures of RPU foams. The addition of both MPP and MC has less effect on the thermal conductivity of RPU foams, especially in the range of loadings studied. Wang et al. (2017) reported synthesis of a phosphorous-nitrogen intumescent flame-retardant, 2,2-diethyl-1,3-propanediol phosphoryl melamine (DPPM), and found that the RPU foam with 25 phr DPPM achieved a UL-94 V-0 rating with a LOI of 29.5%. A novel nitrogen-phosphorus flame retardant, melamine amino trimethylene phosphate (MATMP, Figure 12.4) was synthesized, and the RPU foams containing 15 wt% MATMP passed the UL-94 V-0 test with a LOI of 25.5% and a 34% reduction of the peak HRR compared with untreated RPU foam. Furthermore, the compressive strength of the RPU foams filled with MATMP first increases and then slightly decreases with an increase in the MATMP content compared with those of the untreated foam (Liu et al. 2017a).

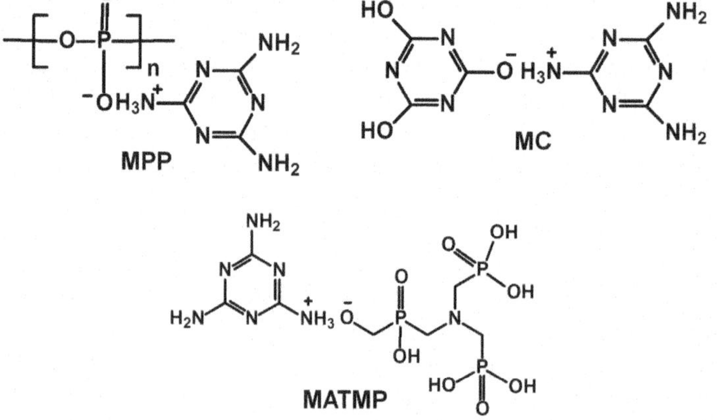

FIGURE 12.4 Structure of melamine polyphosphate (MPP), melamine cyanurate (MC), and amino trimethylene phosphate (MATMP).

12.2.2.3 Organophosphorus Compounds

Organophosphorus compounds, such as phosphines, phosphine oxides, phosphonium compounds, phosphonates, phosphites, and phosphates are an important group of halogen-free flame retardants. Lorenzetti et al. (2011) investigated the gas phase mechanism of dimethylpropanphosphonate (DMPP) and triethylphosphate (TEP) on flame retarded polyurethane foams by means of TG-FTIR, MS, and FTIR spectra. Due to the lower decomposition/volatilization temperatures of TEP and DMPP than those of neat PU foam, they were completely volatilized before polymer decomposition starts and thus no interaction between the flame retardants and the polymer could be expected. The solid residues of PU/TEP and PU/DMPP foams at the end of the TGA analysis showed no significant P content, further confirming that TEP and DMPP had been completely evolved in the gas phase.

9,10-Dihydro-9-oxa-10-phosphaphenanthrene-10-oxide (DOPO) and its derivatives are found to be efficient phosphorus-containing flame retardants and have attracted much attention in recent years. The flame retardant mechanism of DOPO-containing compounds can suppress fire through the release of low molecular weight phosphorous-containing species which are able to scavenge the H· and OH· radicals in the flame. A DOPO-based flame retardant (DOPO-BA) was synthesized from the reaction of benzaldehyde, aniline, and DOPO, and it was found that the LOI value of the RPU foam containing 20 wt% DOPO-BA increased to 28.1% from 20.1% of pure RPU foams (Zhang et al. 2015). In addition, the presence of DOPO-BA also improved the physical properties and thermal stability of RPU foams. Liu et al. (2017b) synthesized a DOPO-based derivative, 6-(2-(4,6-diamino-1,3,5-triazin-2-yl)ethyl) (DTE)-DOPO, and compared it with traditional flame retardant tris(1-chloro-2-propyl) phosphate (TCPP) and a reactive flame retardant (diol) based on oligomeric ethyl ethylene phosphate (PLF140) in RPU foams. They found that DTE-DOPO had not only a lower production of smoke and toxicity, but also an increase in the char residue in RPU foams.

However, some organophosphorus compounds may affect the processing of RPU foams at high loadings. Therefore, many researchers pay more attention to the use of organophosphorus compounds and other flame retardants simultaneously in RPU foams. The effect of the combination of DMMP and EG on flame retardancy of RPU foams was investigated, and the LOI value from 19.2% of the pure RPU foam increased to 33.0% of the RPU foam containing 3.2% DMMP and 12.8% EG (Feng and Qian 2014). Based on LOI results of different DMMP and EG ratios, a synergistic effect between DMMP and EG was confirmed in RPU foams. The DMMP mainly decomposes to gaseous PO_2 fragments which can inhibit the free radical chain reaction in the gas phase, while EG, undoubtedly, shows excellent flame retardant effect in the condensed phase. Therefore, the DMMP/EG system exerts more excellent fire retardant effect than DMMP or EG does alone in the RPU foams. Other researchers also investigated the synergistic effect between pentaerythritol phosphate (PEPA) and EG in RPU foams (Wang et al. 2016a) and between HPCP and EG in RPU foams (Qian et al. 2014).

12.2.2.4 Ammonium Polyphosphate

Ammonium polyphosphate (APP), as an addition-type flame retardant, has wide applications in many polymer materials because it is halogen-free and has a low toxicity and a low cost (Shao et al. 2014). APP usually functions in the condensed phase through catalyzing the char formation (Chen et al. 2015; Zheng et al. 2015). Properties of RPU foams with different types and concentrations of fillers such as APP, borax, aluminum hydroxide (ATH), and MC were compared (Barikani et al. 2010). The results indicated that the density of the filled foams increased with increasing the amount of fillers for most of the used flame retardants. The LOI values of the filled RPU foams with APP were the highest among the used flame retardants at the same loading. The presence of the fillers in the foams resulted in a significant improvement in compressive strength of the filled foams, except for borax. Xu et al. (2015c) investigated the effect of APP, DMMP, or their combination with three nanostructured additives, e.g., zinc oxide (ZnO), zeolite, and montmorillonite (MMT) on flame retardant properties of RPU foams. Their results indicated that FR RPU foams with the nano-additives showed different combustion performances. ZnO and MMT narrowed the heat release peak of the FR RPU foams, but the intensity of the peak did not reduce. The peak HRR of the RPU foam with Zeolite/DMMP/APP was only 91 kW/m^2, which was 56% lower than that of pure RPU foam, and 26% lower than RPU foam only with DMMP/APP. The LOI value of the RPU/8%APP/8%DMMP foam was 29%, and the incorporation of the nano-additives into the above RPU/DMMP/APP foam did lead to the improvement in LOI.

Cellulose whisker (CW) was employed as a novel carbonization agent to build a green and efficient intumescent flame retardant system with APP (Luo et al. 2015). The combination of CW and APP exhibited higher flame retardancy (UL-94 V-0 rating) than that of the foam containing only APP. Moreover, due to the presence of CW, the compressive strength of PU/CW/APP foam was higher than that of PU/APP foam with the same content of flame retardant, which might be attributed to the improvement of compatibility of flame-retardant particles in the PU matrix.

Fly ash (FA) was reported to improve thermal stability and flame retardancy of some polymer materials. Tarakcılar (2015) prepared RPU foams with FA, APP, and pentaerythritol, and found that 5.0 and 7.5 wt% intumescent flame retardant loadings enhanced the thermal stability and improved the flammability retardancy of the foams.

Moreover, there are some publications on the surface-modification of APP in order to improve the compatibility between APP and the PU matrix. For instance, Chen et al. (2017) synthesized a functional surface-modification agent via a reaction between hexachlorocyclo-triphosphazene and aminopropyl triethoxysilane and used it to modify APP. They found that the dispersion of the modified APP (M-APP) in RPU foams was improved, and the addition of 17 wt% modified APP decreased peak HRR by about 51% compared with that of the RPU foam containing the same loading of unmodified APP, and the RPU foams with 17 wt% modified APP or APP reached UL-94 V-0. The influence of polybenzoxazine modified APP (BMAPP) on the thermal behavior and flame retardancy of RPU foams was studied, and the LOI values of RPU foams with BMAPP were all higher than the ones of RPU foams with APP at the same loadings (Luo et al. 2016). Compared with APP, BMAPP can enhance the thermal stability of RPU foams both under air and nitrogen. The microencapsulated APP (MAPP) prepared via in situ polymerization of 4,4-diphenylmethane diisocyanate (MDI) and MEL exhibited a slight increase in LOI and about 20% increase in compressive strength compared with APP in RPU foams (Zheng et al. 2013). The synergistic effect between MAPP/EG or APP/EG in RPU foams is not apparent.

12.2.2.5 Nano-additives

Nano-clay, e.g., MMT is frequently used in many polymer nanocomposites owing to its particular nano-layer structure, which can hinder the mass and heat transfer effectively (Bourbigot et al. 2010; Saha et al. 2008). Zheng et al. (2014) incorporated organically modified montmorillonite (OMMT) as an additional filler in the APP-triphenyl phosphate (TPP)/RPU system to further improve the flame retardancy of RPU foams, and the extraordinary flame retardant performance is attributed to the synergistic effect between the OMMT and the APP-TPP system in forming a mass of integrated, stable, and tight charred layers during the combustion of RPU foams. The effect of clay on the properties of the low density (58–77 kg/m^3) RPU foams was investigated (Cao et al. 2004). In this study, as high as 650% increase in reduced compressive strength was observed in RPU foams with relatively low crosslinking density and urethane content. The compressive modulus and the storage modulus of the RPU foams increased, and the mean cell size decreased with addition of clay, but the hydraulic resistance of the nanocomposite foam was lower than that of the foam without clay. Thirumal et al. (2009) prepared FR RPU foams with organoclay/organically modified nano-clay, and the results showed that the flame retardant properties (LOI and flame spread rate) are improved slightly on the addition (3 phr) of organically modified nano-clay-filled RPU foams. Polyhedral oligomeric silsesquioxane (POSS) consists of a siliceous cage-like core with Si atoms on the vertices and O atoms on the edges. Michałowski et al. (2017) reported the influence of 1,2-propanedi-olisobutyl POSS (PHI-POSS) and octa (3-hydroxy-3-methylbutyldimethylsiloxy) POSS (OCTA-POSS) on the thermal stability and flammability of RPU foams. The results indicated that there is char formation at the surface, and the formed layer acts as an insulating barrier limiting heat and mass transfer with the flame retardant, and thus leading to decreased heat release rate, especially for the foam containing OCTA-POSS. Carbon nanotubes (CNTs) are known to improve the thermal stability of nanocomposites, due to the reduction in the mobility of macromolecules which is correlated with the interfacial interactions (Morgan et al. 2007). Ciecierska et al. (2016) investigated the influence of multi-walled carbon nanotubes (MWCNTs) and graphite on the thermal stability and flammability of RPU foams, and concluded that the increase in LOI for composite foams with MWCNT is probably connected with the formation of a carbon layer on the polymer surface. Graphene, due to its unique two-dimensional layer structure, has exhibited great potential as a nano-filler to enhance the mechanical strengths and thermal properties of polymer foams (Li et al. 2015b, 2016). However, bare graphene has a poor flame retardant effect because of its weak thermal stability during combustion. Some efforts have been devoted to increase the flame retardant efficiency of graphene by surface modification. The flame retardant, mechanical, and thermal properties of flame retardant RPU foams containing DMMP and SiO_2 nanospheres/graphene oxide hybrid (SNGO) were investigated (Wang et al. 2017). The results demonstrated that the combination of DMMP and SNGO enhanced flame retardant and mechanical properties of RPU foams greatly compared with pure RPU foam or DMMP-modified foam. A morphological study indicated that the partial substitution of DMMP with SNGO led to smaller cell sizes and more uniform cell sizes of DMMP-modified RPU foams.

12.2.2.6 Metal Hydroxides

Metal hydroxides (mainly ATH and magnesium hydroxide) are commonly used halogen-free flame retardants. Zhang et al. (2013) introduced ATH and brucite (the mineral form of magnesium hydroxide) into RPU foams and found that the LOI of the RPU foam with 35 wt% fillers increased from 19.8% (pure RPU foam) to 22.6% for ATH and from 19.8% to 21.8% for brucite. Compared to brucite, ATH provided a bit better resistance to combustion in RPU foams because the dehydration temperature of ATH is closer to that of the foams. The compressive strength of the FR RPU foams decreased both with the increase of ATH or brucite content, while the foams filled with brucite showed higher compressive strength than that with ATH. For the RPU foams with 60 wt% hydroxide fillers and 10 wt% DMMP, the LOI values increased to 28.4% for brucite and 32.4% for ATH, respectively. The potential synergy between layered double hydroxide (LDH), EG, and MPP on fire behavior of RPU foams was studied, and the results showed that the LOI value of the RPU foam with a mixture of 10 php (parts per hundred of polyol by weight) EG, 10 php MPP, and 3 php LDH increased from 19.1% (pure RPU foam) to 28.0% (Gao et al. 2013a).

12.3 Flame Retardation of PS Foams

PS foams are generally divided into expanded PS (EPS) foams and extruded PS (XPS) foams according to the type of manufacturing method. EPS foams are produced by two steps. In a first step, the pre-foaming of EPS beads (particles) containing a homogeneously dispersed blowing agent (usually low boiling point hydrocarbons) introduced after completion of the suspension polymerization of styrene is performed in a discontinuous pressurized steam pre-foamer. After the temporary storaging, the pre-expanded beads are molded with steam again to produce EPS foam blocks in the second step. XPS foams are prepared by a continuous foaming extrusion process performed on a screw extruder where a blowing agent (a gas or a low boiling point compound) is introduced into the PS melt.

Nowadays, EPS and XPS foams are widely used as thermal insulation materials for the building and construction sector. These two kinds of foams are very combustible, and thus the flame retardation of XPS and EPS foams is usually needed in this sector.

12.3.1 Flame Retardation of XPS Foams

Traditionally, chlorofluorocarbons (CFCs) due to their low thermal conductivities are used as a blowing agent for XPS foams. However, their use was restricted by the Montreal Protocol in 1987 because

CFCs can deplete the ozone layer. Hydrochlorofluorocarbons (HCFCs) having a low ozone depletion potential (ODP) are now applied in place of CFCs in XPS foams, but they have a high global warming potential (GWP) (Ohara et al. 2004). Therefore, HCFCs will be phased out by 2020. To deal with this problem, some researchers found that low boiling point hydrocarbons (e.g., n-pentane, iso-butane) can be used to substitute for CFCs or HCFCs as a blowing agent in XPS foams. Salehi et al. investigated the effects of operation parameters, e.g., melt temperature and sorption pressure on the foaming process of the PS/n-pentane system and found that the temperature had a great effect on the foaming dynamics and the solubility and diffusivity of n-pentane, as well as melt strength. Moreover, they compared bubble growth behavior of the PS/n-pentane system experimentally and theoretically (Salehi et al. 2016). Compared with CFCs or HCFCs, hydrocarbons such as n-pentane and iso-butane are environmentally friendly blowing agents, however, XPS foams made from the hydrocarbons have two problems: an increase in thermal conductivity and a decrease in flame retardation. To deal with the former problem, people may secure high performance from thermal insulation by means of controlling the cellular structure of the foams. To solve the latter problem, flame retardants are usually incorporated into XPS foams.

The flame retardants used for XPS foams are now mainly halogen-containing flame retardants (e.g., hexabromocyclododecane, HBCD). The HBCD is an efficient flame retardant for both XPS and EPS foams, and antimony trioxide is not applied as a synergist in the HBCD flame retarded PS foams (Wang et al. 2010). According to a 2013/2014 field analytical survey carried out by Jeannerat et al. (2016), the HBCD content in 98 PS foam products was ranging from 0.2% to 2.4% by weight. However, the use of HBCD has led to some serious concerns about its toxicological and environmental impact (Covaci et al. 2006; Messer 2010). Therefore, HBCD was listed as one of the Persistent Organic Pollutants (POPs) by the Stockholm Convention in 2013. China passed the amendment to add HBCD to Annex A of the Stockholm Convention on POPs in 2016 prohibiting its production, use, and import and export, but retains specific exemptions for its production and use in EPS and XPS foams in buildings.

Due to the environmental concern, researchers have been seeking the substitutes for HBCD, especially for their use in flame retarded PS foams in the past decades. Recently the Dow Chemical Company announced the invention and development of a new brominated polymeric flame retardant (Br-PolyFR), and the environmental, health, and safety performances of the new Br-PolyFR were preliminarily evaluated in PS foams (Beach et al. 2013). The researchers from the Dow Chemical Company confirmed that the Br-PolyFR can be used as a direct replacement for HBCD in both XPS and EPS foams, and the foams containing the similar levels of the new flame retardant compared to HBCD have shown no difference in the performances, such as fire retardant, mechanical, and thermal insulating properties. According to the company, the Br-PolyFR is expected to be the "next generation industry standard" flame retardant for use in PS foam insulation applications.

12.3.2 Flame Retardation of EPS Foams

There are two main processes for manufacturing flame retarded EPS foams. The one process is that flame retardants are first introduced into expandable EPS beads (particles) during the polymerization of styrene, and then the flame retardant EPS beads are pre-foamed and molded into flame retarded EPS foams. This method may be suitable for EPS foams with efficient halogen-containing flame retardants, e.g., HBCD, because the introduction of other kinds of flame retardants (e.g., phosphorus-containing compounds) during the polymerization of styrene may influence the molecular weight and properties of PS resin (e.g., T_g) (Levchik et al. 2008).

The other process for manufacturing flame retarded EPS foams is the introduction of flame retardants onto the surface of the pre-foamed EPS beads by means of a binder (e.g., a liquid adhesive) after completion of the polymerization. The flame retardant pre-foamed beads are molded into flame retarded EPS foams. The halogen-free flame retardation of EPS foams is generally carried out by this method.

The binder can be an inorganic adhesive (e.g., soluble silicate) or a polymer adhesive. A thermosetting phenolic resin is usually selected as a binder because the resin has inherent fire retardancy after curing and good compatibility with EPS beads and inorganic fillers (Kandola et al. 2014). For example, Hong et al. (2015) used 10 parts of the mixture containing phenolic resin, aluminum hydroxide, boric acid, and glass fiber, and 1 part of sodium dodecyl sulfate water solution to treat 8 parts of EPS beads to produce composite EPS foams, and they found that the composite foams can melt and retract, but have no dripping when in contact with the flame. We applied the mixture of phenolic resin, curing agent (phosphoric acid and toluene-p-sulfonic acid), APP, and nano-ZrO_2 to treat EPS beads to prepare flame-retarded EPS foams with improved mechanical strengths and good flame retardancy (LOI >31% and UL-94 V-0 rating) (Wang et al. 2016b). Cao et al. (2017) used the mixture (30 g) containing 15 g melamine modified urea formaldehyde (MUF) resin and 15 g intumescent flame retardant (IFR, APP:PER:MEL=3:1:1) to coat 25 g pre-expanded polystyrene particles, and the obtained FR EPS foams can pass UL-94 V-0 test with a LOI of 31.2%. Moreover, they found that the peak heat release rate and maximum smoke density of the FR EPS foams were greatly reduced and thought that their mechanical properties can meet the standard requirements for industrial applications.

Some researchers directly used liquid flame retardants or their combination with other flame retardants to coat pre-expanded PS particles to prepare FR EPS foams. For example, Zhu et al. (2017) synthesized three environmentally friendly FR adhesives, e.g., P(NTMS-phosphoric acid [P(NTMS-PA)], P(NTMS-phosphorous acid) [P(NTMS-POA)], and poly-*N*-beta-(aminoethyl)-gamma-aminopropyltrimethoxysilane [P(NTMS)] (Figure 12.5) based on *N*-β-(aminoethyl)-γ-aminopropyltrimethoxysilane (NTMS) and used them to coat EPS beads to prepare FR EPS foams. Their results indicate that the FR EPS foam with 57 wt% of P(NTMS-PA) coating can pass the UL-94 V-0 rating with a LOI of 31%, the FR foam with 40 wt% of P(NTMS-POA) can reach the V-0 rating with a LOI of 26.5%, and the FR foam with 57 wt% of P(NTMS) gains no rating at all in the UL-94 test. Chen et al. (2014) used a mixture of liquid macromolecular nitrogen–phosphorous compound (MNP) and EG to treat FR EPS beads and found that the modified EPS foam with 20% CPIFR and 10% EG has good flame retardancy (LOI 33.9%, UL-94-V-0).

To sum up, although too many efforts have been made on non-halogenated flame retardation of PS foams, e.g., XPS and EPS foams, they are still not mature enough for large-scale practical applications. The flame retardants used for XPS foams are now mainly halogen-containing flame retardants (e.g., HBCD) in the industry. As for the preparation of FR EPS foams, there are no efficient halogen-free flame retardants to replace HBCD introduced into expandable EPS beads during the polymerization of styrene, which is thought a good way to manufacture FR EPS foams. Non-halogenated FR EPS foams can also be produced with moderate flame retardancy requirements by the surface coating of EPS beads, however some drawbacks exist: (1) flame retardants are not easily distributed evenly on the surface of EPS beads and they easily fall off and (2) the coating by means of adhesives may affect the further formability of the treated EPS particles.

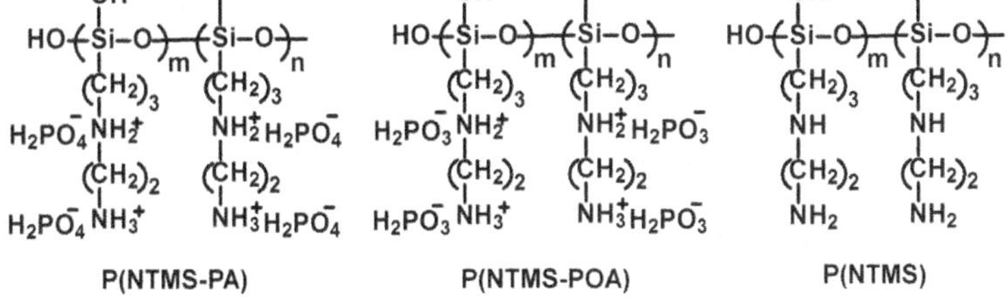

FIGURE 12.5 Structure of P(NTMS-PA), P(NTMS-POA), and P(NTMS).

12.4 Toughening of Phenolic Foams

PF foams are usually prepared by the foaming and curing of a mixture of resol resin, acidic catalyst, surfactant, foaming agent, and other modifiers at moderate temperatures. PF foams have many advantages, such as good heat retardancy, excellent flame retardancy, low smoke and toxic gas release on burning, and good thermal insulation (Lei et al. 2010; Shen et al. 2003). However, the brittleness and friability of PF foams severely restrict their applications, especially in building construction. Therefore, many researchers have been dedicated to improve the toughness of PF foams in the past decades.

There are generally two categories for toughening PF foams: chemical toughening and physical toughening. Chemical toughening is a method to modify the toughness of PF foams by introducing flexible chains into the molecular chain of PF resin, which can be achieved by the partial or full replacement of phenol or formaldehyde with flexible molecules during the preparation of the resin. For example, cardanol or alkyl phenol can be employed as a substitute for phenol, or formaldehyde replaced by glutaraldehyde during the preparation of the PF resin, and then the modified PF resin is used to prepare toughened PF foams. The physical toughening method is carried out by a direct incorporation of toughening agents into the PF foaming mixture by the physical mixing. The commonly used physical toughening agents include polymer resins, fibers, and other modifiers. Some toughening agents, such as epoxy resin, PEG, and polyurethane pre-polymer, however, may deteriorate the excellent flame retardancy of PF foams because those agents are very flammable. Therefore, it is ideal for toughening agents to improve the toughness of PF foams without sacrificing their excellent flame retardancy.

12.4.1 Chemical Toughening

Cardanol is one of the natural phenols, which is a by-product of the cashew nut processing industry. Because of the existence of the long unsaturated chain (C15) in its benzene ring, cardanol is usually employed to partially substitute phenol to prepare a cardanol-phenol-formaldehyde resin, and then the modified PF resin is used to produce the toughened PF foams. For instance, Jing et al. (2014) synthesized bis-phenol by Friedel-Crafts alkylation between cardanol and phenol on the side chain and prepared resol-type prepolymers, and they studied the effects of different amounts of cardanol on the properties of PF foams. They found that when the dosage of cardanol was 10 wt%, the flexural strength and bending modulus of the modified PF foam increased by 22% and 28%, respectively, compared to pure PF foam, indicating that the introduction of cardanol can improve the toughness of PF foams. Liang et al. (2016) first synthesized polyhydroxylated cardanol (PHC) by the reaction of C=C in an unsaturated side chain of cardanol with H_2O_2, and then used it to partly substitute phenol to react with formaldehyde to obtain a modified resol resin (Figure 12.6). When the 20 wt% phenol was replaced by PHC, the compressive strength and flexural strength of the modified PF foam were obviously increased by 57% and 56%, respectively, compared to conventional PF foam.

Lignin and its derivatives are a category of biological and eco-friendly raw materials from natural products, which are also a promising substitute for phenol to prepare modified PF resin (Maldas et al. 1997). Zhuang et al. (2011) fabricated lignin-phenol-formaldehyde (LPF) foam and found that the part replacement of phenol with lignin decreased the brittleness and flexibility of PF foams and slightly increased the LOI. Lee et al. (2002) synthesized a liquefied wood-based resol resin via a reaction of liquefied wood and formaldehyde under alkaline conditions, and the liquefied resin was applied to prepare the foam with improved compressive strength and elastic modulus compared with the conventional PF foam. It was also reported that a phenolated lignosulfonate-modified resol resin was used to prepare the foam with a great enhancement in the compressive properties, e.g., the modulus and compressive strength increasing by 772% and 650%, respectively (Hu et al. 2012). The enhancement of the compressive properties for the modified PF foams may be due to the alkyl side chain in the lignosulfonate structure.

Dicyandiamide can play the same role with phenol to react with formaldehyde in order to increase the toughness of PF resin. It was noted that phenol and dicyandiamide could simultaneously react with paraformaldehyde to prepare phenol-dicyandiamide-formaldehyde (PDF) resin, and the compressive

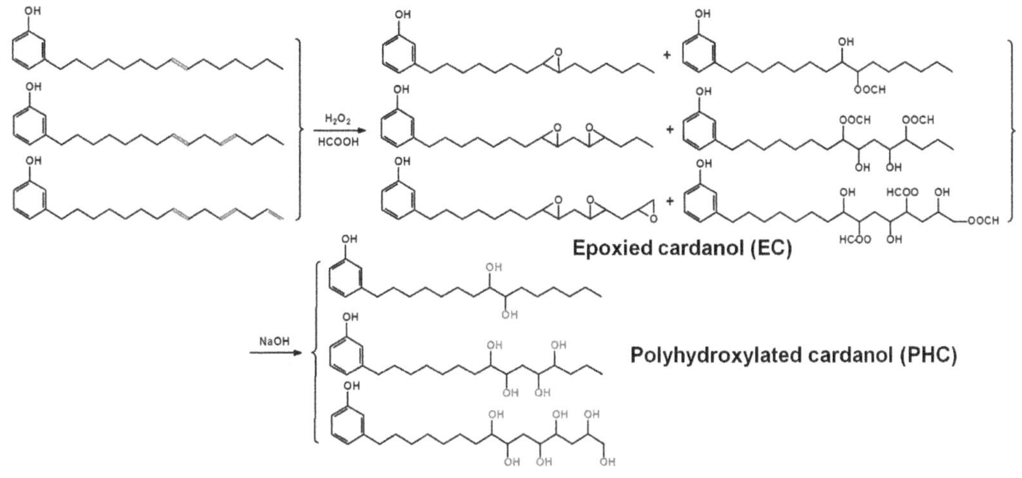

FIGURE 12.6 Synthetic route of polyhydroxylated cardanol (PHC).

strength and impact strength of the foam produced with the PDF resin were dramatically increased by 182% and 43.6%, respectively, compared with those of pure PF foam, while the flame-retardant performance of the modified PF foams were also enhanced (Gao et al. 2016).

12.4.2 Physical Toughening

12.4.2.1 Polymer Resins

Epoxy resin is an effective toughening agent for improving the toughness of PF foams due to its good mechanical property, low shrinkage, and chemical resistance. Because of the etherification reaction between resole methylol, hydroxyl group in PF resin, and ring-opened epoxy group, epoxy resin exhibits a great potential as a toughening agent to reduce the brittleness of PF foams. Many researches have been reported in this aspect. Auad et al. synthesized phenolic foams modified with different types of epoxy resin, and their results indicated that compared with pure phenolic foams, diglycidyl ether of bisphenol A (Epon 826) type and brominated bisphenol A (DER 542) could significantly improve mechanical performance, friability, and resistance to flame of phenolic foams (Auad et al. 2007). Xu et al. (2015a) first synthesized the epoxy-modified phenol-formaldehyde (EPF) resin, and then used the EPF resin to prepare epoxy-modified PF foams. Their results demonstrated that the open-cell foam prepared with 10 wt% epoxy-modified phenolic resin showed greatly improved mechanical strengths (compressive and flexural strengths were 108.195 and 1.64 MPa, respectively), compared to pure PF foam (the values were 0.57 and 0.64 MPa, respectively).

The polypropylene glycol (PPG) or PEG have large numbers of flexible C–O–C units in their molecular structures, therefore they can be employed as the toughening agents in PF foams. It is reported that the compressive and impact strengths of PF foams were dramatically enhanced by the addition of PEG, indicating the excellent toughening effect of PEG (Gao et al. 2013b). In addition, in order to strengthen the interfacial interaction of the PPG with PF resin matrix, the PPG can be modified via the reaction between –NCO and PPG. For example, Niu and Wang (2013) synthesized a series of active polypropylene glycols (APPGs) via the reaction between hexamethylene-1,6-diisocyanate and PPG, and then the APPGs were mixed and reacted with resol resin to obtain the modified PF foams. The results showed that the flexible PPG could be introduced into the crosslink structure of PF foams by the co-crosslinking reaction between –NCO groups in APPGs and –CH$_2$–OH groups in resol resin. When the content of APPG2000 (PPG molecular weight: 2000 g/mol) was 15%, the maximum strain increased by 87%, mass loss (friability) decreased by 52%, and thermal conductivity decreased from 0.033 to 0.028 W/(m·K).

PEG or PPG is a good toughening agent, however, their use usually leads to a decrease in excellent flame retardancy of PF foams. Therefore, many researchers have done some work to prepare flame retardant toughening agents to enhance the toughness while remaining or further increasing flame retardancy of PF foams. For instance, Sui et al. synthesized three kinds of polyethylene glycol phosphates (PEGPs) toughening agents by esterification of phosphorus pentoxide (P_2O_5) with PEG (Sui and Wang 2013). It was found that the incorporation of PEG phosphates not only increased the toughness of PF foams, but also enhanced their fire retardancy and thermal stability. Liu et al. (2015) prepared polyethylene glycol borate (PEG-BAE) toughening agents via the esterification reaction between boric acid and PEG. Compared to pure PF foam, the flexural and compression strength of PEG400-BAE-toughened foam increases by 48.8% and 80%, respectively. Moreover, the flame retardancy of PEG400-BAE-toughened foam was better than that of pure PF foam. Yang et al. (2012) synthesized two kinds of novel phosphorus-containing polyether as toughening agents of PF foams. The results indicated that the addition of 5 wt% toughening agents increased the expansion ratio and promoted the formation of uniform cells. Compared to the pure PF foams, the foam containing 5 wt% PPEG600 had the highest compressive strength which increased by 30.4%. Moreover, the peak heat release rate (PHRR) and total heat release (THR) of the modified foams were reduced significantly.

Polyurethane has flexible chains, and thus it can be incorporated into the molecular structure of phenolic resin, resulting in improving the toughness and reducing the brittleness of PF foams. Yang et al. (2014a) applied post-consumer PU foam scraps (PPU) to strengthen mechanical properties of RPU foam and PF foam, and they observed that the compressive strength of PF foam with 5 wt% PPU (0.48 MPa) increased by 11.6% compared to pure PF foam. Yuan et al. (2013) first synthesized a novel phosphorus-containing polyurethane prepolymer (DOPU, Figure 12.7), and then used DOPU and glass fiber to modify PF foams. The results indicated that the apparent density, compressive strength, and bending strength of the modified phenolic foams tended to increase irregularly with the increasing amount of DOPU, and the LOI value of the modified phenolic foams with 3% DOPU and 0.5% glass fiber still remained at 41.0%. Ding et al. (2015) synthesized a novel phosphorus- and nitrogen-containing polyurethane quasi-prepolymer (PNPUQP) and applied it to toughen PF foams, and they found that the compressive and flexural strengths of the toughened PF foams containing 3 wt% PNPUQP increased by 19.6% and 14.7%, respectively, compared to pure PF foam, meanwhile, the LOI values of the PNPUQP toughened PF foams increased drastically from 36.2% to 43.7% with the increasing loading of PNPUQP from 0 to 8 wt%. Yang et al. (2013a) synthesized a novel phosphorus- and silicon-containing polyurethane prepolymer (PSPUP) by the chemical reaction of phenyl dichlorophosphate with hydroxy-terminated polydimethylsiloxane (HTPDMS) and subsequently with toluene-2,4-diisocyanate. The introduction of PSPUP leads to an increase in the compressive strength and impact strength and a reduction in the pulverization ratio. Moreover, the PSPUP-modified PF foams remained with good flame retardant performances.

12.4.2.2 Fibers

Short fibers reinforcement is an important approach for enhancing the mechanical properties of PF foams. Short chopped glass fibers have been wildly used for the improvement of the strength, toughness, and dimensional stability of PF foams due to their high stiffness property. Nutt et al. did systematic studies on toughening PF foams using fibers (e.g., glass fibers, aramid fibers) (Desai et al. 2008a; Shen and Nutt 2003; Shen et al. 2003). Compared with glass fibers, aramid fibers, such as Kevlar fiber and Nomex fiber exhibit good flexibility and affinity with PF resin, which can obviously increase the toughness without damaging the fire retardancy of PF foams (Shen et al. 2003). The peel and tensile strengths of the PF foams with aramid fibers were obviously higher than that of the foams with glass fibers at the same loading (Shen and Nutt 2003). The PF foams with glass fibers showed higher compression and shear properties than the foams with aramid fibers in parallel direction, and the reinforcement for PF foams from aramid fibers was more effective than that from glass fibers in perpendicular direction. They also found that the PF foams reinforced with glass fibers and aramid fibers exhibited higher shear and compression strengths than the foams with only glass or Nomex fibers. Desai et al. (2008b) also investigated the effect

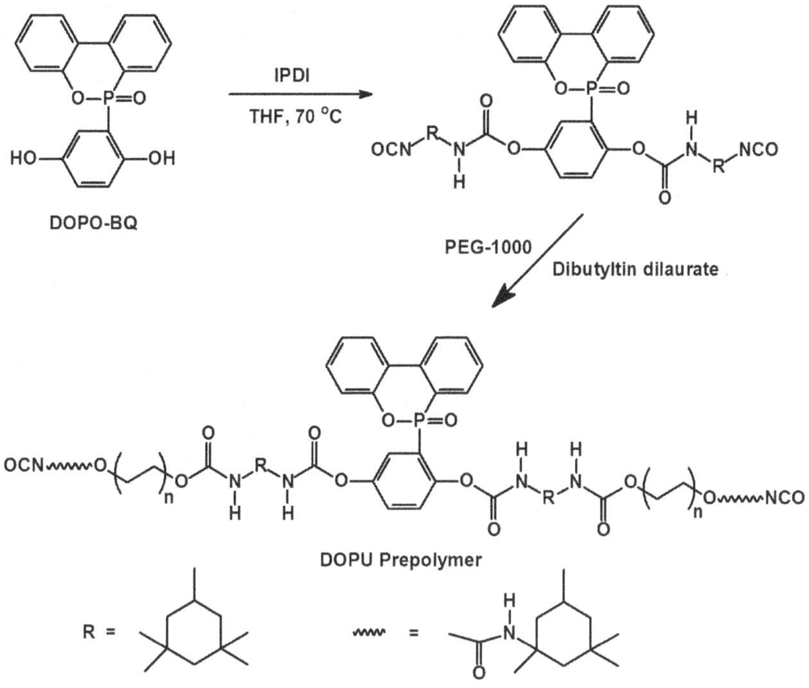

FIGURE 12.7 Synthetic route of DOPU prepolymer.

of glass fibers on the compression properties of PF foam and found that the compressive strength and modulus of the modified PF foams did not depend on the fiber length, and instead these properties were mainly related with the amount of blowing agent and the fiber weight fraction. A statistical predictive model developed by them describes the compressive properties of PF foams reinforced with glass fibers. The conclusion was made that the compressive strength and modulus of the reinforced PF foams did not depend on the fiber length (at least over the range of lengths employed), and instead these properties were mainly related with the amount of blowing agent and the fiber weight fraction (Desai et al. 2008b).

12.4.2.3 Nano-materials

Over the past decades, much attention has been paid to the use of nano-materials as fillers for the improvement of mechanical, thermal, and flame retardant properties of polymer materials. Among nano-materials, nano-clay is well investigated in polymers. It was reported that a sonochemical technique was used to prepare Cloisite clay nanoparticles-modified PF foams. The nanocomposite foam with 2% Cloisite 10A shows about 160% increase in the strength and 182% increase in the modulus, and the dramatic increase may be attributed to the fact that for closed-cell structure foams, the elastic region is controlled by the stretching of the cell faces and edges (Rangari et al. 2007). Moreover, the thermal stability of the clay-modified PF foams was more stable than the neat foam.

Hu et al. (2006) fabricated phenol-urea-formaldehyde foam using nano-clay and glass fiber and proved that nano-clay and glass fiber exhibited good synergistic effects in improving the mechanical strengths and flame retardancy of the composite foam. The impact strength of the composite foam significantly increases with increasing contents of glass fiber and nano-clay, and the pulverization rate of the foam decreases significantly by adding these two kinds of fillers. The incorporation of nano-clay leads to an increase in the thermal stability and a decrease in the maximum heat release rate, total heat release, and total smoke release of the composite foam.

CNTs are a good candidate to reinforce polymers because of their low density, large aspect ratio, excellent mechanical properties, and good thermal stability. There are some reports regarding the use of CNTs to reinforce thermoplastic polymeric foams (Chen et al. 2010, 2011; Zeng et al. 2010). Yang et al. (2013b) reinforced PF foams with functionalized MWCNTs and proved that MWCNTs-reinforced foams have higher compressive strength and thermal stability than the ones of un-reinforced foams and exhibit similar excellent resistance to flame. The carboxyl-modified MWCNTs (MWCNTs-COOH) are the most efficient in raising the compressive strength and specific strength (strength/density) of the reinforced foam among three kinds of MWCNTs (pristine, MWCNTs-COOH, and MWCNTs-NH2). The interfacial compatibility between different MWCNTs and the matrix may play an important role for the external force transfer, resulting in the different reinforcing effect.

Li et al. (2016a) fabricated PF/MWCNTs foam composites by in-situ polymerization. The results indicated that the compressive strength and modulus and the tensile strength and modulus of PF/2wt% MWCNTs nanocomposites dramatically increased by 241.9% and 166.7% and 264.5% and 125.7%, respectively, compared to those of pure PF. Moreover, the pulverization ratios of PF/MWCNTs nanocomposites were all lower than that of pure PF. The thermal conductivity of the nanocomposite with 2 wt% MWCNTs was the lowest.

Graphene, as a two-dimensional carbon layer material, has appealed to many researchers to consider it as a modifier in polymer materials due to its high aspect ratio, chemical stability, and high mechanical strength. Specifically, graphene oxide (GO) has a better compatibility with PF resin because of many oxygen-containing groups such as carbonyl (–COOH), hydroxyl (–OH), and epoxy groups on GO. It was found that the addition of 0.5% GO into PF foam leads to a 16.2% increase in impact toughness and a 26.8% decrease in friability compared with the pure PF foam (Zhou et al. 2014).

The reinforcement of graphene and CNTs on PF foams was compared (Song et al. 2014). The reinforced PF foams were fabricated by means of microwave, and the effects of content of graphene and CNTs particles on the cell morphology, thermal and mechanical properties of PF foams were investigated. The specific compressive strengths of the foams with 0.5% MWCNTs and 1.0% graphene increased by 142% and 157%, respectively, compared to that of the neat PF foam. It was concluded that the reinforcement of graphene on PF foams with a proper resin viscosity and a good dispersion state of the particles before the microwave foaming is a promising way to improve the thermal and mechanical performance of the foams.

However, GO is less thermally stable because there are some labile oxygen-containing functionalities (e.g., –OH, –COOH) on its surface (Yuan et al. 2016a). In order to increase the thermal stability of GO, some thermally stable substances can be doped on its surface. It is noted that the incorporation of the SiO_2/GO hybrid into PF foams greatly improves the mechanical properties, thermal stability, and flame retardancy of PF foams. The flexural and compressive strengths of the PF foam with 1.5 phr SiO_2/GO hybrid increased by 31.6% and by 36.2%, respectively, compared to the pure PF foam, and the addition of SiO_2/GO hybrid into the PF foam also reduced the peak heat release rate and total heat release of the PF foam (Li et al. 2015b). Moreover, the introduction of a small amount of α-zirconium phosphate/graphene oxide (ZGO) hybrid into the PF foam leads to a significant enhancement in the flexural and compressive strengths and a great reduction of pulverization ratio, while the thermal stability and flame retardancy of PF foams are enhanced (Li et al. 2016b).

12.4.2.4 Other Kinds of Nanoparticles

PF foams can also be reinforced by some other nanoparticles, e.g., silica nanoparticles. It was reported that phenolic/silica nanocomposite foams with different silica contents were synthesized (Li et al. 2015b). Phenolic/silica nanocomposites were first synthesized by the introduction of silica sol during the synthesis of the PF resin, and then they were mixed with *n*-pentane, Tween80, and *p*-toluene sulfonic acid to prepare the nanocomposite foams according to the same method for pure PF foam. It was found that the cell sizes of the nanocomposite foams were more uniform and smaller compared to those of pure PF foam, and the compressive strength, compressive modulus, and tensile

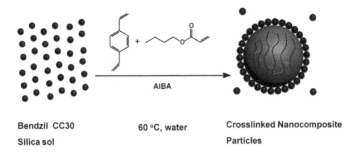

FIGURE 12.8 Preparation of crosslinked poly(n-butyl acrylate)/silica core-shell nanocomposite particles. (Reprinted from Yuan, J. et al., *J. Appl. Polym. Sci.*, 132, 42590–42598, 2015. With permission.)

strength increased by 47.37%, 38.55%, and 57.14%, respectively, when the amount of silica sol was 2 wt%. The nanocomposite foams also exhibited better thermal stability and lower pulverization ratio than those of pure PF foam.

A new type of crosslinked PBA (poly butyl acrylate)/silica core-shell nanocomposite particles were employed as a toughening agent for PF foams (Figure 12.8) (Yuan et al. 2015). The effects of contents of the nanocomposite particles on the structure and properties of the PF foams were investigated. SEM results showed that the addition of a small quantity of the nanocomposite particles obviously enhanced the structural homogeneity of PF foams. The maximum flexural strength and compressive strength of the PF foam modified with only 0.03 phr of the nanocomposite particles increased by 36.0% and 42.9%, respectively, compared to the values of pure PF foam. Moreover, the effect of the nanocomposite particles on the flammability and thermal stability of toughened phenolic foams can almost be neglected.

Recently, the toughening of PF foams using natural materials has also drawn a great deal of attention from many researchers because these kinds of fibers have many advantages, such as low density, biodegradability, recyclability, and low cost (John and Thomas 2008). Del Saz-Orozco et al. (2014) applied wood flour into PF foams to improve their compressive properties and found that the compressive strength and modulus strength of the modified PF foam with 1.5 wt% wood flour reached 154% and 130% of that for the PF foam, respectively. They also investigated the effect of lignin particles and wood flour on PF foams, and their results demonstrated that the addition of lignin particles decreased friability of the PF foam, whereas wood flour increased it (Del Saz-Orozco et al. 2015a, 2015b). Cellulose fiber is another kind of effective natural toughening agent for PF foams. The incorporation of 2 wt% cellulose fiber into PF foams leads to an 18% and 21% increase of the compressive strength and modulus, respectively, compared with those of the unreinforced PF foam.

However, the commonly used polymer toughening agents always sacrifice the excellent fire retardant performance of PF foams. Recently, some FR polymer toughening agents have been prepared as toughening agents for PF foams with less negative effect on flammability. An appropriate amount of solid fillers such as fibers and nano-materials in PF foams can strengthen the cellular walls of the foams and thus lead to an improvement in mechanical strengths and some other performances (e.g., flame retardancy, friability). However, the dispersion of solid fillers, especially some nano-fillers in the foams is of essential importance. The surface modification of solid fillers is usually needed to enhance the compatibility between the filler and the PF resin.

12.5 Conclusion and Perspectives

As for reactive-type flame retardation of RPU foams, the flame retardant diols or polyols in combination with regular diols or polyols are usually applied. However, how the FR diols or polyols participate in or affect the foaming process of RPU foams is still not quite clear. Moreover, the disadvantages of reactive flame retardants, such as complex preparation process and high production cost, restrict their wide application to a certain extent. At present, additive-type flame retardants are commonly used to produce FR RPU foams. Some halogen-free flame retardants (i.e., EG) may affect mechanical properties of RPU foams heavily. The combination of reactive-type flame retardants and additive-type flame retardants may be a better choice for flame retardation of RPU foams in the future.

Although many researchers have paid too much attention to halogen-free flame retardation of PS foams, i.e., XPS and EPS foams, there are no suitable flame retardants to substitute for HBCD for large-scale practical applications nowadays. The use of surface coating of pre-foamed EPS beads by flame retardants themselves or flame retardant adhesives to produce FR EPS foams also has some drawbacks. The coating may affect the further formability of the treated EPS beads.

The commonly used toughening agents for PF foams, such as PEG and polyurethane pre-polymer can greatly improve the toughness of PF foams, but deteriorate the excellent flame retardancy of PF foams. The application of FR polymer toughening agents to toughen PF foams not only increases the toughness, but also increase (or maintain) the excellent fire retardant performance of the foams. Some solid fillers including nano-materials can simultaneously improve mechanical strengths, flame retardancy, and friability. However, their uniform dispersion in the PF resin matrix is of essential importance, and thus their surface modification is usually essential.

References

Al-Homoud, M. S. 2005. Performance characteristics and practical applications of common building thermal insulation materials. *Building and Environment* 40: 353–366.

Auad, M. L., Zhao, L., Shen, H. et al. 2007. Flammability properties and mechanical performance of epoxy modified phenolic foams. *Journal of Applied Polymer Science* 104: 1399–1407.

Barikani, M., Askari, F., and Barmar, M. 2010. A comparison of the effect of different flame retardants on the compressive strength and fire behaviour of rigid polyurethane foams. *Cellular Polymers* 29: 343–358.

Bian, X. C., Tang, J. H., and Li, Z. M. 2008a. Flame retardancy of whisker silicon oxide/rigid polyurethane foam composites with expandable graphite. *Journal of Applied Polymer Science* 110: 3871–3879.

Bian, X. C., Tang, J. H., and Li, Z. M. 2008b. Flame retardancy of hollow glass microsphere/rigid polyurethane foams in the presence of expandable graphite. *Journal of Applied Polymer Science* 109: 1935–1943.

Beach, M. W., Beaudoin, D. A., Beulich, I. et al. 2013. New class of brominated polymeric flame retardants for use in polystyrene foams. *Cellular Polymers* 32: 229–236.

Bourbigot, S., Samyn, F., Turf, T. et al. 2010. Nanomorphology and reaction to fire of polyurethane and polyamide nanocomposites containing flame retardants. *Polymer Degradation and Stability* 95: 320–326.

Cao, B., Gu X. Y., Song, X. H. et al. 2017. The flammability of expandable polystyrene foams coated with melamine modified urea formaldehyde resin. *Journal of Applied Polymer Science* 134: 44423.

Cao, X., Lee, L. J., Widya, T. et al. 2004. Structure and properties of polyurethane/clay nanocomposites and foams. *Chemical Engineering and Materials Science* 2: 1896–1900.

Chen, L., Ozisik, R., and Schadler, L. S. 2010. The influence of carbon nanotube aspect ratio on the foam morphology of MWNT/PMMA nanocomposite foams. *Polymer* 51: 2368–2375.

Chen, L., Schadler, L. S., and Ozisik, R. 2011. An experimental and theoretical investigation of the compressive properties of multi-walled carbon nanotube/poly(methyl methacrylate) nanocomposite foams. *Polymer* 52: 2899–2909.

Chen, X., Jiang, Y., Liu, J. et al. 2015. Smoke suppression properties of fumed silica on flame-retardant thermoplastic polyurethane based on ammonium polyphosphate. *Journal of Thermal Analysis and Calorimetry* 120: 1493–1501.

Chen, X., Liu, Y., Bai, S. et al. 2014. Macromolecular nitrogen-phosphorous compound/expandable graphite synchronous expansion flame retardant polystyrene foam. *Journal of Macromolecular Science: Part D- Reviews in Polymer Processing* 53: 1402–1407.

Chen, Y., Li, L., Wang, W. et al. 2017. Preparation and characterization of surface-modified ammonium polyphosphate and its effect on the flame retardancy of rigid polyurethane foam. *Journal of Applied Polymer Science* 134: 45369–45378.

Cheng, J. J., Shi, B. B., Zhou, F. B. et al. 2014. Effects of inorganic fillers on the flame-retardant and mechanical properties of rigid polyurethane foams. *Journal of Applied Polymer Science* 131: 40253.

Ciecierska, E., Jurczyk-Kowalska, M., Bazarnik, P. et al. 2016. The influence of carbon fillers on the thermal properties of polyurethane foam. *Journal of Thermal Analysis and Calorimetry* 123: 283–291.

Covaci A., Gerecke A. C., Law R. J. et al. 2006. Hexabromocyclododecanes (HBCDs) in the environment and humans: A review. *Environmental Science and Technology* 40: 3679–3688.

Czupryński, B., and Paciorek, J. 1999. The effect of tri(2-hydroxypropyl) borate on the properties of rigid polyurethane—Polyisocyanurate foams. *Polimery* 44: 552–554.

Czupryński, B., Sadowska, J. P., and Liszkowska, J. 2002. Effect of selected boranes on properties of rigid polyurethane-polyisocyanurate foams. *Journal of Polymer Engineering* 22: 59–74.

Del Saz-Orozco, B., Virginia Alonso, M., Oliet, M. et al. 2014. Effects of formulation variables on density, compressive mechanical properties and morphology of wood flour-reinforced phenolic foams. *Composites Part B-Engineering* 56: 546–552.

Del Saz-Orozco, B., Virginia Alonso, M., Oliet, M. et al. 2015a. Lignin particle- and wood flour-reinforced phenolic foams: Friability, thermal stability and effect of hygrothermal aging on mechanical properties and morphology. *Composites Part B-Engineering* 80: 154–161.

Del Saz-Orozco, B., Virginia Alonso, M., Oliet, M. et al. 2015b. Mechanical, thermal and morphological characterization of cellulose fiber-reinforced phenolic foams. *Composites Part B-Engineering* 75: 367–372.

Desai, A., Auad, M. L., Shen, H. et al. 2008a. Mechanical behavior of hybrid composite phenolic foam. *Journal of Cellular Plastics* 44: 15–36.

Desai, A., Nutt, S. R., and Alonso, M. V. 2008b. Modeling of fiber-reinforced phenolic foam. *Journal of Cellular Plastics* 44: 391–413.

Ding, H., Wang, J., Liu, J. et al. 2015. Preparation and properties of a novel flame retardant polyurethane quasi-prepolymer for toughening phenolic foam. *Journal of Applied Polymer Science* 132: 42424–42432.

Duan, H. J., Kang, H. Q., Zhang, W. Q. et al. 2014. Core-shell structure design of pulverized expandable graphite particles and their application in flame-retardant rigid polyurethane foams. *Polymer International* 63: 72–83.

Feng, F., and Qian, L. 2014. The flame retardant behaviors and synergistic effect of expandable graphite and dimethyl methylphosphonate in rigid polyurethane foams. *Polymer Composites* 35: 301–309.

Gao, L., Zheng, G., Zhou, Y. et al. 2013a. Synergistic effect of expandable graphite, melamine polyphosphate and layered double hydroxide on improving the fire behavior of rosin-based rigid polyurethane foam. *Industrial Crops and Products* 50: 638–647.

Gao, M., Wu, W., Wang, Y. et al. 2016. Phenolic foam modified with dicyandiamide as toughening agent. *Journal of Thermal Analysis and Calorimetry* 124: 189–195.

Gao, M., Yang, Y. L., and Xu, Z. Q. 2013b. Mechanical and flame retardant properties of phenolic foam modified with polyethylene glycol as toughening agent. *Advanced Materials Research* 803: 21–25.

Heinen, M., Gerbase, A. E., and Petzhold, C. L. 2014. Vegetable oil-based rigid polyurethanes and phosphorylated flame-retardants derived from epoxydized soybean oil. *Polymer Degradation and Stability* 108: 76–86.

Hong, Y., Fang, X., and Yao, D. 2015. Processing of composite polystyrene foam with a honeycomb structure. *Polymer Engineering and Science* 55: 1494–1503.

Hu, L., Zhou, Y., Zhang, M. et al. 2012. Characterization and properties of a lignosulfonate-based phenolic foam. *Bioresources* 7: 554–564.

Hu, X., Cheng, W., Nie, W. et al. 2016. Flame retardant, thermal, and mechanical properties of glass fiber/nanoclay reinforced phenol-urea-formaldehyde foam. *Polymer Composites* 37: 2323–2332.

Ionescu, M., Wan, X., Bilić, N. et al. 2012. Polyols and rigid polyurethane foams from cashew nut shell liquid. *Journal of Polymers and the Environment* 20: 647–658.

Jeannerat, D., Pupier, M., Schweizer, S. et al. 2016. Discrimination of hexabromocyclododecane from new polymeric brominated flame retardant in polystyrene foam by nuclear magnetic resonance. *Chemosphere* 144: 1391–1397.

Jing, S., Li, T., Li, X. et al. 2014. Phenolic foams modified by cardanol through bisphenol modification. *Journal of Applied Polymer Science* 131: 39942–39949.

John, M. J., and Thomas, S. 2008. Biofibres and biocomposites. *Carbohydrate Polymers* 71: 343–364.

Kandola, B. K., Krishnan, L., and Ebdon, J. R. 2014. Blends of unsaturated polyester and phenolic resins for application as fire-resistant matrices in fibre-reinforced composites: Effects of added flame retardants. *Polymer Degradation and Stability* 106: 129–137.

Lee, S. H., Teramoto, Y., and Shiraishi, N. 2002. Resol-type phenolic resin from liquefied phenolated wood and its application to phenolic foam. *Journal of Applied Polymer Science* 84: 468–472.

Lei, S., Guo, Q., Zhang, D. et al. 2010. Preparation and properties of the phenolic foams with controllable nanometer pore structure. *Journal of Applied Polymer Science* 117: 3545–3550.

Levchik, S. V., and Weil, E. D. 2004. Thermal decomposition, combustion and fire-retardancy of epoxy resins-A review of the recent literature. *Polymer International* 53: 1901–1929.

Levchik, S. V., and Weil, E. D. 2008. New developments in flame retardancy of styrene thermoplastics and foams. *Polymer International* 57: 431–448.

Li, Q., Chen, L., Li, X. et al. 2016a. Effect of multi-walled carbon nanotubes on mechanical, thermal and electrical properties of phenolic foam via in-situ polymerization. *Composites Part A-Applied Science and Manufacturing* 82: 214–225.

Li, Q., Chen, L., Zhang, J. et al. 2015a. Enhanced mechanical properties, thermal stability of phenolic-formaldehyde foam/silica nanocomposites via in situ polymerization. *Polymer Engineering and Science* 55: 2783–2793.

Li, X., Wang, Z., and Wu, L. 2015b. Preparation of a silica nanospheres/graphene oxide hybrid and its application in phenolic foams with improved mechanical strengths, friability and flame retardancy. *RSC Advances* 5: 99907–99913.

Li, X., Wang, Z., Wu, L. et al. 2016b. One-step in situ synthesis of a novel α-zirconium phosphate/graphene oxide hybrid and its application in phenolic foam with enhanced mechanical strength, flame retardancy and thermal stability. *RSC Advances* 6: 74903–74912.

Liang, B., Li, X., Hu, L. et al. 2016. Foaming resol resin modified with polyhydroxylated cardanol and its application to phenolic foams. *Industrial Crops and Products* 80: 194–196.

Liu, L., Fu, M., and Wang, Z. 2015. Synthesis of boron-containing toughening agents and their application in phenolic foams. *Industrial & Engineering Chemistry Research* 54: 1962–1970.

Liu, L., Wang, Z., and Xu, X. 2017a. Melamine amino trimethylene phosphate as a novel flame retardant for rigid polyurethane foams with improved flame retardant, mechanical and thermal properties. *Journal of Applied Polymer Science* 134: 45234.

Liu, X., Salmeia, K. A., Rentsch, D. et al. 2017b. Thermal decomposition and flammability of rigid PU foams containing some DOPO derivatives and other phosphorus compounds. *Journal of Analytical and Applied Pyrolysis* 124: 219–229.

Liu, Y., He, J., and Yang, R. 2016. The synthesis of melamine-based polyether polyol and its effects on the flame retardancy and physical–mechanical property of rigid polyurethane foam. *Journal of Materials Science* 52: 1–13.

Lorenzetti, A., Modesti, M., Besco, S. et al. 2011. Influence of phosphorus valency on thermal behaviour of flame retarded polyurethane foams. *Polymer Degradation and Stability* 96: 1455–1461.

Luo, F., Wu, K., Guo, H. et al. 2015. Effect of cellulose whisker and ammonium polyphosphate on thermal properties and flammability performance of rigid polyurethane foam. *Journal of Thermal Analysis and Calorimetry* 122: 717–723.

Luo, F., Wu, K., and Lu, M. 2016. Enhanced thermal stability and flame retardancy of polyurethane foam composites with polybenzoxazine modified ammonium polyphosphates. *RSC Advances* 6: 13418–13425.

Maldas, D., Shiraishi, N., and Harada, Y. 1997. Phenolic resol resin adhesives prepared from alkali-catalyzed liquefied phenolated wood and used to bond hardwood. *Journal of Adhesion Science and Technology* 11: 305–316.

Meng, X. Y., Ye, L., Zhang, X. G. et al. 2009. Effects of expandable graphite and ammonium polyphosphate on the flame-retardant and mechanical properties of rigid polyurethane foams. *Journal of Applied Polymer Science* 114: 853–863.

Messer, A. 2010. Mini-review: Polybrominated diphenyl ether (PBDE) flame retardants as potential autism risk factors. *Physiology and Behavior* 100: 245–259.

Michałowski, S., Hebda, E., and Pielichowski, K. 2017. Thermal stability and flammability of polyurethane foams chemically reinforced with POSS. *Journal of Thermal Analysis and Calorimetry* 130: 155–163.

Modesti, M., and Lorenzetti, A. 2002a. Halogen-free flame retardants for polymeric foams. *Polymer Degradation and Stability* 78: 167–173.

Modesti, M., and Lorenzetti, A. 2002b. Flame retardancy of polyisocyanurate–polyurethane foams: Use of different charring agents. *Polymer Degradation and Stability* 78: 341–347.

Modesti, M., and Lorenzetti, A. 2003. Improvement on fire behaviour of water blown PIR–PUR foams: Use of an halogen-free flame retardant. *European Polymer Journal* 39: 263–268.

Modesti, M., Lorenzetti, A., Simioni, F. et al. 2002. Expandable graphite as an intumescent flame retardant in polyisocyanurate–polyurethane foams. *Polymer Degradation and Stability* 77: 195–202.

Niu, M., and Wang, G. 2013. The preparation and performance of phenolic foams modified by active polypropylene glycol. *Cellular Polymers* 32: 155–171.

Ohara, Y., Tanaka, K., Hayashi, T. et al. 2004. The development of a non-fluorocarbon-based extruded polystyrene foam which contains a halogen-free blowing agent. *Bulletin of the Chemical Society of Japan* 77: 599–605.

Paciorek-Sadowska, J., Ski, B. C., and Liszkowska, J. 2015a. Boron-containing fire retardant rigid polyurethane-polyisocyanurate foams: Part I—Polyol precursors based on boric acid and di(hydroxymethyl)urea derivatives. *Journal of Fire Sciences* 33: 37–47.

Paciorek-Sadowska, J., Ski, B. C., and Liszkowska, J. 2015b. Boron-containing fire retardant rigid polyurethane-polyisocyanurate foams: Part II—Preparation and evaluation. *Journal of Fire Sciences* 33: 48–68.

Qian, L., Feng, F., and Tang, S. 2014. Bi-phase flame-retardant effect of hexa-phenoxy-cyclotriphosphazene on rigid polyurethane foams containing expandable graphite. *Polymer* 55: 95–101.

Rangari, V. K., Hassan, T. A., Zhou, Y. et al. 2007. Cloisite clay-infused phenolic foam nanocomposites. *Journal of Applied Polymer Science* 103: 308–314.

Saha, M. C., Kabir, M. E., and Jeelani, S. 2008. Enhancement in thermal and mechanical properties of polyurethane foam infused with nanoparticles. *Materials Science and Engineering A* 479: 213–222.

Salehi, M., Rezaei, M., and Hosseini, M. S. 2016. Experimental and theoretical investigation on polystyrene/n-pentane foaming process. *International Journal of Material Forming* 10: 1–14.

Shao, Z. B., Deng, C., Tan, Y. et al. 2014. An efficient mono-component polymeric intumescent flame retardant for polypropylene: Preparation and application. *ACS Applied Materials and Interfaces* 6: 7363–7370.

Shen, H. B., Lavoie, A. J., and Nutt, S. R. 2003. Enhanced peel resistance of fiber reinforced phenolic foams. *Composites Part A-Applied Science and Manufacturing* 34: 941–948.

Shen, H. B., and Nutt, S. 2003. Mechanical characterization of short fiber reinforced phenolic foam. *Composites Part A-Applied Science and Manufacturing* 34: 899–906.

Shi, L., Li, Z. M., Xie, B. H. et al. 2006. Flame retardancy of different sized expandable graphite particles for high-density rigid polyurethane foams. *Polymer International* 55: 862–871.

Shi, L., Li, Z. M., Yang, M. B. et al. 2005. Expandable graphite for halogen-free flame-retardant of high-density rigid polyurethane foams. *Polymer-Plastics Technology and Engineering* 44: 1323–1337.

Singh, H., and Jain, A. K. 2009. Ignition, combustion, toxicity, and fire retardancy of polyurethane foams: A comprehensive review. *Journal of Applied Polymer Science* 111: 1115–1143.

Song, S. A., Chung, Y. S., and Kim, S. S. 2014. The mechanical and thermal characteristics of phenolic foams reinforced with carbon nanoparticles. *Composites Science and Technology* 103: 85–93.

Sui, X., and Wang, Z. 2013. Flame-retardant and mechanical properties of phenolic foams toughened with polyethylene glycol phosphates. *Polymers for Advanced Technologies* 24: 593–599.

Tarakcılar, A. R. 2015. The effects of intumescent flame retardant including ammonium polyphosphate/pentaerythritol and fly ash fillers on the physicomechanical properties of rigid polyurethane foams. *Journal of Applied Polymer Science* 120: 2095–2102.

Thirumal, M., Khastgir, D., Nando, G. B. et al. 2010. Halogen-free flame retardant PUF: Effect of melamine compounds on mechanical, thermal and flame retardant properties. *Polymer Degradation and Stability* 95: 1138–1145.

Thirumal, M., Khastgir, D., Singh, N. K. et al. 2009. Effect of a nanoclay on the mechanical, thermal and flame retardant properties of rigid polyurethane foam. *Journal of Macromolecular Science, Part A: Pure and Applied Chemistry* 46: 704–712.

Wang, C., Wu, Y., Li, Y. et al. 2017. Flame-retardant rigid polyurethane foam with a phosphorus-nitrogen single intumescent flame retardant. *Polymers for Advanced Technologies* 29: 668–676.

Wang, J. Q., and Chow, W. K. 2010. A brief review on fire retardants for polymeric foams. *Journal of Applied Polymer Science* 97: 366–376.

Wang, S., Qian, L., and Xin, F. 2016a. The synergistic flame-retardant behaviors of pentaerythritol phosphate and expandable graphite in rigid polyurethane foams. *Polymer Composites* 39: 329–336.

Wang, X., Pan, Y. T., Wan, J. T. et al. 2014. An eco-friendly way to fire retardant flexible polyurethane foam: Layer-by-layer assembly of fully bio-based substances. *RSC Advances* 4: 46164–46169.

Wang, Z., Jiang, S., and Sun, H. 2016b. Expanded polystyrene foams containing ammonium polyphosphate and nano-zirconia with improved flame retardancy and mechanical properties. *Iranian Polymer Journal* 26: 1–9.

Weil, E. D., and Levchik, S. V. 2004. Commercial flame retardancy of polyurethanes. *Journal of Fire Sciences* 22: 183–210.

Wu, D., Zhao, P., Zhang, M. et al. 2013. Preparation and properties of flame retardant rigid polyurethane foam with phosphorus-nitrogen intumescent flame retardant. *High Performance Polymers* 25: 868–875.

Wu, J. W., Sung, W. F., and Chu, H. S. 1999. Thermal conductivity of polyurethane foams. *International Journal of Heat and Mass Transfer* 42: 2211–2217.

Xu, Q., Gong, R., Cui, M. Y. et al. 2015a. Preparation of high-strength microporous phenolic open-cell foams with physical foaming method. *High Performance Polymers* 27: 852–867.

Xu, Q., Zhai, H., and Wang, G. 2015b. Mechanism of smoke suppression by melamine in rigid polyurethane foam. *Fire and Materials* 39: 271–282.

Xu, W., Wang, G., and Zheng, X. 2015c. Research on highly flame-retardant rigid PU foams by combination of nanostructured additives and phosphorus flame retardants. *Polymer Degradation and Stability* 111: 142–150.

Xu, W. Z., Liu, L., Wang, S. Q. et al. 2015d. Synergistic effect of expandable graphite and aluminum hypophosphite on flame-retardant properties of rigid polyurethane foam. *Journal of Applied Polymer Science* 132: 43842.

Yang, C., Zhuang, Z. H., and Yang, Z. G. 2014a. Pulverized polyurethane foam particles reinforced rigid polyurethane foam and phenolic foam. *Journal of Applied Polymer Science* 131: 39734–39741.

Yang, H., Wang, X., Song, L. et al. 2014b. Aluminum hypophosphite in combination with expandable graphite as a novel flame retardant system for rigid polyurethane foams. *Polymers for Advanced Technologies* 25: 1034–1043.

Yang, H., Wang, X., Yu, B. et al. 2013a. A novel polyurethane prepolymer as toughening agent: Preparation, characterization, and its influence on mechanical and flame retardant properties of phenolic foam. *Journal of Applied Polymer Science* 128: 2720–2728.

Yang, H., Wang, X., Yuan, H. et al. 2012. Fire performance and mechanical properties of phenolic foams modified by phosphorus-containing polyethers. *Journal of Polymer Research* 19: 9831–9841.

Yang, R., Hu, W., Xu, L. et al. 2015. Synthesis, mechanical properties and fire behaviors of rigid polyurethane foam with a reactive flame retardant containing phosphazene and phosphate. *Polymer Degradation and Stability* 122: 102–109.

Yang, R., Wang, B., Han, X. et al. 2017. Synthesis and characterization of flame retardant rigid polyurethane foam based on a reactive flame retardant containing phosphazene and cyclophosphonate. *Polymer Degradation and Stability* 144: 62–69.

Yang, Z., Yuan, L., Gu, Y. et al. 2013b. Improvement in mechanical and thermal properties of phenolic foam reinforced with multiwalled carbon nanotubes. *Journal of Applied Polymer Science* 130: 1479–1488.

Ye, L., Meng, X. Y., Ji, X. et al. 2009. Synthesis and characterization of expandable graphite–poly(methyl methacrylate) composite particles and their application to flame retardation of rigid polyurethane foams. *Polymer Degradation and Stability* 94: 971–979.

Ye, L., Meng, X. Y., Liu, X. M. et al. 2010. Flame-retardant and mechanical properties of high-density rigid polyurethane foams filled with decabrominated dipheny ethane and expandable graphite. *Journal of Applied Polymer Science* 111: 2372–2380.

Yuan, B., Xing, W., Hu, Y. et al. 2016a. Boron/phosphorus doping for retarding the oxidation of reduced graphene oxide. *Carbon* 101: 152–158.

Yuan, H., Xing, W., Yang, H. et al. 2013. Mechanical and thermal properties of phenolic/glass fiber foam modified with phosphorus-containing polyurethane prepolymer. *Polymer International* 62: 273–279.

Yuan, J., Zhang, Y., and Wang, Z. 2015. Phenolic foams toughened with crosslinked poly(n-butyl acrylate)/silica core-shell nanocomposite particles. *Journal of Applied Polymer Science* 132: 42590–42598.

Yuan, Y., Yang, H., Yu, B. et al. 2016b. Phosphorus and nitrogen-containing polyols: Synergistic effect on the thermal property and flame retardancy of rigid polyurethane foam composites. *Industrial and Engineering Chemistry Research* 55: 10813–10822.

Zeng, C., Hossieny, N., Zhang, C. et al. 2010. Synthesis and processing of PMMA carbon nanotube nanocomposite foams. *Polymer* 51: 655–664.

Zhang, A., Zhang, Y., Lv, F. et al. 2013. Synergistic effects of hydroxides and dimethyl methylphosphonate on rigid halogen-free and flame-retarding polyurethane foams. *Journal of Applied Polymer Science* 128: 347–353.

Zhang, L., Zhang, M., Hu, L. et al. 2014. Synthesis of rigid polyurethane foams with castor oil-based flame retardant polyols. *Industrial Crops and Products* 52: 380–388.

Zhang, M., Luo, Z., Zhang, J. et al. 2015. Effects of a novel phosphorus–nitrogen flame retardant on rosin-based rigid polyurethane foams. *Polymer Degradation and Stability* 120: 427–434.

Zhang, X. G., Ge, L. L., Zhang, W. Q. et al. 2011. Expandable graphite-methyl methacrylate-acrylic acid copolymer composite particles as a flame retardant of rigid polyurethane foam. *Journal of Applied Polymer Science* 122: 932–941.

Zhao, B., Liu, D. Y., Liang, W. J. et al. 2017. Bi-phase flame-retardant actions of water-blown rigid polyurethane foam containing diethyl-N, N-bis(2-hydroxyethyl) phosphoramide and expandable graphite. *Journal of Analytical and Applied Pyrolysis* 124: 247–255.

Zheng, X., Wang, G., and Xu, W. 2014. Roles of organically-modified montmorillonite and phosphorous flame retardant during the combustion of rigid polyurethane foam. *Polymer Degradation and Stability* 101: 32–39.

Zheng, Z., Cui, X., and Wang, H. 2015. Preparation of a novel phosphorus-and nitrogen-containing flame retardant and its synergistic effect in the intumescent flame-retarding polypropylene system. *Polymer Composites* 36: 1606–1619.

Zheng, Z., Yan, J., Sun, H. et al. 2013. Preparation and characterization of microencapsulated ammonium polyphosphate and its synergistic flame-retarded polyurethane rigid foams with expandable graphite. *Polymer International* 63: 84–92.

Zhu, Z. M., Xu, Y. J., Liao, W. et al. 2017. Highly flame retardant expanded polystyrene foams from phosphorus-nitrogen-silicon synergistic adhesives. *Industrial and Engineering Chemistry Research* 56: 4649–4658.

Zhuang, X., Li, S., Ma, Y. et al. 2011. Preparation and characterization of lignin-phenolic foam. *Advanced Materials Research* 263–238: 1014–1018.

Zhou, J., Yao, Z., Chen, Y. et al. 2014. Fabrication and mechanical properties of phenolic foam reinforced with graphene oxide. *Polymer Composites* 35: 581–586.

13

Recent Advances in Flame Retardant Textiles

Bin Fei and Bin Yu

13.1 Introduction

Textiles are manufactured using natural and/or man-made fibers. Natural fibers involve plant-based fibers such as cotton, hemp, flax, and jute; protein-based fibers such as silk and wool; and mineral fibers. Man-made fibers involve cellulose-based or regenerated fibers, e.g., rayon, acetate, and triacetate fibers and organic fibers, e.g., polyolefins, polyamide (PA), polyester, and polyacrylonitrile (PAN) (Dolez and Vermeersch 2018). Textiles have a range of applications, such as in apparels, transportation, protective garments, industrial work-wear, military, upholstered furniture, bed linen, and nightwear (Alongi and Malucelli 2012; Nazaré 2009). Being composed of hydrocarbons-rich polymers, most of textiles are easily ignited by an exterior fire and flame spreads very fast, causing huge fires and loss of lives and property. Nowadays, flame retardant (FR) textiles have attracted increasing attention from both the industrial and academic fields (Alongi and Malucelli 2012; Horrocks 2011). Different directives and standards have pushed researchers to explore highly efficient and green methods to solve this issue.

On the basis of the chemical structure and properties, textile fibers and fabrics are endowed with flame retardancy by the use of FRs containing halogen, phosphorus, nitrogen, boron, etc. (Malucelli 2016; Weil and Levchik 2008). These FRs can be directly introduced into fibers during the spinning process, grafted onto the fiber polymer backbone, copolymerized, or deposited on the fiber/fabric surfaces. Chemical modifications for flame retarding fabrics are complex applications of FR agents, e.g., ammonium polyphosphates, FR coatings, e.g., antimony trioxide/halogenated systems, or functional finishing, e.g., organophosphorus- and nitrogen-containing monomers. Functional finishing is particularly advantageous for high-level FR durability as FRs either polymerize within the fiber monomers or react with functional groups in the fiber-forming polymer backbone (Horrocks 2011; Joseph and Tretsiakova-McNally 2013). Surface treatment of fiber/fabric is the conventional method, which can be conducted by finishing and coating (Alongi and Malucelli 2012; Horrocks 2008; Horrocks et al. 2005). The former deals with the immersion of the fabric with an aqueous suspension/solution of FRs, while the latter requires the application of a layer/film on both the surface and back of the fabric (Alongi and Malucelli 2012).

Nowadays, great efforts have been made to achieve surface modifications of textiles with flame retardancy as after-treatments, these treatments are able to impart versatile properties to the textile substrate. Moreover, such after-treatments almost have no adverse effect on the properties of textile substrates. Creating novel FR coatings which are applied to fibers has to be considered. These coatings may be composed of a fully inorganic or organic-inorganic hybrid composition that can be generated by means of different approaches: layer-by-layer (LBL) assembly, sol-gel and dual-cure processes, plasma treatments, ultraviolet (UV)-curing technique, and vapor deposition (Alongi et al. 2013, 2014; Liang et al. 2013).

Flame retardation of textiles mainly involves two FR mechanisms, i.e., condensed phase and gas phase action (Malucelli 2016): to reduce the amount of flammable volatiles, resulting in the formation of more temperature-resistant carbonaceous char layers, which also act as barriers between the textile and the flame; to release reactive free radicals as very efficient flame inhibitors; and to decrease the heat feedback to the polymer to prevent further pyrolysis. FRs that function in condensed phase mechanism are mainly suitable for charring textiles, e.g., cellulose and wool, while FRs in gas phase are applicable to all types of fibers.

This chapter will review the recent advances in halogen-free FR textiles, including natural fibers (such as wool and cotton) and synthetic fibers or textiles (such as polyesters, PAs). New perspectives and innovatory approaches recently achieved in the textile field based on the deposition of novel coatings could confer fire retardancy to the fabrics. Especially, the fire protection of different fabric substrates achieved by exploiting sol-gel, LBL assembly, plasma deposition, and UV-curing techniques is summarized.

13.2 Flame Retardation of Textile Materials

FR textiles can be achieved via four primary approaches, including preparation of FR fibers, FR finishing of textiles, blend of common fibers with FR fibers, and surface-coating approaches.

13.2.1 Preparation of FR Fibers

Common fibers are widely used due to low cost, abundant raw materials, and convenient production. There have been four main methods to achieve FR modification of fibers, namely, copolymerization method, co-blending method, grafting-copolymerization method, and skin-core composite spinning method. The copolymerization method is to introduce FR co-monomers into a polymer chain to modify the molecular structure, and then the modified polymer becomes FR fiber by a simple spinning method. The co-blending method is to mix FRs with the spinning melt/slurry, and then the FR fiber is prepared by a spinning method. Because no chemical bonds between the FRs and the polymers exist and the interfacial interaction is weak, the durability of the FR effect is worse compared to the fiber obtained by copolymerization. Therefore, during the processing, the FRs are often mixed with other additives and resins to produce a master batch to improve the compatibility of the FRs with the matrix, thus improving the FR durability. Grafting-copolymerization method is to graft FR monomers onto the surface of fibers through different copolymerization methods to improve their FR properties. The skin-core composite spinning method refers to FR fibers obtained by the composite spinning process using a FR polymer as the core layer and the normal polymer as the shell.

13.2.2 FR Finishing of Textiles

FR finishing of fabrics means the surface treatments of the fabric in the finishing process, thus conferring the fabrics flame retardancy. Various methods for FR finishing of textiles are involved:

1. Padding-baking approach
   Baking solutions containing FRs are generally used for FR treatment of fiber textile. This approach is the most used for FR finishing of textiles following the processing.

Padding ⟶ Pre-baking ⟶ Baking ⟶ Post-treatment

2. Dipping-drying approach

Dipping ⟶ Drying ⟶ Post-treatment

For example, fabric is dipped into FR solution for a period of time, and then the fabric is removed from the solution, followed by drying and baking. The approach is suitable for FR finishing of hydrophobic fiber fabric.
3. Surface-coating approaches.
FRs and resin as a binder are mixed uniformly and coated on the surface of the textile by the adhering force of the resin. According to the mechanical equipment, it is also divided into the blade-coating, pouring-coating, and rolling-coating methods.
4. Spraying approach
For some certain textiles, such as thick curtains and large carpets that cannot be processed by ordinary equipment, the spraying approach can be used for FR finishing in the last process.
5. Organic solvent approach
FRs dissolved in organic solvents are used for FR finishing. This approach is simple and time-saving, but these organic solvents used are toxic and flammable.

The principal textiles for FR finishing treatments for which effective commercial processes are available include: (1) cotton and blends in which the cellulose fiber ≥50%; (2) natural protein fibers and blends; (3) polyester and polyester-rich blends, most often with cellulosic fibers; and (4) polyamide and blends.

13.2.3 Blend of FR Fibers and Common Fibers

Combining FR fibers and common fibers is another method to achieve flame retardancy of textiles. This method includes FR treatment of fiber, the blend of the treated fiber with common fibers, followed by twisting, and interlacing. The complicated process limits its wide application.

13.2.4 Surface-Coating Approaches

Recently, various surface-coating approaches for deposition of (nano)coatings have been used to impart flame retardancy to textiles. These coatings are generally very thin, even less than 100 nm, and demonstrated as environmentally friendly technologies.

13.2.4.1 UV-Curing Technology

The UV-curing technique has a wide range of applications in industrial FR coatings, due to the following advantages, such as rapid curing process, low energy consumption, marginal environmental impact, superior chemical stability, and low emission of volatile compounds (Chattopadhyay et al. 2005; Liang et al. 2013; Xu and Shi 2005). Liquid, multi-functional monomers or oligomers can be rapidly and efficiently converted to cross-linked polymer networks upon exposure to UV irradiation. To date, various FR monomers containing boron, phosphorus, and silicon, have been applicable to UV-curable coatings for flame retardancy to textiles (Hu et al. 2016; Yuan et al. 2012).

13.2.4.2 Sol-Gel Process

The sol-gel process is a widely used wet-chemical technique for fabrication of fully inorganic or organic-inorganic hybrid FR coatings on textiles. This process primarily involves a two-step hydrolysis/condensation reaction in the presence of (semi)metal alkoxides, such as tetramethoxysilane, aluminium isopropoxide, and titanium tetraisopropoxide (Malucelli 2016). Owing to the combination of organic and inorganic components, the overall properties of the hybrid coatings are enhanced (Altıntaş et al. 2011). Moreover, the possible synergistic effects between organic and inorganic components are at a molecular scale, the coated system is then able to achieve high-level flame retardancy (Alongi et al. 2011b; Hsiue et al. 2001).

13.2.4.3 Layer-by-Layer Assembly Approach

LBL is an eco-friendly, simple technique to fabricate FR (nano)coatings for textiles that are typically less than 1 μm thick (Li et al. 2009; Liang et al. 2013). A substrate can be alternately dipped or sprayed into an oppositely charged polyelectrolyte or inorganic (nano)particle solution or suspension (Alongi et al. 2014; Hu et al. 2016). Bilayer (BL) presents deposition of each positive-negative pair. Electrostatic attractions are the force for LBL assembly coatings, while donor/acceptor interactions, hydrogen-bonding, and covalent bonds were explored to construct multi-layer coatings. This technique is highly desirable to create FR coatings for cotton fabric (Li et al. 2010), polyesters (Alongi et al. 2012a), polyamide (Apaydin et al. 2013), polyurethane foams (Pan et al. 2015), etc.

13.2.4.4 Plasma Technology

The plasma strategy has been widely employed to modify the surface property or deposit nanocoatings onto the surface of various substrates. Different deposition conditions, such as pressure and polymerizable or non-polymerizable gases, can be exploited. This technology has been demonstrated as an eco-friendly alternative for the application of FR coatings on various substrates (Chaiwong et al. 2010; Tsafack and Levalois-Grützmacher 2007). Through this technique, a thin coating can be deposited or grafted on any surface by a corresponding plasma source. The low pressure plasma technique has been used in FR finishing where it imparts highly durable FR performance to textiles (Jama et al. 2001). However, as a result of additional cost and complexity brought from the vacuum operation, atmospheric pressure plasma has been further regarded as a promising alternative to the former process for industrial applications (Tsafack and Levalois-Grützmacher 2006b).

13.2.4.5 Vapor Deposition

Both physical vapor deposition (PVD) and chemical vapor deposition (CVD) have generally been used to treat textile materials and electrical devices due to their inherent merits, such as environmental friendliness and imparting multiple functions and flexibility to functionalized substrates. Sputter coating is one of the most commonly applied PVD techniques to confer flame retardancy to different substrates, such as textile fabrics, nanofibers, or plastic substrates for specific FR applications. The as-prepared coatings on complex surfaces are uniform with excellent coverage and thus offer great potential for FR coating applications. Plasma-enhanced chemical vapor deposition (PEVCD) is commonly used for the deposition of films to achieve flame retardancy.

The combination of surface-coating strategies, e.g., sol-gel plus UV-curing technique (Mülazim et al. 2011), sol-gel plus plasma technique (Totolin et al. 2010), and plasma plus UV-curing technique (Tsafack and Levalois-Grützmacher 2006b), have been developed to construct novel FR coatings.

13.3 Recent Advances in FR Textiles

13.3.1 Newly Developed FR Fibers

13.3.1.1 PANOX® (2017)

PANOX®, developed by the SGL Group, is a fire-retardant industrial textile fiber. It is an oxidized, thermally stabilized PAN fiber. Referring to the information given by the SGL Group, it is produced by thermal stabilization of PAN at 300°C, resulting in the increase of carbon content by approximately 62% in the oxidized textile fiber. It offers excellent chemical resistance and very good electrical insulating properties. PANOX does not burn, melt, soften, or drip because its limiting oxygen index (LOI) is around 50% which is higher than that of other organic fibers. It has not only high thermal stability, but also excellent heat and fire protection. Combining the superior properties, it should be a good insulating material because of its low thermal conductivity. PANOX fibers are produced in the form of continuous fiber tow that can be finished by suitable process for producing staple fibers in different lengths to make non-woven fabric, woven fabric, and knitted fabric.

13.3.1.2 Lenzing FR (2015)

Referring to the website of the Lenzing Group, the newly developed Lenzing FR fibers are an inherently FR fiber. It is on the basis of high wet modulus wood cellulose fiber. With the combination of the unique thermal insulation properties of the fiber and the permanent flame resistance of the finishing, it has superior resistance to all kinds of heat and flame. Since the fibers are made of cellulose, it is biodegradable and environmentally friendly. The blending of Lenzing FR with aramid can highly promote the softness of the clothes. From the official performance test by the Lenzing Group, it claims that the blending of Lenzing FR with aramid fiber can retain less heat than other materials and the breathability of the Lenzing FR blend increases to about 0.33 $W/m^3 \cdot Pa$. The fibers are mostly applied on the firefighters uniform, industrial uniform, and armed forces uniform. Additionally, it is used to make the Defender™ fabric for the US Army combat uniforms.

13.3.1.3 Teijinconex® neo (2015)

Teijinconex neo is registered in 2015 by Teijin Ltd. as a new meta-aramid fiber. In the official brochure, Teijinconex neo has similar FR properties as Teijinconex®. The heat flow of 100% Tenjinconex® fibers is only 0.315 cal/cm^2 sec at 955°C in the heat resistant test. For the flame resistance test, Teijinconex fibers have LOI of around 29%–32%. Also, its smoke concentration is 6% when it reaches the maximum exhaust temperature. From Table 13.1, it can be seen that Teijinconex fibers will only release 536 ppm of CO and 370 ppm of CO_2 in combustion at 400°C that is lower than other pure fibers. For the Teijinconex neo fiber, the emphasized properties are: environmentally friendly in processing, high dyability, and higher softness of the fiber.

13.3.1.4 Fire-Poof® (2016)

The company of Firetect developed a new FR finishing agent called Fire-Poof®. It is an interior non-hazardous, waterborne FR mainly for raw wood and most fabric. Still, it is not effective on 100% nylon, acetate, acrylic, plastic, or surface with water, glue, or stain/water repellent. After application, it should not be heated above 149°C and water washing is not recommended. It passed and got Class A/1 of the surface burning test of ASTM E84.

TABLE 13.1 Comparison of Gas Release of Teijinconex and Other Fiber

	Concentration (ppm)				
	CO	CO_2	NH_3	HCN	H_2S
TEIJINCONEX®	536	370	0	0	0
Polyester	1166	510	0	0	0
Nylon	1030	135	12	0	0
Acrylics	1633	0	499	250	0
PVC	1967	615	0	0	0
Cotton	4000	3300	0	0	0
Wool	1000	1500	600	130	480
Heat temperature: 400°C					

Source: https://www.teijin.com/products/advanced_fibers/aramid/contents/aramid/conex/eng/bussei/conex_bussei_bouen.htm

13.3.1.5 Flame Stop® I (2015)

Flame Stop I was invented by Flame Stop Inc. in 2015, which is a water-base, post treatment, and interior fire retardancy. It is applied on the cellulosic fiber. Under the finishing process, it will penetrate into the material and bond with the cellulose structure of the fiber. The Flame Stop I can self-extinguish when the material is in contact with an open flame. The advantages of Flame Stop I include non-toxic, non-combustible, non-carcinogenic, easy to apply, and contains no PBDEs (polybrominated diphenyl ethers have the problem of bioaccumulation in blood, breast milk, and fat tissues).

13.3.1.6 TexFRon® 9020 (2012)

TexFRon® 9020 produced by ICL group is a washing durable FR finishing chemical for cellulosic fabrics and cellulosic blends. From the official information sheet of the product, TexFRon® 9020 completely penetrates into the fabric during the curing process, resulting in excellent durability. It contains 35% bromine. It is a low melt active component (115°C).

13.3.1.7 Pyrovatex®CP NEW/CP-LF (2012)

Pyrovatex®CP NEW/CP-LF was invented by Huntsman and is a FR finishing agent for cellulosic fiber and fabric blends with a synthetic component up to a maximum of 25%. It is based on a dialkylphos-phonocarboxylic acid amide. With the addition of the methylolamine precondensates, they will react with the hydroxyl groups of the cellulose. As a result, it shows excellent resistance to washing at the boil and dry cleaning. The fabric treated with this finishing agent meets many international standards, for example, ISO 11612 (clothing to protect against heat and flame).

13.3.2 Recent Advances in FR Textiles

13.3.2.1 Cotton Textiles

The LBL assembly technique has been extensively employed to construct FR coatings for cotton textiles. Inorganic nanoparticles-based LBL assembly nanocoatings provide effective barriers on treated substrates by slowing down the rapid release of volatile products and inhibiting the heat transfer during burning. Up to now, LBL-assembled coatings consisting of various nanoparticles and polyelectrolytes have been applied to cotton textiles. Synthetic polyelectrolytes, such as polyethylenimine (PEI), polyallylamine, and polyacrylic acid (PAA), have been used in multi-layer coatings. Inorganic (nano)particles used in LBL assembly coatings have involved nano-clay (Huang et al. 2012a; Li et al. 2010), graphene oxide (Huang et al. 2012b), n-Al_2O_3 (Uğur et al. 2011), α-zirconium phosphate (Fang et al. 2016), polyhedral oligomeric silsesquioxane (POSS) (Chen et al. 2015; Li et al. 2011a), etc.

   Grunlan and co-workers first constructed multi-layer FR nanocoatings on cotton fabric by alternatively depositing negative laponite nanoplatelets and positive branched polyethylenimine (BPEI) (Li et al. 2009). 10-BL-coated fabric had 10 seconds (s) less afterglow time than the uncoated fabric in vertical burning tests (ASTM D6413). After burning, the untreated fabric almost had no residue left on the sample holder, while the coated fabric produced more char residues. The same group reported FR nanocoatings consisting of BPEI and sodium montmorillonite (MMT) on cotton fabric by LBL assembly (Li et al. 2010). The nanocoatings ensured reduced afterglow times and improved fire retardancy. Micro combustion calorimeter (MCC) results revealed that 5-BL BPEI/MMT reduced the total heat release (THR) and heat release capacity (HRC) by 20% and 15%, respectively (Table 13.2).

   POSS represents one group of inorganic-organic hybrid nanomaterials, which is attractive in FR applications due to its superior heat resistance and thermoxidative stability (DeArmitt 2011; Zhang et al. 2017c). Li et al. (2011a) deposited a fully inorganic nanocoating consisting of octa-ammonium (OA)-POSS (positive layer) and OTMA-POSS (negative counterpart) on cotton fabric. 20-BL OA-POSS-OTMA-POSS (octa(3-ammoniumpropyl) octasilsesquioxane octachloride (OA-POSS); octakis(tetramethylammonium) penta-cyclo-[9.5.1.1(3,9).1(5,15).1(7,13)] octasiloxane 1,3,5,7,9,11,13,15-octakis-(cyloxide)hydrate (OTMA-POSS)

TABLE 13.2 MCC Results for Various Coated Fabrics

Sample	Residue (%)	HRC (J/(g·K))	THR (kJ/g)	T_p (°C)
Control	2.88 ± 0.40	273.67 ± 25.38	11.63 ± 0.21	369 ± 0.58
BPEI pH 10/0.2% MMT				
5 BL	6.38 ± 1.50	254.33 ± 25.01	11.23 ± 0.25	374 ± 0.58
20 BL	7.48 ± 0.50	250.33 ± 14.50	11.10 ± 0.36	376 ± 2.65
BPEI pH 7/0.2% MMT				
5 BL	6.75 ± 0.60	260.33 ± 4.04	11.17 ± 0.40	376 ± 2.0
20 BL	6.74 ± 0.20	286.33 ± 8.51	11.90 ± 0.36	369 ± 0.58
BPEI pH 10/1% MMT				
5 BL	10.52 ± 0.30	220.00 ± 6.08	9.87 ± 0.31	382 ± 0.58
20 BL	10.49 ± 0.50	221.30 ± 7.57	10.23 ± 0.06	380 ± 0.58
BPEI pH 7/1% MMT				
5 BL	8.37 ± 0.50	251.30 ± 10.02	10.73 ± 0.25	379 ± 1.00
20 BL	10.54 ± 0.30	240.30 ± 11.37	10.70 ± 0.50	377 ± 2.65

Source: Li, Y.C. et al., *ACS Nano*, 4, 3325–3337, 2010.

was found to reduce a maximum reduction in THR (23%) and peak heat release rate (PHRR) (20%). All coated fabrics had much more residues left after burning than the control fabric (nearly no char left). Increasing BLs on the fabric leads to the production of more residues with the number of BLs increasing from 5% to 20%.

Intumescent FR (IFR) coatings are efficient to improve the flame retardancy of cotton fabrics by generating intumescent char layers during combustion, serving as an effective mass and energy barrier (Chen et al. 2015; Liang et al. 2013). Li et al. (2011b) first constructed a fully organic IFR coating on cotton fabric by utilizing poly(allylamine hydrochloride) (PAH) and poly(sodium phosphate) (PSP) as polymer electrolytes in the LBL assembly process. In this system, PSP, PAH, and a cellulosic substrate functioned as an acid source, a blowing agent, and a carbon source, respectively. The coated cotton with 10 BLs quickly extinguished with a substantial decrease in the THR (−80%) and PHRR (−60%), and the char yield increased from 9.59 wt% for pure cotton to 31.43 wt%. The uncoated control fabric burnt out completely in the vertical flame testing, while the original shape of the coated fabric was maintained. The same group (Laufer et al. 2012) designed an intumescent multi-layer coating made from fully eco-friendly electrolytes, i.e., chitosan (CH) and phytic acid (Figure 13.1). The 30-BL-coated cotton fabric (pH = 4) exhibited the greatest reductions in PHRR and THR of 60% and 76% (MCC results), respectively. The formation of an intumescent phosphorus-containing char layer is responsible for the superior fire retardancy of the modified cotton fabric.

Recently, two-dimensional nanomaterials and graphene oxide (GO) with a unique structure and superior properties have been used as a negative component for the construction of FR nanocoating on cotton fabrics by the LBL assembly process (Huang et al. 2012b; Wang et al. 2017). IFR nanocoatings consisting of alternative polyacrylamide (PAM) and GO were deposited on the surface of cotton fabric by hydrogen-bonding interaction via LBL assembly (Laufer et al. 2012). The IFR-PAM/GO coating was found to improve the thermal stability and flame retardancy of the treated cotton fabric. 20-BL coating increased the time to ignition (TTI) by 23 s and decreased the PHRR by 50%. Ammonium polyphosphate (APP) has been used to create IFR LBL assembly coatings on polyester-cotton blends and cotton with improved fire safety (Alongi et al. 2012a, 2012b; Carosio et al. 2012, 2013a). Carosio et al. (2012) reported two series of assembly coatings consisting of CH-APP and silica nanoparticles-APP for polyester-cotton blends (30/70). For the CH/APP system, CH acted as a charring and foaming agent, while APP functioned as an acid source,

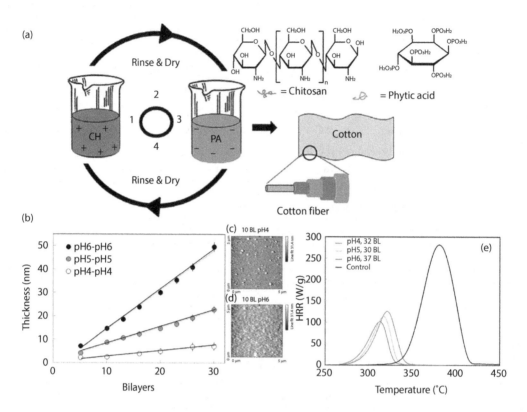

FIGURE 13.1   (a) Schematic of LBL assembly using chitosan (CH) and phytic acid (PA). Steps 1–4 are repeated until the desired number of bilayers are deposited; (b) growth of CH–PA assemblies as a function of deposition solution pH, as measured by ellipsometry. AFM height images of 10-BL assemblies deposited from pH 4 (c) and pH 6 (d) solutions; and (e) heat release rate as a function of temperature for uncoated control and CH–PA-coated cotton fabric. All coated fabrics contained 18 wt% CH–PA. (Reprinted from Laufer, G. et al., *Biomacromolecules*, 13, 2843–2848, 2012. With permission.)

catalyzing the char formation of CH and cellulose. In the silica/APP system, the phosphoric acid from the decomposition of APP induced the dehydration of the cellulose, forming a thermally insulating silica-containing barrier. The 20-BL CH/APP coating significantly decreased the THR and PHRR by ~22% and ~25%, respectively. However, the silica/APP assembly coatings were found to have a worse effect in the reduction of THR and PHRR, but increased the TTI by 9 s. Recently, a hybrid intumescent film made from polyhexamethylene guanidine phosphate (PHMGP)/APP/PHMGP/zirconium phosphate (α-ZrP) was tested on cotton fabric to ensure superior FR efficiency (Fang et al. 2016). The nanobrick wall coating reduced the burning time, eliminated the afterglow behavior, and enhanced the char yield after burning in a vertical configuration. In addition, the presence of the nanobrick wall led to an obvious reduction in PHRR and THR, which provides an improved fire resistance as compared to the PHMGP/APP, PHMGP/α-ZrP, and APP coatings. The flame retardant mechanism is based on the catalytic charring of the IFR and the protection of the unique nanobrick wall as an insulating barrier layer during burning.

Although the LBL technique is very useful to produce various FR coatings for fabrics, the coating durability against washing remains a huge challenge. To improve the washing durability of LBL-coated FR cotton fabric, Pan et al. (2017) modified cotton fabrics by utilizing PEI and sodium alginate (SA) as polymer electrolytes, followed by subsequent crosslinking using barium-, nickel-, and cobalt-containing compounds as crosslinking agents. All the cross-linked coatings on the surface of the cotton fabric led to improved char residue and fire retardancy, especially 28% reduction in PHRR for the cotton-barium sample. More importantly, the washing durability of the coating on the fabric was enhanced after metal ion crosslinking. By contrast, the barium and nickel ion cross-linked coatings prevented the shrinkage during combustion due to the formation of bigger fiber bundles. Among crosslinking coatings, cotton-barium had the most stable char residue, which was responsible for the best fire retardancy (Figure 13.2).

FIGURE 13.2 Photographs of untreated cotton fabric, cotton-blank, cotton-Ba fabric, cotton-Ni, and cotton-Co fabric after the horizontal flame tests. SEM images of cotton-Ba fabric (a,d), cotton-Ni fabric (b,e), and cotton-Co fabric (c,f) after the horizontal tests. Photographs of washed cotton-blank, washed cotton-Ba fabric, washed cotton-Ni fabric and washed cotton-Co fabric after the horizontal flame tests. (Reprinted from Pan, Y. et al., *Carbohydr. Polym.*, 170, 133–139, 2017. With permission.)

Photopolymerization technology has received increasing interest, which exhibits a wide range of applications in paints, coatings, adhesives, and composites (Decker 2002; Schreck et al. 2011). Various multi-functional acrylate and methacrylate monomers, such as polyurethane acrylate, epoxy acrylate, and polyester acrylate have been widely used as the foundation for curable coatings (Miao et al. 2009; Park et al. 2009; Wang et al. 2008). Reactive FR monomers containing phosphorus, nitrogen, and silicon have been explored in UV-cured coatings with excellent fire retardancy (Chen et al. 2011; Cheng and Shi 2010; Liang et al. 2013). Resulting from the synergistic flame retardation, FR formulations of coatings based on phosphorus and nitrogen have been investigated. Through blending triacryloyloxyethyl phosphate (TAEP) and triglycidyl isocyanurate acrylate (TGICA) (Figure 13.3) in various weight ratios, Hu's group formulated a series of UV-cured IFR coatings which were coated on the surface of cotton fabric (Xing et al. 2010, 2011). MCC and LOI results revealed a distinct synergistic effect between TAEP and TGICA. With increasing the loading of FR coating, the LOI value of the treated fabric increased from 21% to 24.5%, and total heat of combustion decreased from 8.8 to 3.6 kJ/g.

Application of the sol-gel process in constructing fully inorganic or organic-inorganic hybrid FR coatings for textiles has been documented in the last few years. Recently, Alongi et al. have given a review on the sol-gel technique for FR textiles (Alongi and Malucelli 2012). Such hybrid architectures function as a thermal insulating barrier to well protect textiles, thus improving the flame retardancy. Various FR units could be selected and introduced into these hybrid systems. Benefiting from synergistic effects between organic and inorganic moieties, the surface coating is usually capable of achieving superior FR efficiency. In the sol-gel process, tetraethyl orthosilicate (TEOS) is the commonly used silica precursor to construct a hybrid FR system. In addition, tetramethyl orthosilicate (TMOS) as an alternative has also been employed to fabricate sol-gel coatings because it is soluble in water (Alongi et al. 2011c). To achieve ideal flame retardancy for cotton fabric with high washing durability, sol-gel reactive conditions such as precursor/water ratio, reaction temperature, and time of thermal treatment were optimized. The optimal performances of the cotton fabric were achieved when the reaction was conducted at 80°C for 15 hours at a 1/1 molar ratio of TMOS/H_2O and pH = 7 (56% increase in TTI; 16% decrease in PHRR). In general, the formation of more uniform coating implies the better FR performance, as observed for cotton, polyester, and their blends. For example, the sol-gel-treated cotton-polyester blend (35 wt% cotton) showed a remarkably increased TTI (up to 98%) and reduced PHRR (up to 34%) (cone colorimeter, 35 kW/m^2) (Alongi et al. 2011d). Following this work, several silica precursors with different hydrolyzable groups have been selected to construct hybrid coatings for cotton fabric, and the impact on the combustion behavior of the treated cotton fabric was evaluated (Alongi et al. 2012c). Specifically, the fire

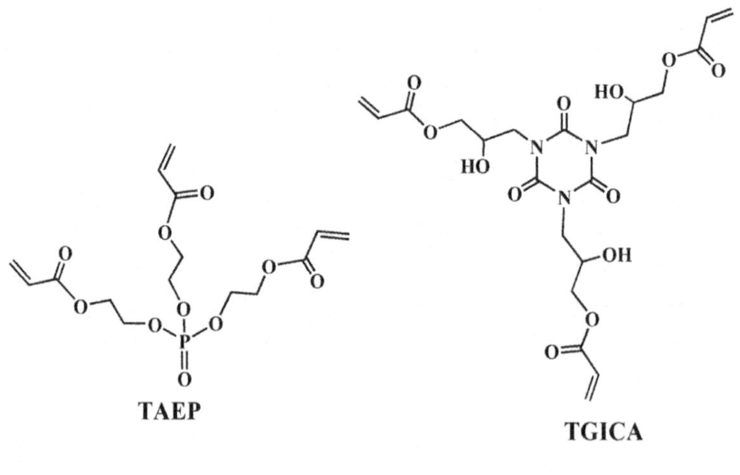

FIGURE 13.3 Chemical structures of triacryloyloxyethyl phosphate (TAEP) and triglycidyl isocyanurate acrylate (TGICA).

retardancy of TMOS-treated cotton fabric was comparatively investigated with those of TEOS-, tetrabutylorthosilicate-(TBOS), diethoxy(methyl)phenylsilane-, 3-aminopropyl triethoxysilane (APTES), 1,4-bis(triethoxysilyl)benzene-, and 1,2-bis(triethoxysilyl)ethane-treated counterparts, respectively. All the precursors facilitated the char formation in air below 360°C and the higher thermal stability was achieved for the hybrid coating derived from the precursors bearing aromatic rings. The presence of silica enabled flame quench and slowed down the burning of cotton upon exposure to a fire. Flame applications twice were necessary to make the treated fabric ignition and the total burning time increase. Compared to the total burning time for the control cotton (40 s), 70, 57, 57, and 50 s were observed for TMOS-, TEOS-, TBOS-, and APTES-treated cotton, respectively. In addition, the residue increased significantly from 10 wt% for the cotton to 48 wt% for TMOS-treated cotton. In contrast, the treated cotton fabric with APTES had an adverse effect on the flame retardancy of the cotton as demonstrated from the increased PHRR (+13%).

Recently, a dual-curing process, e.g., photopolymerization reaction followed by thermal treatment, has been developed to fabricate hybrid organic-inorganic coatings for cotton fabric by the sol-gel process (Alongi et al. 2011a). These hybrid coatings could endow the treated cotton with superior flame retardancy (Alongi et al. 2011a). Formulations involving 30–80 wt% of TMOS in a UV-curable resin (bisphenol A ethoxy diacrylate), 4 wt% of 2-hydroxy-2-methyl-1-phenylpropan-1-one as a photoinitiator, and methacryloyloxypropyltrimethoxysilane as a coupling agent were conducted for dual-cured hybrid coatings. The hybrid coatings prevented the burning of the cotton fabrics. For example, the presence of 60 wt% silica enabled prolonging the total burning time to 150 s for dual-cured-treated fabrics from 40 s for the untreated one, to increase the residue from 8 to 29 wt%, and to postpone the time to ignition from 18 to 34 s during burning tests in a horizontal configuration. More recently, Zhang et al. (2017a) prepared FR and hydrophobic cotton fabric by depositing nanocoatings consisting of APP, sodium MMT, and vinyltrimethoxysilane (VTMS) via LBL self-assembly and sol-gel techniques. High-concentration APP was beneficial to achieving superior hydrophobicity and flame retardation due to its multiple functions as a hydrolysis-condensation catalyst, acid source, and blowing agent in the IFR system. Improved hydrophobicity led to high wash-durability of the formed coatings, but the washing durability of the coated cotton fabric still cannot meet the requirement for practical applications. Other alkoxide precursors have been also exploited to prepare hybrid coatings for FR cotton fabrics. Recently, Alongi et al. (2012d) reported various alkoxide precursors (tetraethylortho-silicate, -titanate, -zirconate, and aluminium isopropylate) for treatment of cotton fabrics by the sol-gel process. These coatings were found to ensure an overall enhancement of the thermal stability and fire retardation of the fabric, in addition to the mechanical properties of the control cotton with respect to tensile strength, deformation, and the abrasion resistance.

Apart from the techniques abovementioned, the plasma technique is a useful tool to deposit FR coatings on the surface of cotton fabrics. Edwards et al. (2012) reported two types of phosphoramidate FR monomers (Figure 13.4), which were applied to cotton textiles via atmospheric-pressure dielectric barrier discharge plasma. The grafted phosphorus-containing polymers were effective to promote

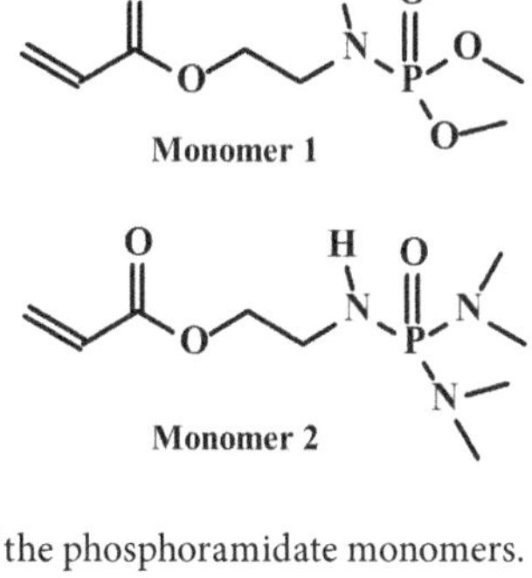

FIGURE 13.4 Chemical structures of the phosphoramidate monomers.

char formation, although the treated fabric did not self-extinguish in vertical burning tests. The grafted phosphorus amount via this strategy was too low to make flame quench, while these compounds exhibit the potential for flame retarding cotton fabrics when the grafting amount of FR is further improved.

13.3.2.2 Natural Protein Fibers (Wool and Silk)

Wool, a cutin fibrous protein, is composed of cysteine, thiocarbamic acid, and cross-linking polypeptides with a helical structure (Forouharshad et al. 2010). As a natural protein fiber, wool has a wide range of applications in apparel, interior textiles, and industrial clothing due to its comfort characteristic and inherent flame retardancy. The wool contains high nitrogen content (16%) and moisture content (10%–14%) (Benisek 1974), ignition temperature (570°C–600°C) (Forouharshad et al. 2010), and LOI value (25%–28%) (Horrocks 1986) as well as low heat release (20.5 kJ/g). Control wool undergoes a complex series of reactions during pyrolysis, yielding plenty of products. Starting from ~230°C, the breakage of the helical structure occurs, and the primary ordered part of the wool protein undergoes a phase change from a solid to liquid (Horrocks and Davies 2000). In the temperature range of 250°C–295°C, sulfur compounds involving hydrogen sulfide are released due to the rupture of the cystine disulfide bonds. Above 250°C, a charring reaction by dehydration occurs (Horrocks and Davies 2000). The initial exothermic reaction depends on the oxidation reaction of the cystine during burning (Horrocks and Davies 2000). Most wool fabrics can pass a horizontal test, but may not pass some 45° or vertical tests. Some special products such as seat coverings, aircraft furnishings and blankets, protective clothing, and carpets of shag pile construction need additional treatment to achieve satisfactory fire retardancy (Cheng et al. 2016).

Zirpro treatment can impart superior flame retardancy to wool, which is based on an electrostatic interaction between negative-charge zirconium or titanium complexes and positive-charge wool under acid conditions (Benisek 1971). The Zirpro treatment is the most commonly used to modify wool with durable fire retardancy (Martini et al. 2010). During the Zirpro process, formic acid and hydrochloric acid are often applied to wool fabrics (Forouharshad et al. 2010, 2011a; Tian et al. 2003). In addition, IFR agents, e.g., tetrakis-hydroxymethyl phosphonium condensates and ammonium phosphates were also capable of enhancing the fire retardation of wool fabrics by forming thermal-insulating char layers (Davies et al. 2000; Horrocks and Davies 2000).

In order to improve FR efficiency, a synergistic system that was composed of zirconium oxychloride, citric acid, and hydrochloric acid was conducted for treating wool (Forouharshad et al. 2011b). The treated wool had increased flame retardancy, as reflected by a 31.9% increase in LOI value. Also, the char residue for the treated wool was increased. This char layer can serve as a shielding layer to protect the underlying substrate from fire attack. In recent work, nano-zirconia and citric acid were used for constructing a multi-functional coating for wool with simultaneous fire retardancy and electromagnetic reflection properties (Gashti et al. 2012). Importantly, the coated sample had reasonable washing durability. Citric acid was used as crosslinking agent to stabilize nano-zirconia by a reaction with free radicals of wool, which led to an electrostatic interaction between Zr^{4+} cations and the carboxylate anions of CA. The excellent flame retardancy can be attributed to a superior heat-insulating effect of nano-zirconia embedded in the coating. Recently, Zhang et al. (2015) prepared a series of boron-doped silica sols using TEOS as a precursor and boric acid, zinc borate, and ammonium borate as boron components, which were applied to the wool fabric based on the sol-gel reaction. The treated wool fabric was found to have superior fire resistance and thermal stability. Especially, the highest FR efficiency is observed for the treated wool fabric with ammonium borate and silica sol. A LOI value of 29.9% and a maximum smoke density value of 99.17 g^{-1} were acquired, respectively (Figure 13.5). Importantly, the coating had an almost adverse effect on the tensile strength and air permeability.

Eco-friendly FR systems have generated tremendous research interest due to increasing environmental pressure. Cheng and co-workers (2016) developed FR wool fabric by adsorption of phytic acid on the surface using an exhaustion technique. 140 wt% loading of phytic acid on the surface of the wool was capable of increasing the LOI value from 23.6% to 35.2%. The fire retardancy of the treated wool

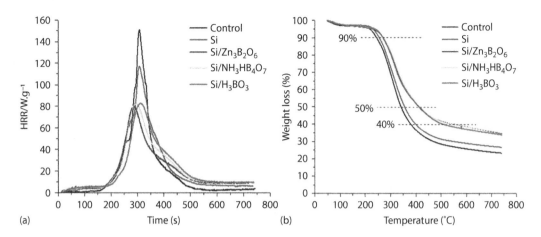

FIGURE 13.5 (a) Heat release rate of wool fabric. (b) TG curves of wool fabric under nitrogen. (Reprinted from Zhang, Q. et al., *Mater. Des.*, 85, 796–799, 2015. With permission.)

depended on the adsorption amount of phytic acid, which was related to the pH of solution, immersing temperature, and initial phytic acid concentration. The surface treatment had almost no adverse effect on the color and mechanical properties of the wool. Moreover, the anti-washing property of the FR fabrics was enhanced.

Flame retardation of silk fabrics can be dated back to 200 years ago by immersing fabrics in the mixture of borax and boric acid solution (Guan and Chen 2006; Zhou 1995). But the surface coating has extremely poor anti-washing properties. In the mid-1980s, Achwal et al. (1987) treated silk fabrics with a urea phosphoric acid salt by a pad-dry process. The treated fabric had an LOI value higher than 28% and the FR finishing was not easily washed out. In the 1990s, organophosphorus FR was utilized to improve the fire safety of silk fabrics (Kako and Katayama 1995). Over the past decade, increasingly, researchers are focused on the FR treatment of silk fabrics. Recently, Guan and co-workers (2009) reported a FR finishing system for silk by combining a hydroxyl-containing organophosphorus oligomer (HFPO) (Figure 13.6) and 1,2,3,4-butanetetracarboxylic acid (BTCA). The presence of HFPO enabled the bonding of BTCA onto the silk by either a BTCA "bridge" between the silk and HFPO or a BTCA-HFPO-BTCA cross-linkage between two silk protein molecules. With the simultaneous existence of 30% HFPO and 5.8% BTCA, the treated silk had a LOI as high as 30.0% (22.8% for control silk) after hand washing once, and it passed the vertical flammability test after 15 hand washing cycles.

FR compounds have been grafted on the surface of silk fabrics with FR durability by a plasma-induced graft polymerization (PIGP) process (Carosio et al. 2013a; Kamlangkla et al. 2011). Kamlangkla et al. (2011) reported FR silk fabric grafted with two kinds of phosphorus containing monomers (Figure 13.7) by means of a two-step process, i.e., argon-induced graft polymerization, followed by a sulphurhexa fluoride (SF 6) plasma treatment. The grafted amount of *ca.* 11% increased the LOI of the treated fabrics to 29.0% and 30.5% for diethyl 2-(acryloyloxyethyl) phosphate (DEAEP) and diethyl 2-(acryloyloxyethyl) phosphoramidate (DEAEPN) from 25.0% for control silk fabric. It is noted that over a 50% grafting yield was achieved at the first step. The FR efficiency of the phosphoramidate monomer for the silk fabrics exceeded the phosphate analogue due to the phosphorus-nitrogen synergistic flame retardation.

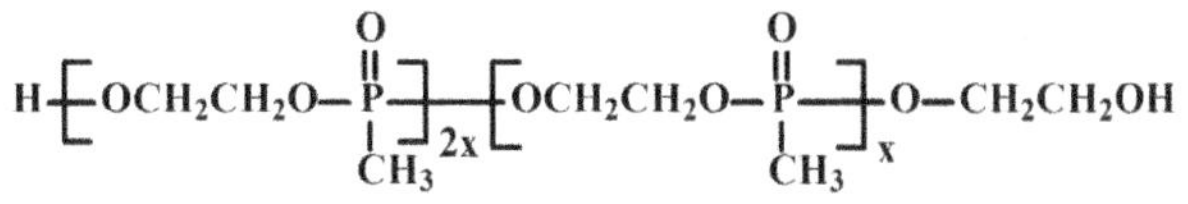

$$H \left[OCH_2CH_2O - \underset{\underset{CH_3}{|}}{\overset{\overset{O}{\|}}{P}} \right]_{2x} \left[OCH_2CH_2O - \underset{\underset{CH_3}{|}}{\overset{\overset{O}{\|}}{P}} \right]_x O - CH_2CH_2OH$$

FIGURE 13.6 Chemical structure of HFPO.

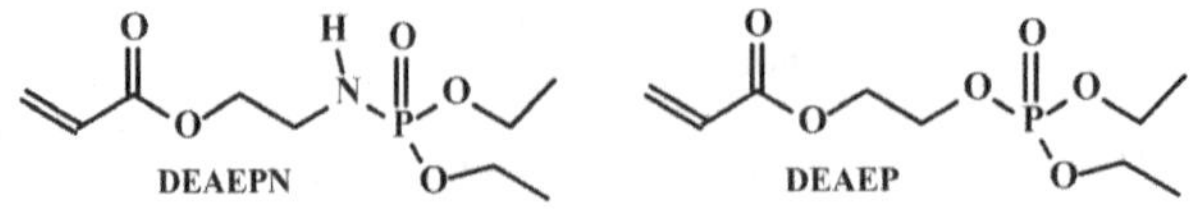

FIGURE 13.7   Chemical structures of DEAEPN and DEAEP.

After 5 minutes of SF 6 plasma treatment, the fabrics became water repellent having a contact angle of 134°, and FR performance remained after 50 cycles of laundering (McSherry method). The second step is critical to fabricate anti-washing multi-functional silk fabrics.

The sol-gel technique is regarded as an effective approach to construct organic-inorganic hybrid coatings on textiles with multi-functional properties. Very recently, various sol-gel-derived architectures have been explored to impart flame retardancy to silk fabrics, including phosphorus-silica (Cheng et al. 2018), boron-nitrogen-silica (Zhang et al. 2017b), and boron-silica (Zhang et al. 2016) hybrid coatings. Zhang et al. (2017b) used a ternary FR finishing system for treating silk fabric containing TEOS, boric acid (H_3BO_3), and urea by the sol-gel reaction. The presence of 24% boric acid and 6% urea (relative to the TEOS) in the sol system achieved the LOI value of 34.6%. Prior to the sol-gel reaction, the use of BTCA for pre-treatment of the silk fabric promoted the washing durability of the coatings. After washing ten times, the LOI of the treated sample reached 30.8% relative to 31.0% for the sample before washing. In addition, this ternary system-treated silk exhibited lower heat release and smoke release than those of the boron-silica binary sol system. In another work, Cheng et al. (2017) developed a pad-dry-cure technique to FR silk fabric using naturally phytic acid, TiO_2 nanoparticles, and BTCA, where BTCA was used to stabilize and improve the adhesion of TiO_2 on the silk surface. The combination of PA, TiO_2, and BTCA improved the flame retardancy of the silk fabric with durability after 25 times of washing. Considering the eco-friendly and water-soluble merits, condensed tannin extracted from dioscorea cirrhosa tubers was applied to treat silk fabric using an impregnation technique in a weak acidic medium (Cheng et al. 2017). An LOI value of over 27% and below 12 cm in the char length in the vertical burning test were observed, even after 20 washing cycles, while 24.7% and 30 cm for the untreated silk, respectively.

Multi-functional FR silk fabrics are highly desirable for diverse applications. GO has been coated onto silk fabrics to improve their fire retardancy and electrical properties by different methods (Cao and Wang 2017; Ji et al. 2017, 2018). Xing and co-workers prepared multi-functional silk fabrics by depositing GO hydrosol onto the silk fabric via a "dry-coating" method, followed by reduction with L-ascorbic acid (Figure 13.8) (Ji et al. 2017). The 19.5 wt% reduced graphene oxide (rGO)-coated silk fabric exhibited a high LOI value of 43.5% as compared to 24.0% for the untreated silk. This value after washing for ten times (42.3%) only had a slight decrease. In addition, shorter time to self-extinguish time (12 s versus 5 s), small damaged length (12.8 cm versus 27.1 cm), and higher residual char (90.1% versus 35.5%) relative to the control silk fabric were obtained. The treated silk fabric with 19.5 wt% rGO had low sheet resistance (0.13 kΩ/sq) and was designed into a FR conductor against 60 s of combustion, which has potential application in fire-fighting field.

13.3.2.3 Synthetic Fabrics

13.3.2.3.1 Polyamide Fabrics

Various FR fillers, such as IFRs and nano-additives have been incorporated into PA fibers to achieve flame retardancy. FRs for PA fibers should have high thermal stability (spinning temperature 250°C–300°C), negligible impact on the mechanical properties of the PA fiber, superior compatibility with the PA fiber, and low migration during service. Two primary approaches, i.e., reactive-type and additive-type are to endow the PA matrix with high fire safety. Reactive-type FRs such as phosphorus-containing polyol and halogenated anhydride can be covalently introduced into primary PA chains. In general, compounds

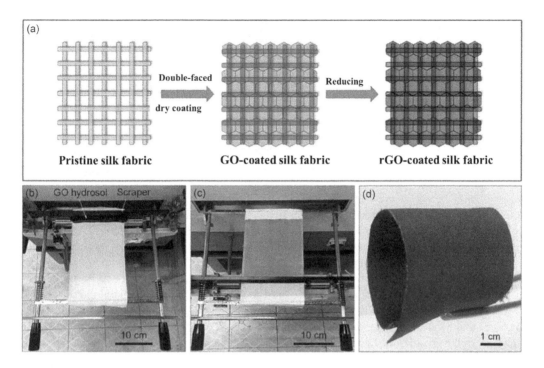

FIGURE 13.8 (a) Illustration of the fabrication process of rGO-coated silk fabric; (b) pristine silk fabric fixed on a sample holder before coating; (c) the silk fabric after coating with GO hydrosol; and (d) digital image showing the as-prepared rGO-coated silk fabric. (Reprinted from Ji, Y. et al., *Mater. Des.*, 133, 528–535, 2017. With permission.)

containing aromatic or heteroaromatic rings are used to increase the chain stiffness, leading to improved thermal stability. Resulting from high cost, this method has limited application in some areas, such as industry and military protective clothing. The nanocomposite technique is regarded as an effective approach to enhance the FR and thermal properties of polymeric materials simultaneously (Laoutid et al. 2009). Bourbigot et al. (2000a, 2000b) first reported FR PA-6 nanocomposite filaments, which were knitted into fabric. 5 wt% nano-clay was able to reduce PHRR by 40% as evaluated by cone calorimeter (35 kW/m²), but the ignition time was shortened. The addition of nano-clay alone cannot reach the acceptable flame retardancy and ignition resistance. In general, to achieve satisfactory flame retardancy of PA textiles, minimal FR loading of 15–20 wt% is required (Levchik and Weil 2000). The synergistic effect of phosphorus-containing FRs is combined for PA 6 and 66 films (~80 mm thick) in the presence of nano-clays (Horrocks et al. 2003, 2004; Horrocks 2011). APP is one of the most efficient FRs for PA, which undergoes thermal degradation in the range of 250°C–300°C matching the melting point of PA 66 (~265°C). The addition of nano-clay enables the reduction in the APP concentration required to achieve a satisfactory level of flame retardancy. For example, to achieve LOI values up to 24% in PA 66, the addition of 2 wt% nano-clay reduced the required level of the APP from approximately 28.5 to 20.1 wt% (Horrocks et al. 2003).

Recently, PA fabrics were endowed with flame retardancy by LBL assembly. Apaydin et al. (2014) designed an IFR coating made from PAH and PSP by LBL assembly to make PA 66 fabric flame retardant. The coating promoted thermal degradation of PA fabrics, and a 40-BL-coated fabric had a maximum PHRR decrease (−36%). Following this work, Apaydin et al. (2015) deposited IFR coating consisting of PAH, PSP, and TiO₂ nanoparticles on PA fabric by LBL assembly. The sample coated with 15 BLs exhibited a maximum reduction in the PHRR (−26%). The presence of TiO₂ enhanced the fire resistance as compared to the (PAH-PSP)ₙ multi-layer coatings. Hu's group reported fully green organic coatings of CH, phytic acid, and oxidized

SA (OSA) on PA 66 fabric (CS-PA-CS-OSA)$_n$ in a quadralayer (QL) process to improve the flame retardancy (Kundu et al. 2017a). The PA 66 fabric treated with 10- and 15-QL coatings could prevent the melt-dripping in a vertical burning test. A maximum reduction (24%) in the PHRR was achieved for the PA 66 fabric with 5-QL depositions (cone calorimeter). The polyelectrolytes promoted the degradation of the control PA 66 fabric, and the char yield was enhanced for all the coated fabrics significantly. The incorporation of OSA could improve the FR durability of the multi-layer nanocoating. In another study, the same group prepared a kind of DOPO-based phosphonamidate monomer (DOPO-diallyl-amine [DAAM]), which was grafted onto PA 66 fabrics by a UV grafting technique (Kundu et al. 2017b). 20 wt% DOPO-diallyl-amine [DAAM]-treated PA 66 fabrics exhibited a 22% reduction in the PHRR without melt-dripping. To improve coating durability, *N,N'*-methylene bisacrylamide (MBAAm) as a photo-cross-linking agent was combined with acrylamide (AM), which was grafted on the PA 66 fabric under UV irradiation during the FR finishing (Liu et al. 2017). In order to evaluate the laundering resistance of AM/MBAAm-g-PA 66 fabric, commercial grade detergent solution (0.5%) was conducted for 50 times. The low MBAAm concentration (0.05–0.1 wt%) led to the improvement in the flame retardancy and tensile properties for AM/MBAAm-g-PA66, while the maximal durable efficiency was observed at the MBAAm concentration of 5.0 wt%.

A plasma technique to graft/deposit a fluorinated acrylate, 1,1,2,2-tetrahydroperfluorodecyl acrylate (AC8), onto the PA 6 surface was demonstrated (Errifai et al. 2004). The treated PA 6 had a 50% decrease in the PHRR. Such a reduction was attributed to the reaction of fluorine-containing radicals with degraded polymer fragments. FR treatments for PA textiles have been conducted by the combination of plasma and nano-technology. Quede et al. (2004) deposited organosilicon thin coatings on PA substrates by polymerization of 1,1,3,3-tetramethyldisiloxane monomer premixed with oxygen through a cold remote nitrogen plasma technique. Improved fire retardancy in the presence of nanocoating (2 wt%) including an increase in LOI (130%) and decrease in the PHRR (41%) and THR (33%) was demonstrated by comparing treated PA 6 with untreated PA fabric. A thin silicon nanoparticles-based nanocoating was deposited onto PA 6 nanofibers and its organic-modified Fe-montmorillonite (Fe-OMT) nanocomposite fibers by a well-controlled sputtering technique (Cai et al. 2008). Thermogravimetric analysis (TGA) results indicated that the silica-coated nanocomposite fibers displayed around 100% increase in char yield, and the PHRR was reduced from 680.7 to 511.1 W/g.

13.3.2.3.2 Polyester Fabrics

Aromatic polyesters are the most important polyesters from which high-performance fibers can be spun due to their high crystallinities and tensile strength. Above 300°C, polyethylene terephthalate (PET) undergoes pyrolysis, producing plenty of highly flammable volatile products, such as acetaldehyde, methane, ethene, and diethyl ether together with less volatile products such as PET oligomers (linear and cyclic), terephthalic acid, and ethylene glycol. PET fabrics have been extensively utilized in civil and industrial fields resulting from their superior comprehensive performances. In general, co-polymerization and co-blending methods are commonly used to modify PET fabrics. As a result of the release of toxic and corrosive gases from halogen-base FR, there is an increasing trend toward using halogen-free FRs. In comparison to the copolymerization method, the co-blending method has been widely used due to its low cost and simple manufacturing process. FRs for PET fabrics with excellent thermal stability and compatibility through a co-blending method are highly desirable. Commonly used co-blending FRs for PET fabrics involve multi-brominated diphenyl ether and phosphorus-containing oligomers.

Recently, the emerging techniques, such as LBL, UV-curing, sol-gel, and plasma techniques have been widely investigated to treat the surfaces of PET or PET blend textiles. LBL coatings consisting of both inorganic nanoparticles and polymer electrolytes have been attempted to endow PET fabrics with flame retardancy. ZrP nanoplateletes (negative layer) coupled with a polydiallyldimethylammonium chloride (PDAC) and two inorganic nanoparticles (POSS and silica) were deposited on PET fabric (Carosio et al. 2011b). As-treated PET exhibited an increased TTI and reduced PHRR. Specifically, 10-BL ZrP/PDAC achieved the highest TTI value, while 5-BL ZrP/POSS ensured the lowest PHRR (20% reduction). The flame retardancy of 5-BL ZrP/silica-treated PET was better than that of 5 silica/silica BL-treated

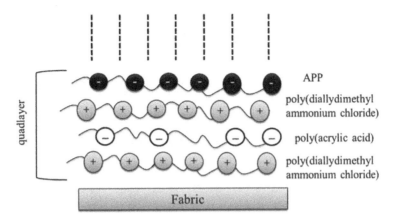

FIGURE 13.9 Schematic representation of the quadralayer architecture. (Reprinted from Alongi, J. et al., *Polym. Degrad. Stab.*, 97, 1644–1653, 2012. With permission.)

PET due to a synergistic effect between ZrP and silica (Carosio et al. 2011c). Another hybrid organic-inorganic architecture composed of PDAC/ PAA/PDAC/APP has been built by QL LBL assembly to thermally protect PET fabric (Figure 13.9) (Alongi et al. 2012a). APP and PAA are used as aromatic char-forming agents. PET fabrics treated with 1, 5, and 10 QL had weight gain of 3, 13, and 33 wt%, respectively. This FR system promoted char formation at low temperature, resulting in the obvious reduction of TTI and PHRR reduction. During burning, the protective coating generated a protective char layer to slow down the rapid release of volatile products and heat transfer (Alongi and Frache 2010). The same QL architectures were further explored for PET-cotton blend (28 wt% PET) under three different heat fluxes (25, 35, and 50 kW/m^2) (Carosio et al. 2013a). The fire performance of the treated fabrics was evaluated by fire tests in horizontal and vertical configurations, and cone calorimeter. The deposition of only one QL led to the formation of the carbonaceous char on the surface of the PET fabric, whereas intumescent-like structures were produced for 5 or 10 QL-treated PET. Overall, these architectures inhibited the flammability of the fabrics, suppressed the afterglow, and eliminated melt dripping and lowered the PHRR during burning. Recently, fully organic LBL coatings consisting of CS, melamine phosphate, and sodium hexametaphosphate were deposited on PET fabric where melamine pre-mixed with CS solution and sodium hexametaphosphate solution were used as cationic and anionic counterparts, respectively (Leistner et al. 2015). The combustion performance of treated PET fabrics was evaluated by vertical flame spread tests and pyrolysis combustion flow calorimeter. Only 15-BL at CS/melamine weight ratio of 0.5/0.9 (w/w) ensured self-extinguishment and substantial reductions in the PHRR.

Constructing versatile coatings by LBL assembly has great potential application in FR PET fabrics, but the current studies are only at the experimental stage. More recently, Carosio et al. (2018) developed hybrid organic-inorganic FR coatings on PET fabrics using OA-POSS and MMT as assembly components at zero and high ionic strengths by LbL assembly on a semi-industrial scale (Figure 13.10). A dyeing padder in an impregnation and exhaustion process was employed to create the hybrid LBL coatings. High ionic strength favors the formation of more uniform and thicker coatings than the reference one. Such coatings prevented the melt dripping behavior completely and reduced the flame spread rate distinctly during the horizontal flame fire tests with only 2 wt% mass add-on. In addition, the flame retardancy was retained after 1 hour washing at 70°C. Cone calorimeter tests indicated that all treated samples showed a 53% reduction in TTI values and decreased PHRR values (202 versus 102 and 107 kW/m^2 for PET, 5 BL and 5 BL 0.10 M NaCl, respectively).

Although dipping is widely employed to construct multi-layer coatings, spray may be more advantageous for its efficiency and feasibility on an industrial scale. Schaaf et al. (2012) explored three different methods, i.e., dipping, vertical, and horizontal spray to construct silica-containing LbL hybrid coatings

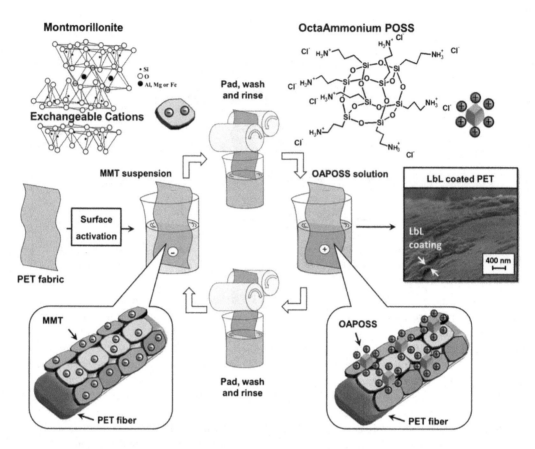

FIGURE 13.10 Schematic representation of the LBL process adopted in the present study. The PET fabric was primed by a BPEI and then alternatively dipped in MMT suspension (negative) and OAPOSS solution (positive) until 5 BL achieved. (Reprinted from Carosio, F. et al., *J. Colloid Interface Sci.*, 510, 142–151, 2018. with permission.)

for flame retarding cotton. The results indicated that the horizontal spray is effective at constructing homogeneous and coherent coatings. Following by this work, the same architecture consisting of 5 BL alumina-coated silica/unmodified silica was deposited on high-density PET fabrics (Carosio et al. 2013b). LbL coating constructed by horizontal spray is more effective than dipping coating at enhancing the flame retardancy (34% versus 3% in terms of the PHRR reduction). As is well known, dipping allows coating to form uniform coatings on the surface, while spray coating follows an island-growth pattern.

A sol-gel approach has been employed to prevent melt dripping and improve fire safety of PET fabrics. Guido et al. (2013) treated PET fabric with 3-glyci-dyloxypropyltrimethoxysilane (GPTMS) sols and/or a synthetic unmodified boehmite. One or two layers of boehmite on PET alone did not have significant influence on their flammability. The presence of GPTMS increased the total combustion time and dripping suppression, while it generated dark and dense smokes. The combination of both boehmite and GPTMS ensured improved fire retardancy of the PET with increased char residues and suppressed dripping behavior.

Plasma technology allows the production of extremely thin coatings with the desired thermal shielding effects on the surface of treated substrates. A number of polymer materials have demonstrated that low pressure, radio frequency discharges plasma treatment in the presence of gaseous (CF_4/CH_4) for flame retardation (Shi 1999). Carosio et al. (2011a) utilized plasma treatment to let nanoparticles adsorb and create a functional coating for PET fabrics with improved thermal stability and FR properties. The presence of the nanoparticles delayed the thermal degradation process of the treated sample in air, as demonstrated by TGA. Cone calorimeter results revealed that plasma-treated fabrics showed a remarkably improved fire safety including increased TTI (up to 104%) and slightly reduced PHRR (ca. 10%) as compared to the original sample. Photo-induced surface grafting of glycidyl methacrylate with a pad-cure FR system consisting of 1-hydroxy ethylidene-1,1-diphosphonic acid and sulfamic acid was applied to PET fabric with improved fire resistance and reduced dripping behavior (Yu et al. 2010). Grafting content of 22.5% was found to have the highest LOI value (25.9%). The treated fabric exhibited lower heat release than the untreated sample. The generation of more residual residues with an intumescent char structure during the burning process is responsible for the improved fire retardancy.

13.3.2.3.3 *PAN Fabrics*

Various approaches, e.g., copolymerization, co-blending, and oxidation modification, have been employed for the FR modification of PAN fibers. To date, most of the industrial FR PAN fibers are prepared by a copolymerization method. The commercial products prepared by copolymerization involve the Dynel fiber produced from the Union Carbide Corporation, the Kanekalon fiber produced from the Kaneka Corporation and the Drlon FLR fiber produced from S. T. Dupont. The phosphorus-silica FR system has been applied to PAN fibers. For example, phosphorus-doped silica coatings prepared by the sol-gel reaction make PAN fibers non-flammable when a 15 s flame was applied, which can also withstand up to ten washing cycles according to TS EN ISO 105-C06-A1S (Yaman 2009). Low pressure plasma-induced graft polymerization (PIGP) is considered as the most promising and versatile technique for industrial application. Tsafack and Levalois-Grützmacher (2006a, 2006b) studied argon PIGP of some acrylic phosphorus-based monomers and revealed their efficiency. An up to 8-unit increase in the LOI value than untreated PAN fibers was achieved, which is dependent on the type of grafted monomer. Tsafack et al. grafted several phosphorus-containing acrylate monomers onto the surface of PAN fabrics by low pressure argon plasma grafting polymerization, including (diethyl(acryloyloxyethyl)phosphate, diethyl-2-(methacryloyloxyethyl) phosphate, diethyl(acryloyloxymethyl)phosphonate, and dimethyl(acryloyloxymethyl)phosphonate (DMAMP) (Tsafack et al. 2004; Tsafack and Levalois-Grützmacher 2006c). The presence of ethyleneglycoldiacrylate (EGDA) as a grafting agent increased the graft yields as high as 28 wt%. The LOI value of the treated PAN fabric acquired 26.5% and was reduced to 21% after laundering.

13.3.3 Flame Retardant Mechanisms for Textiles

Flame retardancy of textiles can be achieved on the basis of gas phase action, condensed phase action, or the combination. The gas phase FR mechanism can be confirmed by detecting evolved species during burning, such as halogen or phosphorus species derived from halogen- and phosphorus-based FRs. These reactive species act as free radical scavengers to inhibit flame in the vapor phase. Commercial phosphorus-containing FR products, e.g., trimethyl phosphate and triphenyl phosphate derivatives have been found to have gas-phase FR action. Some phosphorus species have also been found to have gas phase action, such as PO, HPO_2, and PO_2. These phosphorus species have been confirmed by mass spectrometric analysis of flames. Phosphorus-based FRs produce active low-energy PO·radicals, which can react with high-energy HO or H radicals from the degrading substrates (Eqs 1 and 2) (Granzow 1978; Neisius et al. 2014). Subsequently, HPO reacts with H· and OH· radicals regenerating the PO· species. Thus, this series of reactions result in a lower energy/heat release, cooling down and extinguishing the flame (Eqs 3 and 4):

$$HO_* + PO_* \longrightarrow HPO_2 \qquad (1)$$

$$H_* + PO_* \longrightarrow HPO \qquad (2)$$

$$HPO + HO_* \longrightarrow PO_* + H_2O \qquad (3)$$

$$HPO + H_* \longrightarrow PO_* + H_2 \qquad (4)$$

Condensed phase action is based on the barrier effect of a temperature-resistant char layer derived from the FR system and textiles during burning. Such a protective barrier could prevent the textile against a fire attack. Generally, the FR system promotes the early degradation of textile substrates at lower temperatures by changing the pyrolytic pathways, resulting in the formation of additional char residues. For multi-hydroxyl substrates such as cellulose, phosphorus-containing FRs are degraded to form polyphosphoric acid which reacts with cellulose by releasing water to form a thermal insulating char layer on the surface of the pyrolyzing matrix. The flame retardant efficiency of phosphorous FR in the condensed phase could be further enhanced in the presence of nitrogen-based (e.g., melamine), silica-containing (e.g., TEOS) or nanoscale additives (e.g., graphene). The synergistic FR enhancement was due to possible chemical reactions between degraded products from synergistic agents and the system, resulting in enhancement of the char layer or gas phase action, thereby interrupting the combustion process.

13.4 Summary and Outlook

This chapter reviewed the current effective techniques for imparting the fire safety of natural and synthetic fibers/fabrics by surface engineering. Constructing coatings on the fabric surface through LBL assembly, plasma treatment, sol-gel reaction, UV-curing technique, and dual-curing process has been demonstrated as versatile and efficient tools for making the textile flame retardation. Considering the adverse effect on health and environment, halogen-based coatings systems are being increasingly replaced by eco-friendly FR coating systems. These techniques involve eco-friendly processes which are primarily performed in solvent-free or aqueous systems. Importantly, the formation of extremely thin coatings onto textiles almost has no adverse effect on textile features such as comfort and mechanical properties. Multi-functional treated textiles with improved gas barrier properties, hydrophobicity, and electrical conductivity are achieved via surface-treatment techniques. Although the results reported to date above are impressive, there are still some drawbacks to overcome. Achieving highly efficient coatings with FR durability and large-scale processes for possible industrial application remain huge challenges. Especially for LBL assembly coatings, the weak interaction forces, such as electrostatic interactions or hydrogen bonds between coating and substrate are not enough to withstand the washing process, resulting in the formation of non-durable FR coatings. In contrast, derived architectures onto textiles created by sol-gel and dual-cure processes are demonstrated as a semi-durable or durable treatment because FR systems formed covalent linkage to the textile matrix. Finally, it has become imperative to seek for low-cost, highly efficient synergistic multi-purpose FR systems, which can confer functional properties, such as gas barrier effect, high hydrophobicity, and electrical conductivity, etc. for flexible polyurethane foams (FPUFs).

References

Achwal, W., Mahapatrao, C., and Kaduskar, P. 1987. Flame-retardant finishing of cotton and silk fabrics. *Colourage* 34: 33–37.

Alongi, J., Carosio, F., and Malucelli, G. 2014. Current emerging techniques to impart flame retardancy to fabrics: An overview. *Polymer Degradation and Stability* 106: 138–149.

Alongi, J., Carosio, F., and Malucelli, G. 2012a. Influence of ammonium polyphosphate-/poly (acrylic acid)-based layer by layer architectures on the char formation in cotton, polyester and their blends. *Polymer Degradation and Stability* 97: 1644–1653.

Alongi, J., Carosio, F., and Malucelli, G. 2012b. Layer by layer complex architectures based on ammonium polyphosphate, chitosan and silica on polyester-cotton blends: Flammability and combustion behaviour. *Cellulose* 19: 1041–1050.

Alongi, J., Ciobanu, M., and Malucelli, G. 2011a. Cotton fabrics treated with hybrid organic–inorganic coatings obtained through dual-cure processes. *Cellulose* 18: 1335–1348.

Alongi, J., Ciobanu, M., and Malucelli, G. 2011b. Novel flame retardant finishing systems for cotton fabrics based on phosphorus-containing compounds and silica derived from sol–gel processes. *Carbohydrate Polymers* 85: 599–608.

Alongi, J., Ciobanu, M., and Malucelli, G. 2011c. Sol–gel treatments for enhancing flame retardancy and thermal stability of cotton fabrics: Optimisation of the process and evaluation of the durability. *Cellulose* 18: 167–177.

Alongi, J., Ciobanu, M., and Malucelli, G. 2012c. Sol–gel treatments on cotton fabrics for improving thermal and flame stability: Effect of the structure of the alkoxysilane precursor. *Carbohydrate Polymers* 87: 627–635.

Alongi, J., Ciobanu, M., and Malucelli, G. 2012d. Thermal stability, flame retardancy and mechanical properties of cotton fabrics treated with inorganic coatings synthesized through sol–gel processes. *Carbohydrate Polymers* 87: 2093–2099.

Alongi, J., Ciobanu, M., Tata, J., et al. 2011d. Thermal stability and flame retardancy of polyester, cotton, and relative blend textile fabrics subjected to sol–gel treatments. *Journal of Applied Polymer Science* 119: 1961–1969.

Alongi, J., and Frache, A. 2010. Flame retardancy properties of α-zirconium phosphate based composites. *Polymer Degradation and Stability* 95: 1928–1933.

Alongi, J., Horrocks, A. R., Carosio, F., et al. 2013. *Update on Flame Retardant Textiles.* Shawbury, UK: Smithers Rapra.

Alongi, J., and Malucelli, G. 2012. State of the art and perspectives on sol–gel derived hybrid architectures for flame retardancy of textiles. *Journal of Materials Chemistry* 22: 21805–21809.

Altıntaş, Z., Çakmakçı, E., Kahraman, M., et al. 2011. Preparation of photocurable silica–titania hybrid coatings by an anhydrous sol–gel process. *Journal of Sol-Gel Science and Technology* 58: 612–618.

Apaydin, K., Laachachi, A., Ball, V., et al. 2013. Polyallylamine–montmorillonite as super flame retardant coating assemblies by layer-by layer deposition on polyamide. *Polymer Degradation and Stability* 98: 627–634.

Apaydin, K., Laachachi, A., Ball, V., et al. 2014. Intumescent coating of (polyallylamine-polyphosphates) deposited on polyamide fabrics via layer-by-layer technique. *Polymer Degradation and Stability* 106: 158–164.

Apaydin, K., Laachachi, A., Ball, V., et al. 2015. Layer-by-layer deposition of a TiO2-filled intumescent coating and its effect on the flame retardancy of polyamide and polyester fabrics. *Colloids and Surfaces A: Physicochemical and Engineering Aspects* 469: 1–10.

Benisek, L. 1971. Use of titanium complexes to improve natural flame retardancy of wool. *Journal of the Society of Dyers and Colourists* 87: 277.

Benisek, L. 1974. Communication: Improvement of the natural flame-resistance of wool. Part I: Metal-complex applications. *Journal of the Textile Institute* 65: 102–108.

Bourbigot, S., Devaux, E., and Flambard, X. 2002a. Flammability of polyamide-6/clay hybrid nanocomposite textiles. *Polymer Degradation and Stability* 75: 397–402.

Bourbigot, S., Devaux, E., Rochery, M., et al. 2002b. Nanocomposite textiles: New route for flame retardancy. *47th International SAMPE Symposium and Exhibition 2002*, pp. 1108–1118.

Cai, Y., Wu, N., Wei, Q., et al. 2008. Structure, surface morphology, thermal and flammability characterizations of polyamide6/organic-modified Fe-montmorillonite nanocomposite fibers functionalized by sputter coating of silicon. *Surface and Coatings Technology* 203: 264–270.

Cao, J., and Wang, C. 2017. Multifunctional surface modification of silk fabric via graphene oxide repeatedly coating and chemical reduction method. *Applied Surface Science* 405: 380–388.

Carosio, F., Alongi, J., and Frache, A. 2011a. Influence of surface activation by plasma and nanoparticle adsorption on the morphology, thermal stability and combustion behavior of PET fabrics. *European Polymer Journal* 47: 893–902.

Carosio, F., Alongi, J., and Malucelli, G. 2011b. α-Zirconium phosphate-based nanoarchitectures on polyester fabrics through layer-by-layer assembly. *Journal of Materials Chemistry* 21: 10370–10376.

Carosio, F., Alongi, J., and Malucelli, G. 2012. Layer by layer ammonium polyphosphate-based coatings for flame retardancy of polyester–cotton blends. *Carbohydrate Polymers* 88: 1460–1469.

Carosio, F., Alongi, J., and Malucelli, G. 2013a. Flammability and combustion properties of ammonium polyphosphate-/poly(acrylic acid)-based layer by layer architectures deposited on cotton, polyester and their blends. *Polymer Degradation and Stability* 98: 1626–1637.

Carosio, F., Di Blasio, A., Cuttica, F., et al. 2013b. Flame retardancy of polyester fabrics treated by spray-assisted layer-by-layer silica architectures. *Industrial & Engineering Chemistry Research* 52: 9544–9550.

Carosio, F., Di Pierro, A., Alongi, J., et al. 2018. Controlling the melt dripping of polyester fabrics by tuning the ionic strength of polyhedral oligomeric silsesquioxane and sodium montmorillonite coatings assembled through layer by layer. *Journal of Colloid and Interface Science* 510: 142–151.

Carosio, F., Laufer, G., Alongi, J., et al. 2011c. Layer-by-layer assembly of silica-based flame retardant thin film on PET fabric. *Polymer Degradation and Stability* 96: 745–750.

Chaiwong, C., Tunma, S., Sangprasert, W., et al. 2010. Graft polymerization of flame-retardant compound onto silk via plasma jet. *Surface and Coatings Technology* 204: 2991–2995.

Chattopadhyay, D., Panda, S. S., and Raju, K. 2005. Thermal and mechanical properties of epoxy acrylate/methacrylates UV cured coatings. *Progress in Organic Coatings* 54: 10–19.

Chen, L., Song, L., Lv, P., et al. 2011. A new intumescent flame retardant containing phosphorus and nitrogen: Preparation, thermal properties and application to UV curable coating. *Progress in Organic Coatings* 70: 59–66.

Chen, S., Li, X., Li, Y., et al. 2015. Intumescent flame-retardant and self-healing superhydrophobic coatings on cotton fabric. *ACS Nano* 9: 4070–4076.

Cheng, X. E., and Shi, W. F. 2010. Synthesis and thermal properties of silicon-containing epoxy resin used for UV-curable flame-retardant coatings. *Journal of Thermal Analysis and Calorimetry* 103: 303–310.

Cheng, X. W., Guan, J. P., Chen, G., et al. 2016. Adsorption and flame retardant properties of bio-based phytic acid on wool fabric. *Polymers* 8: 122.

Cheng, X. W., Guan, J. P., Yang, X. H., et al. 2017. Improvement of flame retardancy of silk fabric by bio-based phytic acid, nano-TiO2, and polycarboxylic acid. *Progress in Organic Coatings* 112: 18–26.

Cheng, X. W., Liang, C. X., Guan, J. P., et al. 2018. Flame retardant and hydrophobic properties of novel sol-gel derived phytic acid/silica hybrid organic-inorganic coatings for silk fabric. *Applied Surface Science* 427: 69–80.

Davies, P. J., Horrocks, A. R., and Miraftab, M. 2000. Scanning electron microscopic studies of wool/ intumescent char formation. *Polymer International* 49: 1125–1132.

DeArmitt, C. 2011. Polyhedral oligomeric silsesquioxanes in plastics. In *Applications of Polyhedral Oligomeric Silsesquioxanes*, ed. C. Hartmann-Thompson, 209–228. Basel, Switzerland: Springer Nature.

Decker, C. 2002. Kinetic study and new applications of UV radiation curing. *Macromolecular Rapid Communications* 23: 1067–1093.

Dolez, P., and Vermeersch, O. 2018. Introduction to advanced characterization and testing of textiles. In *Advanced Characterization and Testing of Textiles*, ed. P. Dolez, O. Vermeersch, and V. Izquierdo, 3–21. Cambridge, UK: Woodhead Publishing.

Edwards, B., El-Shafei, A., Hauser, P., et al. 2012. Towards flame retardant cotton fabrics by atmospheric pressure plasma-induced graft polymerization: Synthesis and application of novel phosphoramidate monomers. *Surface and Coatings Technology* 209: 73–79.

Errifai, I., Jama, C., Le Bras, M., et al. 2004. Elaboration of a fire retardant coating for polyamide-6 using cold plasma polymerization of a fluorinated acrylate. *Surface and Coatings Technology* 180: 297–301.

Fang, F., Tong, B., Du, T., et al. 2016. Unique nanobrick wall nanocoating for flame-retardant cotton fabric via layer-by-layer assembly technique. *Cellulose* 23: 3341–3354.

Forouharshad, M., Montazer, M., Bameni Moghadam, M., et al. 2010. Flame retardancy of wool fabric with zirconium oxychloride optimized by central composite design. *Journal of Fire Sciences* 28: 561–572.

Forouharshad, M., Montazer, M., Moghadam, M., et al. 2011a. Preparation of flame retardant wool using zirconium acetate optimized by CCD. *Thermochimica Acta* 520: 134–138.

Forouharshad, M., Montazer, M., Moghadam, M. B., et al. 2011b. Flame retardant wool using zirconium oxychloride in various acidic media optimized by RSM. *Thermochimica Acta* 516: 29–34.

Gashti, M. P., Almasian, A., and Gashti, M. P. 2012. Preparation of electromagnetic reflective wool using nano-ZrO2/citric acid as inorganic/organic hybrid coating. *Sensors and Actuators A: Physical* 187: 1–9.

Granzow, A. 1978. Flame retardation by phosphorus compounds. *Accounts of Chemical Research* 11: 177–183.

Guan, J., Yang, C. Q., and Chen, G. 2009. Formaldehyde-free flame retardant finishing of silk using a hydroxyl-functional organophosphorus oligomer. *Polymer Degradation and Stability* 94: 450–455.

Guan, J. P., and Chen, G. Q. 2006. Flame retardancy finish with an organophosphorus retardant on silk fabrics. *Fire and Materials* 30: 415–424.

Guido, E., Alongi, J., Colleoni, C., et al. 2013. Thermal stability and flame retardancy of polyester fabrics sol–gel treated in the presence of boehmite nanoparticles. *Polymer Degradation and Stability* 98: 1609–1616.

Horrocks, A. R. 1986. Flame-retardant finishing of textiles. *Coloration Technology* 16: 62–101.

Horrocks, A. R. 2011. Flame retardant challenges for textiles and fibres: New chemistry versus innovatory solutions. *Polymer Degradation and Stability* 96: 377–392.

Horrocks, A. R. 2008. Flame retardant/resistant textile coatings and laminates. In *Advances in Fire Retardant Materials*, ed. A. R. Horrocks, and D. Price, pp. 159–187. Cambridge, UK: Woodhead Publishing.

Horrocks, A. R., and Davies, P. J. 2000. Char formation in flame-retarded wool fibres. Part 1. Effect of intumescent on thermogravimetric behaviour. *Fire and Materials* 24: 151–157.

Horrocks, A. R., Kandola, B. K., Davies, P., et al. 2005. Developments in flame retardant textiles– A review. *Polymer Degradation and Stability* 88: 3–12.

Horrocks, A. R., Kandola, B. K., and Padbury, S. 2003. The effect of functional nanoclays in enhancing the fire performance of fibre-forming polymers. *Journal of the Textile Institute* 94: 46–66.

Horrocks, A. R., Kandola, B. K., and Padbury, S. A. 2004. Effectiveness of nanoclays as flame retardants for fibres. *Proceedings of Flame Retardants 2004, Interscience Communications*, London, UK: Interscience Publications Ltd., pp. 97–108.

Hsiue, G. H., Liu, Y. L., and Liao, H. H. 2001. Flame-retardant epoxy resins: An approach from organic–inorganic hybrid nanocomposites. *Journal of Polymer Science Part A: Polymer Chemistry* 39: 986–996.

Hu, Y., Yu, B., and Song, L. 2016. Novel fire-retardant coatings. In *Novel Fire Retardant Polymers and Composite Materials*, ed. D. Y. Wang, pp. 53–91. Cambridge, UK: Woodhead Publishing.

Huang, G., Liang, H., Wang, X., et al. 2012a. Poly (acrylic acid)/clay thin films assembled by layer-by-layer deposition for improving the flame retardancy properties of cotton. *Industrial & Engineering Chemistry Research* 51: 12299–12309.

Huang, G., Yang, J., Gao, J., et al. 2012b. Thin films of intumescent flame retardant-polyacrylamide and exfoliated graphene oxide fabricated via layer-by-layer assembly for improving flame retardant properties of cotton fabric. *Industrial & Engineering Chemistry Research* 51: 12355–12366.

Jama, C., Quédé, A., Goudmand, P., et al. 2001. Fire retardancy and thermal stability of materials coated by organosilicon thin films using a cold remote plasma process. In *Fire and Polymers*, ed. G. L., Nelson, C. A. Wilkie, pp. 200–213. Washington, DC: ACS Publications.

Ji, Y., Chen, G., and Xing, T. 2018. Rational design and preparation of flame retardant silk fabrics coated with reduced graphene oxide. *Applied Surface Science*. doi:10.1016/j.apsusc.2018.03.120

Ji, Y., Li, Y., Chen, G., et al. 2017. Fire-resistant and highly electrically conductive silk fabrics fabricated with reduced graphene oxide via dry-coating. *Materials & Design* 133: 528–535.

Joseph, P., and Tretsiakova-McNally, S. 2013. Chemical modification of natural and synthetic textile fibres to improve flame retardancy. In *Handbook of Fire Resistant Textiles*, ed. F. S., Kilinc, pp. 37–67. Cambridge, UK: Woodhead Publishing.

Kako, T., and Katayama, A. 1995. Performance properties of silk fabrics flameproofed with organic phosphorous agent. *The Journal of Sericultural Science of Japan* 64: 124–131.

Kamlangkla, K., Hodak, S. K., and Levalois-Grützmacher, J. 2011. Multifunctional silk fabrics by means of the plasma induced graft polymerization (PIGP) process. *Surface and Coatings Technology* 205: 3755–3762.

Kundu, C. K., Wang, W., Zhou, S., et al. 2017a. A green approach to constructing multilayered nanocoating for flame retardant treatment of polyamide 66 fabric from chitosan and sodium alginate. *Carbohydrate Polymers* 166: 131–138.

Kundu, C. K., Yu, B., Gangireddy, C. S. R., et al. 2017b. UV grafting of a DOPO-based phosphoramidate monomer onto polyamide 66 fabrics for flame retardant treatment. *Industrial & Engineering Chemistry Research* 56: 1376–1384.

Laoutid, F., Bonnaud, L., Alexandre, M., et al. 2009. New prospects in flame retardant polymer materials: From fundamentals to nanocomposites. *Materials Science and Engineering: R: Reports* 63: 100–125.

Laufer, G., Kirkland, C., Morgan, A. B., et al. 2012. Intumescent multilayer nanocoating, made with renewable polyelectrolytes, for flame-retardant cotton. *Biomacromolecules* 13: 2843–2848.

Leistner, M., Abu-Odeh, A. A., Rohmer, S. C., et al. 2015. Water-based chitosan/melamine polyphosphate multilayer nanocoating that extinguishes fire on polyester-cotton fabric. *Carbohydrate Polymers* 130: 227–232.

Levchik, S. V., and Weil, E. D. 2000. Combustion and fire retardancy of aliphatic nylons. *Polymer International* 49: 1033–1073.

Li, Y. C., Mannen, S., Morgan, A. B., et al. 2011b. Intumescent all-polymer multilayer nanocoating capable of extinguishing flame on fabric. *Advanced Materials* 23: 3926–3931.

Li, Y. C., Mannen, S., Schulz, J., et al. 2011a. Growth and fire protection behavior of POSS-based multilayer thin films. *Journal of Materials Chemistry* 21: 3060–3069.

Li, Y. C., Schulz, J., and Grunlan, J. C. 2009. Polyelectrolyte/nanosilicate thin-film assemblies: Influence of pH on growth, mechanical behavior, and flammability. *ACS Applied Materials & Interfaces* 1: 2338–2347.

Li, Y. C., Schulz, J., Mannen, S., et al. 2010. Flame retardant behavior of polyelectrolyte-clay thin film assemblies on cotton fabric. *ACS Nano* 4: 3325–3337.

Liang, S., Neisius, N. M., and Gaan, S. 2013. Recent developments in flame retardant polymeric coatings. *Progress in Organic Coatings* 76: 1642–1665.

Liu, W., Zhang, S., Sun, J., et al. 2017. An improved method for the durability of the flame retardant PA66 fabric. *Journal of Thermal Analysis and Calorimetry* 128: 193–199.

Malucelli, G. 2016. Surface-engineered fire protective coatings for fabrics through sol-gel and layer-by-layer methods: An overview. *Coatings* 6: 33.

Martini, P., Spearpoint, M., and Ingham, P. 2010. Low-cost wool-based fire blocking inter-liners for upholstered furniture. *Fire Safety Journal* 45: 238–248.

Miao, H., Cheng, L., and Shi, W. 2009. Fluorinated hyperbranched polyester acrylate used as an additive for UV curing coatings. *Progress in Organic Coatings* 65: 71–76.

Mülazim, Y., Kahraman, M. V., Apohan, N. K., et al. 2011. Preparation and characterization of UV-curable, boron-containing, transparent hybrid coatings. *Journal of Applied Polymer Science* 120: 2112–2121.

Nazaré, S. 2009. Environmentally friendly flame-retardant textiles. in: *Sustainable Textiles*, ed. R. S. Blackburn, 339–368. Cambridge, UK: Woodhead Publishing.

Neisius, N. M., Lutz, M., Rentsch, D., et al. 2014. Synthesis of DOPO-based phosphonamidates and their thermal properties. *Industrial & Engineering Chemistry Research* 53: 2889–2896.

Pan, H., Wang, W., Pan, Y., et al. 2015. Formation of layer-by-layer assembled titanate nanotubes filled coating on flexible polyurethane foam with improved flame retardant and smoke suppression properties. *ACS Applied Materials & Interfaces* 7: 101–111.

Pan, Y., Wang, W., Liu, L., et al. 2017. Influences of metal ions crosslinked alginate based coatings on thermal stability and fire resistance of cotton fabrics. *Carbohydrate Polymers* 170: 133–139.

Park, Y. J., Lim, D. H., Kim, H. J., et al. 2009. UV-and thermal-curing behaviors of dual-curable adhesives based on epoxy acrylate oligomers. *International Journal of Adhesion and Adhesives* 29: 710–717.

Quede, A., Mutel, B., Supiot, P., et al. 2004. Characterization of organosilicon films synthesized by N2-PACVD. Application to fire retardant properties of coated polymers. *Surface and Coatings Technology* 180: 265–270.

Schaaf, P., Voegel, J. C., Jierry, L., et al. 2012. Spray-assisted polyelectrolyte multilayer buildup: From step-by-step to single-step polyelectrolyte film constructions. *Advanced Materials* 24: 1001–1016.

Schreck, K. M., Leung, D., and Bowman, C. N. 2011. Hybrid organic/inorganic thiol–ene-based photo-polymerized networks. *Macromolecules* 44: 7520–7529.

Shi, L. 1999. Investigation of surface modification and reaction kinetics of PET in RF CF_4-CH_4 plasmas. *Journal of Polymer Engineering* 19: 445–455.

Tian, C., Li, Z., Guo, H., et al. 2003. Study on the thermal degradation of flame retardant wools. *Journal of Fire Sciences* 21: 155–162.

Totolin, V., Sarmadi, M., Manolache, S. O., et al. 2010. Atmospheric pressure plasma enhanced synthesis of flame retardant cellulosic materials. *Journal of Applied Polymer Science* 117: 281–289.

Tsafack, M. J., Hochart, F., and Levalois-Grützmacher, J. 2004. Polymerization and surface modification by low pressure plasma technique. *The European Physical Journal-Applied Physics* 26: 215–219.

Tsafack, M. J., and Levalois-Grützmacher, J. 2006a. Flame retardancy of cotton textiles by plasma-induced graft-polymerization (PIGP). *Surface and Coatings Technology* 201: 2599–2610.

Tsafack, M. J., and Levalois-Grützmacher, J. 2006b. Plasma-induced graft-polymerization of flame retardant monomers onto PAN fabrics. *Surface and Coatings Technology* 200: 3503–3510.

Tsafack, M. J., and Levalois-Grützmacher, J. 2006c. Plasma-induced graft-polymerization of flame retardant monomers onto PAN fabrics. *Surface and Coatings Technology* 200: 3503–3510.

Tsafack, M. J., and Levalois-Grützmacher, J. 2007. Towards multifunctional surfaces using the plasma-induced graft-polymerization (PIGP) process: Flame and waterproof cotton textiles. *Surface and Coatings Technology* 201: 5789–5795.

Uğur, Ş. S., Sarışık, M., and Aktaş, A. H. 2011. Nano-Al_2O_3 multilayer film deposition on cotton fabrics by layer-by-layer deposition method. *Materials Research Bulletin* 46: 1202–1206.

Wang, F., Hu, J., and Tu, W. 2008. Study on microstructure of UV-curable polyurethane acrylate films. *Progress in Organic Coatings* 62: 245–250.

Wang, W., Wang, X., Pan, Y., et al. 2017. Synthesis of phosphorylated graphene oxides based multi-layer coating: Self-assembly method and application for improving the fire safety of cotton fabrics. *Industrial & Engineering Chemistry Research* 56: 6664–6670.

Weil, E. D., and Levchik, S. V. 2008. Flame retardants in commercial use or development for textiles. *Journal of Fire Sciences* 26: 243–281.

Xing, W., Jie, G., Song, L., et al. 2011. Flame retardancy and thermal degradation of cotton textiles based on UV-curable flame retardant coatings. *Thermochimica Acta* 513: 75–82.

Xing, W., Song, L., Hu, Y., et al. 2010. Combustion and thermal behaviors of the novel UV-cured intu-mescent flame retardant coatings containing phosphorus and nitrogen. *e-Polymers* 10: 684–694.

Xu, G., and Shi, W. 2005. Synthesis and characterization of hyperbranched polyurethane acrylates used as UV curable oligomers for coatings. *Progress in Organic Coatings* 52: 110–117.

Yaman, N. 2009. Preparation and flammability properties of hybrid materials containing phosphorous compounds via sol-gel process. *Fibers and Polymers* 10: 413–418.

Yu, L., Zhang, S., Liu, W., et al. 2010. Improving the flame retardancy of PET fabric by photo-induced grafting. *Polymer Degradation and Stability* 95: 1934–1942.

Yuan, H., Xing, W., Zhang, P., et al. 2012. Functionalization of cotton with UV-cured flame retardant coatings. *Industrial & Engineering Chemistry Research* 51: 5394–5401.

Zhang, D., Williams, B. L., Shrestha, S. B., et al. 2017a. Flame retardant and hydrophobic coatings on cotton fabrics via sol-gel and self-assembly techniques. *Journal of Colloid and Interface Science* 505: 892–899.

Zhang, Q. H., Chen, G. Q., and Xing, T. L. 2017b. Silk flame retardant finish by ternary silica sol containing boron and nitrogen. *Applied Surface Science* 421: 52–60.

Zhang, Q. H., Gu, J., Chen, G. Q., et al. 2016. Durable flame retardant finish for silk fabric using boron hybrid silica sol. *Applied Surface Science* 387: 446–453.

Zhang, Q., Zhang, W., Huang, J., et al. 2015. Flame retardance and thermal stability of wool fabric treated by boron containing silica sols. *Materials & Design* 85: 796–799.

Zhang, X., Akram, R., Zhang, S., et al. 2017c. Hexa(eugenol)cyclotriphosphazene modified bismaleimide resins with unique thermal stability and flame retardancy. *Reactive and Functional Polymers* 113: 77–84.

Zhou, H. 1995. Now and future of flame retardancy of fabrics. *Guangxi Textile Science Technology* 24: 35–39.

14

Flame Retardant Polymer Materials Design for Wire and Cable Applications

Christian Lagreve,
Laurent Ferry,
and Jose-Marie
Lopez-Cuesta

14.1 Introduction

Cables are present everywhere in our technological and domestic environments. Due to their main function to transport electric currents for power transmission, power distribution, communication, and instrumentation, and the presence of polymers used as insulating or jacketing materials, cables represent a main source of fire risk due to the ignition of combustible material or propagation. Statistical studies carried out in various European countries have shown that electrical equipment represent the main source of domestic fires (Netherland Institute for Safety 2009). For example, in France, from the Ministry of the Interior, 25% of fires are originated in electrical equipment (INPES 2009). Yang et al. (2013) reported that in China, electrical facilities account for 30% of the causes of fires. The rapid social and economic development in emerging countries has led to urban building complexes combining multi-type enclosure spaces and various urban functions, such as transportation hubs, in which cables are employed extensively and laid intensively. Cables are a common cause of fire, not only in transportation and industries, but also in power plants since they are widely used in power plants and particularly in nuclear ones. Table 14.1 reports different fire incidences that occurred due to cable failure and propagation, in power plants or transportation places.

TABLE 14.1 Fire Incidences Occurred Due to Cable Failure and Other Sources

San Onofre (USA)	1968	Self-ignited cable fire due to variation in cable layout
Browns Ferry (USA)	1975	Fire in cable spreading room (reactor building)
Greifswald (Germany)	1975	Large switchgear and cable fire
Beloyarsk (Russia)	1978	Cable fire in turbine hall spreading to other plant areas
Zaporoshye (Ukraine, USSR)	1984	Cable fire
Kalinin (Russia)	1984	Turbine hall fire with several pilot fires at power cable
Ignalina (Lithuania)	1988	Self-ignition of cable fire
Waterford (USA)	1995	Switchgear fire propagating via vertical cables
Dusseldorf airport (Germany)	1996	Fire propagating via cables
London Holborn tube station	2016	Cable fire and underground propagation

Source: Chaudhary, A. et al. *Procedia Earth Planet. Sci.*, 11, 376–384, 2015.

According to their various fields of applications, cables operate different functions. Hence, the range of fire behavior properties required can be different. The resistance of fire propagation and resistance to ignition are always among the first order criteria for the behavior. Other properties such as, for example, ability to self-extinguish, total energy released, smoke emission, and release of corrosive degradation products may be prioritized. In addition, for all areas of use of cables, specific mechanical properties are required for the polymers with emphasis on elongation at maximal tensile stress and at break. Hence, glass temperature transition for the selected polymers has also to be considered.

The fire risk related to the use of cables depends on the nature of the polymers and their associated flame retardant (FR) systems, but also on the construction of cables. Polymers are used as insulators, but also as jacketing material. Metals are not only used as electric conductors, but also as metallic layers used to protect some components such as optical fibers or to play a role of armor toward external electromagnetic fields. It appears that the assembly of the various polymer and metal components of the cable significantly governs its fire behavior. Hence, the characterization of fire requires specific fire tests, as well as modeling, since the study of the specific behavior of the polymer materials used in insulation and jacketing does not bring enough information on the cable fire behavior. Among the main materials used for both kinds of components, poly(vinyl chloride) (PVC) and polyolefins play the main role. The selection of other materials such as polydimethylsiloxanes, fluorinated polymers, polyurethanes, and various polymer alloys correspond to particular conditions of use, for example, the contact with specific chemicals or the resistance to elevated temperatures during use.

In the following sections, after an overview of the main standards and fire tests for cables, the different polymers used as well as the usual FR systems for these materials will be detailed. A specific focus will be made on the advantages and limitations of such systems in order to show the interest of nanotechnology to improve the fire performance. The last part will be devoted to environmental issues since they become inescapable for the development of new FR systems as well as concerning the end-of-life of cables.

14.2 Standards and Tests for Cables

14.2.1 Standards and Sectors

Cables and wires can be found in a great number of applications for energy transport, signal, or data transmission. The specifications that they have to fulfill in terms of fire behavior depend on the region of the world where they are traded and on the sector in which they are used (buildings, railways, maritime, aeronautics, etc.). As examples, in the European Union, cables used in buildings have been included in the classification system under the European Construction Products Regulation (CPR) defined in EN 13501-6 standard (EN 2014a). Thus, the specifications and testing methods for power, control, and communication cables have been defined in the harmonized product standard EN 50575 published in 2016 (EN 2016). Cables used in the railway sector have to respect the specifications of the EN 45545 standard that defines fire testing applicable to railway components (EN 2013). Cables used in ships shall comply with the provisions determined by the International Maritime Organization (IMO) in its fire testing procedure code (IMO 2010).

Whatever the sector, all the fire tests that are applied to cables can be divided into two categories: reaction to fire tests that assess the behavior of cables as a fuel source and fire resistance tests that assess the evolution of the physical properties of cables during a fire.

14.2.2 Reaction to Fire Tests

Standards dedicated to reaction to fire of cables assess mainly four parameters: (i) flame propagation, (ii) heat release, (iii) smoke production, and (iv) emission of corrosive.

14.2.2.1 Flame Propagation

IEC 60332-1 is the international standard test characterizing the vertical flame propagation for a single insulated wire or cable (IEC 2004a). In this test, a 1 kW and 125 mm high pre-mixed flame from a propane burner is applied 100 mm over the bottom of a 600 mm vertically oriented cable (Figure 14.1). The burner is oriented at 45° from the vertical cable, and the flame is applied during a period ranging from 60 to 480 seconds (s) depending on the overall diameter of the cable. The test is deemed passed if, after the flame has been removed, the burning cable extinguishes itself and the fire damage is at least 50 mm from the upper mounting clamp. In North America, the VW-1 test (UL 2006) is used to characterize flame propagation on a single wire or cable (UL 2006). In principle, this test is analogous to IEC 60332-1 even if some differences exist in operating conditions: energy of the source, orientation of the burner, and application of the flame. The test is passed if self-extinguishing occurs within 1 minute after removing the flame and dropping material does not ignite cotton lying beneath.

IEC 60332-3 is the international standard test for the vertical flame spread of vertically mounted bunched wires and cables (IEC 2004b). In this test, a series of 3.5 m long cables are mounted on a tubular

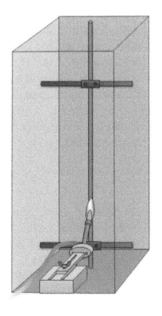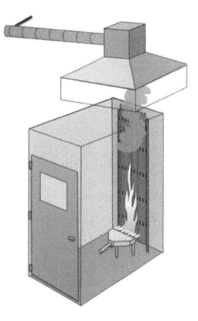

FIGURE 14.1 IEC 60332-1 and IEC 60332-3.

steel ladder (Figure 14.1). The ignition source is a propane burner whose flame-producing surface consists of a flat metal plate drilled by an array of holes, and its power is 70,000 BTU/h (20.5 kW). The flame is applied in the center of the bunch at least 500 mm over the bottom of the cables. The number of pieces attached to the ladder as well as the flame application time depends on the category that is targeted. For the most severe category (category A), 7 L of combustible (non-metallic material) per meter must be used and the flame application time is 40 minutes (min). Depending on the cross-sectional area of conductors, cables have to be tested in a spaced configuration or in a touching configuration. The test is deemed passed if after the burner has been removed, the fire damage does not cover more than 2.5 m measured from the bottom cable end. The analogous test in North America is the UL 1685 or CSA FT4 test (CSA 2001; UL 2007).

14.2.2.2 Heat Release

The simplest criterion assessing heat release by a product is the determination of its gross heat of combustion. In the European EN 50575 standard relative to construction materials, the best classification A can be obtained if the product exhibits a gross heat of combustion lower than 2 mJ/kg and is considered as incombustible (EN 2016). This is never the case for cables, and other parameters have to be assessed. The rate at which heat is generated is a critical parameter to characterize a fire. Hence, heat release rate (HRR) measurement was integrated into the EN 50399 standard that described the common test methods for cables under fire conditions (EN 2011). EN 50399 is based on IEC 60332-3 and utilizes the same equipment with additional measurement of HRR by oxygen consumption using an oxygen analyzer. Three parameters are considered for cable classification: the total heat release after 1200s (THR_{1200s}), the peak of HRR, and the fire growth rate (FIGRA) which is defined as the maximum value of the function (HRR)/(elapsed test time).

14.2.2.3 Smoke Production

One of the main testing methods in the cable industry to assess the smoke production potential in the case of a fire is described in the IEC 61034-1 and -2 standards (IEC 2005, 2006). Cables are placed above a tank containing a specific fire source, mainly ethanol, in a 27 m^3 cabin. When the cable materials burn, smoke is released, reducing the light intensity in the test chamber. This attenuation is measured by a photocell placed in front of a white light beam at a fixed distance: path length of 3 m. The final result is given in terms of transmittance, It (%). Without any other specification, the normal pass/fail criterion is 60% of light transmittance. The smoke production was also integrated into the EN 50399 standard using the same measurement principle (EN 2011).Transmittance is converted into an extinction coefficient that enables determining the smoke production rate knowing the air flow rate and the smoke temperature. Three parameters are considered for cable classification: the transmittance It, the total smoke production after 1200s (TSP_{1200s}), and the peak of smoke production rate. The UL-1685 standard can be considered as the corresponding North American test evaluating smoke production (UL 2007). A small difference can be noted with EN 50399 concerning the threshold criteria in char height, total smoke released, and peak of smoke release.

14.2.2.4 Emission of Corrosive

The emission of an acidic substance during a fire may have dramatic consequences on equipment in industrial plants. Acidity classification for cable materials can be determined according to the EN 60754 standard (EN 2014b) and is taken into account in the Euroclass cable classification. A static sample is decomposed in a stream of flowing air in the middle of the tube furnace, above 935°C, and the effluent is collected in bubblers prior to analysis. The acidity of the effluent solution is quantified in terms of pH and conductivity.

14.2.3 Fire Resistance

In terms of fire resistance, the purpose of standards is to assess the functional integrity of cables during a fire. In the IEC 60331-11 international standard (IEC 1999), a flame (T = 850°C) is applied across the entire horizontal length of a single, 120 cm cable for a period of 90 min. The test is deemed passed if the cable continues to conduct electricity without shorting throughout the 90 min flame application and for a subsequent 15 min cooling period. This test is generally only passed by cables and wires with special, flame-retardant glass or mica wrapping enclosing the individual cores as well as the entire cable bundle. BS 6387 category A, B, C, or S adopt a similar test set up, but different flame temperatures are tested (650°C, 750°C, and 950°C) and the test duration is also different (180 or 20 min). BS 6387 category W is a version of the preceding test including water spray with sprinkler (BS 2013).

The EN 50200/IEC 60331-2 test is close to IEC 60331-11, but this time the cable is mounted in a "U" on a fireproof frame and a mechanical shock is applied to the frame every 5 min (Figure 14.2) (EN 2015a). The test is deemed passed if the circuit integrity is ensured. Different classifications can be obtained depending on flame exposure time (15, 30, 60, and 120 min). BS 6387 category X, Y, & Z standards also assess cable functional integrity by mechanically shocking the frame on which the cable is mounted. However, the cable is mounted in a "Z," the shocking frequency is higher (every 30 s), and different flame temperatures are applied (650°C–950°C). The EN 50362/IEC 60331-1 standard is similar to EN 50200, but is dedicated to large diameter cables (EN 2003). The main difference lies in the fact that the cable is mounted on a metallic ladder.

Fire resistance can be assessed in still more severe conditions by combining mechanical shocks and water spray. These conditions correspond to EN 50200 (annex E) and BS 8434-2 standards (BS 2009). The two standards are very close, but differ in the flame temperature (850°C/950°C) and the duration of the test (30/120 min).

Larger-scale tests can be also used to assess the fire resistance of cables. These tests are generally carried out in a furnace and involve horizontal cable trays (DIN 4102-12 [DIN 1998], NBN 713-020 [NBN 1994], EN 50577 [EN 2015b]).

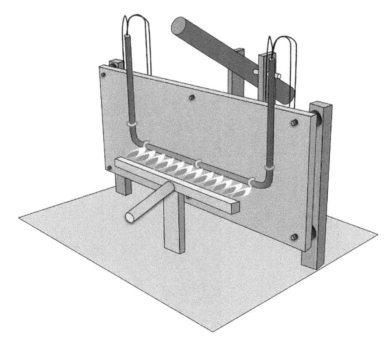

FIGURE 14.2　EN 50200/IEC 60331-2.

14.2.4 Correlations between Standard Tests and Lab Tests

Before being rated through standard tests, cables and more especially sheath materials are often tested using small-scale lab tests. The most common of these tests are the limiting oxygen index (LOI), cone calorimeter, and microscale calorimeter (MCC). The question that is immediately raised is the correlation between classification tests and small-scale tests.

In the early 1990s, Weil et al. (1992) investigated correlations between LOI and other fire tests. With regard to wire and cable tests, they reported a common statement that a LOI value of 30 for the cable sheathing enables it to meet most cable fire tests. However, they concluded that no good correlation could be observed due to the fact that cable tests depend on other parameters than the sheath properties (i.e., inner layer, cable layout, etc.). The Fire Propagation Apparatus developed by Tewarson and co-workers (1992) can be considered as one of the first set ups enabling assessment of the fire behavior of cables at a small-scale with the aim to simulate large-scale fire conditions. Its use was later defined in the ASTM-2058 standard. Using this test, Khan et al. (2006) studied the Fire Propagation Index (FPI) of different cables and evidenced this can be a good predictor for larger-scale tests like UL-910, IEEE 1202, or NFPA 262.

The cone calorimeter is another piece of bench-scale equipment that was proven to be adaptable to predict full-scale tests for various kinds of products (Hirschler 2015). This technique was considered by Babrauskas et al. (1991) as exhibiting good correlation with large-scale tests when testing wires and cables. Hirschler (1994) and thereafter Barnes et al. (1996a, 1996b) studied a series of 21 electric cables identical in structure, but shown to be differing by the composition of their sheath or insulation through a large-scale test (ASTM D5424, FT4 protocol) and cone calorimeter test (ISO 5660). It was shown that there was a good correlation between peak heat release rate (PHRR) data of both tests for cables that pass the large test, but a wide scattering for cables that fail as highlighted in Figure 14.3. It was also evidenced that smoke obscuration in large-scale tests can be reasonably related to the smoke factor in the cone calorimeter (Figure 14.4).

Nakagawa (1998) compared the cone calorimeter data obtained on pieces of different electric cables to LOI experiments performed on the sheath materials, the insulated wires, and the complete cables. The peak HRR was evidenced to vary as the inverse of LOI of cables or LOI of sheaths with a high determination coefficient especially for the tests performed at a low irradiance (20 kW/m^2). Meinier et al.

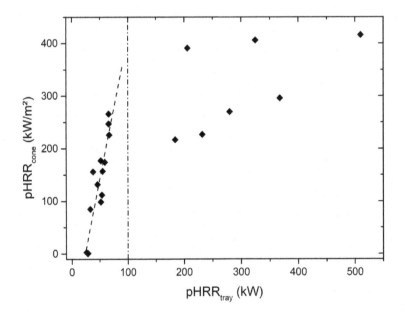

FIGURE 14.3 PHRR in the cone calorimeter (20 kW/m^2) as a function PHRR of cable tray rate.

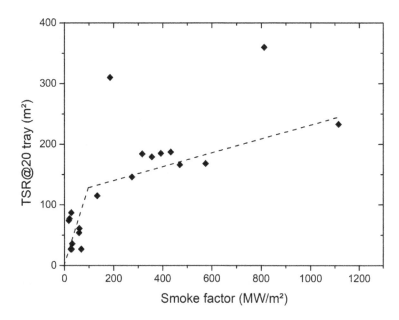

FIGURE 14.4 Total smoke release (TSR) after 20 min for cable tray as a function of smoke factor after 5 min in the cone calorimeter (40 kW/m²).

(2018) studied the influence of external heat flux and spacing between cables on the fire response using the cone calorimeter. The results indicated that in halogen-free flame retardant, electric cable ignition is mainly governed by the properties of the sheath material. A transition heat flux depended on cable composition, and structure was identified as the utmost parameter controlling the development of fire and thus the heat release by the combustion.

MCC is an interesting technique since it enables testing polymer flammability using only a few milligrams of material. Thus, MCC was used to assess some basic burning properties of materials that compose cable sheath (Xie et al. 2010; Yang et al. 2013; Sonnier et al. 2012). Correlations between MCC data and conventional flammability test results for wire and cable compounds have been also investigated (Lin et al. 2007). Cogen et al. (2009) studied correlations between MCC, cone calorimetry, limiting oxygen index, and flame spread tests performed on seven halogen-free FR compounds used for cable sheathing. A poor correlation was found between LOI and the other tests, and therefore the authors discourage using this test as a selecting tool. On the other hand, some cone data (time to PHRR, FIGRA) and pyrolysis-combustion flow calorimetry (PCFC) data (temperature at PHRR) exhibited good correlation with wire burn time as measured by a MS-8288 test especially when considering only the ranking of cables. When considering absolute values, the degree of correlation is lesser as shown in Figure 14.5.

In the multi-year program called Cable Heat Release, Ignition, and Spread in Tray Installations during FIRE (CHRISTIFIRE), burning properties of electric cables were characterized from microscale (using MCC) via bench-scale (using cone calorimeter) to large-scale through a burning test on horizontal cable trays. No correlations between the different tests were attempted, but a model (called Flashcat) was proposed to simulate the fire behavior of cable trays from data collected at lower-scale (McGrattan et al. 2010).

14.2.5 New Tests or Measurement Systems

Mangs et al. (2013) developed a new test to study flame spread on cables. The originality of this test compared to IEC 60332 is to enable the control of the ambient temperature surrounding the cable. This was evidenced to significantly impact the spread velocity as illustrated in Figure 14.6. In the same vein,

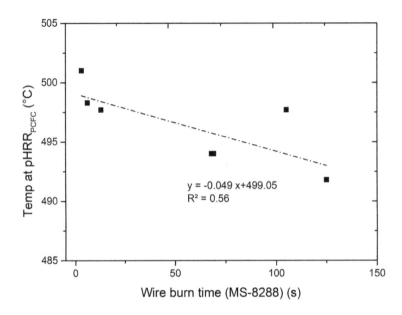

FIGURE 14.5 Temperature of the peak heat release rate (PHRR) in pyrolysis-combustion flow calorimetry (PCFC) versus wire burn time in MS-8288 test.

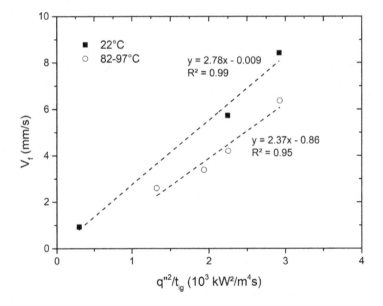

FIGURE 14.6 Flame spread velocity from cone calorimeter versus flame spread velocity from 2-m apparatus.

Courty and Garo (2017) proposed a bench-scale test allowing the studying of spontaneous or piloted ignition of cables as a function of ambient temperature using a device in which the cables are placed into a furnace. In order to investigate the simultaneous effect of external radiation and oxygen concentration on the flammability of wire insulation, Miyamoto et al. (2016) developed a new test inspired from the ASTM limiting oxygen test in which the burning limit of a wire placed in a controlled atmosphere and exposed to a radiant heat flux is measured. The application of an external heat flux tends to decrease the limiting oxygen concentration and may change the ranking between several insulation materials.

Polansky and Polanska (2015) modified the IEC 60331 test by adding an insulation resistance meter likely to monitor the actual state of the cable insulation during the fire test. This system enabled highlighting the influence of different phenomena (e.g., polymer melting, flame retardant decomposition, or formation of a silica layer) on the insulation resistance.

The full-scale EN 50399 tests required for cables to be used in buildings is very tricky to set up and is also time consuming. Gallo et al. (2017) proposed an alternative bench-scale fire test for faster and cheaper cable screening. The new test used the cone calorimeter equipment in which balance, cone heater, and sample holder were replaced by a burner and a portion of the cable tray in the vertical position, while the complete gas analysis of the cone calorimeter monitoring oxygen consumption, CO, and smoke production was preserved. The new set up was shown to efficiently predict the results of the full-scale CPR test, the main properties (PHRR, FIGRA, flame spread) being linked by a factor of around 5.

With similar motivations, Girardin et al. (2016) developed a novel bench-scale fire test. In this method, a flat specimen of the material constituting the external cable sheathing was placed in a vertical position and submitted to the flame of a methane burner. HRR as well as flame spread were measured. The authors found a good correlation between the fire performance obtained at reduced- and large-scale for five benchmark materials.

Sarazin et al. (2017) developed a small-scale fire test aimed at investigating the influence of the electric current flowing through the cable on the flame spread. Cables can be assessed in a vertical or horizontal position. Ignition is generated by the application of a Bunsen burner delivering a calibrated flame. The current in the conductor can be varied from 0 to 60 A. The flame spread and time to delamination are the two main parameters measured. It was highlighted that the electric current induces an additional heating causing an increase of the flame propagation rate. Moreover, it lowers the viscosity of the intumescent char and provokes its delamination during the test.

Chatenet et al. (2017) designed a bench-scale apparatus working in different controlled atmospheres with the aim to assess the fire behavior of sheath materials at different fire scenario stages. As illustration of this, it was shown that the gaseous compounds released during the first decomposition step of ethylene-vinyl acetate (EVA)/aluminum trihydroxide (ATH) composition decreases as the atmosphere is enriched in oxygen.

14.3 Main Polymers and Flame Retardants Used in the Cable Industry

14.3.1 Main Polymers Used and Their Characteristics

PVC, low density poly (ethylene) (LDPE), and its copolymers such as EVA are the two main polymer types used for wire and cable insulation, from Babrauskas (2006).

The heat release capacity and LOI of the materials which can be used in cables are very different, showing *a priori* the interest of PVC and fluorinated polymers in comparison with polyolefins (Table 14.2) (Hirschler 2017).

PVC has corresponded to about two-thirds of the insulation used in building cables in the United States due to its interesting profile for the following properties: dielectric strength, electrical resistivity, chemical and water resistance, good processability, and also intrinsic fire behavior (Kaufman 1985; Kotak 1983).

When PVC is used for wire/cable insulation, it is not used as a pure polymer. PVC used in wire/cable insulation or jacketing needs to be flexible through the use of plasticizers. Wickson (1993) indicates that typical wire/cable formulations contain 52%–63% PVC resin, 25%–29% plasticizer, 2%–4% stabilizer, 0.2%–0.3% wax, less than 0.1% of antioxidants, and small amounts of lubricants and colorants. Moreover, a variable amount of mineral fillers is often added, such as $CaCO_3$ and kaolin, but other ones can act as components of flame retardant systems. The plasticizers are selected according to the required service temperature. They are typically phthalates (e.g., diisodecyl phthalate, ditridecylphthalate) or

TABLE 14.2 Heat Release Capacity and Limiting Oxygen Index of Different Polymers

Polymer	Heat Release Capacity (J/g·K)	Limiting Oxygen Index (%)
High density polyethylene	1450	17.0
Polypropylene	1106	17.1
Polystyrene	1088	17.7
Polycarbonate	578	26.2
Polyamide 6.6	565	25.1
Poly(methyl methacrylate)	480	17.9
Poly(ethylene terephtalate)	366	20.0
Polyether ether ketone	345	–
Poly(vinylidene fluoride)	309	22.6
Polyphenyl sulfone	219	31.1
Polyoxymethylene	200	15.7
Polyether imide	197	36.5
PVC (rigid)	157	47
PVC flexible non FR	–	23–30
Fluorinated ethylene propylene	82	–
Polytetrafluoroethylene	–	95

Source: Hirschler, M. M., *Fire Mater.*, 41, 993–1006, 2017.

trimellitates, e.g., tris(2-ethylhexyl) trimellitate). Phosphate esters or halogenated phthalates can also be incorporated in order to combine plasticizing and flame retardant effect. Moreover, arylphosphates contribute to reduce smoke generation (Moy 2004).

Alternative solutions to phthalates particularly result from environmental grounds (Krauskopf 2003).

The use of plasticizers allows the brittleness temperature to be decreased as low as −60°C (Kaufman 1985), nevertheless, it leads to a strong decrease of the LOI which can attain 15% (Coaker 2003; Hirschler 2017) and needs the use of the highest amounts of flame retardants other than rigid PVC. About stabilizers, environmental requirements such as the European Directive of the Restriction of Hazardous Substances in electrical and electronic equipment (RoHS 2002) have led to the phase out of Pb and Cd compounds. Systems based on Ca-Zn, Mg-Zn, Ca-Zn-Al (Kaseler 1993) are now predominant and allow PVC to play again a major role in the cable industry. Co-stabilizers are also frequently added, such as zeolites (e.g., 4A variety) or hydrotalcites which act as HCl absorbers (Grossman 2000) to avoid smoke toxicity, irritancy, and particularly corrosivity, which can represent a main concern in the telecommunication industry (Gibbons and Stevens 1989).

Thermal degradation of PVC and the migration of stabilizers are among the main issues related to the use of this chlorinated polymer. Moreover, among the aging mechanisms leading to a loss of electrical resistivity, oxidation and absorption of water have to be considered (Quenehen et al. 2015).

As it is well known, thermal degradation of PVC is a two-stage process (Wu et al. 1994) with an initial mechanism of dehydrochlorination, leading to HCl release. Babrauskas (2006) reports that various studies have shown that measurable amounts of HCl can be detected, even at 100°C and also that from 180°C, there is no induction period for HCl release. Van Luik (1974) reported that submicron particles were released from PVC at 143°C. It has been proven that the dehydrochlorination reaction is autocatalytic (Hjertberg and Sorvik 1978). In consequence, the role of stabilizers is to counteract this process by increasing the starting temperature of HCl release and to scavenge the HCl produced (Fisch and Bacaloglu 1999; Montaudo and Puglisi 1991). However, the effectiveness of stabilizers is often limited by the adverse effect of the plasticizers (Zaikov et al. 1998). But a loss of plasticizer can occur even at moderate temperatures. Ekelund et al. (2007) have shown that at service temperature ranges, such as (20°C–50°C), the deterioration of a PVC jacketing is dominated by the loss of plasticizer by migration.

At higher temperatures, it has been shown that a plasticized PVC with dioctyl phthalate (DOP) exposed during 30 hours at 100°C resulted in a loss of 30% of the plasticizer (Titow 1990). The migration of a plasticizer from a PVC jacket can also significantly alter the transmission characteristics of the cable by increasing the dissipation factor for high frequency range (Loadholt and Kaufman 1982). Arrhenius extrapolation methods based on constant activation energy are currently used to predict plasticizer migration (Linde and Gedde 2014). Among the solutions to prevent migration, the development of macromolecular additives such as polymeric polyester plasticizers obtained by reacting a dibasic carboxylic acid with a glycol or a mixture of different dibasic carboxylic acids with one or more glycols has been presented by Svoboda as an alternative route (Svoboda 1991).

Even if PVC exhibits better intrinsic flammability than polyolefins, these last ones are frequently used as insulation and jacketing materials. Different materials of this category are used, regarding differences in fire behavior, electrical (particularly the dielectric loss factor), and mechanical properties, as well as their resistance to various solvents (Table 14.3). But, for many applications, LDPE is replaced by its copolymers, mainly ethylene-vinyl acetate (EVA), but also ethylene ethyl acrylate (EEA), ethylene methyl acrylate (EMA), and ethylene propylene diene monomer (EPDM). This last one is frequently crosslinked to increase its fire performance as well as its resistance to solvents. LDPE is also crosslinked using radiations leading to noticeable modifications of properties, as shown by Mo et al. (2013).

The use of poly(propylene) (PP) allows better tensile strength to be obtained, nevertheless, due to its tertiary carbon in the polymer chain; it is well known that its resistance to thermo- or photo-chemical aging is lower than this of LDPE. The use of copolymers such as EVA, EEA, or EMA stems from the better ability for these polymers to char compared to LDPE, and particularly to enable high loadings of inorganic compounds which can play a role of flame retardants, without too high a loss of mechanical properties. The thermal degradation pathway of the copolymers is different from LDPE since elimination of the side group of the co-monomer can occur prior to the chain breaking in the backbone. For instance, in the case of EVA, the release of acetic acid leads to less fuel emission than hydrocarbon degradation products. Moreover, the carbon double bonds formed can react through Diels-Alder reactions to promote char structures, and cohesion and thermal stability can be reinforced using specific fillers (see infra).

In some cases, the use of PVC or polyolefins is not able to face drastic requirements regarding fire performance. Fluoropolymers and silicones are more expensive materials, but present a very low flammability. Silicones exhibit specific fire behavior due to their ability to form an expanded ceramic structure

TABLE 14.3   Comparison of PE, XLPE, and PVC Plastics Performance

Property	PVC	PE	XLPE
Density (g/cm^{-3})	1.4	0.92	0.92
Insulation thickness (mm)	1.0	0.7	0.7
Max. temperature (°C)	60–70	75	95
Softening temperature (°C)	120	105–115	127
Instantaneous short-circuit temperature (°C)	135	150	250
Insulation resistance (mΩ·K·m^{-1})	20	1000	1000
Volume resistivity (Ω·cm)	1012–1015	1017	1017
Dielectric strength (KV·cm^{-1})	20–35	20–35	35–50
Dielectric constant (60 Hz)	6–8	2.3	2.3
Dielectric loss tangent (60 Hz)	0.1	10^{-4}	10^{-4}
Weathering resistance	Excellent	Reduced	Common
Aging resistance	Reduced	Common	Excellent
Oil resistance	Reduced	Common	Excellent
Embrittlement at low temperature	Reduced	Common	Excellent

Source: Mo, S. J. et al., *Procedia Eng.*, 52, 588–592, 2013.

TABLE 14.4   LOI of Partially Fluorinated Polymers

Polymer	LOI (%)
Polyvinyl fluoride	22.6
Copolymer of ethylene and tetrafluoro ethylene (ETFE)	30.0
Polyvinylidene fluoride (PVDF)	44.0
Copolymer of ethylene and chlorofluoro ethylene (ECTFE)	≥ 52

Source: Lin, S.C., and Kent, B., Flammability of fluoropolymers, Chapter 17, in *Fire and Polymers V*, ACS Symposium Series, American Chemical Society, Washington, DC, 2009.

when it thermally decomposes. The cohesion of this ceramic structure made of silica and silicon carbides regarding the fire parameters (temperature increase, ventilation) (Camino et al. 2001) can be improved using mineral fillers in order to prevent short circuits (Hamdani et al. 2010). Concerning fluoropolymers, the main concern is the risk of corrosive release in case of fire, nevertheless, recent industrial compounds for wires and cables can scavenge acids during combustion (Kiddoo 2007). Among the fluoropolymers, around 65% of fluorinated ethylene propylene (FEP) is used in cable insulation. Its interest lies in its low smoke emission in comparison with other polymers. So, it can be considered as an excellent solution for confined and dense spaces of buildings (Gardiner 2015). Ethylene chlorotrifluoro ethylene (ECTFE) and ethylene trifluoroethylene (ETFE) copolymers are also often used for transportation cable and wires. PVDF presents one of the highest LOI values (Table 14.4), nevertheless, its use is restricted to low voltage applications, due to its too high dielectric loss factor (Lin and Kent 2009).

Other polymers correspond to minor uses in cables and wires, such as thermoplastic polyurethanes. TPU is extensively used due to its excellent physical properties, such as flexibility at low temperature and abrasion resistance. However, TPU presents some drawbacks, such as low thermal stability and mechanical strength (Tabuani et al. 2012), low ability to self-extinguish (LOI of 21.5) (Wang et al. 2016). Moreover, the thermal decomposition or combustion of the polymer produces large amounts of toxic gases, such as CO, HCN, and NO_x.

It appears that the thermal decomposition of PUs is complex and takes place via any of the following routes or more likely through a combination of them: (1) random-chain scission, (2) chain-end scission, i.e., unzipping, and (3) crosslinking. The composition of the decomposition products leading to a complex char depends on the structure of the initial PU material.

14.3.2 Usual Flame Retardants in the Cable Industry: Advantages and Limitations

A particular feature of the cable industry is the important use of mineral fillers which can play a role of extenders or flame retardants. The use of calcium carbonate is frequent in PVC for cable applications as well as other ones, but generally without specific functional properties except the ability to act on possible HCl release. Conversely, calcined kaolin is used for mechanical reinforcement (Zazycny and Matuana 2005). The main mineral fillers used as flame retardants in PVC are antimony trioxide, aluminum trihydrate, zinc borate, zinc stannate, and hydroxystannate. Table 14.5 presents the most used flame retardants in flexible PVC (Shen 2006).

Many of these FRs are able to act through specific mechanisms which can interact with the presence of chlorine in the PVC macromolecules. In the case of Sb_2O_3, the mechanism is similar to the use of halogenated flame retardants in various polymers. The intrinsic flame retardancy of antimony oxide is very low (Laachachi et al. 2004), but it can act as a synergist with the Cl·free radicals released from the decomposition of PVC. $2ZnO·3B_2O_3·3.5H_2O$ zinc borate (ZB) has been used in various polymers as a

TABLE 14.5 Mostly Used Flame Retardants in PVC

Flame Retardant or FR Synergist	Chemical Formula	Mode of Action
Antimony trioxide and pentoxide	Sb_2O_3/Sb_2O_5	Gas phase
Zinc borates	$2ZnO{\cdot}3B_2O_3{\cdot}3.5H_2O$ $2ZnO{\cdot}3B_2O_3{\cdot}3H_2O$	Mainly condensed phase
Zinc hydroxy stannate	$ZnSn(OH)_6$	Condensed phase
Ammonium octamolybdate	$(NH_4)_4Mo_8O_{26}$	Condensed phase
Melamine molybdate	–	Condensed phase
Zinc oxide and magnesium oxide complex	$ZnO{\cdot}MgO$	Condensed phase
Zinc molybdate	$ZnMoO_4$	Condensed phase
Phosphate esters	Various	Condensed/gas phase
Chlorinated paraffin	Various	Gas phase

flame retardant and acts by forming a glassy structure able to protect the PVC residue from heat sources (Shen 2006).

Zinc stannate (ZS) and hydroxystannate (ZHS) have proven higher efficiency than ATH regarding fire performance (Wu et al. 2008; Xu et al. 2005). They act by catalyzing the dehydrochlorination of PVC through the formation of $ZnCl_2$ and $SnCl_2$ Lewis acids which promote the building of a cohesive charred structure (Qu et al. 2008, 2009). ZHS was also used as a coating of metallic hydroxides (Figure 14.7) such as aluminum and magnesium hydroxides (Cusack and Hornsby 1999; Qu et al. 2008), but from Cross et al. (2003), it seems more effective as a filler than as a coating. Other types of metallic hydroxystannates have been used as coatings of ATH by Jiao and Xu (2009) and Qu et al. (2009).

A synergistic effect was found, but not explained. Moreover, coatings of ZHS on $CaCO_3$ were also tested interestingly by Yang et al. (2007), in order to increase the fire performance. ZHS particles form a close packing at the $CaCO_3$ surface, leading to a good charring ability.

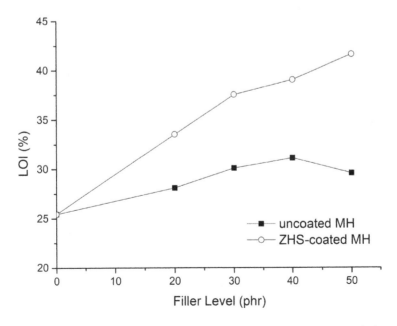

FIGURE 14.7 Effect of the filler level on the LOI of flexible PVC containing magnesium hydroxide (MH) and ZHS-coated MH.

As ATH has been used as the main flame retardant component in PVC for decades, it is of prime interest to find compounds which can act as intrinsic FRs or only as synergists with PVC. Consequently, various ternary compositions were proposed to improve fire retardancy of flexible PVC. ATH has been successfully associated with Sb_2O_3, ZB, or ZHS to increase the fire performance of PVC (Shen 2006), leading to industrial developments in the cable industry and materials for building.

ATH, magnesium hydroxide, and hydromagnesite ($5MgO\cdot4CO_2\cdot5H_2O$), a hydrated magnesium carbonate (Hollingbery and Hull 2010; Toure et al. 1996) are the main FRs or components of flame retardant systems used in polyolefin cables. Whereas in PVC, the ATH amount is less than 30 wt% (Shen 2006), on the whole, the lesser intrinsic fire performance of polyolefins requires higher amounts of fillers (up to 65 wt%) (Rothon 2001). From Hull et al. (2011), who investigated the mode of action of ATH using thermal analysis and literature data, 55% of the fire performance effect would be due to the endothermal decomposition of the mineral filler, 23% to the effect of water vapor in gaseous phase, 13% ascribed to the formation of a barrier layer, and 9% to the absorption of heat by the filler. It can be noticed that the contribution of the barrier layer formed by the remaining dehydrated alumina is relatively low due to its low mechanical cohesion as in the case of the two other hydrated fillers abovementioned (Delfosse et al. 1989). Various efforts have been made to improve the cohesion of the residual mineral layer by replacing a part of the hydrated fillers by other compounds. Synergistic effects were expected at constant loading as well as the possibility to improve mechanical properties by reducing the whole loading at constant fire performance. The interest of borates such as ATH or magnesium hydroxide (MH) synergistic agent has been shown by various authors (Carpentier et al. 2000; Durin-France et al. 2000; Genovese and Shanks 2007). These last authors highlighted the formation of a magnesium orthoborate to account for the synergy between MH and zinc borate. The interest of the use of high aspect ratio talcs in combination with MH has been showed by Clerc et al. (2005). The ability for talc to contribute to the cohesive character of the final residue was ascribed to its high specific surface area allowing catalytic effects to entail charring. During these last 15 years, various kinds of particles having a huge quantity of interface with polyolefins were tested, most of them were nanoparticles.

14.4 Nanotechnology, Environmental Issues, and Future Trends

14.4.1 Use of Nanoparticles in Cables to Improve Fire Performance

The cable industry can be considered as a pioneer industry regarding the extent of nanoparticles used from the beginning of the new century. The main objective was to improve the global fire performance of flame retarded polymers through specific mechanisms involving nanoparticles without the loss of functional properties. About mechanical properties, it was expected, for instance, to reduce the detrimental effect of high filler loadings on ultimate properties such as elongation at break or tear resistance, particularly in the case of polyolefins.

The main strategy is to substitute a limited fraction of metallic hydroxides (typically 3–5 wt% of the global composition) by nanoparticles. The main categories of nanoparticles used were the different families of phyllosilicates (or layered silicates) and mainly organo-modified montmorillonites using alkyl ammonium ions, allowing exfoliated or intercalated morphologies. The modes of action of the organo-modified layered silicates (OMLS) in polyolefins and other thermoplastic polymers have been highlighted by various authors and involve both physical and chemical mechanisms (Gilman et al. 2005; Lewin et al. 2006; Morgan et al. 2002; Zanetti et al. 2002). Physical ones correspond to the building of a barrier made of the lamellar particles. It results from the thermal ablation of the polymer, but also from the migration toward the surface exposed to radiant flux due to viscosity and surface tension gradients. This barrier formation limits the heat and mass transfer of combustible volatiles and oxygen through the residual material.

Chemical mechanisms are related to the catalytic effect of layered silicates on the degradation pathway of polymers. This action leads to the formation of charred structures or to the increase of the amount of

FIGURE 14.8   Expanded and porous residue obtained at cone calorimeter for MH55/MMT5/EVA45 (wt%) compositions.

char formed for polymers charring intrinsically. The presence of the lamellar particles at the surface is also able to reinforce the charred structure. Moreover, in the case of metallic hydroxides/OMLS combinations, it has been proven that the nanoparticles could increase the cohesion of the mineral residue formed at the surface after polymer ablation. It has been shown by Ferry et al. (2005) for MH/organo-modified montmorillonite (MMT) in EVA that the formation of a cohesive barrier at the surface of the material could generate a kind of intumescent and porous mineral structure due to the hindered release of decomposition gaseous products of EVA (Figure 14.8).

More recently, a similar combination in crosslinked PE (XLPE) was tested by Liu (2014) with a total filler amount of 52 wt%. PE grafted with maleic anhydride (PEgMA) was also used in substitution of PE as compatibilizer to obtain an exfoliated morphology. The LOI values of XLPE composites enhanced when 4 wt% of MH was replaced with clay. The increase in the char residue of nanocomposites indicated that the thermal stability was improved.

ATH/OMMT using alkylammonium ion compositions were also investigated. Beyer (2001) studied these combinations in EVA for a filler total loading of 65 wt% with a substitution rate of 5 wt% for OMLS. Thermogravimetric analysis performed demonstrated a clear increase in the thermal stability of the layered silicate-based nanocomposites. Moreover, char formation in the case of the nanocomposites has been improved and is responsible for the better flame retardancy.

Zhang and Wilkie (2005) investigated also ATH/montmorillonite compositions in PE. Nevertheless, a lauricoligomerically modified clay was used in this study. Various PE/ATH ratios (with a global loading up to 70 wt%) and percentages of OMLS were investigated. The combination of polyethylene with 2.5% inorganic clay and 20% ATH led to a 73% reduction in the PHRR, similar to that obtained with 40% ATH used alone.

More recently, Chang et al. (2014) prepared EVA/PE/EVA compositions with a filler total loading of 50 wt% and percentages of OMLS from 1 to 3 wt%. Intercalated morphologies were noticed using X-ray diffraction. The best tensile resistance, elongation at break, and fire resistance (LOI = 28) were obtained at 3 wt% montmorillonite and 47 wt% ATH.

Hydromagnesite has been frequently used as a flame retardant filler (Hollingberry and Hull 2010) and can play a similar role as ATH and MH. Its decomposition leads to H_2O, but also CO_2 release, with a higher mass loss than ATH and MH (54 wt%), but the release of CO_2 occurs during a higher temperature range up to 550°C than this of H_2O, limiting its interest. Hydromagnesite (HM) has been combined with organo-modified montmorillonite (55 wt% HM and 5 wt% montmorillonite) by Laoutid et al. (2006a) in EVA and Haurie et al. (2007) in LDPE/EVA equimassic blends. Synergistic effects were noticed between HM and organo-modified montmorillonite. The residues formed at the cone calorimeter tests showed the formation of magnesium silicate ($MgSiO_4$) when either MH or HM were combined with MMT. Finally, a better stability of charred and ceramized residue was noticed.

Layered double hydroxides (LDHs) are often considered as anionic clays. Their diversity of compositions allows them similar improvement of properties and particularly fire reaction to be imparted as with organo-modified layered silicates (Manzi-Nshuti et al. 2009).

Zhang et al. (2007) and Ye et al. (2007) have associated ultrafine MH with LDH in EVA and EVA/LDPE blends. Exfoliated morphologies are obtained with less than 5 wt% of LDH, whereas intercalated ones occur for highest amounts. From these authors, LDH allows a better dispersion of MH particles in EVA and EVA/LDPE. Synergistic effects were highlighted on the LOI values and ascribed to a better dispersion of MH.

Among the categories of nanoparticles able to reinforce the cohesion of polyolefins containing ATH or MH, silica represents an economic solution for the cable industry. Fu and Qu (2004) have highlighted the synergistic action of fumed silica in EVA containing MH. The results showed that the addition of increasing percentages of fumed silica in substitution of MH up to 8 wt% increased the LOI value while keeping the V-0 rating at UL-94 test (Table 14.6). The cone calorimeter tests showed that silica allowed the peak of heat release rate and mass loss to be reduced, but also the smoke release in comparison with that of EVA/MH compositions.

Natural silicas such as diatomites can also improve the fire performance of hydrated fillers in polyolefins. Cavodeau et al. (2016) have shown the synergistic action between diatomite and ATH in EVA (5–15 wt% of diatomite for a total loading of 60 wt%) with the formation of an expanded layer during the cone calorimeter tests. This layer presents a complex structure which better insulates the sample and reduces the HRR. Rheological measurements have suggested an influence of the viscosity of the sample melt, imparted by the porous texture of diatomite. Better residue cohesion was also obtained in the presence of diatomite.

Synergistic effects of different types of nanofillers with metallic hydroxides were investigated. Laoutid et al. (2006b) have used micro-indentation in order to characterize the increase of cohesion of the residue obtained with MH combined with organo-modified montmorillonite (55 wt% MH and 5% OMMT) by replacing a fraction of the layered silicate by various types of commercial fumed silica. The presence of silica of high specific surface area improved the self-extinguishing ability, thanks to a better thermal stability at higher temperatures, and reduced the PHRR in the cone calorimeter tests.

TABLE 14.6   LOI values and UL-94 rating of EVA containing MH and fumed silica

Formulation	EVA (wt%)	MH (wt%)	SiO_2 (wt%)	LOI (%)	UL-94
A	100	0	0	18	Fail
B	40	60	0	35	V-0
C	40	58	2	37	V-0
D	40	55	5	39	V-0
E	40	52	8	39	V-0
F	40	48	12	37	V-1

Source: Fu, M.Z., and Qu, B.J., *Polym. Degrad. Stab.*, 85, 633–639, 2004.

The interest of carbon nanotubes (CNTs) to improve the fire reaction of polyolefins has also been high-lighted, and mechanisms based on the formation of thermal shields as well as trapping of free radicals released by the polymer decomposition were proposed (Kashiwagi et al. 2005; Peeterbroeck et al. 2007).

Combinations between organo-modified montmorillonites and CNT, and then combined with ATH, for cable manufacturing applications were tested by Beyer (2003, 2005a, 2005b). 2.5 phr of both OMMT and CNT in EVA were tested by cone calorimeter (35 kW/m^2) and a synergistic effect was highlighted. It was concluded that OMMT plays a significant role in the formation of a compact char, whereas CNTs, due to their high aspect ratio, could reinforce it. Furthermore, from the production of insulated wire compositions tested by cone calorimeter, it was shown that the above synergistic effect was maintained for compositions containing both nanoparticles associated with ATH.

In another study, Gao et al. (2005) investigated the detailed mechanisms leading to the CNT/OMMT synergy. Due to the presence of CNTs, a low permeable char containing graphitic carbon is formed, and this effect is enhanced when CNT and OMMT are associated. Another interpretation of the synergistic effect was proposed by Marosfoi et al. (2006) who noted that CNTs tend to promote deacetylation of EVA, whereas OMMTs could promote crosslinking and char formation.

Nanoparticles can also be used in reactive systems devoted to improve the fire retardancy of polyole-fins devoted to cable and wire applications. Some low voltage insulation systems are based on ethylene-butyl acrylate copolymers (EBA). Combinations of silicone powder and $CaCO_3$ (typically 5 and 35 wt%, respectively) are able to generate the formation at high temperature (close to 1000°C) of Ca_2SiO_4, CaO, $Ca(OH)_2$, as well as silicon oxycarbide (Hermansson et al. 2003). These commercial compositions known as *CASICO* systems in the industrial sector were modified by different authors to introduce nanoparticles in order to produce synergistic effects. Karlsson et al. (2011) have investigated the effect of nanofillers on the flame retardant properties of a *CASICO* system. The authors found that the addition of OMMT enabled the reduction of dripping and heat release rate. This was related to an increase in melt viscosity, and the authors concluded that the flame retardant properties of these systems were dependent on the formation of a protective surface skin. More recently, Realinho et al. (2016) completed a *CASICO* system using silica associated with zinc borate micro-sized particles (4 wt%) and montmo-rillonite associated with graphene nanoplatelets (2 wt%). Significant fire behavior improvements were observed for this composition with a lowest value of PHRR, EHC (effective heat of combustion) and FIGRA in comparison with the reference *CASICO* system.

The interest of nanotechnology to improve flame retardancy of cable and wires is not restricted to the polyolefins for which the control of filler loading is of prime importance, regarding the other func-tional properties. Various works were carried out to introduce different types of nanoparticles in PVC and polyurethane compositions. Tabuani et al. (2012) have developed fire resistant TPU formulations containing different amounts of melamine cyanurate (MC) and OMLS (18 or 25 wt% MC and 5 wt% OMMT). Synergistic effects on the LOI and heat release rate were noticed and the improvement of fire retardancy was not achieved at the expense of the mechanical properties.

Beyer (2007, 2008) has studied the interest of the use of different types of OMLS (montmorillon-ites and hectorites) in PVC. However, OMLS initiated accelerated HCl release from PVC resulting in unwanted color changes of the compounds. So, a method was proposed to assess the basic character of OMLS to avoid this effect. Nevertheless, only reductions in PHRR of around 20% in compari-son with pure plasticized PVC could be noticed by introducing 5 phr OMMT into the composition. Consequently, the use of pristine Montmorillonite-Na$^+$in PVC by Wang et al. (2002) could appear more interesting, since higher PHRR reduction (around 40%) was achieved by introducing 15 wt% of clay in a plasticized PVC.

In addition, Zhao et al. (2006) studied both PVC/(sodium clay) and PVC/(organically modified clay) nanocomposites containing different percentages of plasticizer and clay. The cone calorimeter tests indi-cated that PVC nanocomposites have a similar or higher HRR and a lower mass loss rate when com-pared with those of the virgin polymer.

Conversely to OMLS containing alkylammonium ions, layered double hydroxides seem more advantageous to improve the fire performance of PVC. Hydrotalcite is a hydrated magnesium–aluminum (Mg–Al) hydroxycarbonate with a lamellar structure similar to that of brucite.

The general formula is $Mg_6Al_2(OH)\cdot16CO_3\cdot4H_2O$. Wang and Zhang (2004) investigated the effect of hydrotalcite on the thermal stability, mechanical properties, and rheology and flame retardance of PVC. TG analysis showed that the thermal stability of PVC was improved significantly only in the presence of a complex of hydrotalcite and the organotin stabilizer. Hydrotalcite also improved flame retardancy and reduced smoke emission with the LOI value of more than 28.7 and UL-94 V-0 ranking at a PVC/hydrotalcite weight ratio of 70/30.

Bao et al. (2008) have prepared PVC/hydrotalcite(HT) nanocomposites through vinyl chloride suspension polymerization with HT nanoparticles surface modified with alkyl phosphate (AP). TG analysis showed that thermal stability increased as the weight fraction of HT increased up to 5.3 wt%. The maximum smoke density decreased about 50% in the presence of 5.3 wt% nano-sized HT.

Since tin compounds have proven their effectiveness as synergists in flame retardant systems for plasticized PVC, LDH-containing tin was synthesized by Zheng and Cusack (2013) using a co-precipitation method. Sn-LDH compounds were compounded into PVC, and their fire performance has been evaluated using the LOI and cone calorimeter techniques. PHRR and smoke release could be reduced by 64% and 81%, respectively, through a substitution of 10 wt% of ATH by Sn-LDH, while keeping the total fire-retardant loading at 100 phr (with 40 wt% plasticizer). Moreover, TG analysis indicated that Sn-LDH is an effective char promoter for PVC.

14.4.2 Environmental Issues and Future Trends

The main environmental issues related to the use of cables containing flame retardant systems are the use of less toxic and more environmentally friendly components, the limitation of toxic degradation products in case of fire, and the waste management of end-of-life cables.

14.4.2.1 Toward More Environmentally Friendly Flame Retardant Systems

Since the nineties, the objective to reduce the use of halogenated flame retardants (HFR) has driven the research on fire retardancy of polymers and particularly in the cable industry to find new solutions allowing the equivalent level of performance to be achieved. In parallel, regulations and directives on the use and waste management on HFR were more restrictive owing to the accumulation of evidences about their toxicological and eco-toxicological effects (Lomakin and Zaikov 2003; Yu et al. 2016). Despite the evolution of regulations in many industrial countries, outside Europe is very limited, the Restriction of Hazardous Substances (RoHS) directive was enforced in 2006 in the European Union. This directive restricts the use of six substances: lead, mercury, cadmium, hexavalent chromium, polybromobiphenyls (PBB), and polybromodiphenylethers (PBDE) in the production of various types of electric and electronic equipment. Since PBB were banned for many years, due to previous European Union risk assessment, the main consequences of the RoHS directive for the cable and wire industry concerned the use of PBDE, mainly in polyolefins and the lead stabilizers in PVC. RoHS directive was associated with the Registration, Evaluation, Authorization, and Restriction of Chemicals (REACH) regulation. REACH was enforced in 2007 and is devoted to the production and use of chemical substances and potential impacts on human health and environment. It can be noticed that this complex legislation is particularly able to act on the design of flame retardant systems intended to be incorporated in products exported to the European market. The last update of REACH for 2018 stipulated that tonnages from 1T/year are concerned by the registration of chemicals. Nevertheless, since there is no specific REACH regulation about nanomaterials, some chemicals or nanoparticles used in very low quantity in flame retardant systems could escape from this regulation.

As mentioned above, nanotechnology has already proven that it could be part of alternative solutions for the use of HFR in cables. The use of minor fractions of usual nanoparticles such as silicas, OMLS, or LDH, in combination with metallic hydroxides is able to improve their fire performance at a lower cost than other alternative solutions such as intumescent FR systems. Another route leading to greener flame retarded cables is to use bio-based FR components, that is a new challenge for future developments.

14.4.2.2 Assessment of the Potential Risks Caused by the Degradation Products

The development of analytic techniques aiming to improve the performance of fire testing devices or to investigate flame retardant mechanisms can also quantify and assess the potential toxic character of released decomposition products: gases and particles in aerosols released. For example, novel set ups were developed by coupling a cone calorimeter with a multi-stage cascade impactor system and condensation nuclei counter (Motzkus et al. 2012).

Fire toxicity of combustion of polymers and its assessment has been previously summarized by Stec and Hull (2011) and different set ups have been described including flow-through tests. Hull et al. (2008) have carried out a comparison of toxic product yields of burning cables (PVC, PE, and crosslinked PE) in bench- and large-scale experiments. This shows only marginal sensitivity to the differences in toxic product yield (CO, HCl, and smoke) between the tube furnace [steady-state tube furnace (SSTF) method (IEC 60695-7-50, Purser furnace)] and the large-scale test, suggesting that using the burning data from a suitable fire scenario with toxic product yields from the Purser furnace could produce data needed for toxic hazard assessment from burning cables. In another study, Hull et al. (2003) have carried out investigations about the decomposition and burning behavior of EVA containing ATH and OMMT using the Purser furnace. Little difference in combustion toxicity from CO release was observed for the composition containing OMMT in comparison with pristine EVA and this with only ATH, but under stoichiometric conditions a higher than normal yield of CO was noticed. The large decrease of the rate of thermal decomposition, particularly at 420°C in the tube furnace, suggests that the clay particles were able to reinforce the protective charred layer formed by the crosslinked polyene formed after deacetylation of EVA.

The development of nanotechnology in cable and wires has also entailed new issues about the fate of nanoparticles used as components of FR systems regarding accidental fires or incineration (Lopez-Cuesta et al. 2014). In a research project funded by the French agency of Environment (ADEME), the particle release from EVA containing 50 wt% ATH, 5% OMMT, and 3 wt% silica to simulate cable compositions has been investigated by Ounoughene (2015) using both a tubular furnace and a cone calorimeter, both devices being coupled with on-line particle analyzers. It has been shown that a tubular furnace was more representative of industrial conditions of incineration. Moreover, the protective layer that formed very quickly at the surface of the material residue limited the release of nanometric and ultrafine particles conversely to the composition without nanoparticles.

14.4.2.3 Recycling and Other Industrial Issues and Trends for the Cable Industry

Recycling of cables and wires is often governed by the interest in recovering valuable materials and mainly the metallic conductors. Polymeric materials from waste cables are often disposed into landfills or incinerated, since they have only lower value in recycling recovery channels. Nevertheless, some studies such as this of Suresh et al. (2017) showed that polymers from end-of-life cable could be reprocessed successfully.

The waste management of plastics fraction was not taken into account in a large extent before the enforcement of specific regulations concerning the end of life of waste electric and electronic equipment. The WEEE (Waste Electrical and Electronic Equipment) European directive sets collection, recycling, and recovery targets for end-of-life products and particularly aims to reduce the toxic waste. The WEEE directive imposes separate collection of FR plastics containing HFR. Nevertheless, many plastic parts from printed circuits, cables, and wire are too small after shredding and grinding to be sorted correctly using the current separation techniques including automatic sorting using near infra-red spectroscopy (Beigbeder et al. 2013). Miniaturization of electric systems, including cables and wire to meet reductions of space and weight requirement leads to concerns about polymer recycling. Furthermore, high levels of flame retardancy are more difficult to achieve for thin cable insulation and jacketing. Finally, other issues and challenges could be mentioned for the cable industry such as:

- The control of the microstructure of more and more complex flame retardant systems containing, for example, various types of micronic and nanometric particles for which particle size distribution and dispersion have to be controlled
- The need to use flame retarded polymers able to resist higher temperatures under the hood in cars or miniaturized equipment without significant increase of costs.

14.5 Conclusions

According to the applications, the geography and since cables are most frequently used in confined space, fire behavior is governed by multiple standards and related tests. Hence, new developments are in progress through innovative set ups, analytic tools, methodologies, and modeling to give a better understanding of the fire behavior of cables. In particular, many studies aim to improve the transfer between laboratory-scale tests and real-scale ones.

To meet the various requirements resulting for the multiple functional properties and constraints for the plastics used for cable insulation and jacketing, a large range of materials and related costs are available. Nevertheless, the demand of materials able to resist to increasing temperatures in confined spaces will likely expand the use of fluoropolymers or silicone at the expense of PVC and polyolefins.

The contribution of nanotechnology is essential, particularly to upgrade FR systems based on hydrated fillers, but addresses new issues concerning the control of the microstructures resulting from the processing of more and more complex FR systems as well as the fate of the nanoparticles at the end-of-life of cables. Environmental concerns are one of the main drivers of the development of new FR systems. The emergence of new regulations in developed countries aims to phase out toxic products and is prone to create new challenges to develop environmentally friendly FR systems able to meet high levels of flame retardancy. Nevertheless, a lot of research and industrial developments need to be carried out to promote bio-based materials and FR systems devoted to cable applications.

References

Babrauskas, V. 2006. Mechanisms and modes for ignition of low-voltage, PVC-insulated electrotechnical products. *Fire and Materials* 30: 151–174.

Babrauskas, V., Peacock, R. D., Braun, E., et al. 1991. Fire performance of wire and cable: Reaction-to-fire tests: A critical review of the existing methods and of new concepts, NIST Technical Note 1291.

Bao, Y. Z., Huang, Z. M., Li, S. X., et al. 2008. Thermal stability, smoke emission and mechanical properties of poly(vinyl chloride)/hydrotalcite nanocomposites. *Polymer Degradation and Stability* 93: 448–455.

Barnes, M. A., Briggs, P. J., Hirschler, M. M., et al. 1996a. A comparative study of the fire performance of halogenated and non-halogenated materials for cable applications. Part I tests on materials and insulated wires. *Fire and Materials* 20: 1–16.

Barnes, M. A., Briggs, P. J., Hirschler, M. M., et al. 1996b. A comparative study of the fire performance of halogenated and non-halogenated materials for cable applications. Part II tests on cable. *Fire and Materials* 20: 17–37.

Beigbeder, J., Perrin, D., Mascaro, J. F., et al. 2013. Study of the physico-chemical properties of recycled polymers from waste electrical and electronic equipment (WEEE) sorted by high resolution near infrared devices. *Resources, Conservation and Recycling* 78: 105–114.

Beyer, G. 2001. Flame retardant properties of EVA-nanocomposites and improvement by combination of nanofillers with alumina trihydrate. *Fire and Materials* 25: 193–197.

Beyer, G. 2003. Progress on nanostructured flame retardants. *Proceedings of the 14th BCC Conference on Flame Retardancy*, Business Communication Co. edition, Norwalk, USA.

Beyer, G. 2005a. Flame retardancy of nanocomposites based on organoclays and carbon nanotubes with aluminium trihydrate. *Polymers for Advanced Technologies* 17: 218–225.

Beyer, G. 2005b. Filler blend of carbon nanotubes and organoclays with improved char as a new flame retardant system for polymers and cable applications. *Fire and Materials* 29: 61–69.

Beyer, G. 2007. Flame retardancy of thermoplastic polyurethane and polyvinyl chloride by organoclays. *Journal of Fire Sciences* 25: 65–78.

Beyer, G. 2008. Organoclays as flame retardants for PVC. *Polymers for Advanced Technologies* 19: 485–488.

BS 6387: Test method for resistance to fire of cables required to maintain circuit integrity under fire conditions, London, UK, 2013.

BS 8434-2: Methods of test for assessment of the fire integrity of electric cables. Test for unprotected small cables for use in emergency circuits. BS EN 50200 with a 930° flame and with water spray, London, UK, 2009.

Camino, G., Lomakin, S., and Lazzari, M. 2001. Polydimethylsiloxane thermal degradation. Part 1. Kinetic aspects. *Polymer* 42: 2395–2402.

Carpentier, F., Bourbigot, S., Foulon, M., et al. 2000. Charring of fire retarded ethylene vinyl acetate copolymer—magnesium hydroxide/zinc borate formulations. *Polymer Degradation and Stability* 69: 83–92.

Cavodeau, F., Otazaghine, B., Sonnier, R., et al. 2016.Fire retardancy of ethylene-vinyl acetate composites—Evaluation of synergistic effects between ATH and diatomite fillers. *Polymer Degradation and Stability* 129: 246–259.

Chang, M. K., Hwang, S. S., Liu, S. P. 2014. Flame retardancy and thermal stability of ethylene-vinyl acetate copolymer nanocomposites with alumina trihydrate and montmorillonite. *Journal of Industrial and Engineering Chemistry* 20: 1596–1601.

Chatenet, S., Gay, L., and Authier, O. 2017. Reaction to fire of electrical cables in underventilated conditions, in *Proceedings of the Fire Retardant Polymeric Materials 2017 conference.*

Chaudhary, A., Gupta, S. K., Gupta, A., et al. 2015. Burning characteristics of power cables in a compartment. *Procedia Earth and Planetary Science* 11: 376–384.

Clerc, L., Ferry, L., Leroy, E., et al. 2005. Influence of talc physical properties on the fire retarding behaviour of (ethylene–vinyl acetate copolymer/magnesium hydroxide/talc) composites. *Polymer Degradation and Stability* 88: 504–511.

Coaker, A. W. 2003. Fire and flame retardants for PVC. *Journal of Vinyl and Additive Technology* 9: 109–115.

Cogen, J. M., Lin, T. S., and Lyon, R. E. 2009. Correlations between pyrolysis combustion flow calorimetry and conventional flammability tests with halogen-free flame retardant polyolefin compounds. *Fire and Materials* 33: 33–50.

Courty, L., and Garo, J. P. 2017. External heating of electrical cables and auto-ignition investigation. *Journal of Hazardous Materials* 321: 528–536.

Cross, M. S., Cusack, P. A., and Hornsby, P. R. 2003. Effects of tin additives on the flammability and smoke emission characteristics of halogen-free ethylene-vinyl acetate copolymer. *Polymer Degradation and Stability* 79: 309–318.

CSA FT4: Vertical flame test: Cables in cable trays, Toronto, Canada, 2001.

Cusack, P. A., and Hornsby, P. R. 1999. Zinc stannate-coated fillers: Novel flame retardants and smoke suppressants for polymeric materials. *Journal of Vinyl and Additive Technology* 5: 21–30.

Delfosse, L., Baillet, C., Brault, A., et al. 1989. Combustion of ethylene-vinyl acetate copolymer filled with aluminium and magnesium hydroxides. *Polymer Degradation and Stability* 23: 337–347.

DIN 4102-12: Fire behavior of building materials and elements—Fire resistance of electric cable systems required to maintain circuit integrity—Requirements and testing, Berlin, Germany, 1998.

Durin-France, A., Ferry, L., Lopez-Cuesta, J. M., et al. 2000. Magnesium hydroxide/zinc borate/talc compositions as flame-retardants in EVA copolymer. *Polymer International* 49: 1101–1105.

Ekelund, M., Edin, H., and Gedde, U. W. 2007. Long-term performance of poly(vinyl chloride) cables. Part 1: Mechanical and electrical performances. *Polymer Degradation and Stability* 92: 617–629.

EN 13501-6: Fire classification of construction products and building elements—Part 1: Classification using data from reaction to fire tests, Brussels, Belgium, 2014a.

EN 50575: Power, control and communication cables - Cables for general applications in construction works subject to reaction to fire requirements, Brussels, Belgium, 2016.

EN-45545-2: Fire testing o railway component, Brussels, Belgium, 2013.

EN 50399: Common test methods for cables under fire conditions—Heat release and smoke production measurement on cables during flame spread test—Test apparatus, procedures, results, Brussels, Belgium, 2011.

EN 60754-1: Test on gases evolved during combustion of materials from cables. Determination of the halogen acid gas content, Brussels, Belgium, 2014b.

EN 50200: Method of test for resistance to fire of unprotected small cables for use in emergency circuits, Brussels, Belgium, 2015a.

EN 50362: Method of test for resistance to fire of larger unprotected power and control cables for use in emergency circuits, Brussels, Belgium, 2003.

EN 50577: Electric cables. Fire resistance test for unprotected electric cables (P classification), Brussels, Belgium, 2015b.

Ferry, L., Gaudon, P., Leroy, E., et al. 2005. Intumescence in ethylene-vinylacetate copolymer filled with magnesium hydroxide and organoclays, in *Fire Retardancy of Polymers: New Applications of Mineral Fillers*, pp. 302–312. London, UK: Royal Society of Chemistry.

Fisch, M. H., and Bacaloglu, R. 1999. Mechanism of poly(vinyl chloride) stabilization. *Plastics, Rubber and Composites* 28: 119–124.

Fu, M. Z., and Qu, B. J. 2004. Synergistic flame retardant mechanism of fumed silica in ethylene-vinyl acetate/magnesium hydroxide blends. *Polymer Degradation and Stability* 85: 633–639.

Gallo, E., Stocklein, W., Klack, P., et al. 2017. Assessing the reaction to fire of cables by a new bench-scale method. *Fire and Materials* 41: 768–778.

Gao, F., Beyer, G., and Huan, Q. 2005. A mechanistic study of fire retardancy of carbon nanotube/ EVA copolymers and their clay composites. *Polymer Degradation and Stability* 89: 559–564.

Gardiner, J. 2015. Fluoropolymers: Origin, production, and industrial and commercial applications. *Australian Journal of Chemistry* 68: 13–22.

Genovese, A., and Shanks, R. A. 2007. Structural and thermal interpretation of the synergy and interactions between the fire retardants magnesium hydroxide and zinc borate. *Polymer Degradation and Stability* 92: 2–13.

Gibbons, J. A. M., and Stevens, G. C. 1989. Limiting the corrosion hazard from electrical cables involved in fires. *Fire Safety Journal* 15: 183–190.

Gilman, J. W., Kashiwagi, T., and Morgan, A. B. 2005. NISTIR 6531, National Institute of Standards Report.

Girardin, B., Fontaine, G., Duquesne, S., et al. 2016. Fire tests at reduced scale as powerful tool to fasten the development of flame-retarded material: Application to cables. *Journal of Fire Sciences* 34: 240–264.

Grossman, R. F. 2000. Acid absorbers as PVC co-stabilizers. *Journal of Vinyl and Additive Technology* 6: 4–6.

Hamdani, S., Longuet, C., Lopez-Cuesta, J. M., et al. 2010. Calcium and aluminium-based fillers as flame-retardant additives in silicone matrices. I. Blend preparation and thermal properties. *Polymer Degradation and Stability* 95: 1911–1919.

Haurie, L., Fernandez, A. I., Velasco, J. I., et al. 2007. Thermal stability and flame retardancy of LDPE/EVA blends filled with synthetic hydromagnesite/aluminum hydroxide/montmorillonite mixtures. *Polymer Degradation and Stability* 92: 1082–1087.

Hermansson, A., Hjertberg, T., and Sultan, B. 2003. The flame retardant mechanism of polyolefins modified with chalk and silicone elastomer. *Fire and Materials* 27: 51–70.

Hirschler, M. M. 1994. Comparison of large-and small-scale heat release tests with electrical cables. *Fire and Materials* 18: 61–76.

Hirschler, M. M. 2015. Flame retardants and heat release: Review of traditional studies on products and on groups of polymers. *Fire and Materials* 39: 207–231.

Hirschler, M. M. 2017. Poly(vinyl chloride) and its fire properties. *Fire and Materials* 41: 993–1006.

Hjertberg, T., and Sorvik, E. M. 1978. On the Influence of HCl on the thermal degradation of poly(vinyl chloride). *Journal of Applied Polymer Science* 22: 2415–2426.

Hollingberry, L. A., and Hull, T. R. 2010. The fire retardant behaviour of huntite and hydromagnesite—A review. *Polymer Degradation and Stability* 95: 2213–2225.

Hull, T. R., Lebek, K., Pezzani, M., et al. 2008. Comparison of toxic product yields of burning cables in bench and large-scale experiments. *Fire Safety Journal* 43: 140–150.

Hull, T. R., Price, D., Liu, Y., et al. 2003. An investigation into the decomposition and burning behaviour of ethylene-vinyl acetate copolymer nanocomposite materials. *Polymer Degradation and Stability* 82: 365–371.

Hull, T. R., Witkowski, A., and Hollingbery, L. 2011. Fire retardant action of mineral fillers. *Polymer Degradation and Stability* 96: 1462–1469.

IEC 60332-1: Test for vertical propagation for a single insulated wire or cable—Procedure for 1 kW premixed flame, Geneva, Switzerland, 2004a.

IEC 60332-3: Test for vertical flame spread of vertically-mounted bunched wires and cables, Geneva, Switzerland, 2004b.

IEC 61034-1: Measurement of smoke density of cables burning under defined conditions—Part 1 test apparatus, Geneva, Switzerland, 2005.

IEC 61034-1: Measurement of smoke density of cables burning under defined conditions—Part 2 test procedures, Geneva, Switzerland, 2006.

IEC 60331-11: Tests for electric cables under fire conditions—Circuit integrity—Part 11: Apparatus—Fire alone at a flame temperature of at least 750°C, Geneva, Switzerland, 1999.

IMO. International code for application of fire test procedures, 2010.

INPES Campagne Nationale de Prévention des Incendies Domestiques, 2009. http://inpes.santepubliquefrance.fr/30000/actus2009/031.asp. Accessed May 25, 2019.

Jiao, Y., and Xu, J. Z. 2009. Flame-retardant properties of magnesium hydroxystannate and strontium hydroxystannate coated calcium carbonate on soft poly(vinyl chloride). *Journal of Applied Polymer Science* 112: 36–43.

Karlsson, L., Lundgren, A., Jungqvist, J., et al. 2011. Effect of nanofillers on the flame retardant properties of a polyethylene–calcium carbonate–silicone elastomer system. *Fire and Materials* 35: 443–452.

Kaseler, T. G. 1993. Non-lead stabilizer systems for PVC wire and cable extrusion. *Journal of Vinyl and Additive Technology* 15: 196–201.

Kashiwagi, F., Du, F., Winey, K. J., et al. 2005. Flammability properties of polymer nanocomposites with single-wall carbon nanotubes: Effects of nanotube dispersion and concentration. *Polymer* 46: 471–481.

Kaufman, S. 1985. PVC in communication cable. *Journal of Vinyl Technology* 7: 107–111.

Khan, M. M., Bill, R. G., and Alpert, R. L. 2006. Screening of plenum cables using a small-scale fire test protocol. *Fire and Materials* 30: 65–76.

Kiddoo, D. B. 2007. Cable component material innovations for stringent fire safety and environmental compliance requirements. *Proceedings of the 56th IWCS International Wire & Cable Symposium*.

Kotak, J. 1983. PVC wire and cable compound—An overview, *Journal of Vinyl Technology* 5: 77–78.

Krauskopf, L. G. 2003. How about alternative to phthalate plasticizers. *Journal of Vinyl and Additive Technology* 9: 187–192.

Laachachi, A., Cochez, M., Ferriol, M., et al. 2004. Influence of Sb2O3 particles as filler on the thermal stability and flammability properties of poly(methyl methacrylate) (PMMA). *Polymer Degradation and Stability* 85: 641–646.

Laoutid, F., Gaudon, P., Taulemesse, J. M., et al. 2006a. Study of hydromagnesite and magnesium hydroxide based fire retardant systems for EVA containing organo-modified montmorillonite. *Polymer Degradation and Stability* 91: 3074–3082.

Laoutid, F., Ferry, L., Leroy, E., et al. 2006b. Intumescent mineral fire retardant systems in EVA copolymer: Effect of silica particles on char cohesion. *Polymer Degradation and Stability* 91: 2140–2145.

Lewin, M., Pearce, E. M., Levon, K., et al. 2006. Nanocomposites at elevated temperatures: Migration and structural changes. *Polymers for Advanced Technologies* 17: 226–234.

Lin, S. C., and Kent, B. 2009. Flammability of fluoropolymers, Chapter 17, In *Fire and Polymers V, ACS Symposium Series*, American Chemical Society, Washington, DC.

Lin, T. S., Cogen, J. M., and Lyon, R. E. 2007. Correlations between microscale combustion calorimetry and conventional flammability tests for flame retardant wire and cable compounds. *Proceedings of 56th IWCS*, pp. 176–185.

Linde, E., and Gedde, U. W. 2014. Plasticizer migration from PVC cable insulation-the challenges of extrapolation methods. *Polymer Degradation and Stability* 101: 24–31.

Liu, S. P. 2014. Flame retardant and mechanical properties of polyethylene/magnesium hydroxide/montmorillonite nanocomposites. *Journal of Industrial and Engineering Chemistry* 20: 2401–2408.

Loadholt, J. T., and Kaufman, S. 1982. Development of a flame-retardant noncontaminating PVC jacket for coaxial cable. *Journal of Vinyl and Additive Technology* 4: 128–132.

Lomakin, S. M., and Zaikov, G. E. 2003. Polyhalogenated flame retardants and dioxins. *Journal of Environmental Protection and Ecology* 4: 95–119.

Lopez-Cuesta, J. M., Longuet, C., and Chivas-Joly, C. 2014. Thermal degradation, flammability and potential toxicity of polymer nanocomposites, Chapter 13. In *Health and Environmental Safety of Nanomaterials*, Njuguna J., Pielichowsly K. and Zhu H. (Eds.), Woodhead Publishing.

Mangs, J., and Hostikka, S. 2012. Vertical flame spread on charring materials at different ambient temperatures. *Fire and Materials* 37: 230–245.

Manzi-Nshuti, C., Songtipya, P., Manias, E., et al. 2009. Polymer nanocomposites using zinc aluminum and magnesium aluminum oleate layered double hydroxides: Effects of LDH divalent metals on dispersion, thermal, mechanical and fire performance in various polymers. *Polymer* 50: 3564–3574.

Marosfoi, B. B., Marosi, G. J., Szep, A., et al. 2006. Complex activity of clay and CNT particles in flame retarded EVA copolymer. *Polymers for Advanced Technologies* 17: 255–262.

McGrattan, K., Lock, A., Marsh, N., et al. 2010. Cable heat release, ignition, and spread in tray installations during fire (CHRISTIFIRE) Phase 1: Horizontal Trays. Office of Nuclear Regulatory Research, NUREG/CR-7010, Vol. 1.

Meinier, R., Sonnier, R., Zavaleta, P., et al. 2018. Fire behavior of halogen-free flame retardant electrical cables with the cone calorimeter. *Journal of Hazardous Materials* 342: 306–316.

Miyamoto, K., Huang, X., Hashimoto, N., et al. 2016. Limiting oxygen concentration (LOC) of burning polyethylene insulated wires under external radiation. *Fire Safety Journal* 86: 32–40.

Mo, S. J., Zhang, J., Liang, D., et al. 2013. Study on pyrolysis characteristics of cross-linked polyethylene material cable. *Procedia Engineering* 52: 588–592.

Montaudo, G., and Puglisi, C. 1991. Evolution of aromatics in the thermal degradation of poly(vinyl chloride): A mechanistic study. *Polymer Degradation and Stability* 33: 229–262.

Morgan, A. B., Harris, R. H., Kashiwagi, T., et al. 2002. Flammability of polystyrene layered silicate (clay) nanocomposites: Carbonaceous char formation. *Fire and Materials* 26: 247–253.

Motzkus, C., Chivas-Joly, C., Guillaume, E., et al. 2012. Aerosols emitted by the combustion of polymers containing nanoparticles. *Journal of Nanoparticle Research* 14: 687–704.

Moy, P. 2004. Aryl phosphate ester fire-retardant additive for low-smoke vinyl applications, *Journal of Vinyl and Additive Technology* 10: 109–115.

Nakagawa, Y. 1998. A comparative study of bench-scale flammability properties of electric cables with different covering materials. *Journal of Fire Sciences* 16: 179–205.

NBN 713-020: Fire fighting—Fire performance of building materials and products—fire resistance of building materials, Brussels, Belgium, 1994.

Netherland Institute for Safety 2009. Consumer fire safety: European statistics and potential fire safety measures, https://www.ifv.nl/kennisplein/Documents/09-06-24_rapport_consumer_fire_safety_pdf1.pdf. Accessed May 25, 2019.

Ounoughene, G. 2015. Study on emissions from thermal decomposition of nanocomposites: The application of incineration. PhD dissertation, University of Nantes, France.

Peeterbroeck, S., Laoutid, F., Swoboda, B., et al. 2007. How carbon nanotubes crushing can improve flame retardant behavior in polymer nanocomposites. *Macromolecular Rapid Communications* 28: 260–271.

Polansky, R., and Polanska, M. 2015. Testing of the fire-proof functionality of cable insulation under fire conditions via insulation resistance measurements. *Engineering Failure Analysis* 57: 334–349.

Qu, H., Wu, W., Wei, H., et al. 2008. Metal hydroxystannates as flame retardants and smoke suppressants for semirigidpoly(vinyl chloride). *Journal of Vinyl and Additive Technology* 14: 84–90.

Qu, H., Wu, W., Xie, J., et al. 2009. Zinc hydroxystannate-coated metal hydroxides as flame retardant and smoke suppression for flexible poly vinyl chloride. *Fire and Materials* 33: 201–210.

Quenehen, P., Royaud, I., Seytre, G., et al. 2015. Determination of the aging mechanism of single core cables with PVC insulation. *Polymer Degradation and Stability* 119: 96–104.

Realinho, V., Antunes, M., and Velasco, J. I. 2016. Enhanced fire behavior of Casico-based foams. *Polymer Degradation and Stability* 128: 260–268.

RoHS. 2002. Directive no 2002/95/CE, http://eur-lex.europa.eu/legal-content/EN/TXT/?uri =CELEX:32011L0065. Accessed May 25, 2019.

Rothon, R. N. 2001. *Particulate Fillers for Polymers*. Smithers Rapra Technology, Shrewsbury, UK.

Sarazin, J., Bachelet, P., and Bourbigot, S. 2017. Fire behavior of simulated low voltage intumescent cables with and without electric current. *Journal of Fire Sciences* 35: 179–194.

Shen, K. K. 2006. Overview of flame retardancy and smoke suppression of flexible PVC. In *Proceedings SPE Vinyltech Conference*, Atlanta, GA.

Sonnier, R., Viretto, A., Taguet, A., et al. 2012. Influence of the morphology on the fire behavior of a polycarbonate/poly(butylene terephthalate) blend. *Journal of Applied Polymer Science* 125: 3148–3158.

Stec, A., and Hull, R. 2011. Fire toxicity and its assessment, Chapter 17. In *Fire Retardancy of Polymeric Materials*, Wilkie C. A. and Morgan A. (Eds.), CRC Press: Boca Raton, FL.

Suresh, S. S., Mohanty, S., and Nayak, S. K. 2017. Composition analysis and characterization of waste polyvinyl chloride (PVC) recovered from data cables. *Waste Management* 60: 100–111.

Svoboda, R. D. 1991. Polymeric plasticizers for higher performance flexible PVC applications. *Journal of Vinyl Technology* 13: 130–133.

Tabuani, D., Bellucci, F., Terenzi, A., et al. 2012. Flame retarded thermoplastic polyurethane (TPU) for cable jacketing application. *Polymer Degradation and Stability* 97: 2594–2601.

Tewarson, A., and Khan, M. M. 1992. A new standard test method for the quantification of fire propagation behavior of electrical cables using Factory Mutual Research Corporation's small-scale flammability apparatus. *Fire Technology* 28: 215–227.

Titow, W. V. 1990. *PVC Plastics: Properties, Processing, and Applications*. Elsevier: London, UK.

Toure, B., Lopez-Cuesta, J. M., Gaudon, P., et al. 1996. Fire resistance and mechanical properties of a huntite/hydromagnesite/antimony trioxide/decabromodiphenyl oxide filled PP-PE copolymer. *Polymer Degradation and Stability* 53: 371–379.

UL 1581: Reference Standard for Electrical Wires, Cables, and Flexible Cords, Northbrook, USA, 2006.

UL 1685: Vertical-Tray Fire-Propagation and Smoke-Release Test for Electrical and Optical-Fiber Cables, Northbrook, USA, 2007.

Van Luik, F. W. 1974. Characteristics of invisible particles generated by precombustion and combustion. *Fire Technology* 10: 129–139.

Wang, B. B., Sheng, H. B., Shi, Y. Q., et al. 2016. The influence of zinc hydroxystannate on reducing toxic gases (CO, NOx and HCN) generation and fire hazards of thermoplastic polyurethane composites. *Journal of Hazardous Materials* 314: 260–269.

Wang, D., Parlow, D., Yao, Q., et al. 2002. Melt blending preparation of PVC-sodium clay nanocomposites. *Journal of Vinyl and Additive Technology* 8: 139–150.

Wang, X., and Zhang, Q. 2004. Effect of hydrotalcite on the thermal stability, mechanical properties, rheology and flame retardance of poly(vinyl chloride). *Polymer International* 53: 698–707.

Weil, E. D., Patel, N. G., Said, M. M., et al. 1992. Oxygen index: Correlations to other fire tests. *Fire and Materials* 16: 159–167.

Wickson, E. J. 1993. *Handbook of Polyvinyl Chloride Formulating.* Wiley: New York.

Wu, C. H., Chang, C. Y., Hor, J. L., et al. 1994. Two-stage pyrolysis model of PVC. *Canadian Journal of Chemical Engineering* 72: 644–650.

Wu, W., Qu, H., Li, Z., et al. 2008. Thermal behavior and flame retardancy of flexible poly(vinyl chloride) treated with zinc hydroxystannate and zinc stannate. *Journal of Vinyl and Additive Technology* 14: 10–15.

Xie, Q. Y., Zhang, H. P., and Tong, L. 2010. Experimental study on the fire protection properties of PVC sheath for old and new cables. *Journal of Hazardous Materials* 179: 373–381.

Xu, J., Zhang, C., Qu, H., et al. 2005. Zinc hydroxystannate and zinc stannate as flame-retardant agents for flexible poly(vinyl chloride). *Journal of Applied Polymer Science* 98: 1469–1475.

Yang, H., Fu, Q., Cheng, X. D., et al. 2013. Investigation of the flammability of different cables using pyrolysis combustion flow calorimeter. *Procedia Engineering* 62: 778–785.

Yang, L., Hu, Y., You, F., et al. 2007. A novel method to prepare zinc hydroxystannate-coated inorganic fillers and its effect on the fire properties of PVC cable materials. *Polymer Engineering and Science* 47: 1163–1169.

Ye, L., Ding, P., Zhang, M., et al. 2007. Synergistic effects of exfoliated LDH with some halogen-free flame retardants in LDPE/EVA/HFMH/LDH nanocomposites. *Journal of Applied Polymer Science* 107: 3694–3701.

Yu, G., Bu, Q., Cao, Z., et al. 2016. Brominated flame retardants (BFRs): A review on environmental contamination in China. *Chemosphere* 150: 479–490.

Zaikov, G. E., Gumargalieva, K. Z., Pokholok, T. V., et al. 1998. PVC wire coatings: Part 1-Ageing process dynamics. *International Journal of Polymeric Materials and Polymeric Biomaterials* 39: 79–125.

Zanetti, M., Kashiwagi, T., Falqui, L., et al. 2002. Cone calorimeter combustion and gasification studies of polymer layered silicate nanocomposites. *Chemistry of Materials* 14: 881–887.

Zazycny, J. M., and Matuana, L. M. 2005. Fillers and reinforcing agents, Chapter 7. In *PVC Handbook*, edited by C.E Wilkie, C.A. Daniels and J.W. Summers, Hanser Publishers, Munich, Germany.

Zhang, G., Ding, P., Zhang, M., et al. 2007. Synergistic effects of layered double hydroxide with hyperfine magnesium hydroxide in halogen-free flame retardant EVA/HFMH/LDH nanocomposites. *Polymer Degradation and Stability* 92: 1715–1720

Zhang, J., and Wilkie, C. A. 2005. Fire retardancy of polyethylene-alumina trihydrate containing clay as a synergist. *Polymers for Advanced Technologies* 16: 549–553.

Zhao, Y., Wang, K., Zhu, F., et al. 2006. Properties of poly(vinyl chloride)/wood flour/montmorillonite composites: Effects of coupling agents and layered silicate. *Polymer Degradation and Stability* 91: 2874–2883.

Zheng, X., and Cusack, P. A. 2013. Tin-containing layered double hydroxides: Synthesis and application in poly(vinyl chloride) cable formulations. *Fire and Materials* 37: 35–45.

15

Flame Retardant Epoxy Resin Formulations for Fiber-Reinforced Composites

Manfred Döring,
Sebastian Eibl,
Lara Greiner, and
Hauke Lengsfeld

15.1 Introduction

Fiber-reinforced epoxy resins are widely used in automotive and aircraft construction as lightweight materials with optimized mechanical properties. The following section describes composite materials (Section 15.1.1), the use of epoxy resins in fiber-reinforced composites (Section 15.1.2), and basic parameters of incorporating flame retardants (FRs) into the matrix (Section 15.1.3).

15.1.1 Composite Materials

For many applications, neat epoxy resins do not provide the required strength or stiffness. Mechanical properties can be increased by incorporating fibers into the resin matrix, yielding fiber-reinforced composites. In composite materials, two or more different, but compatible materials are combined in order to achieve new properties that could not be obtained by a single component. The matrix as a binding material holds the fibers in place, transfers forces into and between the fibers, and is responsible for corrosion resistance, chemical resistance, fatigue, and the fire performance of the composite. The fibers as a reinforcement component, provide strength and stiffness to the composite. As a consequence, the material has a high strength to weight ratio and can replace conventional materials in load bearing and light weight applications. Developing fiber-reinforced materials requires the consideration of many variables. The resin type as well as the type, orientation, and amount of fibers are parameters influencing the performance of the material for the designated application. Different fiber types are used in fiber-reinforced materials. Glass fibers are most commonly applied, as they are affordable, have good mechanical properties, corrosion and water resistance. For higher requirements concerning mechanical properties, e.g., in aerospace applications, carbon fibers are of great importance. They offer a very high strength and modulus, while keeping the weight of the composite very low. Other common fiber types are aramid (Kevlar®) or natural fibers. As epoxy resins are thermosets, the shape of a component is achieved during the hardening process. Fiber-reinforced epoxy composites can be produced by different procedures. Hand lay-up is used for small quantities and complex geometries, e.g., for boats used in sports. Vacuum-assisted processes (VAP) and resin transfer molding (RTM) demand higher investments, but produce larger quantities per time. RTM is used for the production of vehicle and aircraft components and sporting goods. Vacuum infusion (rotor blades for wind engines), pultrusion (e.g., tubes, beams), and filament winding (tubular structures, e.g., pressure vessels) are alternative processes. For the aerospace industry, the prepreg/autoclave technique is important (Mazumdar 2002).

15.1.2 Epoxy Resins and Composite Materials

Epoxy resins represent a highly versatile class of thermosetting materials. In the case of a poly addition reaction, they typically consist of molecules with two or more epoxy groups and a curing agent. Typical curing agents are diamines, anhydrides, or phenolic resins. The concentration of the curing agent can vary from 5% to over 50% in the formulation. Mechanical properties in the cured state can thereby be adjusted in a wide range, e.g., from brittle to tougher flexible to highly ridged. Furthermore, processing properties like viscosity, reactivity, and latency can be adapted to manufacturing technologies used, e.g., liquid resin infusion, high pressure RTM, or for prepreg manufacture. In all cases, an adapted reactivity is an important characteristic for processing the material. The curing of epoxy resins (poly addition) is an exothermic process. The exothermic behavior as well as the kinetically determined initial temperature for the curing process are influenced by steric and electronic effects in the hardener. Those variables have to be adjusted for the technology (e.g., oven, autoclave, hot press) used to avoid polymer degradation, e.g., by temperature spikes as well as porosity generation in the component. Either of both is not desirable due to the negative impact on the mechanical properties of the final product.

Epoxy resins typically exhibit very good wetting of fibers and textiles, e.g., carbon, glass, or other fibers (with appropriate sizing of the fibers) used for fiber-reinforced composites. When selecting a matching fiber sizing/epoxy resin combination, due to these wettability characteristics, suitable adhesion of fiber and matrix for excellent mechanical properties as well as tensile strength or interlaminar shear strength is obtained. With the points stated above, the application of epoxy resin systems is as widely spread as their versatile properties. Numerous highly sophisticated applications of epoxy-based composites can be found in space and aerospace (mainly in the fuselage, wing and tail fin components, control surfaces, and doors) requiring the highest mechanical performance, as well as in transport, automotive, wind energy, and electronics —each application with its specific requirements in terms of handling, processing, and properties of cured material. Finally, yet none the less importantly, epoxy resin can provide well-balanced solutions at an attractive price level due to their aforementioned versatility. Table 15.1 provides a few examples of non-flame-retarded systems currently used alongside their application.

Common prepreg systems—as shown in Table 15.1—are, especially in aerospace, formulated with 3- or 4-functional epoxy resins providing high glass transition temperatures (T_g) and good mechanical properties. Bifunctional epoxies of various molecular weights are often used to balance properties like viscosity—respectively tack—which is often an important parameter in composite manufacturing, and resin flow during cure. Fracture toughness is typically improved by using amounts of up to 20–25 wt% of high molecular weight polyether sulfone (PES) or PEI (polyether imide) dissolved in the epoxy resin mixture. Although the PES-containing systems are not explicitly designed for flame retardancy, they show some improved flame-retardant behavior compared to other epoxy resin formulations. Therefore, the sulfone in the thermoplastic modifier already shows some flame-retardant effect. This does not include RTM systems, which do not contain any thermoplastic modifiers nor flame retardants and are entirely composed of 3- or 4-functional epoxy resin and aromatic amine hardeners. All remaining epoxy systems from Table 15.1 with a chemical backbone mainly based on Bisphenol-A and Bisphenol-F resins typically are providing no intrinsic fire, smoke, and toxicity (FST)-properties. A typical example for a wet

TABLE 15.1 Examples of Non-Flame-Retarded Epoxy Resin Systems and Their Applications

Resin System	Curing Conditions	Properties	Market/ Application	Configuration, Manufacturing Technology	Manufacturer and Example
3- and 4-Functional aromatic epoxy resins, 4-functional aromatic amine hardeners, high temperature thermoplastics	Autoclave, 2 hours 180°C, 7–10 bar	Very high T_g (180°C–200°C), high stiffness and high fracture toughness	Aerospace structural parts	Prepreg, automatic tape laying, or fiber placement	**HexPly® 6376 and M21, Hexcel® Cycom® 977-2, Cytec Solvay Group®**
3- and 4-Functional aromatic epoxy resins, 4-functional aromatic amine hardeners	Autoclave, open, or closed mold, 2 hours 180°C, 7–10 bar	Very high T_g (180–200°C), high stiffness and medium fracture toughness	Aerospace structural parts	Liquid resin system, liquid resin infusion, or RTM technology	**HexFlow® RTM6. Hexcel® Cycom® 977-20, Cytec Solvay Group®**
2-Functional aromatic epoxy resins, 2-3 functional aliphatic amine hardeners	Oven, open, or closed mold, 2–4 hours 130°C	High T_g (100°C–150°C), good stiffness and facture toughness	Industrial composites		**Biresin CR135, Sika®**
		High T_g (100°C–130°C), good stiffness and facture toughness	Industrial composites	Liquid resin system, vacuum infusion technology	**Biresin CR131, Sika®**
	Oven, open, or closed mold, 24 hours room temperature (RT) + 24 hours 180°C	Very high T_g (180°C–200°C), good stiffness	Composite tooling		**EPIKOTE™ RIM R 9000, Hexion™**
	Oven, open mould, 24 hours RT + 24 hours 80°C	Medium T_g (80-90°C), good stiffness and fracture toughness	Structural parts for small planes and gliders	Liquid, wet laminating	**EPIKOTE™MGS™ LR285, Hexion™**
2-Functional cycloaliphatic epoxy resins, aliphatic anhydride hardener	Oven, open, or closed mold, 2 hours 120°C		Industrial composites		**PRIME' 180, Gurit®**

laminating system based on 2-functional epoxy resins and liquid aminic hardeners is Hexion® LR285®, which is approved for small airplanes and gliders manufacture. Especially because of the outstanding material properties, it is one of the most commonly used systems in this section of the aircraft industry today. For this application, good FST properties were not required by airworthiness regulations. All other systems mentioned above can be used for any composite part production depending on required mechanical and thermo-mechanical properties of the application.

15.1.3 Incorporation of Flame Retardants—Basic Parameters

The most serious disadvantage of epoxy-based resins and their composites is their insufficient fire resistance. Many applications like electric and electronic devices demand exigent fire-resistant requirements. For the incorporation of flame retardants, numerous variables besides the fire performance and smoke requirements have to be considered.

The manufacturing process of fiber-reinforced composites is crucial when incorporating flame retardants into the matrix resin. For example, infusion techniques like RTM or LRI (liquid resin infusion require—in addition to specific viscosity of the modified resin—a homogeneous matrix resin system. Otherwise the fiber preform placed in the closed mold may filter solid phase additive particles when the resin is injected. This leads to a flame-retardant gradient in the material and consequently to an uneven fire and material performance. The particle distribution during resin infusion/injection depends on many parameters (Laurenzi and Marchetti 2012). If the particles are small enough and there is a homogeneous and stable suspension of the particles in the resin formulation, the use of techniques like RTM or LRI is possible under certain further conditions (Fiedler et al. 2006; Thostenson et al. 2001). Hand lay-up in contrast leads to no filtration and is less sensitive to increased viscosities, but the achievable maximum fiber content is lower and it is less productive. Mechanical and thermal property requirements have to be considered as well. In most cases, these properties are deteriorated by incorporating flame retardants into epoxy resins and composite materials as the crosslinking density of the cured epoxy-based materials is decreased. With a few exceptions, non-reactive, organic flame-retardant additives act as plasticizers and lead to a decrease of the glass transition temperature, another crucial parameter of epoxy thermosets. The mentioned parameters among others determine the resin, the flame-retardant additives, and the fiber type as well as what architecture are suitable for the application.

Depending on the end-use application for the flame-retarded part, fire and smoke performance must fulfill different requirements. In the United States, there is the ASTM E-84 test and in the European Union, the Single Burning Item (SBI) test for building and construction applications. For aerospace and ship hulls, Federal Aviation Regulations (FAR) and International Maritime Organization codes apply.

15.2 Influence of the Epoxy Resin Formulation on the Performance of Flame Retardants

15.2.1 Fire Behavior of Different Epoxy Resin Formulations

For each area of application and processing technology, different kinds of epoxy compounds and curing agents have been developed. Most neat epoxy-based thermosets are easily inflammable. However, the chemical structures of the epoxy components and curing agents as well as the crosslinking density strongly influence whether a resin burns vigorously or slowly, and in some cases, they might even render a system self-extinguishing. The most widely used epoxy compounds are diglycidyl ether of bisphenol A (DGEBA, Figure 15.1) and its oligomers. Due to their rather low network density, most DGEBA-based materials show strong burning and dripping after they have been exposed to a flame. A higher functional epoxy resin is tetraglycidyl methylene dianiline (TGMDA, Figure 15.1) that is used in aviation applications, this generates higher network densities. Important commercial curing agents are amines, anhydrides, or novolacs. Due to the high aromatic content, novolac-cured epoxies show the best intrinsic

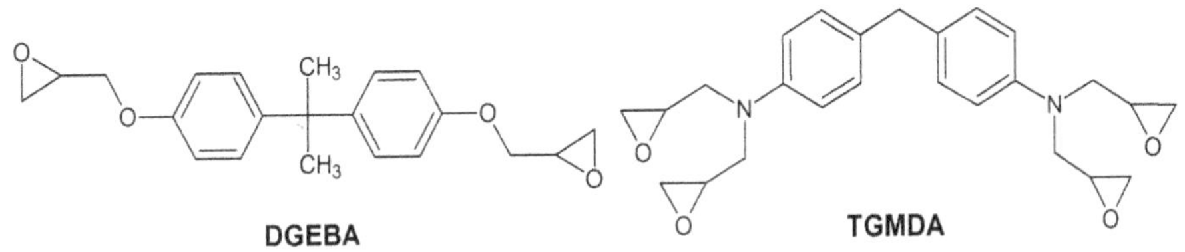

FIGURE 15.1   Molecular structures of the epoxy resins: DGEBA and TGMDA.

flame-retardant performance. At comparable crosslinking densities acid- or anhydride-cured epoxy resins are more flammable than epoxy resins cured with amines, as the latter produce more char (Martin and Price 1968). Aliphatic diamines or polyamines are used as curing agents especially if homogeneous systems with low viscosities are required. As the flammability is higher for aliphatic components than for aromatic components, these resin formulations are highly inflammable. Therefore, highly aromatic diamines help to improve the flame-retardancy. If high glass transition temperatures above 200°C are requested, 4,4′-diaminodiphenylsulfone (4,4′-DDS) is used as a curing agent. Besides the aromatic groups, it contains a sulfonic group, and the flame-retardant effect of heteroatoms induces a moderate improvement of the fire behavior (Braun et al. 2006). A combination of TGMDA and 4,4′-DDS is relevant for aviation applications. The commercial formulation M 18/1 contains only a moderate amount of a boron-based flame retardant and shows self-extinguishing behavior. Also dicyanamide and phenol novolacs are widely used to cure epoxy resins especially for printed circuit boards. Formulations thereof have a lower content of aliphatic substructures and show a better fire behavior, but are not self-extinguishing.

Among the special epoxy resins that are inherently flame-retarded are glycidyl-functionalized novolac-based-cured resins. These poly functional epoxy compounds with a high content of aromatic units form dense and thermally stable networks. No dripping occurs after samples are exposed to an ignition source. Basically, compounds that contain novolac derivatives including aromatic groups in the main chain have far higher flame-retardancy than that of epoxy resin compounds without aromatic groups (Iji and Kiuchi 2001a, 2001b). The use of special epoxy monomers containing highly aromatic bisphenols (e.g., phenolphthalein, bisphenol-fluorenone) (Arada et al. 1979; Chen et al. 1982a; Lin and Pearce 1979), bisphenol-anthrone and tetraphenol-anthracene (Chen et al. 1982b), or double bonds able to undergo the Diels-Alder reaction (e.g., mono-, di- or trihydroxystyrylpyridine) (Yan and Pearce 1984) may increase the char yield. Also epoxy resins and curing agents based on condensed aromatic hydrocarbons—such as naphthalene and biphenyl—were developed. When both components are combined, the flammability is even more decreased than for epoxy novolac resins (Song et al. 2013). Several suppliers (DIC Corporation, Huntsman) offer combinations of naphthalene-based epoxy resins and special novolac hardeners. The cured samples show self-extinguishing behavior and may be classified V-0 in UL-94 tests, so additional flame retardants are unnecessary.

15.2.2 Performance of Flame Retardants

The performance of flame retardants depends on the resin composition. High performance resins used by the aerospace industry are primarily composed of aromatic structures. Therefore, there is a lower heat release potential to start with, and they are easier to equip with a flame protected. The most common flame retardant used for epoxy resins is tetrabromobisphenol A (TBBPA, Figure 15.2). This molecule contains two phenolic hydroxyl groups that react with epoxy groups in a fusion process (or so-called preformulation reaction) and it is thereby incorporated into the network without decreasing the network density. While the epoxy novolac resin D.E.N.™ 438 (Dow Chemical, DEN438) cured with 2-Cyanoguanidine (DICY)/fenurone requires only a loading of 13.6 wt% TBBPA to achieve UL-94 V-0 classification, DGEBA cured with the same hardener requires a loading of 33.5 wt% (Döring et al.

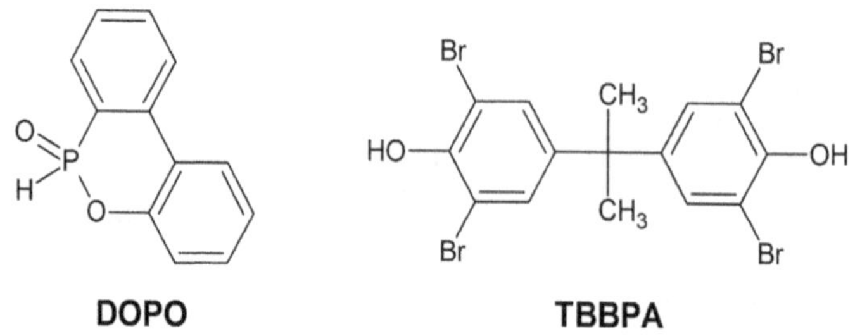

DOPO   **TBBPA**

FIGURE 15.2   Chemical structures of DOPO and TBBPA.

2012). Halogen-free, less toxic, and ecologically friendly alternatives to bromine-based flame retardants are a current research topic. Especially organophosphorous compounds like the commercially available flame-retardant 9,10-dihydro-9-oxa-10-phosphaphenanthrene-10-oxide (DOPO) (Figure 15.2) or its derivatives have been designed. DOPO itself is a mono-functional reactive flame retardant and therefore decreases the network density when it is added to epoxy groups and becomes a part of the network backbone. Consequently, novel DOPO derivatives should have no negative effect on the epoxy network density. For example, DOPO additive flame retardants do not react with epoxy groups and will not be an integrative part of the network. Also, hardeners and epoxy resins already containing phosphorous species were designed. Other trends are oligomeric and polymeric flame retardants, silicon-based flame retardants, and synergistic flame-retardant formulations (Ciesielski et al. 2017).

The epoxy novolac resin DEN438 could be protected by moderate loadings of different mono- and/or bifunctional gas-phase active phosphorous compounds based on DOPO, diphenylphosphine oxide, 5,5-dimethyl-2-oxid O-1,3,2-dioxaphosphinan, and 2,8-dimethyl-phenoxyphosphine-10-oxide, whereas DGEBA failed to achieve a UL-94 classification with all investigated mono-functional compounds in a study by Döring et al. (Ciesielski et al. 2009; Döring et al. 2012). This can partly be explained by the higher aromaticity of DEN438 compared to DGEBA. Non-reactive phosphorous-containing flame retardants in different matrices were investigated by Schartel et al. and Döring et al. among others. Perret et al. (2011a) described the synthesis of the new flame-retardant adduct of DOPO and tris-[(acryloyloxy)] ethyl isocyanurate (DOPI) (Figure 15.3). This molecule contains DOPO units that are linked to a star-shaped aliphatic molecule. The resin formulations used for the examination of flame retardancy were RTM6 [HexFlow®RTM6 (Hexcel®) consisting of TGMDA and aromatic amine hardeners] and a combination of DGEBA and 3,3'-dimethyl-4,4'-diamino-dicyclohecylmethane (DMDC). Samples without any flame-retardant additives served as references. In this study, the impact of fiber reinforcement was tested as well (see Section 15.3.2). Whereas the neat resin formulations did not reach any UL-94-classifications, 2 wt% of phosphorous in RTM6 was enough to reach V-0 classification. However, DGEBA/DMDC did not reach any classification with the same amount of phosphorous. The fire behavior with respect to actions in the gas phase and the condensed phase for the two flame-retarded epoxy resins was different. Schartel et al. proposed different decomposition pathways for the resin formulations based on the results of mass loss, evolved gas analysis, and residue analysis. DOPI is primarily supposed to act in the gas phase. DGEBA/DMDC/DOPI did not show an enhanced residue and therefore no condensed-phase mechanism is expected. RTM6/DOPI showed an enhanced amount of residue determined by thermogravimetry and phosphorus species in the gas phase monitored in evolved gas analysis, consequently, there are indications for gas and condensed-phase mechanisms. So not only the higher network density of RTM6 leads to a decreased flammability in comparison to DGEBA, also the mechanism of added flame retardants is influenced by chemical structures of the resin components.

Döring et al. (Ciesielski et al. 2010; Zang et al. 2011) synthesized and investigated two star-shaped *tris*(2-hydroxyethyl) isocyanurate (THIC)-bridged derivatives of DOPO and 5,5-dimethyl-1,3,2-dioxaphosphorinan-2-one (DDPO) (Figure 15.3) incorporated in DEN438 and DGEBA, both cured with DICY/fenurone. UL-94 V-0-classifications were reached by DEN438/DICY/fenurone/THIC-(DDP-O)$_3$

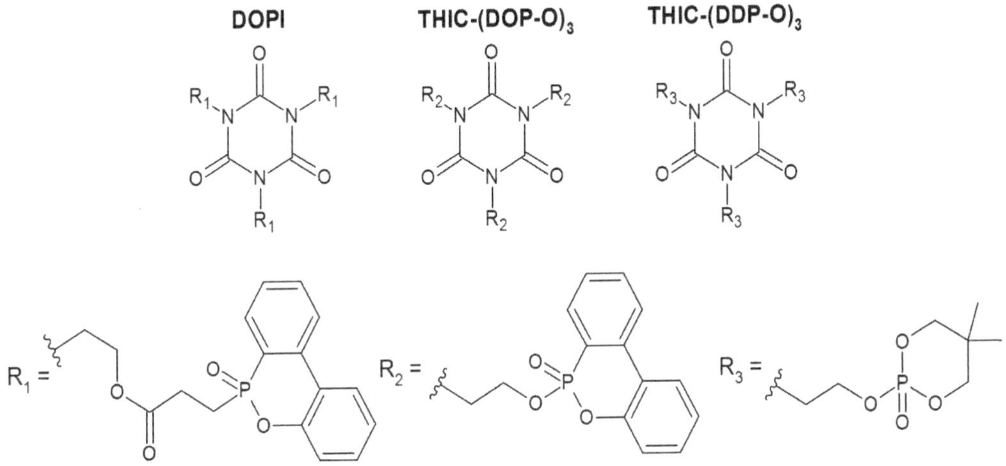

FIGURE 15.3 Star-shaped phosphorous-containing flame-retardant additives.

and DGEBA/DICY/fenurone/THIC-(DDP-O)$_3$ formulations if 2.5 wt% phosphorous were incorporated. THIC-(DOP-O)$_3$ could achieve the same rating in DEN438 if 1.0 wt% phosphorous were incorporated, but only a V-2-classification in DGEBA for 2.5 wt% phosphorous. The corresponding resorcinol-bridged derivatives with a rod-like geometry were investigated for comparison. DOPO derivatives which act primarily via gas-phase mechanism were found to be preferable for epoxy resins with a high amount of aromatic subunits (DEN438), whereas DDPO derivatives are condensed-phase active and most suitable for epoxy resins with a high amount of aliphatic hydrocarbons (DGEBA).

15.3 Impact of Fiber Reinforcement on the Fire-Retardant Behavior of Epoxy-Based Composites

15.3.1 Influence of the Reinforcement on the Flame Retardancy of Composites

As shown in Section 15.2.1, the single components of epoxy resin formulations have a great impact on the fire behavior in terms of flame spread and smoke. There are additional impacts on the burning characteristics that have to be discussed if there is fiber reinforcement. The fiber type and amount has to be considered. Combustible reinforcing fibers significantly contribute to the burning behavior of composites (Brown et al. 1994). Glass fibers instead act as inert fillers and thermal isolators as well as carbon fibers if a threshold of 600°C is not reached (Eibl 2017a). They show a diluting effect (Le Bras et al. 1998; Levchik et al. 1996) and have an impact on the velocity of combustion through formation of a protective layer (Hume 1992). This protective layer acts as a barrier in transportation processes and has an impact on the combustion mechanism. If the sample is exposed to heat radiation that penetrates the matrix, the matrix degrades and gaseous degradation products are produced. They migrate through the laminates and through possibly formed char to the burning surface. This process is hindered by the barrier. Another point is that occurring delamination often causes unsteady burning of the composite as delaminated layers burn fast and fierce. It was shown that already a small increase of the volume fraction of carbon fibers in a carbon fiber-reinforced epoxy resin from 56% to 59% leads to a sharp decrease of the heat release rate and total heat release (Dao et al. 2016). Also dimensional parameters influence the fire properties like the specimen thickness (Fateh et al. 2017). An increase in the thickness of a sample leads to a later ignition because of the heat conduction into the bulk material (Scudamore 1994) and an increase in the heat emitted during combustion, but the specific heat released per volume unit stays the same (Gibson and Hume 1995). The incident heat flux influences the flammability for a given composite of defined thickness. This heat flux may vary in real fires. Cone calorimetric measurements (Scudamore 1994) showed that the thermally thin to thick effect decreases as the external heat flux increases. Thick samples (9.5 mm) ignited less easy than thin samples (3 mm) at 35 and 50 kW/m² as expected, but at 75 and 100 kW/m² both samples behaved thermally thin. The fiber architecture has an impact on the burning behavior as well. Eibl (2012, 2017b) did different studies on carbon fiber-reinforced composites and was able to show by cone calorimetric measurements and the observation of temperature distributions through the laminate that the velocity and degree of combustion are dominated by fiber orientation of continuous filaments for a given resin. If this resin is less thermally stable, it is expected that the effects are less distinct. Simultaneously, it was investigated how the influence of fiber orientation on reaction-to-fire properties of a composite depends on the heat flux (15–80 kW/m²) and the specimen thickness (0.25–8 mm). In general, a quasi-isotropic fiber orientation leads to faster ignition, because of preferred delamination, but retards combustion processes more effectively than a unidirectional lay-up. With increasing thickness the effect is stronger. For quasi-isotropic samples with a technically relevant thickness, a better performance in case of fire is foreseen. The second study showed flame retarding effects by out of plane-oriented fibers in carbon fiber-reinforced plastic materials. Samples with different angles for the out of plane orientation were investigated (0°–90°). Cone calorimetric measurements (60 kW/m²) showed that with increasing angles (0°: typical unidirectional lay-up), the barrier effect of the fibers is reduced until an out of plane angle of 15°. Higher angles do not show significant differences in the heat release rate and the flame retardation by the carbon fiber. The heating of the sample on the backside was taken as an indicator for the velocity for the migration of the pyrolysis zone through the sample. Also a comparison between fabric and roving fiber reinforcement was done (Eibl 2017b). For identical resin types and carbon fibers, the fabric provides a better retardation of the pyrolysis than a roving. Delamination occurs easier in a fabric-reinforced sample due to resin rich areas between the fiber plies, resulting in a more unsteady combustion. As shown, the impact of the fiber reinforcement on the burning behavior of epoxy resins depends on many parameters, such as fiber type and architecture as well as fraction, matrix type, additives, and possible synergism or antagonism between different components.

Carbon fiber reinforcement may cause an additional hazard next to smoke and toxic gases produced by the combustion of the matrix resin as the fibers degrade significantly if a threshold temperature of about 600°C is reached. The degradation increases with irradiation time and heat flux (Eibl 2017a). Fiber fragments of certain shapes can be sharp enough to puncture human skin or small enough to be respirable and therefore may cause cancer. In Section 15.5, this phenomenon as well as possible prevention methods will be discussed.

15.3.2 Composites with Incorporated Flame Retardants

There are many published studies on the flame retardancy of epoxy resins. Mainly phosphorous-based flame retardants as an alternative for halogenated flame retardants are investigated. The impact of fiber reinforcement on the fire behavior of epoxy-based materials that are halogen-free flame protected is discussed less often. The results of these few studies showed that a condensed-phase action is inefficient in composites containing high amounts of carbon fibers (70 wt%). The surface area compared to the volume is increased greatly by the incorporation of fibers so the shielding gained by char formation is less effective (wick effect) (Perret et al. 2011a, 2011b). The efficiency of the flame-retardant additives adduct of DOPO and tetra-[(acryloyloxy)ethyl] pentarythrit (DOPP) and DOPI, shown in Figure 15.4, and their mode of action in neat and carbon fiber-reinforced materials were investigated by Schartel et al. DOPP and DOPI are based on DOPO units linked to a star-shaped aliphatic molecule. DOPO derivatives act primarily via gas-phase mechanism. The matrix materials used were RTM6 and a combination of DGEBA and DMDC. Samples without any flame-retardant additives served as references. The investigations of these samples showed for both carbon fiber-reinforced matrixes enhanced limiting oxygen index (LOI) values compared to the neat resins. RTM6 went from 25.0% without carbon fiber reinforcement to 33.2% with a fiber amount of 70 wt%. The DGEBA/DMDC matrix showed a LOI of 20.4% without and 27.0% with carbon fiber reinforcement (60 wt%). UL-94 measurements showed no classification for samples without any flame retardant as well as the DGEBA/DMDC/DOPI formulation with 2 wt% phosphorous.

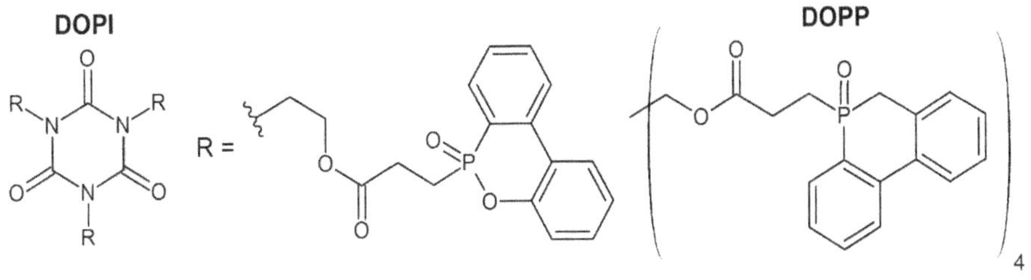

FIGURE 15.4 Star-shaped phosphorous-containing flame-retardant additives DOPI and DOPP.

The comparable reinforced sample reached a V-0-rating. Overall, the UL-94 ratings of the materials containing flame-retardant additives were improved by the reinforcement, so that all flame-retarded carbon fiber-reinforced samples achieved a V-0-rating. The RTM6/DOPP sample without fibers was only classified V-1. In conclusion, the fiber reinforcement of either formulation decreased the fire risk. Cone calorimetric tests (50 kW/m²) showed an extended time to ignition for both matrix resin systems and gave information about the mode of action of the flame retardants. The total heat evolved was reduced, and the char formation was suppressed by the addition of carbon fibers. The authors concluded that in the neat resin a condensed-phase action and in the composite material a gas-phase action is predominant, so the fiber reinforcement has an influence on the mode of action of the flame retardant.

Müller (2013) used 2,4,6-Trioxo-1,3,5-triazinane-1,3,5-triyl)tris(ethane-2,1-diyl) tris(3-(dimethoxyphosphoryl) propanoate (DMPI) and 2,2-Bis(((3-(dimethoxyphosphoryl)propanoyl)oxy)methyl)propane-1,3-diyl bis(3-(dimethoxy phosphoryl)propanoate (DMPP) (structures shown in Figure 15.5) as flame retardants that act mainly in the condensed phase, and confirmed these observations. The structures are quite similar to DOPP and DOPI, but instead of DOPO, dimethyl phosphonate units were incorporated. V-0-ratings were reached for DGEBA/DMDC samples with DMPI (2 wt% phosphorous) and DMPP (3.2 wt% phosphorous) and for RTM6 samples with DMPI (1.5 wt% phosphorous) and DMPP (1.0 wt% phosphorous). In this case, the impact of the carbon fiber reinforcement led to worse UL-94 results. No UL-94 classification was reached even with 3 wt% phosphorous contents in both matrixes. In conclusion, a condensed-phase action is inefficient in composites containing high amounts of carbon fibers (70 wt%). The surface area of the condensed phase is increased by the fibers so that the shielding by the char is not effective.

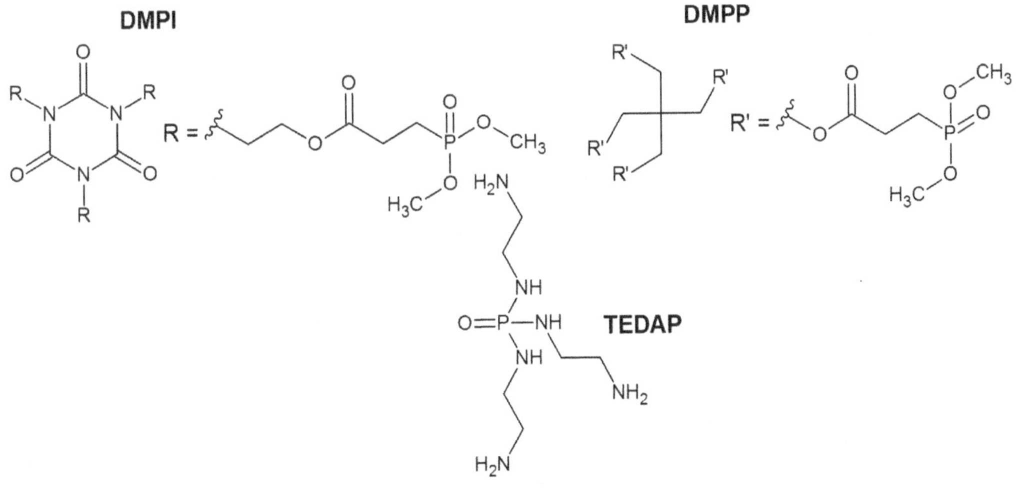

FIGURE 15.5 Molecular structures of DMPI, DMPP, and TEDAP.

The impact of lower amounts of fibers (40 wt%) on the flame retardancy of carbon fiber-reinforced epoxy resins was investigated by Toldy et al. (2011). The authors compared the flame retardancy of aliphatic and aromatic epoxy resin matrices and composites flame-retarded by the phosphorous-containing crosslinking amine N, N′,N″-*tris*(2-aminoethyl) phosphoric triamide (TEDAP, see Figure 15.5). As epoxy components, a system based on DGEBA was compared to a system based on the glycidyl ether of pentaerythritol. Only the LOI of the neat resin was enhanced by the fiber reinforcement. Contrary to the observations of Schartel et al., the LOI (31%–33%) and the UL-94 results (V-0-ratings) of the flame-retarded samples were not influenced by the fibers. Cone calorimetric measurements (50 kW/m²) showed a delay of the ignition for fiber-reinforced samples as well as data to discuss the flame-retardant mode of action. The carbon fibers tend to hinder condensed phase action or intumescence of the flame retardant and only the gas-phase activity remains. In conclusion, these results confirm that carbon fiber reinforcement reduces the efficiency of condensed-phase active phosphorous-based flame retardants.

Not only is the impact of carbon fibers important to investigate, also glass fibers are widely used as reinforcement in epoxy resins. Becker et al. (2012) investigated the flame-retardant effect of intercalated layered double hydroxides (LDHs) in a DGEBA/triethylenetetramine formulation as well as in glass fiber-reinforced samples thereof. LDHs are nanofillers and also known as hydrotalcite-like compounds or anionic clays. The overall flame retardancy of LDH is created by CO_2 and H_2O formation as well as the metal oxide residue. UL-94 horizontal burning tests showed a decrease in the burning rate when glass fibers are present. But overall, there was only a rather low impact of the glass fiber reinforcement onto the fire behavior in flame-retarded and non-flame-retarded samples as no change of the vertical UL-94 test classifications was observed.

The impact of fiber reinforcement on the fire-retardant behavior of epoxy-based composites is not completely investigated. The discussed studies showed that the impact depends on a variety of parameters. The type and amount of fibers, the structural type of the matrix components, and the flame retardant itself interact in a complex way so that predictions can only be done by investigating similar systems. By considering the mentioned studies, it can be concluded that a gas-phase action of flame retardants is generally more effective in those composites than charring or intumescence mechanisms.

15.4 Selected Examples for Flame-Retardant Epoxy Resin Formulations for Fiber-Reinforced Composites

Due to the combustible nature of epoxy resins and their widely spread use in all kinds of applications, there are various flame-retarded systems or flame-retardant additives available in the market to meet the required level of flame retardancy. The risk has to be associated with the application. All products have to fulfill certain specified regulations for particular applications. While the FST requirements in electronic applications can be fulfilled by meeting the certification criteria for, e.g., UL-94 V-0, public transport (rail, ship) and aerospace applications require a much higher level of FST properties, especially in terms of time to ignition, smoke density, and HRR (heat release rate), to meet specification, e.g., FAR 25.873. Hence, the resin formulations very much depend on criteria such as application and required FST properties, pricing, as well as environmental aspects. Flame retardancy of epoxy resin systems has a long history and with the change of some of the aforementioned requirements over time more modern FR systems were developed, while some older ones became less popular or even vanished from the market. Table 15.2 presents some examples of epoxy-based resins containing different flame retardants showing the broad variety of FR additives.

Prepreg systems like HexPly® M26 and M92 of Hexcel Corporation were developed during the last 20 years and are designed for aerospace sandwich and interior applications, i.e., that next to thermomechanical characteristics, fracture toughness and flame retardancy, also manufacturing parameters like tackiness and resin flow are important criteria. An older system is the HexPly® F155 which still uses antimony trioxide and tetrabromobisphenol-A. Both flame retardants have shown strong

TABLE 15.2 Examples of Commercial Epoxy-Based Resins Containing Different Flame Retardants

Resin system	Market/application	Requirement	Flame retardant	Configuration, manufacturing technology	Manufacturer and example
2- and 4-Functional aromatic epoxy resins, 4-functional aromatic amine hardeners, high temperature thermoplastics	Aerospace interior	FAR 25.873, ABD0031	Red phosphorus	Prepreg, automatic tape laying, autoclave	HexPly® M92, Hexcel®
			Zinc borate	Prepreg, hand layup, autoclave / hot press	HexPly® M26, Hexcel®
			Antimony trioxide, tetrabromo-bisphenol A (TBBA)		HexPly® F155, Hexcel®
2-Functional aromatic epoxy resins, 2-3 functional aliphatic amine hardeners	Industrial composites, electronics	UL-94 V-0, V-1, V-2	Organophosphorus, inorganic filler	Liquid, wet laminating	CR134FR, Sika®
	Aerospace interior	FAR 25.873	Organophosphorus	Liquid, wet laminating	EPOLAM2500, Axxon®
	Rail and aerospace interior, automotive	FAR 25.873, UL-94 V-0	Organophosphorus, inorganic filler		Polyphlox 3742 & 3760, Struktol®

environmental disadvantages in the past, but due to the fact that this system was qualified to aerospace specifications for parts being manufactured, this system is still in use. As mentioned before, such prepreg systems typically are formulated with 3- or 4-functional epoxy resins and amine hardeners for high temperature cure (above 120°C) providing high glass transition temperatures and good mechanical properties, by using high loadings of a thermoplastic modifier (PES) resin mixture. The sulfonyl groups in the thermoplastic modifier provide some additional improvement of the flame retardancy next to the flame retardant (Kandola et al. 2010).

For wet laminating and/or resin infusion, there are some systems available on the market as well. Sika®CR134FR resin system, made for wet laminating, is modified with organophosphorus compounds and inorganic fillers. These fillers are not suitable for resin infusion due to the filtering effect in the fabric. Axson® EPOLAM2500 and Struktol® Polyphlox 3760 provide this feature. These systems use pure organophosphorus modifiers either dissolved in the epoxy resin or, in the case of Polyphlox 3760, preformulated with the resin. In fact, the latter system of Struktol® is a highly phosphorus-containing epoxy functional additive which can dissolve in any epoxy resin. Due to the level of flame retardancy and characteristics under burning conditions, both systems can be used for interior aerospace resp. rail application.

15.5 Formation of Respirable Fibers in Carbon-Fiber-Reinforced Epoxy Resins during Combustion and Prevention Thereof

Carbon fiber-reinforced plastic materials (CFRP) are widely used, especially in aircraft, automotive, and ship construction. Several mishaps, crashes, and disasters accompanied by fires have recently increased the interest in assessing an additional health risk due to the formation of respirable carbon fiber dust.

15.5.1 Formation of Respirable Fiber Fragments by Thermal Degradation during Fire

Most commercial carbon fibers are manufactured of polyacrylonitrile (PAN) fibers by a temperature treatment beyond 2000°C in nitrogen (AVK—Industrievereinigung Verstärkte Kunststoffe e. V. 2014; Flemming et al. 1995). Pitch-based fibers and various natural fibers play a minor role as precursors.

Manufacturing is carried out by a three step process, at which the final graphitization temperature determines chemical properties. For example, higher temperatures beyond 2000°C increase the orientation of the graphite micro structure responsible for high tensile strength (Ehrenstein 2006). All types of carbon fibers consist of at least ca. 90% carbon (typically more than 95%), with the rest being mainly nitrogen, which makes them prone to oxidation reaction. A significant oxidation reaction is observed above ca. 600°C in air within minutes, slightly influenced by heating rate and strongly influenced by oxygen concentration (Eibl 2017a). Thermo-oxidative degradation is expressed in principal by two phenomena: a continuous decrease of fiber diameter and the formation of surface defects, which end up in a porous structure with severe damage. Figure 15.6 depicts a series of scanning electron microscopic images of two crossed carbon fibers during prolonged heating at 650°C. Fibers thinner than 3 µm are formed after 10 minutes.

According to the definition of the World Health Organization (WHO), fibers are considered to be respirable, if they are thinner than 3 µm and longer than 5 µm. Fibers with these dimensions are thin enough to penetrate the deep lung areas (alveola) and too long to be easily exhaled (World Health Organization [WHO] 1999). Additionally, the length to diameter ratio has to be higher than 3. Commercially available carbon fibers are typically thicker than 5 µm and therefore uncritical with respect to inhalation, as long as CFRP material has not undergone high temperatures during fire or other degradation processes. A fiber diameter below 3 µm is reached faster and at lower temperatures for initially thinner fibers than for thicker fibers, when thermally treated, whereas the fiber types such as high tensile (HT) or intermediate modulus (IM) etc. only play a minor role. Figure 15.7 gives the average fiber diameters of initially 5.2 µm (IM) and 7.2 µm (HT) thick fibers after prolonged thermal treatment in air between 600°C and 800°C. Thin fibers can easily break and form respirable fragments according to the WHO-definition.

It is well known that carbon fibers may form occupational safety risks under several conditions due to respirable fiber fragments (Hertzberg 2005; Holt and Horne 1978; Jones et al. 1982; Seibert 1990). A risk of the formation of respirable fiber fragments is increased for prolonged fires at elevated

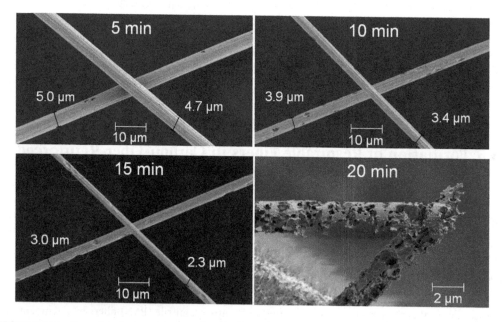

FIGURE 15.6 Scanning electron micrographs of a fiber pair after thermal treatment at 650°C (initial diameter 5.2 µm). (Reprinted with permission from Eibl, S., *Fire Mater.*, 41, 808–816, 2017a.)

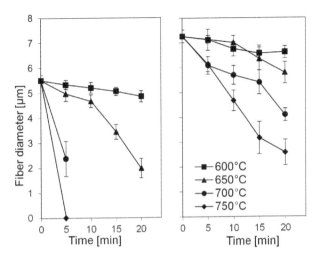

FIGURE 15.7 Fiber diameter of A: IM7 and B: HT fiber after thermal treatment in an oven. (A diameter of 0 μm corresponds to a complete decomposition of the sample). (Reprinted with permission from Eibl, S., *Fire Mater.*, 41, 808–816, 2017a.)

temperatures as in the case of fuel fires. Critical temperatures are typically not reached in the absence of an external heat load. Therefore, not all composite fires have to be considered to form respirable fiber fragments. Even at high enough temperatures, respirable fibers are transported by thermal uplift during a fire and diluted outdoors without critical concentrations. Currently, no data are available which provide emission rates of respirable fibers in a composite fire. Especially, handling of combusted composite material was identified to pose a risk of significant exposition according to several health regulations (TRGS 521 2008; TRGS 905 2008), as fiber dust is released due to mechanical impact. Critical fiber concentrations were reported for large-scale fire tests and collecting the flight recorder of a crashed helicopter (Eibl and Scholz 2014; Eibl et al. 2014). Personal protection is necessary for personnel in repeated contact with combusted CFRP. For a reduction of fiber exposition after a fire, fiber binding agents may be applied (Eibl 2016). However, on the side of passive measures of fire protection, it is necessary to improve the thermal properties of composites with respect to the release of respirable fibers after a fire.

15.5.2 Flame Retarding Principles to Protect Carbon Fibers

Typical flame retardants act physically, by cooling or dilution, or chemically, by quenching radical reactions in the flame or by forming barriers for the transportation of heat or oxygen (Troitzsch 2004). Preventing the polymer matrix from combustion can be achieved by flame retardants acting at lower temperatures than necessary for fiber degradation. However, efficient carbon fiber protection requires retarding mechanisms beyond ca. 600°C. For example, an epoxy-based composite with initially 7.2 μm thick carbon fibers was observed to form fibers with a diameter of ca. 3 μm after 20 minutes at a maximum temperature of ca. 800°C (Eibl 2017a). Figure 15.8 shows this reached mean diameter of carbon fibers close to the surface of CFRP panels after combustion in a cone calorimeter at 60 kW/m² for 20 minutes. An epoxy-based composite containing up to 25% magnesium hydroxide in the resin exhibits a reduction of the fiber diameter to 3.2 μm, as pronounced as for the composite without a flame retardant and nearly independent of its content. Magnesium hydroxide retards resin combustion by the formation of water beginning from ca. 350°C, but is not able to efficiently protect the fiber from degradation for a prolonged exposition of the composite at ca. 800°C (60 kW/m²). In contrast, zinc borate provides efficient fiber protection, even at low concentration indicated by a minimum fiber diameter of 5.0 μm. Zinc borate forms a glassy layer during a fire, which covers the fibers and prevents oxygen

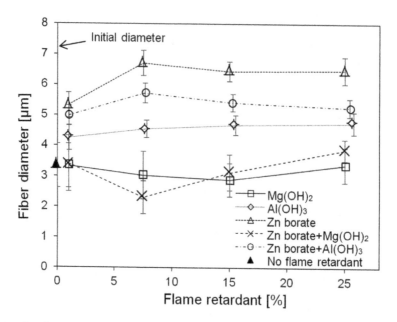

FIGURE 15.8 Fiber diameter at the surface of the irradiated side of 2 mm thick epoxy-based composite panels reinforced with a HT carbon fiber and modified with various flame retardants after combustion in a cone calorimeter irradiated at 60 kW/m² for 20 minutes, measured by scanning electron microscopy. (Reprinted with permission from Eibl, S., *FireMater.*, 41, 808–816, 2017a.)

from accessing them. This layer is thermally resistant enough to withstand high temperatures during prolonged irradiation. Adding magnesium hydroxide to zinc borate as flame retardant limits the fiber protection by negatively affecting the glassy barrier layer. The reached fiber diameters are even roughly the same as for the unprotected composite. The same effects are observed for aluminum hydroxide. However, it provides a limited fiber protection by itself and does not interfere with the zinc borate as pronounced as magnesium hydroxide.

Therefore, an efficient way to protect carbon fibers is to impregnate them with a flame retardant forming a barrier layer. However, fiber to matrix adhesion must not be harmed, this is what makes it difficult to find flame retardants compatible with the fiber sizing. A strategy to reduce this problem is to add a flame retardant to the resin matrix, which forms a protection layer on the fiber surface during combustion (Eibl 2017c). Appropriate chemistry of flame retardants forming in situ protection layers is a topic of ongoing research. Basically, all additives forming glass or ceramics at high temperatures such as silanes promise efficient fiber protection. Other strategies follow the use of glass formers such as phosphates, borates, and low melting glasses such as vanadates (Malmendier and Sojka 1974), etc. In order to increase the concentration of the in situ flame retardant close to the fiber, it can be impregnated with the matrix containing the flame retardant in a first step and curing the CFRP with the pure matrix in a second step.

15.5.3 Combination of Flame Retardancy, Fiber Protection, and Processability

The last sections focused on fiber protection, but it is important that flame retardancy and processability are taken into account as well. As mentioned above, especially, flame retardants with gas-phase activity are efficient for epoxy-based composites, whereas condensed-phase active compounds are needed for fiber protection. Latest results (Greiner et al. 2019) showed that the incorporation of char-forming or intumescent phosphorous-containing flame retardants into the matrix resin leads to a significant retardation of the formation of respirable fibers. In contrast to zinc borate, which is a substance of very high concern, these compounds are homogenous miscible with epoxy resins, so infusion processes like RTM are still possible. A combination of gas-phase activity and condensed-phase activity unified in one molecule is possible and therefore flame retardancy, fiber protection, and processability are considered. Polymeric or oligomeric compounds with less influence on material parameters and good solubility in the resin are a favored solution.

References

Arada, B., Lin, S. C., and Pearce, E. M. 1979. Epoxy resins IV: The stability of the epoxy-triethanolamine borate system. *International Journal of Polymeric Materials* 7(3–4): 167–184.

AVK—Industrievereinigung Verstärkte Kunststoffe e. V. 2014. *Handbuch Faserverbundkunststoffe/ Composites*. 4. Hrsg. Wiesbaden, Germany: Springer Vieweg.

Becker, C. M., Dick, T. A., Wypych, F. et al. 2012. Synergetic effect of LDH and glass fiber in the properties of two- and three-component epoxy composites. *Polymer Testing* 31(6): 741–747.

Braun, U., Balabanovich, A. I., Schartel, B. et al. 2006. Influence of the oxidation state of phosphorous on the decomposition and fire behavior of flame-retarded epoxy resin composites. *Polymer* 47(26): 8495–8508.

Brown, J. R., Fawell, P. D., and Mathys, Z. 1994. Fire hazard assessment of extended-chain polyethylene and aramid composites by cone calorimetry. *Fire and Materials* 18(3): 167–172.

Chen, C. S., Bulkin, B. J., and Pearce, E. M. 1982a. New epoxy resins. I. The stability of epoxy-trialkoxy-boroxinestriaryloxyboroxine system. *Journal of Applied Polymer Science* 27(14): 1177–1190.

Chen, C. S., Bulkin, B. J., and Pearce, E. M. 1982b. New epoxy resins. II. The preparation, characterization, and curing of epoxy resins and their copolymers. *Journal of Applied Polymer Science* 27(9): 3289–3312.

Ciesielski, M., Altstädt, V., Diederichs, J. et al. 2010. Novel phosphorus-based flame-retardants for epoxy resins and their carbon fibre reinforced composites in automotive and aviation industry. In: *14th European Conference on Composite Materials*, Budapest, Hungary.

Ciesielski, M., Burk, B., Heinzmann, C. et al. 2017. Fire-retardant high-performance epoxy based materials. In: *Novel Fire Retardant Polymers and Composite Materials*, Wang D-Y. (ed.), Cambridge, UK: Elsevier Ltd., Woodhead Publishing, pp. 3–51.

Ciesielski, M., Diederichs, J., Döring, M. et al. 2009. Advanced flame-retardant epoxy resins for composite materials. In: *Fire and Polymers V*. Washington, DC: American Chemical Society, pp. 174–190.

Dao, D., Rogaume T., Luche, J. et al. 2016. Thermal degradation of epoxy resin/carbon fiber composites: Influence of carbon fiber fraction on the fire reaction properties and on the gaseous species release. *Fire and Materials* 40(1): 27–47.

Döring, M., Ciesielski, M., and Heinzmann, C. 2012. Synergistic flame retardant mixtures in epoxy resins. In: *Fire and Polymers VI: New Advances in Flame Retardant Chemistry and Science*, Morgan A. B., Wilkie C. A. and Nelson G. L. (eds.), Washington, DC: American Chemical Society, pp. 295–309.

Ehrenstein, G. W. 2006. *Faserverbund-Kunststoffe: Werkstoffe—Verarbeitung—Eigenschaften*. München, Germany: Carl HanserVerlag.

Eibl, S. 2012. Influence of carbon fiber orientation on reaction-to-fire properties of polymer matrix composites. *Fire and Materials* 36(4): 309–324.

Eibl, S. 2016. Faserbindelackbei CFK-Brandschäden (Fiber bonding lacquer with CFRP fire damage). Germany, Patentnr. DE 102014010465 A1.

Eibl, S. 2017a. Potential for the formation of respirable fibers in carbon fiber reinforced plastic materials after combustion. *Fire and Materials* 41(7): 808–816.

Eibl, S. 2017b. Flame retarding effects by variously orientated fibers in carbon fiber reinforced plastic materials. *16th European Meeting on Fire Retardant Polymeric Materials (FRPM)*, Manchester, UK.

Eibl, S. 2017c. Flammgeschützte Kohlenstofffaser (Flame retardant carbon fiber). Germany, Patentnr. DE 102015010001 A1.

Eibl, S., Reiner, D., and Lehnert, M. 2014. Gefährung durch lungengängige Faserfragmente nach dem Abbrand Kohlenstofffaser verstärkter Kunststoffe (Health hazards by respirable fiber fragments after combustion of carbon fiber reinforced plastic material). Gefahrsoffe—Reinhaltung der Luft, 74(7/8), p. 285.

Eibl, S., and Scholz, N. 2014. Besondere Gefährdung beim Abbrand von Carbon-Kunststoffen (Particular hazards related to combustion of carbon fiber reinforced plastic material). Brandschutz, Deutsche Feuerwehr-Zeitung, pp. 423–427.

Fateh, T., Zhang, J., Delichatsios, M. et al. 2017. Experimental investigation and numerical modeling of the fire performance for epoxy resin carbon fibre composites of variable thicknesses. *Fire and Materials* 41(4): 307–322.

Fiedler, B., Gojny, F. H., Wichmann, M. H. G. et al. 2006. Fundamental aspects of nano-reinforced composites. *Composites Sciences and Technology* 66(16): 3115–3125.

Flemming, M., Roth, S., and Zeigmann, G. 1995. *Faserverbundbauweisen —Fasern und Matrices*. Berlin, Germany: Springer Verlag Heidelberg.

Gibson, A. G., and Hume, J. 1995. Fire performance of composite panels for large marine structures. *Olastics, Rubbers & Composites Processing and Applications* 23(3): 175–183.

Greiner, L., Kukla, P., Eibl, S., and Döring, M. 2019. Phosphorus Containing Polyacrylamides as Flame Retardants for Epoxy-Based Composites in Aviation. *Polymers* 11(2): 284.

Hertzberg, T. 2005. Dangers relating to fires in carbon-fiber based composite material. *Fire and Materials* 29(4): 231–248.

Holt, P. F., and Horne, M. 1978. Dust from carbon fiber. *Environmental Research* 17(2): 276–283.

Hume, J. 1992. Assessing the fire performance characteristics of GRP composites. *International Conference on Materials and Design against Fire*, London, UK, pp. 11–15.

Iji, M., and Kiuchi, Y. 2001a. Flame-retardant epoxy resin compounds containing novolac derivatives with aromatic compounds. *Polymers for Advanced Technologies* 12(7): 393–406.

Iji, M., and Kiuchi, Y. 2001b. Self-extinguishing epoxy molding compound with no flame-retarding additives for electronic components. *Journal of Materials Science-Materials in Electronics* 12(12): 715–723.

Jones, H. D., Jones, T. R., and Lyle, W. H. 1982. Carbon fiber results of a survey of process workers and their environment. *The Annals of Occupational Hygiene* 26(8): 861–867.

Kandola, B. K., Biswas, B., Price, D. et al. 2010. Studies on the effect of different levels of toughener and flame retardants on thermal stability of epoxy resin. *Polymer Degradation and Stability* 95: 144–152.

Laurenzi, S., and Marchetti, M. 2012. Advanced composite materials by resin transfer molding for aerospace applications. In: *Composites and Their Properties*. Rijeka, Croatia: InTech Open, pp. 197–226.

Le Bras, M., Bourbigot, S., Mortaigne, B. et al. 1998. Comparative study of the fire behaviour of glass-fibre reinforced unsaturated polyesters using a cone calorimeter. *Polymers & Polymer Composites* 6: 535–539.

Levchik, S. V., Camino, G., Costa, L. et al. 1996. Mechanistic study of thermal behaviour and combustion performance of carbon fibre-epoxy resin composites fire retarded with a phosphorous-based curing system. *Polymer Degradation and Stability* 54(2–3): 317–322.

Lin, S. C., and Pearce, E. M. 1979. Epoxy resins. II. The preparation, characterization, and curing of epoxy resins and their copolymers. *Journal of Polymer Science Polymer Chemistry Edition* 17(10): 3095–3119.

Malmendier, J., and Sojka, J. 1974. Low melting vanadate glasses. US, Patent no. US 3837866 A.

Martin, F. J., and Price, K. R. 1968. Flammability of epoxy resins. *Journal of Applied Polymer Science* 12(1): 143–158.

Mazumdar, S. 2002. *Composites Manufacturing—Materials, Product, and Process Engineering.* Boca Raton, FL: CRC Press LLC.

Müller, P. 2013. *Dissertation—Mehrfunktionelle phosphorhaltige Flammschutzmittel für Epoxidharze,* Heidelberg, Germany: Ruprecht-Karls-Universität Heidelberg.

Perret, B., Schartel, B., Stöß, K. et al., 2011a. A new halogen-free flame retardant based on 9,10-dihydro-9-oxa-10-phosphaphenanthrene-10-oxide for epoxy resins and their carbon fiber composites for the automotive and aviation industries. *Macromolecular Materials and Engineering* 296(1): 14–30.

Perret, B., Schartel, B., Stöß, K. et al. 2011b. Novel DOPO-based flame retardants in high-performance carbon fibre-epoxy composites for aviation. *European Polymer Journal* 47(5): 1081–1089.

Scudamore, M. J. 1994. Fire performance studies on glass-reinforced plastic laminates. *Fire and Materials* 18(5): 313–325.

Seibert, J. F. 1990. Composite fibre-hazards, Report 90-EIOO178MGA, Texas: Air Force occupational and environmental health laboratory.

Song, T., Li, Z., Liu, J. et al. 2013. Synthesis, characterization and properties of novel crystalline epoxy resin with good melt flowability and flame retardancy based on an asymmetrical biphenyl unit. *Polymer Science, Series B* 55(3–4): 147–157.

Thostenson, E. T., Ren, Z., and Chou, T.W. 2001. Advances in the science and technology of carbon nanotubes and their composites: A review. *Composites Science and Technology* 61(13): 1899–1912.

Toldy, A., Szolnoki, B., and Marosi, G. 2011. Flame retardancy of fibre-reinforced epoxy resin composites for aerospace applications. *Polymer Degradation and Stability* 96(3): 371–376.

TRGS 521. 2008. Technical Rules for Hazardous Substances (TRGS 521): Demolition, reconstruction and maintenance work with biopersistent mineral wools, Germany: Committee on Hazardous Substances—BundesanstaltfürArbeitsschutz und Arbeitsmedizin.

TRGS 905. 2008.Technical Rules for Hazardous Substances (TRGS 905): Index of substances which can cause cancer, genetic changes or limit reproductive capability, Germany: Committee on Hazardous Substances—BundesanstaltfürArbeitsschutz und Arbeitsmedizin.

Troitzsch, J. 2004. *Plastics Flammability Handbook—Principles, Regulations, Testing and Approval.3rd Hrsg.* Munich, Germany: Carl HanserVerlag.

World Health Organization. 1999. Hazard prevention and control in the work environment: Airborne dust, WHO/SDE/OEH/99.14. http://www.who.int/occupational_health/publications/en/oehairbornedust3.pdf. Accessed December 18, 2017.

Yan, H. J., and Pearce, E. M. 1984. Stryrylpyridine-based epoxy resins: Synthesis and characterization. *Journal of Polymer Science Polymer Chemistry Edition* 22(11): 3319–3334.

Zang, L., Wagner, S., Ciesielski, M. et al. 2011. Novel star-shaped and hyperbranched phosphorus-containing flame retardants in epoxy resins. *Polymers for Advanced Technologies* 22(7): 1182–1191.

Index

Note: Page numbers in italic and bold refer to figures and tables, respectively.